W9-BHZ-559

ELEMENTARY FUNCTIONS

THE INTEXT SERIES IN MATHEMATICS

under the consulting editorship of

RICHARD D. ANDERSON   Louisiana State University

ALEX ROSENBERG   Cornell University

ELEMENTARY FUNCTIONS

AN ALGORITHMIC APPROACH

Theodore C. Burrowes
COLLEGE OF WOOSTER

and

Sharon K. Burrowes

Intext Educational Publishers · NEW YORK and LONDON

Copyright © 1974 by Intext, Inc.
All rights reserved. No part of this book may be reprinted,
reproduced, or utilized in any form or by any electronic,
mechanical, or other means, now known or hereafter invented,
including photocopying and recording, or in any information
storage and retrieval system, without permission in writing
from the Publisher.

Library of Congress Cataloging in Publication Data

Burrowes, Theodore C.
    Elementary functions.
    (The Intext series in mathematics)
    1. Functions. 2. Electronic data processing—Func-
tions. I. Burrowes, Sharon K., joint author.
II. Title.
QA331.3.B87   515   73–20118
ISBN 0–7002–2453–X

Intext Educational Publishers
257 Park Avenue South
New York, New York 10010

Text design by Barbara Rasmussen

To Bobby, Pop, and Mom: Inspirations in the living of life!

CONTENTS

PREFACE

This book is intended for courses which immediately precede a course in calculus. It has been our observation that most teachers of such courses hope not only to communicate a body of factual information prerequisite to a study of calculus but also to develop in their students an appreciation for mathematics as seen and done by mathematicians, and a capacity to participate in the activities of discovery, derivation, and proof. In writing this book we developed several tactics that we hope will foster a lively and stimulating context for the student's learning and make it easier for the student to achieve these goals. Among these tactics are

1. A conversational and developmental method of presentation,
2. An algorithmic approach together with opportunities for computer involvement at several levels,
3. An unusually wide spectrum of exercises.

Let us elaborate briefly on each of these points:

1. The student's natural curiosity, intuition, and skepticism are recognized and used to help understanding. In the pull between intuition and rigor, we have tried to make constructive use of intuition while at the same time developing skepticism toward it as the final word. The role and techniques of proof are discussed and many proofs are included in the text and among the problems. We have also attempted to broaden the student's awareness that mathematics is an art and a language as well as a valuable tool.

2. At the heart of our algorithmic approach is the *concept of an algorithm*. Knowledge of computers is not presumed, nor is use of computers required, though various opportunities for computer involvement are provided.

We make use of flowcharts and computer outputs where they will aid the credibility of a conjecture, the visualization of a process, or assist understanding in some other way. We have avoided the inclusion of computer-related topics and methods for their own sake, seeking to keep them always in the service of the mathematics at hand.

Perhaps the single greatest benefit of an algorithmic approach appears when students create their own algorithms. This follows the time-honored principle that the best way to learn a subject is to teach it. The creation of an algorithm involves careful examination and clear communication which enhances learning. Opportunities for such student activity are made available by the inclusion of a preliminary section on flowchart language and flowcharting techniques and by the inclusion of flowchart exercises among the problems at the end of each chapter.

If both time and computing hardware are available, students can make use of the computer with the programs of Appendix C without any need for knowing or learning a programming language. The programs can be used as computational tools and/or apparatus in a mathematics laboratory.

Finally, opportunities for active student programming and the accompanying benefits to learning are made available through the flowchart exercises, and instruction in both Basic and Fortran adequate for the needs of this book are made available in Appendix B.

3. Exercises in the book are divided into three categories: Practice Exercises, Problems, and Projects.

The Practice Exercises are designed to allow the student straightforward drill on specific problem types. They are arranged by chapter section and subsection and are numerous enough for the student to gain confidence in his/her capacity to do such problems. These are located at the end of each chapter rather than after each section in order to maintain a greater sense of continuity and unity within each chapter.

The Problems, also located at the end of each chapter are varied in thrust, difficulty, and length and are randomly arranged to avoid student's thinking being too confined ("It's after Section 3 so it must use that idea" or "It's the last problem in the section so it must be too hard for me").

The Projects, found toward the end of several chapters, are designed to provide guided but independent experiences with a mathematical idea, offering not only a more complete "experience" with mathematics but also review and extension of important aspects of the chapter.

The book has three other features which we hope students will find helpful. First is an algebra review (Appendix A). We have found that most students discover occasional weaknesses in their working knowledge of algebra but that these often require no more than the kind of reminder which this review makes available to the student whenever the need arises. Second, marginal glosses are provided as a reference guide more handy than the Table of Contents and Index. And third, each chapter contains a Summary of Important Concepts and Notation which serves as a kind of thumbnail sketch of the chapter as well as providing definitions which do not depend on a context of discussion as is sometimes the case in the text development.

This book has matured over several years of class use and contains more material than can reasonably be covered in one term. Our experience is that in a term of 50 class hours, giving the attention to flowcharts but not programming, essentially all the material on sets, functions and relations, polynomial functions, and circular functions can be covered without haste. In this context algebra is assumed and only reviewed informally in class as problems arise. A course in plane geometry is also assumed.

Many other patterns of topic selection are also possible, and we believe that most two term courses would find adequate material in this book for their needs.

We pause here to accept the responsibility for the existence of those tenacious errors that have adeptly avoided detection by so many sets of eyes throughout the revisions and rereadings preceding final publication, and we would welcome your calling our attention to any you may find.

Finally, our gratitude extends to various people associated directly or indirectly with the writing of this book. To Melcher Fobes, Ruth Smyth, Edward Assmus, Richard Koch, and David Moursund for their contributions as teachers to our understanding and love of what mathematics is all about; to the students who grew with the prepublication versions, to the staff of Office Services of the College of Wooster for their skill and valuable suggestions in producing the prepublication version, and to David, Bethanne, Scott, and Bonnie whose patience exceeded their understanding of why Mommy and Daddy couldn't always sit down and play with them!

LIST OF SYMBOLS

ELEMENTARY FUNCTIONS

PRELIMINARIES

1. INTRODUCTION

This chapter is intended to provide some background in two topics that will facilitate your work in the rest of the book: sets and flowcharts.

Sets are discussed in Section 2. It is assumed that the reader has an understanding of the fundamental concepts of set theory. The primary purpose of the section is to establish a general vocabulary for the use of sets.

Section 3 deals with the concepts of algorithms and flowcharts. Here it is assumed that the reader has no prior exposure to these topics. The coverage of this section depends on the extent to which students will be expected to develop algorithms and write flowcharts. If little or no such activity is planned, a brief reading of the section should provide adequate background to understand and benefit from the algorithmic treatment of topics later in the book. If extensive flowcharting and/or programming is planned, the reader should be sure that he can work a variety of the problems provided for this section at the end of the chapter.

For those readers and/or classes that have adequate time to include direct computer involvement as part of their learning about the elementary functions, an appendix is provided with an introduction (albeit brief)

1

to the programming languages Basic and Fortran. This appendix can serve as a review and reference section for those who already know one or both languages. It can also serve as a base for learning either language in conjunction with Section 3, for those who have no prior programming experience. Some explanation and enough examples are given to permit learning the basics of either language "from scratch" without the need of outside sources other than user's manuals at the local installation.

2. SETS

SET **2a. Introduction**

You have no doubt already encountered the idea of a set. You are also probably familiar with a variety of operations that may be performed on sets, as well as the properties that these operations have. In this section, we shall review most of these ideas briefly. In places, we shall add a few words of explanation or advice in order to indicate what will be assumed as common vocabulary and notation.

DEFINITION 1
Set

By the word *set*, we shall simply mean any collection of objects† which enjoys the property that every object is either in or not in the collection. Right away, we have run into one of the traits of mathematicians that often annoy students but which is an essential part of mathematics: the insistence on precision. Ordinarily, sets may be considered to be collections of objects. But consider the following example.

In a small rural community, Maryanne makes clothes for *each* person in the community who does not make his own clothes. This sounds reasonable and it seems that one could discuss the set of all people for whom Maryanne makes clothes, call it set A. But does Maryanne belong to this set or not? As the set was formulated, she can neither belong nor fail to belong to A.

If Maryanne were in set A, then she would be making her own clothes— but she sews only for people who do not make their own clothes (and so could not be in A).

If Maryanne were not in set A, then she would not be making her own clothes; but the definition of her job is making clothes for *everyone* who does not make his own clothes (so she would *have* to be in A).

Confused? Don't be surprised if you are. Go back and reread it a number of times. The example is not an easy one, but it is important because what *seemed* to be a set in this case is not really clearly defined. While this example is admittedly off the beaten track, it does point out the possibility of trouble if the definition is not precise.‡ Meanwhile, do not be misled— at the heart of the matter is still the idea that a set is, essentially, any collection of objects.

---

† We use the word *object* in this book to mean *object* or *idea* or *concept*.
‡ This is one form of the famous Russell Paradox formulated by Bertrand Russell (1872–1970). The reader interested in such logical paradoxes is referred to Volume 3 of *The World of Mathematics* edited by James R. Newman (New York: Simon and Schuster), pp. 1949–1953.

2b. Basic Concepts and Notation

We ordinarily write sets in one of two forms : The *roster* form lists all the elements between two braces with elements separated by commas. As elsewhere in mathematics, the symbol "$\cdots$" is used to indicate a repeated pattern.

DEFINITION 2
Roster

Thus

$$\{a, e, i, o, u\}$$
$$\{a, b, c, d, \cdots, x, y, z\}$$
$$\{1, 2, 3, 4, \cdots\}$$

EXAMPLE 1

and

$$\{\cdots, -3, -2, -1, 0, 1, 2, \cdots\}$$

are all legitimate set descriptions.

The *set builder* form exhibits a property that is characteristic of all elements of the set and of no other objects, and so specifies the elements of the set by a kind of remote control. The notation we use is

DEFINITION 3
Set Builder

$$\{x \,|\, P(x)\},$$

where $P(x)$ is shorthand for the property that an object x must have to be an element of the set.

Thus, in set builder notation, the four sets above become

EXAMPLE 2

$$\{x \,|\, x \text{ is a vowel in our alphabet}\}$$
$$\{x \,|\, x \text{ is a letter in our alphabet}\}$$
$$\{x \,|\, x \text{ is a positive integer}\}$$
$$\{x \,|\, x \text{ is an integer}\}.$$

The variable x in the set builder notation is a "dummy variable"—it has no intrinsic meaning; i.e., $\{x \,|\, p(x)\} = \{\alpha \,|\, p(\alpha)\} = \{A \,|\, p(A)\}$ are all notations for the same set.

Many students find it easy to confuse the property $p(x)$, which characterizes the elements of the set, with a property of the set as a separate entity. You should learn to see the difference between $\{x \,|\, x$ is a state in the USA$\}$ and $\{x \,|\, x$ is the USA$\}$. The former is $\{$Alaska, $\cdots$, Wyoming$\}$, while the latter is a set containing only one element: the country USA. To reiterate, the defining property is to apply to one element at a time, not to the whole.

Finally, two sets are said to be *equal* if they contain precisely the same elements. You will recall that the order in which the elements are listed makes no difference and that one disregards repetitions of elements. Thus

DEFINITION 4
Equality of Sets

$$\{s, p, o, t\} = \{t, o, p, s\} = \{p, o, s, t, s\}.$$

Although other definitions would be possible (and several possibilities do warrant enough attention to be given other names than equality),† this is the definition agreed upon within the mathematical community.

---

† Set equivalence, equality of vectors, and equality of sequences are some examples.

DEFINITION 5
Universe

In most contexts, there is an underlying set called the *universe of discourse* or just *universe*. It is understood by the people who are communicating about sets so that a completely precise statement for $p(x)$ in $\{x|p(x)\}$ is not needed. Thus $\{x|x$ is prime$\}$ is interpreted in the context of whole numbers and so is understood to mean $\{x|x$ is a whole number and x is prime$\}$. Recall that prime numbers are positive integers whose only divisors are one and itself. Further, the number 1 is not a prime. Thus, we do *not* mean prime rib, prime objectives, or prime ministers! (Only prime *numbers*.)

In other words, the universe of discourse is a set consisting of all the objects of interest to the current discussion. If and when it overtly enters a discussion, it is often denoted by U.

DEFINITION 6
Element and
Subset

One frequently encounters two different types of containment when dealing with sets, exemplified in the distinction between being a member of a committee and being a subcommittee of a committee. The former is an example of an element's being a member of a set. We use the symbol "$\in$" to denote such membership, and so say $e \in \{a, e, i, o, u\}$. The latter is an example of a set's being a subset of another set. We use the symbol $\subseteq$ to denote this kind of containment. Thus $\{a, e, o\} \subseteq \{a, e, i, o, u\}$. More formally we say $x \in A$ (read x is an element of A) if x is an element in the set A, and $B \subseteq A$ (read B is a subset of A) if B is a set, each of whose elements is also an element of A. As an imprecise rule of thumb, then, subsets are always sets, while elements are usually "individuals."

This distinction may seem obvious enough, but many students find it more subtle than it first appears, especially when dealing with subsets containing only one element. To return to the committee example, if the committee is $C = \{$Linda, Mark, Sue, Andrea, Tony$\}$ and if Sue is a one-member subcommittee in charge of advertising, we should write $\{$Sue$\} \subseteq C$. We may also write Sue $\in C$, but $\{$Sue$\} \notin C$ and Sue $\nsubseteq C$.

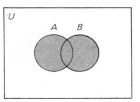

FIGURE 1. Shaded area represents $A \cup B$.

Do you see the distinction? If not, do not shrug your shoulders. Instead reread the paragraph and try to create some examples of your own. If understanding still has not been reached, formulate as precise a question as you can to ask a friend or your instructor. Such an effort in itself may provide a clue you needed or clarify your thinking so an answer is more likely to make sense to you. (Such a strategy is wise to follow whenever you encounter difficulties in this or any other mathematics book.) †

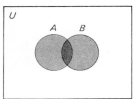

FIGURE 2. Heavily shaded area represents $A \cap B$.

There are three familiar set operations: *union*, *intersection*, and *complementation*, denoted, respectively, by $\cup$, $\cap$, and $'$. Union and intersection are binary operations; i.e., they act on two sets at once. You will probably recall the definitions and corresponding Venn diagrams in Figures 1 and 2.

DEFINITION 7
Union and
Intersection

$$A \cup B = \{x|x \in A \quad \text{or} \quad x \in B\}; \qquad A \cap B = \{x|x \in A \quad \text{and} \quad x \in B\}.$$

EXAMPLE 3 As examples, if $A = \{a, b, c, d, e\}$ and $B = \{a, e, i, o, u\}$, then

$$A \cup B = \{a, b, c, d, e, i, o, u\}, \qquad A \cap B = \{a, e\}$$

---

† Occasionally it is useful to distinguish between *proper* and *improper* subsets. A proper subset of a set B is any subset of B other than B itself, while B is called an improper subset of itself. In this book we shall use the word subset to mean both proper *and* improper subset.

Some readers will already have observed that the only difference between the definitions of $A \cup B$ and $A \cap B$ are the words *or* and *and*. These two words occur repeatedly in logical statements and set descriptions. Their use in these settings is not always the same as their use in ordinary parlance.

For example, in ordinary parlance a college official concerned with off-campus housing might be concerned with the students who are 21 years old or older, and with students who are married (perhaps as a minimal group of persons who would be permitted to live off campus). One might try to write this group in set notation as $\{x \mid x \text{ is} \geq 21 \text{ years old and is married}\}$. But careful examination of this set shows it to be different from what we had intended—it includes only those students who are simultaneously married and at least 21. Married 18 year olds and single 30 year olds are *not* included. What is needed is the set $\{x \mid x \text{ is} \geq 21 \text{ years old } or \text{ } x \text{ is married}\}$. The "and" of the ordinary parlance construction became an "or" in the precise mathematical setting.

The use of a construction that might be correctly interpreted in conversation but which is technically incorrect can destroy the delicate structure of a mathematical statement. Care is therefore called for in using these words in mathematical settings and the unpracticed student is advised to think carefully before he uses them.

The operation of complementation is a unary operation—it acts on only one set, but its action is in the context of the universal set. Certainly, the familiar definition

$$A' = \{x \mid x \notin A\}$$

does not intend to include all of creation except A, but only those objects currently being studied which are not in A. This is also shown in the Venn diagram in Figure 3.

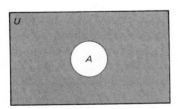

FIGURE 3. Shaded area represents A'.

This matter of context can occasionally be ambiguous. Therefore, we shall find it useful to include another form of complementation, which you may not have encountered before: a binary operation called the relative complement, denoted by $-$. Its definition is

$$A - B = \{x \mid x \in A \quad \text{and} \quad x \notin B\}$$
$$= A \cap B'.$$

Thus, if

EXAMPLE 4

$$A = \{1, 2, \cdots, 9, 10\}$$
$$B = \{0, 2, 4, 6, \cdots, 48, 50\}$$

then

$$A - B = A \cap B' = \{1, 2, \cdots, 9, 10\} \cap \{1, 3, 5, \cdots, 47, 49\}$$
$$= \{1, 3, 5, 7, 9\}.$$

The Venn diagram of the relative complement is shown in Figure 4.

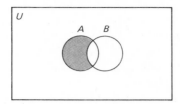

FIGURE 4. Shaded area represents
$A - B$.

2c. Some Special Sets

There is one special set we avoided in the previous sections, which you have probably seen before. Try to answer the following questions:

What is U'?
What is $A \cap B$ if $A = \{1, 3, 5, 7, \cdots\}$ and $B = \{2, 4, 6, 8, \cdots\}$?
What is $\{x \mid x$ is a whole number and $x^2 = -1\}$?

In each case, what is called for is a set that contains no elements.

DEFINITION 10
Empty Set
This new set, called the *empty set* or *null set*, is the set that has no elements and is denoted either by $\emptyset$ or $\{\ \}$. Of course, this set will occur in a wide variety of disguises, such as in the examples above; but when one wishes to refer specifically to the empty set, one of the two notations just suggested is usually used.

Because it contains no elements, $\emptyset$ can often be tricky to deal with rigorously. Of particular concern in this regard is showing that $\emptyset \subseteq A$ for all sets A. The definition of subset we currently use is of little help because there are no $x \in \emptyset$ to test. Instead, we will try an *equivalent definition of subset*, which is $A \subseteq B$ if there are no elements of A that fail to be members of B.

DEFINITION 11
Subset

In other words, if we check each element of B' we shall find that *none* of them is an element of A, that is, no elements of A fail to be in B. For example, if $A = \{a, e, i, o, u\}$ and $B = \{a, b, c, d, \cdots, x, y, z\}$ then, using our new definition, we say that $A \subseteq B$ because no vowel fails to be a letter (read that again). Or, using Venn diagrams (Figure 5), we observe that none of A is outside of B.

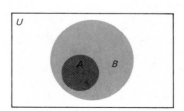

FIGURE 5

Now, using this definition, since $\emptyset$ has no elements at all, it has no **EMPTY SET** mischievous ones (i.e., ones that are "outside" of B), so $\emptyset$ is a subset of B no matter what set B is.

This is an example of what mathematicians often call *degenerate cases*, **NOTE** cases that meet the letter of the law but whose spirit seems to be questionable.

Degenerate cases often result in proofs like the one above, which leave one feeling disatisfied at first, even though the proof is completely rigorous. Your understanding and even acceptance of these cases will come with additional experience. For now, reread the above argument and think about it for awhile.

One of the circumstances giving rise to the need for the empty set **DEFINITION 12** Disjoint was that in which we asked about $A \cap B$ when, in fact, A and B had no elements in common. This circumstance occurs frequently and is termed *disjoint*: we say two sets A and B are disjoint if $A \cap B = \emptyset$.

For example, if $X = \{1, 2, 3\}$ and $Y = \{5, 6, 7\}$, $X \cap Y = \emptyset$ and X and Y **EXAMPLE 5** are said to be disjoint.

There are several sets of numbers to which we shall frequently refer, **SPECIAL SETS OF NUMBERS** so we give these sets special notations. There are many other symbols in use for these same ideas. We choose the ones below both because of their wide usage and their distinctiveness—they are not likely to be confused with other notations we shall use.

$\mathbb{N}$ will denote the set of natural numbers, i.e., $\mathbb{N} = \{1, 2, 3, 4, 5, \cdots\}$.

$\mathbb{Z}$ will denote the set of integers $\{\cdots, -3, -2, -1, 0, 1, 2, 3, \cdots\}$.

$\mathbb{Q}$ will denote the set of rational numbers;† $\{x \mid x$ can be expressed as p/q where p and q are integers$\}$.

$\mathbb{R}$ will denote the set of all real numbers. Recall that this set includes $\mathbb{Q}$ together with the irrational numbers (e.g. $\sqrt{2}$, π, etc.). A precise definition requires more subtle "machinery" than we have available. Intuitively, $\mathbb{R}$ is the set of all numbers on the number line.

The $\mathbb{Z}$ for the integers is motivated by the German word *Zahl*, for number; the $\mathbb{Q}$ can be thought of as coming from the word quotient, since rational numbers are quotients of integers. If you take a few minutes now to learn these notations, you will save both page turning and confusion later.

We shall find that in dealing with real numbers, sets such as $\{x \mid x \in \mathbb{R}$ **INTERVALS** and $x < 5\}$, $\{x \mid x \in \mathbb{R}$ and $x \geq 0\}$, and $\{x \mid x \in \mathbb{R}$ and $-2 < x \leq 10\}$ occur with such frequency that more succinct notations are needed.

Such sets as those mentioned above are called intervals. We use several specific examples to introduce the necessary notation, and then provide a complete catalogue of the eight varieties of intervals.

$(3, 4)$ means $\{x \mid x > 3$ and $x < 4, x \in \mathbb{R}\}$; in other words, all the real **EXAMPLE 6** numbers between 3 and 4 but not including 3 and 4.

† The word *rational* here has as its root the word *ratio*. Rational numbers are "ratio-numbers" or quotients. You should not think of the word *rational* as connoting "sensible," although the reference to nonrational numbers as irrational often suggests this erroneous idea. Irrational numbers are just as sensible as rational numbers.

[3, 4] means $\{x\,|\,x \ge 3$ and $x \le 4, x \in \mathbb{R}\}$; in other words, all the real numbers between 3 and 4, including 3 and 4.

(3, 4] means $\{x\,|\,x > 3$ and $x \le 4, x \in \mathbb{R}\}$; in other words, all the real numbers between 3 and 4, including 4 but not 3.

By now you should see that the "(" means that the number next to the parenthesis is not included in the interval and "[" means that the number next to the bracket *is* included in the interval. Thus [3, 4) means include 3 but not 4, remembering that all the real numbers between 3 and 4 are also in the interval. These numbers next to the parentheses and brackets are called *endpoints*. In the example above, [3, 4), then, the endpoint 3 is included and the endpoint 4 is not included.

A problem arises with sets as $\{x\,|\,x \ge 5, x \in \mathbb{R}\}$. This set is written $[5, \infty)$ (or $[5, +\infty)$) with $+\infty$ meaning that the set goes on indefinitely in the positive direction. We always use parentheses, not brackets, with the ∞ symbol because one cannot include an endpoint that one never reaches and, indeed, does not even exist. Thus

$$\{x\,|\,x \le 5, x \in \mathbb{R}\} = (-\infty, 5]$$

and

$$\{x\,|\,x > 5, x \in \mathbb{R}\} = (5, +\infty).$$

DEFINITION 13
Interval

We summarize this discussion with the following list of all eight kinds of intervals along with graphical representations of each. Read through the list and try to understand each one. Can you tell, for example, that [3, 4) is of type (2) and $\{x\,|\,x \in \mathbb{R}$ and $x > 7\}$ is of the type (5)?

1. $(a, b) = \{x\,|\,x \in \mathbb{R}$ and $a < x < b\}$.
2. $[a, b) = \{x\,|\,x \in \mathbb{R}$ and $a \le x < b\}$.
3. $(a, b] = \{x\,|\,x \in \mathbb{R}$ and $a < x \le b\}$.
4. $[a, b] = \{x\,|\,x \in \mathbb{R}$ and $a \le x \le b\}$.
5. $(a, \infty) = \{x\,|\,x \in \mathbb{R}$ and $x > a\}$.
6. $[a, \infty) = \{x\,|\,x \in \mathbb{R}$ and $x \ge a\}$.
7. $(-\infty, b) = \{x\,|\,x \in \mathbb{R}$ and $x < b\}$.
8. $(-\infty, b] = \{x\,|\,x \in \mathbb{R}$ and $x \le b\}$.

Types 1, 5, and 7 are called *open intervals*; types 4, 6, and 8 are called *closed intervals*, and types 2 and 3 are called *half-open intervals*.

Convention usually dictates that one should not consider as intervals such things as (2, 2). After all, if it were allowed, it would be $\varnothing$. Also, such intervals as [2, 2] $(= \{2\})$ while legal, are generally to be avoided.

Although the interval (a, b) could be confused with the ordered pair (a, b) as a point in the plane, the context is ordinarily sufficient to remove any ambiguity. Otherwise, one must specify which he means.

3. PROBLEM ANALYSIS AND FLOWCHARTS

3a. Introduction

What do we mean by "an algorithmic approach" to the elementary functions? To answer that, one needs to know what an algorithm is.

For most purposes, it suffices to define an algorithm as any set of unambiguous instructions designed to accomplish a specific task. A more precise definition, one acceptable as a mathematical definition, requires a few additional provisions. The instructions must be linked in a definite order so that completion of the instructions is accomplished in finitely many steps, and the instructions need to be specific enough so that exactly the same activities and results occur each time the same starting conditions occur. These restrictions rule out such things as attempts to get an exact, complete decimal representation of $\frac{1}{3}$ and rule out such instructions as "Pick a number between 1 and 10."

DEFINITION 14
Algorithm

You are already familiar with algorithms to add, subtract, multiply, and divide real numbers, to factor trinomials, and to solve various kinds of equations. You may also know algorithms for computing square roots, completing the square, constructing various geometric figures, and finding missing angles or sides in certain triangles. Indeed, the solutions to most problems met in mathematical (and many other) settings can be expressed as sets of directions.

For example, one solves the linear equation $ax + b = c$ by (1) subtracting b from both sides of the equation, (2) dividing both sides by a, and (3) observing the number that results on the right-hand side of the last equation.

EXAMPLE 7a

Our algorithmic approach, then, consists primarily of an attempt to emphasize and make clear the careful thinking that goes into most mathematical procedures, and to nurture your ability to work and communicate at such a level. Thus the purpose of the algorithmic approach is to encourage understanding that goes beyond a mere recognition and repetition of facts and sample problems.

3b. Flowcharts

Flowcharts are algorithms in a diagrammatic form. More specifically, a flowchart consists of various shaped boxes containing instructions and arrows connecting the boxes in the order in which the instructions are to be carried out. The purpose of a flowchart is to make the logic or flow of the algorithm more visible, and hence more readily understood than is the case if one merely writes down a list of the instructions. A flowchart, therefore, is nothing more than a conceptual tool.

DEFINITION 15
Flowchart

As an example, let us write a flowchart for the equation-solving algorithm mentioned in the introduction to this section (Figure 6).

EXAMPLE 7b

Several points may be apparent to you immediately. For one, the algorithm is short and simple enough so that the flowchart may not really add anything to your understanding. If that is the case, you need only be patient for a few pages until we encounter more substantive algorithms, where the flowchart will prove of benefit. Also, we have used three different shapes for the boxes. The use of different shaped boxes for different kinds of instructions lends additional visual impact to the attempt in the flowchart to display geometrically the logical structure of the algorithm. The shapes we shall use are adopted from the field of computer science (although there is not, as yet, total standardization within that field). There are two reasons for this choice. First, these shapes are as convenient as any, and second, should you decide to pursue further education in computer science or even only a bit of computer programming, the practice gained from this book will be more easily transferred.

The shapes introduced in the above example are :

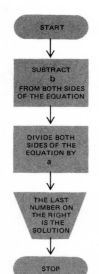

An *oval* for start and stop instructions. These are worthy of high-
lighting because, in a more complicated flowchart, they can be
as hard to find as the ends of a tangled piece of yarn.
A *rectangle* for " process " instructions. These are the usual " do this,"
" do that, " kinds of instructions.
A *trapezoid* for instructions that identify answers or insert new
information. We will refer to these activities as *output* and *input*,
respectively. These words are suggestive, and have a natural carry-
over into a computer setting.

We shall shortly encounter a few other shapes for boxes in flowcharts.

Still another observation you may already have made is that the flow-
chart of Figure 6 has some built-in assumptions. It assumes that the equation is set up in a particular way; it assumes that one understands the roles of the letters b and a in the equation; and it also assumes that one knows how to do arithmetic. In the context of our earlier formulation of the problem, these are all reasonable assumptions to make. But in a more general environment the author of a flowchart needs to be very aware of the context in which the flowchart will be read. For example, if an algorithm you were writing involved completing the square, " Complete the square " would be sufficient instructions to a mathematics professor, while a very detailed description of how to do all the arithmetic involved would be required for a student who had never learned any algebra at all. One of the issues to which we shall repeatedly return in this section is that of this context-dependence of a flowchart. In particular, we shall try to establish a general context for the flowcharts to be used and written in this course. Particular details of this, however, will depend on the structure of your class and so must come from class discussions and activities.

FIGURE 6

EXAMPLE 8   Now let us move to another example and, simultaneously, a new type of box. Again we choose a short example, an algorithm describing the activities of a hypothetical working person upon waking up (see Figure 7).

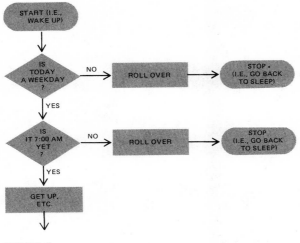

FIGURE 7

The next box shape is a *diamond* for "decision" instructions. These are usually for questions that admit a yes or no answer.† The "fork in the road" diagram that results is one of the more important visual improvements of a flowchart over an ordinary list of instructions. Notice that we may use more than one "stop" if we wish, or we could have noted the similarity of the two rightward branches of the flowchart and written it in the form shown in Figure 8. This has the advantage of showing at a glance that the same course of action results from two different conditions.

The next example brings us back into a more mathematical context: a flowchart of the usual definition of absolute value (see Figure 9). That definition is

EXAMPLE 9

$$|x| = \begin{cases} x & \text{if} \quad x \geq 0 \\ -x & \text{if} \quad x < 0. \end{cases}$$

Notice that this definition does have an algorithmic flavor—absolute values are "produced" by the definition, so a flowchart is not out of order. Indeed some of you may find the flowchart adds perceptibly to your ability to visualize and understand absolute value.

---

† While almost any decision can be dissected into a sequence of yes-no type decisions, one does find occasions when multibranched decision instructions are valuable. Adaptations of the diamond box are appropriate in such cases. For example, you will recall that, in the quadratic formula, the nature of the set of solutions depends on whether the discriminent $b^2 - 4ac$ is negative, zero, or positive. A flowchart for this formula might advantageously include the following decision:

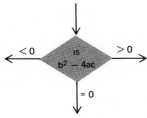

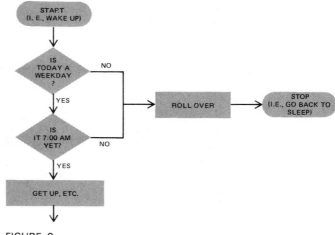

FIGURE 8

The two flowcharts in Figure 9 communicate the same algorithm, but the one labeled (b) has new notations inside some of the boxes. We shall use these notations frequently from now on because of the nature of the algorithms to be encountered in this book. Because much of our work will entail establishing numerical values for quantities, such an activity ought to be given a suggestive form in our flowcharts. We distinguish between two kinds of assignment (though classifying a particular assignment as one or the other type may not always be obvious). When an algorithm accepts numbers as raw materials to be worked with (such as x in the absolute value example, or a, b, and c in the $ax + b = c$ equation example), we consider these as "input" and register their arrival on the scene by an input instruction.

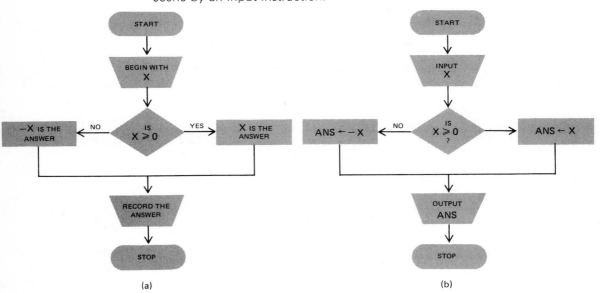

FIGURE 9

When an algorithm generates a number, whether an answer or only an intermediate result, then we think of assigning this value to another variable and indicate the assignment by a leftward arrow. This occurs within a "process" box, since it is a routine "doing" instruction.

FLOWCHART NOTATION

As individual examples of these "assignment statements," we could have **EXAMPLE 10**

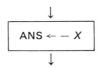

$$Z \leftarrow 5$$

which would assign Z the specific value 5 (i.e., Z then is a shorthand for 5) or

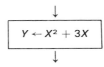

$$ANS \leftarrow -X$$

as in the flowchart above, or

$$Y \leftarrow X^2 + 3X$$

which would mean to take some previously established value represented by the variable X, perform the arithmetic indicated on the right side of the arrow, and then call the result Y.

As for the output instruction, it is used to signify the desire to report some information, whether a final answer(s) or intermediate results of some kind.

It is worthy of explicit attention to observe that flowcharts make use of variables much as is done in algebra. A symbol is used to represent a number, because the algorithm is communicating a general process that can be used on many different numbers, and one wishes to communicate the general process without becoming entangled in specific examples. Unlike algebra, however, we often let variables in flowcharts be words or abbreviations, rather than single letters. This is helpful in reminding the reader of the flowchart what the variable represents.

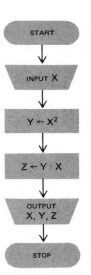

FIGURE 10

Now, consider another example to clarify some of these ideas. Suppose **EXAMPLE 11** we had the flowchart shown in Figure 10 and wished to "feed" the algorithm the number 5.

Then as we obey

INPUT X

X would become a symbol for the number 5.

Next

$$Y \leftarrow X^2$$

tells us to square 5 (yielding, of course, 25) and call this number Y.
Then,

$$Z \leftarrow Y \cdot X$$

tells us to take 5 and 25, multiply (we get 125) and call this number Z.

The output instruction tells us to report (whether by writing on a piece of paper, calling out to a listener, or some other way) the numbers 5, 25, and 125, in that order.

We may immediately return to this algorithm with the number $-\frac{1}{2}$, in which case X now represents $-\frac{1}{2}$, Y becomes $\frac{1}{4}$, and Z becomes $-\frac{1}{8}$. Therefore, the variable names in flowcharts have the same kind of generality as do x and y in the algebraic statement $(x + y)^2 = x^2 + 2xy + y^2$.

DEFINITION 16
Tracing

The process of reading a flowchart and performing the algorithm yourself with a specific example is called *tracing* the flowchart. It is highly recommended that whenever you write or encounter a flowchart, you trace it to be sure that it does what you intended and to get a "hands on" feel for the algorithm being described.

DEFINITION 17
Loop

Many algorithms require repeatedly performing some task or tasks until some condition is met. Flowcharts are adept at picturing such repetitions and motivating the following definition: Any set of instructions that may be executed more than once in succession is called a *loop*.

As our first example of an algorithm with a loop, we shall work with a nonmathematical problem. We shall also use this as our first example of the evolution, or creation, of a flowchart from the statement of the problem through the initial form of solution to the finished form.

EXAMPLE 12

Suppose that a motorist stopped and asked us for directions to the nearest hamburger stand. In solving this problem, we would assume that the motorist knew the basic directional instructions and basic vocabulary associated with driving a car. Our solution might be the following:

Drive to the end of this block and turn right. Drive past several traffic lights until you reach one on the same corner as a theater, then turn left. Drive $3\frac{1}{2}$ blocks further and turn right into the hamburger stand's parking lot.

This would be fine for most motorists, but for illustrational purposes we go into more detail. We suppose that the motorist can absorb and carry out only one instruction at a time and that instructions as complicated as that given in the second sentence in the above solution are too complex for him to understand. If we separate compound sentences and dissect our thinking more carefully, we could come up with a new set of instructions:

1. Drive to the end of this block.
2. Turn right.

3. Drive to the next traffic light.
4. If there is a theater at this corner go to instruction 5; otherwise return to instruction 3.
5. Turn left.
6. Drive $3\frac{1}{2}$ blocks.
7. Turn right into parking lot.

FLOWCHART NOTATION

These directions are very much in the spirit we shall follow. Each instruction is fairly simple, and the instructions are linked in a clear sequence. They are then easily translated into the flowchart of Figure 11.

Before leaving this example, however, a few more words about loops are appropriate. It is very important that *exit* from a loop be provided, that one go through the loop a certain number of times but still be sure of eventually going on to other tasks (i.e., be sure of exiting from the loop). Merely providing an arrow that leads out of the loop, however, may not suffice. An instruction that seems to be an exit may not always turn out to be so. For example, in our directions to the hamburger stand, if our memory had been wrong and the theater was actually on a different street, the motorist following our directions might never (in theory) find the conditions necessary to exit from his loop of stopping at traffic lights to look for a theater. If instead we had been able to tell him how many traffic lights to go through before turning, our directions would have been less likely to result in an infinite loop. Therefore, be sure your loops have good exit criteria.

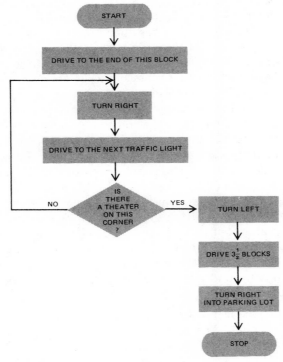

FIGURE 11

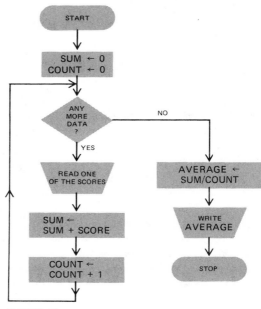

FIGURE 12

EXAMPLE 13   It is now time to become more actively involved with flowcharts; with this in mind, we urge you to take pencil and paper and actually trace the algorithm of Figure 12. It is designed to input a set of test scores. We suggest that you make up three or four test scores; find the average yourself; then make columns labeled SUM, COUNT, SCORE, and AVERAGE, in which to keep track of the most recent value assigned to each variable; and finally, use the flowchart with your data in a very mechanical fashion doing exactly what the instructions say, no more and no less.

If you had any difficulty in tracing the algorithm, try to find exactly where and then formulate a specific question. This is a very important activity in reading mathematics books and learning mathematics in general, but one that most students do not *make* time for. However, the investment of your time and effort will pay off in a growing ease of reading and learning mathematics. (The same comments apply also to doing examples as suggested!) The activity of trying to isolate and define your difficulty will often tell you where to get the information to resolve the difficulty yourself, and thereby provide a substantive instance of self-learning. Otherwise, it will enable you to get more directed help from a friend or instructor and again *your* learning will be aided because *you* are more in control of it.

EXAMPLE 14   Our next to the last example will be the construction of an algorithm to compute factorials. The factorial of an integer N is defined to be the product $N(N-1)(N-2) \cdots (3)(2)(1)$, and is denoted $N!$. Thus $4! = 4(3)(2)(1) = 24$, while $7! = 7 \cdot 6 \cdot 5 \cdot 4 \cdot 3 \cdot 2 \cdot 1 = 5040$.

Our initial view of the problem might produce a flowchart such as that in Figure 13.

This flowchart restates the definition given above, but for many of your classmates, reading this flowchart without reading the explanation just preceding it would not have clearly communicated the meaning of N ! More detailed explanation of the "$\cdots$" is called for.

Flowcharts we shall write should ordinarily be devoid of any notations that require thinking on the part of the reader. As a guideline, it may help to think of writing your flowchart for an obedient but firmly rooted litera-list,† who knows how to do arithmetic, compare two numbers as well as input and output numbers, but who has a very short memory (so that one needs to keep record of numbers through the use of variables) and who has essentially no common sense.

The amplification we need for our N ! flowchart comes in the form of a loop. Each time we go through the loop, we wish to multiply by the next number in the list of factors. Notice that different numbers as input will require a different number of trips through the loop, so that our exit criterion or criteria (often there is more than one appropriate way out of a loop) will need to reflect this variability.

It is true for most algorithms that different people will create different-looking flowcharts which are entirely equivalent in their end results. So do not be alarmed if your flowcharts trace correctly yet appear different from your classmates' flowcharts or flowcharts in the "Answers." Two different flowcharts would result here by starting the multiplication with N and "multiplying down" to 1, or by starting with 1 and "multiplying up" to N.

We choose to start multiplying with 1 and work up until we reach N (in this way we shall not be tempted to change the value of N in the process of finding N ! and so end up with both N and N ! when we are done), but the other is equally valid. If you try to modify the flowchart in Figure 13 yourself, you will discover the need to store both the partially computed answer and the variable used for the next factor in the list. One possible version of the flowchart is shown in Figure 14.

Now trace this flowchart for $N = 4$ and $N = 7$ to see how it generates the values 24 and 5040 claimed earlier. Notice that this leaves no instruction open to either the need or possibility of interpretation by a reader. Also, ANS does not in fact represent the answer until the last execution of the loop. The moral of that story is do not believe that a variable is what its name suggests until you have verified if and when it actually is, whether or not you are the author.

And now to a final, more complicated example.

EXAMPLE 15

The problem we shall solve is to grade a five-question multiple choice test for a whole class of students.

If you were to do the grading yourself, you would probably write an answer key and then take each student's answers, compare them with the

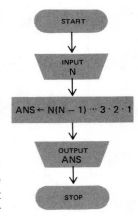

FLOWCHART NOTATION

START

INPUT
N

ANS ← N(N − 1) ⋯ 3 · 2 · 1

OUTPUT
ANS

STOP

FIGURE 13

---

† You know the kind—the one who really *does* put on his shoes and socks—in that order!

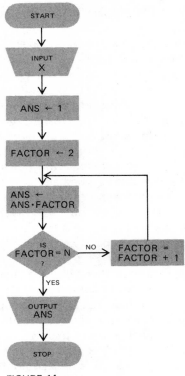

FIGURE 14

key, one at a time, add the number of correct answers and this would be the grade. As with the N! algorithm just completed, such an initial analysis provides a point of departure. It often suggests major subdivisions of the algorithm that can be of value, since " divide and conquer" is often a very useful strategy; i.e., break a problem into smaller (easier) subproblems and solve these subproblems one at a time. At the very least, an analysis at this level gets one started in actively solving the problem.

One might well write an initial, very crude flowchart that describes the algorithm in these broad terms. Although our example is not too involved, Figure 15 shows such a flowchart for this problem.

As a first refinement of this flowchart, let us try to analyze the second rectangular box further. Considering one student at a time, we shall compare answers with the key and add the number correct, then output. " One student at a time" suggests the use of a loop, so our first refinement of the second rectangular box might look like Figure 16. Note the use in that figure of an amorphous form called a *blob* to remind us of jobs left undone at such an intermediate stage. It is wise to be liberal in the use of such notes and reminders. Let us formalize this new (and quite unstandard) flowchart shape : a *blob* is used for a wide variety of reminders in incomplete flowcharts.

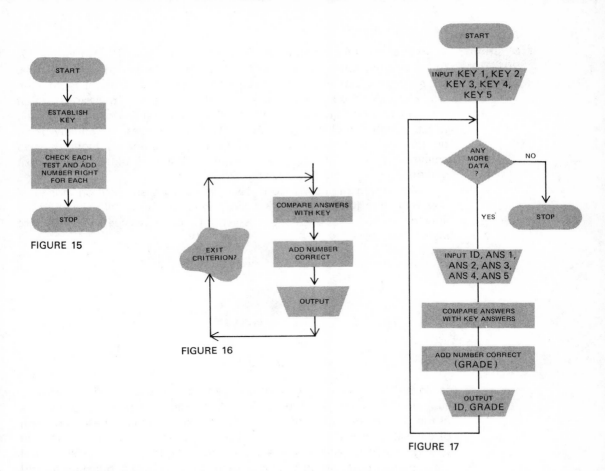

FIGURE 15

FIGURE 16

FIGURE 17

Before we can dissect this segment further, we need to know more about the data for this problem. For this, we turn our attention to input and output. We shall assume that the answers to the test are coded by numbers, not letters, and we shall plan on inputting all of one students' answers at a time. In inputting several numbers at once, the first number encountered is assigned to the first variable in the list to be read, the second number to the second variable, and so on.

Before constructing any input instructions, it is also wise to consider the output needs. In our example, it would be sensible to output not only the scores, but also who got which score. We shall need to input such identifying information, then, even though it is not used in the computations at all. Let us identify students by student number in our flowchart.

So far, then, we shall need to input a student number and list of test answers for each student, and output the student number and number correct for each student. Our initial flowchart (Figure 15) reminds us that the key needs to be considered, too. It will have to be read in before we can do any grading. At this stage, we might wish to write an intermediate flowchart for the whole problem, which includes all the analyses so far, but including sections that still need further elaboration. Figure 17 shows such a flowchart.

Notice the incorporation of the segment of Figure 16 into Figure 17. In its original form, the segment did not include any input provisions, but as we analyzed other facets of the problem, we found this an appropriate place for input. Rough flowcharts are likely to have such omissions so that the student who follows this pattern of constructing flowcharts is advised to use rough flowcharts as guides, not finished pieces, continually being on the lookout for omissions and rearranging steps as more detailed analysis suggests. Notice, too, in Figure 17, the resolution of the exit criterion question left unresolved in Figure 16.

All that remains now is to dissect the actual grading of an individual test. Since our friend the literalist is of too limited a mental capacity to be able to look at the key and set of answers all at once, we need a sequence of instructions such as

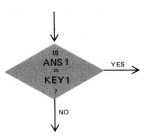

to do the comparison. But what are we to do for each of the paths from such an instruction? In grading the tests by hand, we make a mental note of the correct answers or of the wrong answers, summing these at the end of each test, or else make marks on the paper to keep track of right and wrong answers, tallying these marks at the end. Our literalist cannot make mental note, but in the latter case, we are really suggesting a rudimentary form of counting. The most sensible procedure for our literalist is to keep a running count of the number of correct answers. Then no separate addition would be needed at the end (again our earlier version of the flowchart would have been misleading had we adhered to its exact structure, instead of using it as a reminder of tasks to be done). In other words, if the answer to the question in the above decision instruction is YES we add one to the running count and continue to the next instruction; if the answer is NO we merely continue to the next instruction. In putting this into our flowchart, we should also see the need to reset the running count to zero as we start grading each student's test. If we call our running count variable GRADE, Figure 18 gives a final flowchart describing our algorithm to grade the multiple choice test. You should make up a few sets of data and trace this flowchart to gain more insight into its structure.

In Figure 18, the two small circular boxes with the number "1" inside are called *remote connectors*. They are used to replace long arrows, which would complicate a flowchart. When one encounters a remote connector in tracing a flowchart, one searches for the remote connector with the same number that has an arrow leading out of it, rather than into it (there may be several of the same number with arrows coming into them but only

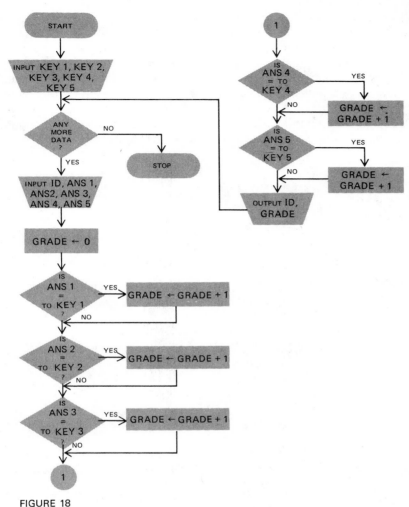

FIGURE 18

one with an arrow going out—different numbers are used to distinguish different such interrupted pathways).

Furthermore, notice how our use of flowcharts in this whole example has been in a kind of symbiotic relationship with our understanding of the problem and its solution. Each new version of the flowchart provided a foundation on which to build new understanding, which in turn enabled us to write a more complete flowchart. The possibility of such assistance in thinking by use of flowcharts is one of the reasons we encourage you to approach problems and procedures from a flowchart orientation.

We can carry such a symbiosis further in this particular problem to get a more compact flowchart.

Notice that the flowchart just completed (Figure 18) has five similar groups of instructions to perform the actual grading. This kind of repetition can be very tedious and suggests the use of subscripted notation combined with a loop to shorten the flowchart. Figure 19 exhibits the flowchart of Figure 18 with this modification. One frequently encounters circumstances in which the use of a loop like this one, which uses a counter, is of great value; therefore, an early mastery of this technique is recommended.

This test grading example has been worked out in detail to suggest strategies and methods for attacking problems and constructing algorithms. It should be noted, however, that individuals will differ in their approach.

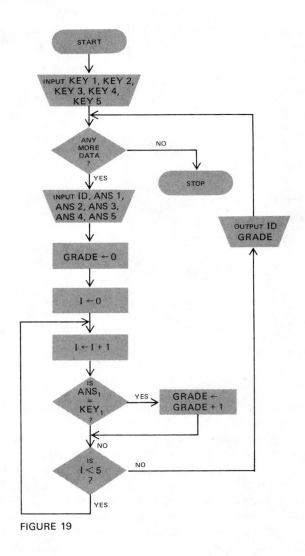

FIGURE 19

SUMMARY OF IMPORTANT CONCEPTS AND NOTATION

This summary, found at the end of each chapter, is intended to help the reader find and review major concepts that were covered in the chapter.

SET: (Definition 1) A set is any collection of objects with the property that every object either is in or is not in the collection.

ROSTER: (Definition 2) The roster form of writing sets lists all the elements between two braces with the elements separated by commas.

SET BUILDER: (Definition 3) The set builder form exhibits a property that is characteristic of all elements of the set and no other objects. The notation used is $\{x|P(x)\}$, where $P(x)$ represents the property that an object x must have to be an element of the set.

EQUALITY OF SETS: (Definition 4) Two sets are said to be equal if they contain precisely the same elements.

UNIVERSE (of discourse): (Definition 5) The universe of discourse is the set consisting of all the objects of interest to the current discussion.

ELEMENT (of a set): (Definition 6) The object x is an element of A (written $x \in A$) if x is one of the objects in the set A.

SUBSET: (Definitions 6 and 11) The set B is a subset of the set A (written $B \subseteq A$) if each element of set B is also an element of set A.

UNION: (Definition 7) $A \cup B$ is the set of all elements that are either elements of A or elements of B. Notationally,

$$A \cup B = \{x|x \in A \quad \text{or} \quad x \in B\}.$$

INTERSECTION: (Definition 7) $A \cap B$ is the set of all elements that are both elements of A and elements of B. Notationally,

$$A \cap B = \{x|x \in A \quad \text{and} \quad x \in B\}.$$

COMPLEMENTATION: (Definition 8) A' is the set of all elements that are not in set A. Notationally, $A' = \{x|x \notin A\}$.

RELATIVE COMPLEMENT: (Definition 9) $A - B$ is the set of elements that are contained in set A but are not contained in set B. Notationally,

$$A - B = \{x|x \in A \quad \text{and} \quad x \notin B\}.$$

EMPTY SET (null set): (Definition 10) The empty set is the set that has no elements and is denoted either by $\emptyset$ or $\{\}$.

DISJOINT: (Definition 12) Two sets A and B are disjoint if $A \cap B = \emptyset$.

SPECIAL SETS OF NUMBERS: (Section 2c)

$\mathbb{N} = $ natural numbers $= \{1, 2, 3, 4, 5, \cdots\}$

$\mathbb{Z} = $ integers $= \{\cdots, -3, -2, -1, 0, 1, 2, 3, \cdots\}$

$\mathbb{Q} = $ rational numbers $= \{x|x$ can be expressed as p/q, where p and q are integers$\}$.

$\mathbb{R} = $ real numbers $= \mathbb{Q} \cup \{x|x$ is irrational$\}$.

INTERVALS: (Definition 13) The interval $[a, b]$ is defined to be $\{x|x \in \mathbb{R}$ and $a \le x \le b\}$. (Similar definitions hold for each of the seven other types of intervals, as given in Definition 13.)

ALGORITHM: (Definition 14) An algorithm is a set of unambiguous instructions that

1. Are linked in a definite order.
2. Are completed in finitely many steps.
3. Accomplish a specific task.
4. Produce the same activities and results each time the same initial conditions occur.

FLOWCHART: (Definition 15) A flowchart consists of various shaped boxes containing instructions and arrows connecting the boxes in the order in which the instructions are to be carried out.

FLOWCHART NOTATION (Section 3b)

Start-stop: ⬭

Process: ▭

Input/output: ⬦

Decision: ◇

Assignment: ⟵——

Reminders: ♡

Remote connectors: ◯

TRACING: (Definition 16) Tracing is the process of reading a flowchart and performing the algorithm yourself with a specific example.

LOOP: (Definition 17) A loop is any set of instructions in an algorithm that may be executed more than once in succession.

PRACTICE EXERCISES

Exercises in this book will be divided into two categories: "Practice Exercises" and "Problems." Practice Exercises will consist of routine and drill-type problems to cement ideas or techniques discussed in the text; they will be identified by sections and subsections so that you may easily test and reinforce your understanding of a portion of the text before proceeding to the next. The Problems will contain a wide variety of problems which may be more involved than the practice exercises, or bring together ideas from various sections and/or chapters, or extend some ideas beyond the text material.

Generally, they will ask you to work as a mathematician in applying all your expertise. The problems will not ordinarily be arranged in any special sequence by either difficulty or content.

Section 2b

1. Write the following sets in roster form.
   (a) The set of courses you are taking
   (b) The set of states in the United States
   (c) The set of planets in our solar system
   (d) The set of positive odd numbers less than 20
   (e) The set of odd numbers greater than 20
   (f) The set of whole numbers that are perfect squares
   (g) The set of letters in the word roster
   (h) $\{x \mid x$ is a positive multiple of $5\}$
   (i) $\{x \mid x$ is a perfect square$\}$
   (j) $\{x \mid x$ is a prime number$\}$
   (k) $\{x \mid x$ is an integer and $10 \leq x \leq 15\}$
   (l) $\{x \mid x$ is an integer and $x \geq 15$ or $x \leq 10\}$
   (m) $\{x \mid x$ is a state in the USA$\}$
   (n) $\{x \mid x$ is a planet in our solar system$\}$
2. Write the following sets in set builder form
   (a) The set of books in your school's library
   (b) The set of students in your mathematics class
   (c) The set of U.S. senators currently in office
   (d) The set of all fractions between 0 and 1
   (e) The set of stars in our galaxy
   (f) The set of all geometric figures having exactly five straight sides
   (g) The set of all sets that contain precisely three elements
   (h) $\{1, 2, 3, \cdots, 20\}$
   (i) $\{10, 20, 30, 40, \cdots\}$
   (j) $\{3, 6, 9, 12, \cdots\}$
   (k) $\{-1, -2, -3, -4, \cdots\}$
   (l) $\{a, I\}$
   (m) $\{$Australia, Asia, Europe, North America, Antarctica, South America, Africa$\}$
   (n) $\{$Mercury, Venus, $\cdots$, Neptune, Pluto$\}$
3. Express the following in ordinary English.
   (a) $\{x \mid x$ is now the president of your college$\}$
   (b) $\{z \mid z$ is a member of your immediate family$\}$
   (c) $\{t \mid t$ is a bill currently pending in the U.S. Congress$\}$
   (d) $\{w, h, y\}$
   (e) $\{a, A, b, B, \cdots, y, Y, z, Z\}$
   (f) $\{10, 11, 12, \cdots, 98, 99\}$
   (g) $\{x \mid x > 0\}$ (h) $\{x \mid x \geq 0\}$
   (i) $\{x \mid x < 0\}$ (j) $\{x \mid x \leq 0\}$

(k) {stop, pots, tops, spot, post}

(l) {4, 16, 36, 64, 100, $\cdots$}

(m) {acute, scalene, obtuse, right, isosceles, equilateral, equiangular}

4. In each of the following problems, determine whether or not the two sets are equal.

(a) {a, b, c} and {x, y, z}

(b) {2, 4, 8} and {2, 2^3, 2^2}

(c) {1, 2, 5, 7} and {7, 5, 1, 2, 5, 1}

(d) {3, 5, 16} and {3, $\frac{32}{2}$, 5, 16}

(e) {a, b, c, d} and {1, 2, 3, 5, 9}

(f) {x, y, z, w} and {a, b, c, d}

(g) {0, 1, 2} and {2, 1, 0}

(h) {a, b, c, d, e} and {a, e, i, o, u}

(i) {$m, a, t, h, e, m, a, t, i, c, s$} and {$m, a, t, h, e, i, c, s$}

(j) {2, 4, 6, 8, $\cdots$} and {1, 3, 5, 7, $\cdots$}

(k) {e, a, s, y} and {h, a, r, d}

(l) {1, 4, 9} and {$-3, -2, -1, 1, 2, 3$}

5. If $A = \{a, e, i, o, u\}$; $B = \{a, b, c, \cdots, x, y, z\}$; $C = \{m, a, t, h, e, i, c, s\}$; and $D = \{g, h, i, j, k\}$, which of the following statements are true? If false, suggest a symbol that would produce a correct statement.

(a) $A \in B$ (b) $a \in B$ (c) $\{a\} \in A$

(d) $C \subseteq B$ (e) $e \subseteq C$ (f) $C \in B$

(g) $c \in B$ (h) $A \subseteq B$ (i) $D \subseteq B$

(j) $\{a, b, c\} \in B$ (k) $\{a, b, c\} \subseteq B$ (l) $D \subseteq B$

(m) $B \subseteq D$ (n) $g \in D$ (o) $x \subseteq B$

6. Find all the subsets of each of the following sets.

(a) {a, b}   (b) {1, 2, 3}   (c) {*, ?, ", @}

(d) {1, 2, 3, 4, 5}

7. If $A = \{$red, orange, yellow, green, blue$\}$, $B = \{$yellow, gold, black, green$\}$, $C = \{$red, green, blue, lavender, brown$\}$, and $D = \{$blue, brown, black$\}$, what are each of the following (in roster form)?

(a) $A \cap B$ (b) $A \cup B$

(c) $(A \cap B) \cup C$ (d) $(A \cup B) \cap C$

(e) $(A \cap B) \cap C$ (f) $A \cap B \cap C \cap D$

(g) $A \cup B \cup C$ (h) $C \cup D$

(i) $C \cap D$ (j) $B \cap D$

(k) $A \cup D \cup B$ (l) $A \cap (B \cap D)$

8. Find the following and express in either roster or set builder notation.

(a) B' if $B = \{1, 2, 3\}$ and $U = \{1, 2, 3, \cdots, 9, 10\}$

(b) A' if $A = \{a, b, c\}$ and $U = \{a, b, c, : \cdots, x, y, z\}$

(c) R' if $R = \{a, b, c\}$ and $U = \{a, b, c, \cdots, r, s\}$

(d) Q' if $Q = \{2, 4, 6, 8, \cdots\}$ and $U = \{x \mid x$ is a positive whole number$\}$

(e) X' if $X = \{1, 5, 10\}$ and $U = \{1, 2, 3, \cdots, 14, 15\}$

(f) Y' if $Y = \{1, 2, 3, \cdots, 13, 15\}$ and $U = \{1, 2, 3, \cdots, 14, 15\}$

(g) Z' if $Z = \{$red, orange, yellow$\}$ and
$U = \{$red, orange, yellow, blue, purple, brown, black$\}$

(h) K' if $K = \{\text{green}\}$ and
    $U = \{\text{red, orange, yellow, green, blue, purple, brown, black}\}$
(i) L' if $L = \{\text{brown, blue, black}\}$ and
    $U = \{\text{red, orange, yellow, green, blue, purple, brown, black}\}$
(j) P' if $P = \{a\}$ and $U = \{a, b\}$
(k) W' if $W = \{x \mid x \text{ is a week day}\}$ and
    $U = \{x \mid x \text{ is a day of the week}\}$
(l) T' if $T = \{y \mid y \text{ is a state with a Republican governor}\}$
    and $U = \{z \mid z \text{ is a state in the USA}\}$
(m) M' if $M = \{w \mid w \text{ is a word with three or fewer syllables}\}$
    and $U = \{x \mid x \text{ is an English word}\}$
(n) D' if $D = \{t \mid t \text{ is a math book}\}$ and
    $U = \{b \mid b \text{ is a book now in print}\}$

9. If $U = \{1, 2, 3, 4, 5, 6, 7, 8, 9, 10\}$, $W = \{2, 4, 6, 8, 10\}$,
   $Y = \{1, 2, 3, 4, 5\}$, and $Z = \{3, 5, 7, 9\}$, what are each of the following
   (in roster form) ?

(a) W'
(b) Y'
(c) $W \cap Y$
(d) $W \cup Y$
(e) $Y \cap Z$
(f) $W \cup Y \cup Z$
(g) $(W \cup Z) \cap Y$
(h) Z'
(i) $(W \cup Y)'$
(j) $(W \cap Y)'$
(k) $W' \cap Y'$
(l) $W' \cup Y$
(m) $(W \cap Y')'$
(n) $(W \cup Y')'$
(o) $(W' \cup Y) \cup W$
(p) $(Z \cap Y)' \cup W$
(q) $Y - W$
(r) $W - Y$
(s) $Z - (W \cup Y)$
(t) $(Z - W) \cup Y$
(u) $(W - Y)'$
(v) $W' - Y'$
(w) $W - W'$
(x) $W - (Y - Z)$
(y) $(W - Y) - Z$
(z) $(W' - Y) \cap Z$

10. If $U = \{x \mid x \text{ is a student at your college now}\}$
    $K = \{x \mid x \text{ is a mathematics major}\}$
    $L = \{x \mid x \text{ is male}\}$
    $M = \{x \mid x \text{ is a sophomore}\}$
    Describe each of the following sets in set builder notation.

(a) $K \cap L'$
(b) $L \cap M$
(c) $(K \cap L) \cap M$
(d) $K' \cap L' \cap M'$
(e) $(L' \cup M) \cup L$
(f) $K - L$
(g) $M - K$

Section 2c

11. (a) Give three representatives of the empty set such as the " disguises "
        in the text, but other than those mentioned.
    (b) Give three pairs of disjoint sets other than those in the text.
12. Write in interval notation (let the universe be $\mathbb{R}$).

(a) $\{x \mid x > 7\}$
(b) $\{x \mid x \geq 3 \text{ and } x < \frac{25}{19}\}$
(c) $\{x \mid x < 0\}$
(d) $\{x \mid x \leq 4 \text{ and } x \geq -4\}$
(e) $\{x \mid x \geq -16\}$
(f) $\{x \mid x < 29\}$
(g) $\{x \mid -1 \leq x < \frac{4}{5}\}$
(h) $\{y \mid -1 < y \leq \frac{4}{5}\}$
(i) $\{t \mid -1 \leq t \leq \frac{4}{5}\}$
(j) $\{z \mid z > 100\}$

13. Write the following intervals as sets of real numbers using set builder
    notation.

(a) (2, 3)   (b) (4, $+\infty$)   (c) [3, 21]

(d) [$\sqrt{2}$, 5)   (e) ($-\infty$, 3]   (f) (-20, $+\infty$)

(g) (-20, 30) (h) ($-\infty$, -15) (i) [10, $15\frac{1}{2}$)

(j) [$-\frac{3}{2}$, $\sqrt{13}$]

14. Which of the following are not valid intervals? Explain why not.

(a) (3, 7)   (b) (-4, -7)   (c) (0, ∞]

(d) [0.1, 0.2]   (e) ($-\infty$, 4)   (f) [10, 5)

(g) (2, 2)   (h) [2, 2]

Section 3b

For each of the problems 15–22,

 (a) Trace the following flowcharts with the given data and give the resulting output.

 (b) Hypothesize the purpose of the algorithm that the flowchart represents, explaining why you came to your conclusions.

*15.†

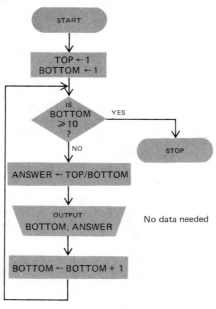

No data needed

---

† A starred problem (in the preliminary chapter) means that the problem is suited to programming in either Basic or Fortran (see Appendix 2).

*16. Data : 100, 0.05, 5.

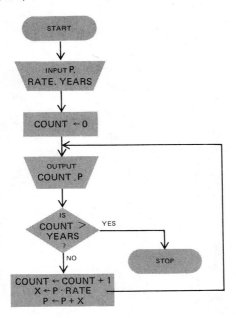

*17.

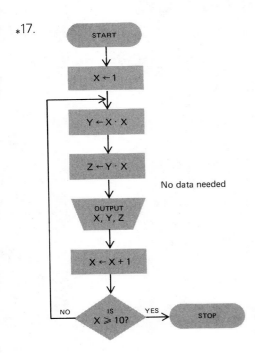

No data needed

*18. Data : 4, 9, 2.

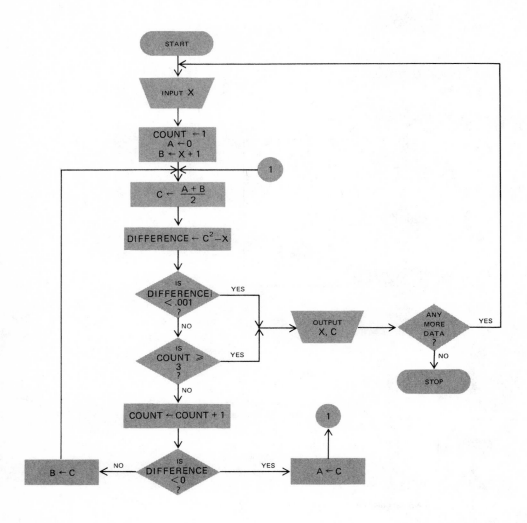

*19. Data : 3, 10, 6, 1, —2, 5.

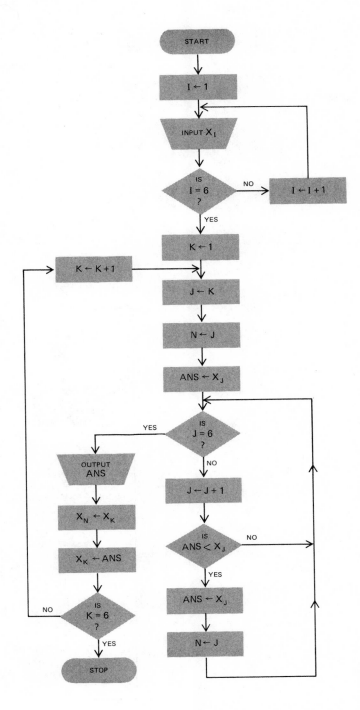

*20. Data: 77289, 77452, 13.2.

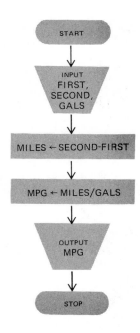

*21. Data: 70, 80, 90.

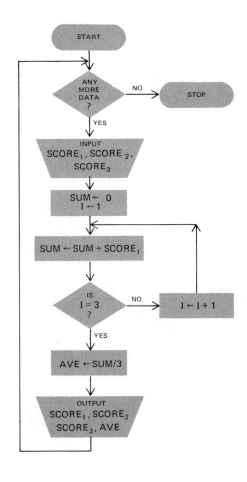

*22. Data: 9200, 3, 0.15.

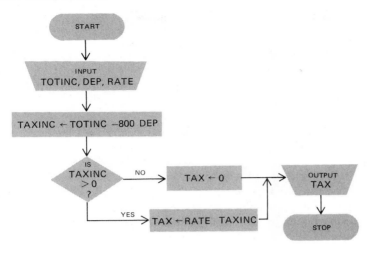

PROBLEMS

23. If $A = \{$lemon, licorice, orange, chocolate, lime$\}$,
    $A \cap B = \{$chocolate, licorice$\}$, and
    $A \cup B = \{$grape, licorice, orange, lime, raspberry, lemon, mint, chocolate, cherry$\}$,
    what is B?
24. There are 100 dogs in a particular dog show. If 40 are black, 35 have short hair, and 10 are both black and short-haired, how many are
    (a) neither black nor short-haired?
    (b) black but not short-haired?
25. Is it possible for there to be sets A, B, and C so that $A \in B$ and $B \in C$ but $A \notin C$? If not, prove so; if so, give an example.
26. Discuss the solutions to the following equations. (Hint: How many solutions are there for each? How might one solve such an equation?)
    (a) $A \cap X = A$ (b) $A \cap X = U$
    (c) $A \cup X = A$ (d) $A \cup X = U$
27. Which of the following relationships are true? Explain why not for for those that are false.
    (a) $(1, 2) \subseteq (3, 4)$ (b) $(1, 2) \in \mathbb{R}$
    (c) $[-1, 1] \cap [0, \infty) = [0, 1]$ (d) $(1, 2) \cap (2, 3) = \varnothing$
    (e) $(1, 2) \cup (2, 3) = (1, 3)$ (f) $3 \in (0, 10)$
    (g) $3 \subseteq (0, 10)$ (h) $(10, 20) \subseteq (0, 30]$
    (i) $[17, 20] \subseteq (15, 21)$ (j) $(1, 2) = \varnothing$
28. Carefully explain the difference between $\{\,\}$ and $\{\{\,\}\}$.
29. Write all the subset relationships that occur among $\mathbb{N}$, $\mathbb{Z}$, $\mathbb{Q}$, and $\mathbb{R}$.
30. Given a set S, we may associate with S another set called the *power set* of S, which is denoted $\mathscr{P}(S)$ and is defined by

$$\mathscr{P}(S) = \{A \mid A \text{ is a subset of } S\}.$$

(a) Write out the power set for {algebra, analysis}.

(b) Write out the power set for {a, b, c}.

(c) Write out the power set for {1, 2, 3, 4}.

(d) Having done parts (a), (b), and (c), suggest a formula for the number of elements in the power set of a set containing n elements. Can you justify or prove your formula?

31. The set builder notation is sometimes modified for efficiency. Rewrite each of the following examples of this in the standard set builder form (these do not exhaust the varieties of modification that are possible).

(a) $\{x > 0 \mid x^2 = y\}$ (b) $\{x \in \mathbb{R} \mid x^2 < 5\}$

(c) $\{n^2 \mid n \in \mathbb{N}\}$ (d) $\{x \in \mathbb{R} \mid x \notin \mathbb{Q}\}$

(e) $\{p/q \in \mathbb{Q} \mid 0 \le p < q\}$

32. It is a valid mathematical concern to inquire into the properties of the set operations we have at our disposal (and we shall give attention to similar questions about operations with functions in Chapter 1). We consider a few such inquiries in this problem. Recall that an operation $*$ is said

to be commutative if $a * b = b * a$

to be associative if $a * (b * c) = (a * b) * c$ and   } for all

to distribute over the operation $\square$ if $a * (b \square c) =$ } a, b, c

     $(a * b) \square (a * c)$

For each of problems (a)–(j), answer first with the aid of Venn diagrams, and then try to make a more formal proof.

(a) Is set union commutative?

(b) Is set intersection commutative?

(c) Is relative complement commutative?

(d) Does union distribute over intersection?

(e) Does intersection distribute over union?

(f) Does union distribute over relative complement?

(g) Is union associative?

(h) Is intersection associative?

(i) Is relative complement associative?

(j) (DeMorgan's Laws) Derive expressions for $(A \cup B)'$ and $(A \cap B)'$ that do not involve parentheses. (Be careful.)

33. Some additional, interesting algebraic properties of sets are stated below. Prove that each is true.

(a) $A \cup U = U$ for all $A \subseteq U$ (b) $A \cap U = A$ for all $A \subseteq U$

(c) $A \cap A = A$ for all sets A (d) $A \cup A = A$ for all sets A

(e) $A \cup \varnothing = A$ for all sets A (f) $A \cap \varnothing = \varnothing$

(g) $A \cap A' = \varnothing$ (h) $U' = \varnothing$

(i) $\varnothing' = U$

34. Looking at Problem 33, $\varnothing$ behaves for union as zero behaves for addition in the real numbers, while the universe behaves for intersection as the number one behaves for multiplication. Investigate this similarity a little further. (See next page.)

(a) Does $\emptyset$ behave for intersection as zero does for multiplication? (Explain carefully.)

(b) Does the universe behave for union as one does for addition? (Explain carefully.)

(c) Write down any further similarities or dissimilarities that you can find.

35. In Problems a–f, we use a new set operation called symmetric difference. It is defined by

$$A \bigtriangleup B = \{x \mid x \text{ is an element of } A \text{ or of } B \text{ but not both}\}.$$

For (c)–(f), answer first by using Venn diagrams; then try to make a more formal proof.

(a) Define $A \bigtriangleup B$ using A, B, intersection, union, and complement only.

(b) Draw a Venn diagram showing $A \bigtriangleup B$.

(c) Is $\bigtriangleup$ commutative?

(d) Is $\bigtriangleup$ associative?

(e) Does $\cap$ distribute over $\bigtriangleup$?

(f) Does $\cup$ distribute over $\bigtriangleup$?

36. Still another operation that one can perform on two sets A and B is called the *Cartesian product* of A and B. It is quite different from the operations studied so far in that the new set formed is a completely new set which has no subset relationship with either of the original sets. It is denoted by $A \times B$ (read "A cross B"), and is defined by:

$$A \times B = \{(a, b) \mid a \in A \quad \text{and} \quad b \in B\},$$

where (a, b) is an ordered pair (i.e., a set of two elements in which the order does make a difference—see p. 40). Thus, if $A = \{1, 2\}$ and $B = \{a, b, c\}$, then

$$A \times B = \{(1, a), (1, b), (1, c), (2, a), (2, b), (2, c)\}.$$

(a) Write out $A \times B$ if $A = \{a, b\}$ and $B = \{1, 2, 3\}$.

(b) Write out $A \times B$ if $A = \{$Joe, George, Mike$\}$ and $B = \{$Sue, Kathy, Barbara$\}$.

(c) Is $\times$ a commutative operation? Prove your claim.

(d) Does $\times$ distribute over union? Prove your claim.

(e) Does $\times$ distribute over intersection? Prove your claim.

(f) Give a geometric interpretation of $[1, 3] \times [1, 2]$. (Hint: You need a geometric interpretation for ordered pairs of real numbers.)

Problems such as 37, 38, and 39 occur with surprising frequency in many areas of data processing, although the abstract setting of these problems does not suggest this.

*37. Write a flowchart that will input two sets A and B, each of which consists of five numbers and output $A \cup B$ without writing the same element twice. (If $A = \{1, 2, 5, 8, 10\}$ and $B = \{2, 11, 7, 5, 6\}$, the input would consist of ten data: 1, 2, 5, 8, 10, 2, 11, 7, 5, 6, while the output should be the list 1, 2, 5, 8, 10, 11, 7, 6.)

*38. Write a flowchart as in Problem 37, but which outputs $A \cap B$ instead of $A \cup B$.

*39. Write a flowchart as in Problem 37, but which inputs a universal set of 10 numbers and a set A of as many as 10 elements, and outputs A'.

*40. Write a flowchart that will
    (a) Output a number and its square.
    (b) Output a number and its square root (you need not describe how to compute square roots, $y \leftarrow \sqrt{x}$ will suffice.)
    (c) Output a number, its square, and its square root. (Note: Once you have found the square of a number you can wait to output it until you have found its square root.)
    (d) Output a table of squares and square roots for the whole numbers between 0 and 100. (There are two general ways of doing this, by inputting each number and by using a counter to generate the numbers. Can you do both? The latter is much more efficient.)

*41. Write a flowchart that will
    (a) Cube a number.
    (b) Find three times the cube of a number.
    (c) Add seven to a number.
    (d) Find $3x^3 + 7$.

*42. Write a flowchart which, if given a set of monthly bills with the names of the people who owe the bills, will
    (a) Output a list of those less than $5.
    (b) Output a list of 10% of each bill that is greater than or equal to $5.
    (c) Output a list containing the customer's name, total bill, and, in a third column, either the total bill (if less than $5) or 10% of the bill (if greater than or equal to $5). This third column might be a listing of a minimum monthly payment on a credit card bill.

*43. Write a flowchart that will input a number and
    (a) Double it five times.
    (b) Double it N times, where the value of N is also read in.
    (c) Double it N times; handle seven such problems in succession before stopping.
    (d) Double it N times; handle M such problems in succession before stopping, where the value of M is read in, too.

*44. Write a flowchart that inputs the three coefficients A, B, and C of a quadratic equation $Ax^2 + Bx + C = 0$ and outputs the roots (use the quadratic formula and have the flowchart "discard" a problem having complex roots).

45. Write a flowchart for a beginning algebra student that describes how to complete the square. Assume that the student is familiar with signed number arithmetic, variables, literal constants, and the basic arithmetic of algebraic expressions.

46. Write a flowchart that describes the process of adding two numbers. For concreteness, suppose that the numbers are no more than five digits long.

*47. Redo the flowchart of Figure 14 so that the multiplication starts with N instead of with 1.

*48. Write a flowchart that accepts as input a price in dollars and a payment in dollars (each $\leq$ \$10) and output the "efficient change," i.e., the minimum number of coins and bills that total to the correct change.

*49. Write a flowchart that will input a value for x; compute $3x^2 - 5x + 2$; and output the result.

*50. Write a flowchart that will input three numbers and output the largest of the three. (Assume that one can compare only two numbers at a time.)
  (a) Assume that the numbers are distinct (i.e., all different).
  (b) Make provisions for possible repeated values.

*51. Write a flowchart that inputs a natural number N, finds the sum of the first N natural numbers (i.e., $1 + 2 + 3 + \cdots + N$), and outputs the result. (If you know a short formula for this result, do not use it.)

*52. (a) On a certain planet, a population explosion is predicted to be imminent. The population started with two individuals, and each decade the population has doubled. Write a flowchart that will output the population at the end of each decade so that someone looking at the total output can see the population growth and determine when population growth should be stopped. The flowchart should stop when the population has reached 10 billion.
  (b) Redo this problem with both the original population and the rate of increase per decade read in.

*53. Write a flowchart that inputs the number of hours a man works in a week, his hourly rate of pay, and percentage for deductions, and then computes and outputs his total wages, deductions, and net pay. The man earns time-and-one-half for the first 10 hours over 35 hours and double time for time beyond that.

*54. Write a flowchart that will find the largest number in a large set of numbers without storing the whole set. (Hint: Think of how you would determine the largest if you turned over cards one by one and could not go through the deck more than once.)

*55. Write a flowchart that inputs the test scores of 20 students, computes the mean score, and outputs the mean and all scores above the mean. (Hint: Use subscripts and loops to shorten your flowchart.)

*56. Expand and modify the flowchart of Figure 19 so that it will grade a 50-question multiple choice test.

*57. Expand and modify the flowchart of Figure 19 so that it will grade an n-question multiple choice test, where n is an integer that may vary on different runnings of the algorithm.

*58. Write a flowchart to compute a loan repayment schedule. It should input the amount of the loan, the interest rate *per month*, and the monthly payment. The output should consist of a table, each line of which gives the payment number, the interest accumulated during the previous month, the payment due, and the new balance after the payment is made. (The first payment comes 1 month after the loan is granted, and interest is computed each month on the unpaid balance of the loan, not the full original amount.) The last line of the table should show a zero balance; i.e., the last payment should be exact although it might well be less than a standard payment.

*59. Election day is drawing near, and as head of a nationally known polling organization, you have just received reports from your field workers indicating voter preferences in various parts of the country. You are to write a flowchart that will input the reports and determine the national preferences for the candidates in terms of percents. In particular, suppose that there are two persons running for offices and you have also recorded those who are undecided or favor minor candidates in a third category. Each field worker submits a card with the number for candidate A, then the number for candidate B, and then the number of others.

CHAPTER 1

RELATIONS AND FUNCTIONS

1.1. INTRODUCTION

The keystone of modern mathematics is the concept of function. The vocabulary and notation of functions pervade all of mathematics to the point that some areas consist almost entirely of the study of certain special sets of functions. For example, algebraic topology, sometimes referred to as "rubber-sheet geometry," is essentially the study of special functions called homeomorphisms. Even in those areas of mathematics in which the emphasis on functions is less total, the concept remains a common cord tying these areas to the mainstream of mathematics.

There are several ways of viewing the material to be covered in the remainder of this book. In certain respects, the contents fall into the category of mathematics in which there is only a moderate emphasis on functions. Many of the topics we shall study were originally developed without the modern functional approach. Although they could be learned in their original form, the use of functions will bring to light similarities that would otherwise remain hidden and will permit more rapid assimilation of the ideas involved.

In other respects, the contents of this book fall into the mainstream of mathematics in the sense that the elementary functions are the patriarchs of the family of functions. They also provide the framework on which the functional approach to mathematics is built and the substance from which many other functions are constructed. For example, some students may have encountered Fourier series, which are built up using circular functions.

If one views mathematics as a means for solving problems and describing phenomena, the elementary functions are also of great importance. When used in this context, they are called *elementary*, since they are the functions one encounters first and most often in applications. Although rather complicated, "nonelementary," functions exist which yield more precise and sophisticated results and descriptions (you may have encountered examples such as the Schroedinger wave equation or solutions to some differential equations), the elementary functions are neither trivial, nor useless, nor to be plowed through only for the sake of reaching a realm beyond. In both the theoretical and applied views, the elementary functions are extremely useful, interesting, and beautiful in their own right.

1.2. RELATIONS AND FUNCTIONS

1.2a. Relations

RELATIONS

One of mankind's most common activities is that of associating different objects or ideas. A physicist associates with each liquid such characteristics as its specific heat, freezing point, and viscosity. A librarian associates a call number with each book. A sports fan may associate with the umpire a string of not altogether complimentary epithets. With each course, a student associates its difficulty, the price of the textbook(s), the attractive members of the opposite sex in the class, and the value of the course for his intellectual and professional growth. Many disciplines other than those cited above seek to associate cause and effect, and so on.

Often sets of similar associations are grouped together in some way. The library card catalogue, a student's transcript of courses and grades, and the telephone book are examples of such groupings. In most cases, such sets of associations can be distilled further, to the point that they appear as sets of pairings. A particular object may be paired with many others. Thus one might associate with a particular woman her job, her address, her age, the name of her husband, her social security number, etc. But the set of associations may nevertheless be viewed as a set of pairings. In the above-mentioned examples, we might have the pairings Zedia Spif-accountant, Zedia Spif-807 College, Zedia Spif-35, Zedia Spif-George, Zedia Spif-133 55 8765, and so on. Often, there is a "directionality" to such pairings, too. For example, the physicist is likely to start with a set of liquids and associate with them their boiling points rather than start with the boiling points.

Mathematicians have a simple object that facilitates discussions of such pairings: the ordered pair. No doubt, you have already encountered ordered pairs and are familiar with the ordered pair (a, b) as a specialized set containing exactly two elements, the specialization being that order is now important, i.e., (a, b) is *not* the same as (b, a) unless a = b, whereas {a, b} = {b, a} for any choices of a and b. More generally, we define, the equality of ordered pairs by (a, b) = (s, t) if and only if a = s and b = t.

Using ordered pairs, the set of associations that we have been considering can be thought of as sets of ordered pairs. Again using the above example, we would have the set {(Zedia Spif, accountant), (Zedia Spif, 807 College), (Zedia Spif, 35), (Zedia Spif, George), (Zedia Spif, 133 55 8765)}. The order within each pair imparts the directionality of the associations: the first element—Zedia Spif—is associated with the second element—occupation, etc. If for any or all pairs (a, b), the order really makes no difference, we need only include both of the ordered pairs (a, b) and (b, a) to include the associations in both directions. For example, if we had the set of associations of numbers {(2, 3), (4, 5), (6, 7), (8, 9)} and it really made no difference which way the associations went, we would only need to include the reversed ordered pairs, and get {(2, 3), (4, 5), (6, 7), (8, 9), (3, 2), (5, 4), (7, 6), (9, 8)}.

DEFINITION 1.1
Equality of Ordered Pairs

This, then, is the starting place in our investigation of the elementary functions, sets of ordered pairs. We define a *relation* to be any set of ordered pairs.

DEFINITION 1.2
Relation

Such a general definition seems to open the door to many unruly and previously unwelcome individuals because there is no restriction that the sets of ordered pairs must have any sense, unifying theme, or purpose. But sense, theme, and purpose, like beauty, lie in the eye of the beholder. Whereas you might see no rhyme or reason to the set of ordered pairs {(1, 3), (2, 3), (7, 5), (5, 4), (27, 11)}, we might then point out that this associates with the numbers 1, 2, 7, 5, and 27 the number of letters in the English word for that number. While this particular relation may not be profound, it is, at least, sensible. So we will not try to put a condition in the definition that requires meaning or sense. Generally, though, in this book, we shall consider relations that, to us, do have some sense or reasonableness.

Although we are discussing a special kind of set, and sets are ordinarily denoted with capital letters, convention dictates that relations are denoted by any kind of letter—upper or lower case, English or Greek—and sometimes by other symbols. Thus we could have the relation $f = \{(x, y) \mid x^2 + y^2 = 25\}$ or the relation $\sigma = \{(s, t) \mid s$ is a state of the U.S.A. and t is a U.S. senator from that state}.

1.2b. Domain and Range

When dealing with relations, we often want to consider the set of first elements from all ordered pairs in the relation and the set of second elements from all ordered pairs in the relation. If we have the relation $r = \{(x, 1), (y, 2), (z, 3)\}$, the set of first elements would be $\{x, y, z\}$ and the set of second elements would be {1, 2, 3}.

DOMAIN AND RANGE

At times, these two sets have a semi-independent life. For example, the relation that pairs each freshman with his dormitory room might be visualized more dynamically as a set of freshmen arriving and then being divided among a set of rooms. The set of freshmen is the set of first elements from the ordered pairs making up the relation, while the set of rooms is the set of second elements.

At other times, one wishes to note that the rule has built-in restrictions and so needs to refer to those sets of first and second elements of the ordered pairs. For example in the relation $\{(x, y) \mid y = \sqrt{x}, x \in \mathbb{R}\}$, x must be limited to $[0, \infty)$ for y to be a real number† and y also turns out to be limited to $[0, \infty)$, not all of $\mathbb{R}$.

At other times, one wishes to limit the application of the rule that defines a relation, such as $\{(m, n) \mid m^2 = n, m \in \mathbb{N}\}$. Here, although we could have talked about squares of any real number, we wish to consider only the squares of natural numbers. Now that we have seen some reasons for considering the sets of first and second elements, let us look at the more precise definitions of these sets as well as the terminology we need to discuss them.

DEFINITION 1.3a
Domain

DEFINITION 1.3b
Range

If T is a relation, we let D_T denote the set of all first elements in the ordered pairs that comprise T, and R_T denote the set of all second elements in the ordered pairs that comprise T. The set D_T is called the *domain* of T, R_T the *range* of T. More succinctly, these definitions are

$$\text{domain of the relation } T = D_T = \{x \mid (x, y) \in T \quad \text{for some } y\}$$
$$\text{range of the relation } T = R_T = \{y \mid (x, y) \in T \quad \text{for some } x\}.$$

The specification "for some y" is to emphasize that for x to be an element of D_T, one does not need to know the exact identity of the corresponding y value(s), only that at least one does exist. The underlying idea in Definition 1.3 is that to be in the domain, an object must *be connected to* some element in the range; similarly, for the phrase "for some x" in the set definition of R_T.

Somewhat picturesquely, one can think of a relation as being a computer program. One inputs an element of the domain and receives, as output, each corresponding element from the range. One such program might output the numbers 9 and 27 when the number 3 was inputted. This would mean that (3, 9) and (3, 27) are members of that relation. In the context of this analogy, the domain of a relation is the set of all permissible input values, while the range is the set of all possible output values. The values, of course, need not be numerical.

EXAMPLE 1

Consider the following relations:
$$a = \{(1, 5), (2, 4), (3, 12), (4, 1)\}$$
$$b = \{(x, y) \mid y = x^2\}$$
$$c = \left\{(w, s) \mid s = \frac{1}{w}\right\}$$
$d = \{(t, z) \mid t$ is a letter of the alphabet and z is an English word that starts with $t\}$

---

† Remember if $x < 0$, i.e., negative, then $\sqrt{x}$ is an "imaginary" or complex number. For further discussion of this, see Chapter 7.

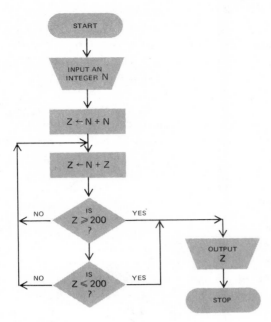

```
                    START

                    │
                    ▼
              INPUT AN
              INTEGER N
                    │
                    ▼
              Z ← N + N
                    │
                    ▼
              Z ← N + Z
                    │
                    ▼
    NO          IS           YES
◄──────      Z ⩾ 200      ──────►
               ?
                    │                  OUTPUT
                    ▼                    Z
    NO          IS           YES
◄──────      Z ⩽ 200      ──────►      STOP
               ?
```

FIGURE 1-1

$e = \{(x, y) \mid x$ is a valid input number for the flowchart in Figure 1.1 and y is one of the numbers outputted as a result$\}$.

Then the domains and ranges of these five relations are

$$D_a = \{1, 2, 3, 4\}$$
$$R_a = \{5, 4, 12, 1\}$$
$$D_b = \{x \mid x \text{ is a number that may be squared}\} = \mathbb{R}, \text{ since any}$$

real number can be multiplied by itself;

$$R_b = \{y \mid y \text{ is a possible value for } x^2\} = [0, \infty), \text{ since}$$

$x \cdot x \geq 0$ for all real numbers x, and each positive real is the square of its own square root $(y = \sqrt{y} \cdot \sqrt{y})$. Also,

$$D_c = \{w \in \mathbb{R} \mid w \neq 0\},$$

since one may divide 1 by any real number except 0;

$$R_c = \{s \in \mathbb{R} \mid s \neq 0\},$$

since any nonzero real number s is the reciprocal of its reciprocal $\left(\text{i.e., } \dfrac{1}{1/s} = s\right)$;

$$D_d = \{a, b, c, \cdots, x, y, z\}$$
$$R_d = \{B \mid B \text{ is an English word}\}$$
$$D_e = \{N \mid N \text{ is a permissible input value for the flowchart of Figure 1.1}\} = \mathbb{Z}$$

$R_e = \{Z \mid Z$ is a number obtainable from the flowchart of Figure 1.1$\}$.

(Can you find a more explicit description of the last set?)

A few words of caution are in order at this point. First, note that the domain and range of a relation are merely sets of elements and do not determine the rule that defines the relation. Although one may need to consult such a rule to determine the domain or range, the rule is not specified by the domain and range. Thus, knowing that $D_f = \mathbb{R}$ and $R_f = \mathbb{N}$ tells one nothing about how f associates elements in D_f with elements in R_f.

Second, one needs to distinguish clearly between the domain (or range) and the elements of the domain (or range). Thus one refers to 2 as an element of the domain of a in the example above but does not say that 2 is a domain of a.

Last, finding domains and ranges is usually a matter of careful observation. Go back over the previous examples to be sure you understand the why involved in each one. In general terms, given a set of ordered pairs (x, y) comprising a relation, the questions one usually asks in finding domain and range are: "What x values are 'legal' and produce legal values for y?" and "What are all the possible y values I can get if all possible legal values of x are used?" These questions will not always have easily found answers, but they should help once you understand their meaning.

1.2c. Functions

The concept of relation is a very broad one. It provides a place to start when one is concerned with sets of associated objects, but one ordinarily restricts his attention to specific kinds of relations and develops more structure and understanding within these more limited confines. One might expect the mathematician to restrict his attention first to relations that deal with numbers. Although the concern of this book is primarily with such relations, mathematics as a whole makes no such restrictions at all. Indeed, a great deal of mathematics has nothing to do with numbers. Instead, the restriction of most value and concern for the mathematician is the requirement that no two different ordered pairs in the relation have the same first element. Many people initially feel that this restriction does not isolate a meaningful set of relations. To help relieve this feeling, let us return to the computer analogy, where we see that such a relation corresponds to a program that outputs one and *only* one item for each element inputted.

Such a program might be one that outputted a customer's total bill when his name was inputted. If records are updated once each month, then on, say, October 5th, one would never get both $100.10 and $500.25 as bills for Z. K. Smith; i.e., one could *not* get the ordered pairs (Z. K. Smith, $100.10) and (Z. K. Smith, $500.25) from the relation represented by the program described above.

Such relations present no ambiguity when the question "What [single] range (second) element corresponds to this domain (first) element?" is asked. Mathematicians (and many others) want to keep their lives from being cluttered with too much information.

This restriction does leave open the possibility of several ordered pairs having the same second element. For example, the relation $c = \{(x, y) \mid x$ is a natural number and y is the number of letters in the English word for that number$\}$ contains no two ordered pairs with the same first element but does contain many with the same second element, e.g., (1, 3), (2, 3), (6, 3), and (10, 3).

Such "single-valued" relations are called functions. Thus, technically, a *function* is a set of ordered pairs no two of which have the same first element. Because of the restriction, we may say that the (single) range element y corresponds to a given domain element x, and we write $y = f(x)$ to express this correspondence between x and y in the function f. This is read "y equals f of x."

DEFINITION 1.4a
Function

We may rephrase the definition of function in a more self-contained form that is easier to understand. If some rule assigns to each element x of one set X a single element from a second set Y, then the set of ordered pairs (x, y) is called a *function f*, and one writes $y = f(x)$. The set X is called the domain of f; the set Y is called the range.

DEFINITION 1.4b
Function

Because the concept of function is central to the remainder of this book (and subsequent mathematics), you should stop now long enough to be sure you understand the preceding two paragraphs. Do you see, for example, that the two definitions convey the same concept? If not, try to formulate a specific question or objection. Or for another example, can you explain to a friend why $\{(1, 2), (2, 3), (3, 4)\}$ is a function but $\{(1, 2), (1, 3), (2, 4)\}$ is not?

Another notation that is commonly used with functions emphasizes the directionality of the ordered pairs and the general nature of the range. If $D_f = X$ and $R_f \subseteq Y$, one may speak of the function f from X to Y and write $f : X \to Y$, where Y is called the *codomain* of f. For example, we might wish to speak of $\psi : \mathbb{R} \to \mathbb{R}$ with the defining rule $\psi(x) = 3x^4 - 7x^3 + 2$. Now, $D_\psi = \mathbb{R}$, but the exact identity of R_ψ is not obvious. We do know, however, that $\psi(x)$ is always a real number, and our notation conveys this knowledge.

DEFINITION 1.5
Codomain

Along with this notation, we sometimes use a diagram reminiscent of Venn diagrams. If $g : A \to W$, we draw

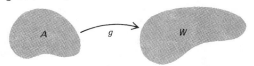

To illustrate the general idea of codomain, we can draw

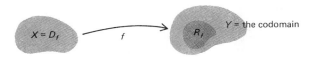

FUNCTIONS Related diagrams allow us to "watch" specific ordered pairs. If $k = \{(x, y) | y = x^2 + 2\}$, then $k : \mathbb{R} \to \mathbb{R}$ and, for ordered pairs $(-1, 3)$, $(0, 2)$, and $(2, 6)$, we would have

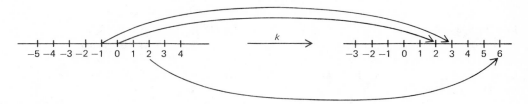

The $y = f(x)$ notation is sometimes altered to permit another way of defining a function: one writes $f : y =$ (some rule), e.g., $f : y = 3x^5 - 7x^3 + 2$.

Returning now to the $y = f(x)$ notation, we should carefully note that $f(x)$ is quite different from f. Using the function c on p. 45 as an example, we note that $c(x)$ refers to a specific range element: the one corresponding to the domain element x in the ordered pair $(x, c(x))$. Thus $c(3)$ is 5 (five letters in three), which corresponds to the ordered pair $(3, c(3))$ or $(3, 5)$. However, c alone refers to the entire function.

Nevertheless, the notation is often abused and one speaks of the function $f(x)$ or the function $y = f(x)$ instead of the function f. As long as the context generates no ambiguity and as long as we are willing to return to more precise statements whenever there is any need to do so, we shall allow such looseness for the sake of convenience.

DEFINITION 1.6
Argument and
Value

Corresponding to the words *domain* and *range*, but on the level of individual elements, are the words *argument* and *value*. An argument of a function is any element of the domain of the function, while a value of a function is any element of the range of the function. Thus, in reference to the function c above, one says that corresponding to the argument 3 is the value 5.

Functions are often specified with the help of an equation. For example, $g = \{(x, y) | y = x^3\}$ is a function that associates with each number its cube. Because functions are so frequently defined using equations, one is inclined to gloss over the difference and speak of the function $y = x^3$ instead of $\{(x, y) | y = x^3\}$. When we use $y = x^3$, we are really only stating a relationship between some number (x) and its cube $(x^3 = y)$. However, when we say the "function $y = x^3$," we mean "the set of ordered pairs that satisfy the equation $y = x^3$." This last statement is exactly what the more precise $\{(x, y) | y = x^3\}$ says.†

As the following examples indicate, the actual computation of functional values is a routine matter, even though your assimilation of the vocabulary developed above will probably require careful study.

EXAMPLE 2

1. Find $f(3)$ if $f = \{(x, y) | y = 4x^2\}$. Here, $f(3) = 4(3)^2 = 4 \cdot 9 = 36$. Thus $f(3) = 36$.

---

† Do you see that this issue is similar to the distinction made between f and $f(x)$? In fact, we usually write $f = \{(x, y) | y =$ (some equation)$\}$ and $f(x) = y =$ (some equation).

2. Find $g(-5)$ if $g(x) = 2x + x^2$. Here $g(-5) = 2(-5) + (-5)^2 =$ $2(-5) + (25) = -10 + 25 = 15$. Thus $g(-5) = 15$.

3. Find $h(0)$ if

$$h(x) = \frac{2x + 3}{4}. \text{ Here } h(0) = \frac{2 \cdot 0 + 3}{4} = \frac{0 + 3}{4} = \frac{3}{4}.$$

Thus $h(0) = \frac{3}{4}$.

In general, then, just "plug in" the value given wherever the argument appears on the right-hand side of the equation defining the function.

We have covered a good deal of ground in this section—it may take several readings to digest it all. Conceptually, the key point to remember is that functions are just special relations, ones in which each domain element has only one range element corresponding to it. Notationally, the following example summarizes the major features of this section.

The function $f = \{(x, y) \mid y = \sqrt{x + 7}\}$ can also be specified by the rule $f(x) = \sqrt{x + 7}$, or $f : y = \sqrt{x + 7}$, or even just by $y = \sqrt{x + 7}$. We can write $f : X \to Y$, where $X =$ the domain of $f = D_f = [-7, \infty)$ and Y, the codomain, may be $\mathbb{R} = (-\infty, \infty)$. The range of f, R_f, is a subset of Y, and is, in particular, $[0, \infty)$.

1.2d. Some Special Functions

Let us look at some particular functions as examples because these functions are ones that you shall encounter from time to time in this book and/or in your later studies. Note: Not all of the names we use are standardized.

$i = \{(x, x) \mid x \in \mathbb{R}\}$. This is called the *identity function*. It associates each real number with itself; D_i and R_i are both $\mathbb{R}$. Some sample pairs from i would be $(2, 2)$, $(3.65, 3.65)$, (π, π), etc. **EXAMPLE 3**

$h = \{(x, y) \mid y = \sqrt{x}\}$. This is called the *square root function*. Its domain is $[0, \infty)$, since square roots are defined only for nonnegative numbers. Its range is also $[0, \infty)$ because, by convention, $\sqrt{x}$ means the non-negative number whose square is x. (The negative square root is denoted $-\sqrt{x}$.) Thus $y = \sqrt{4}$ means the ordered pair $(4, 2)$, and $z = -\sqrt{4}$ means the ordered pair $(4, -2)$. Note: $(4,2) \in h$; $(4, -2) \notin h$. **EXAMPLE 4**

$k = \{(x, 0) \mid x \in \mathbb{R}\}$. This is called the *zero function*, since $k(x) = 0$ no matter what x is. Also, $D_k = \mathbb{R}$, while $R_k = \{0\}$. Sample ordered pairs are $(-6, 0)$, $(0.321, 0)$, $(10^{20}, 0)$, etc. **EXAMPLE 5**

For c, a fixed real number, let $p_c = \{(x, c) \mid x \in \mathbb{R}\}$. Each p_c is called a *constant function*. Then $p_2 = \{(x, 2) \mid x \in \mathbb{R}\}$ with such ordered pairs as $(3, 2)$, $(\pi, 2)$, $(0.6, 2)$ and $p_{9.1} = \{(x, 9.1) \mid x \in \mathbb{R}\}$ with such ordered pairs $(2, 9.1)$, $(-60, 9.1)$, $(\frac{1}{2}, 9.1)$, etc. (In fact, Example 5 is just p_0.) The domain, D_{p_c}, is $\mathbb{R}$, while the range, R_{p_c}, is $\{c\}$. **EXAMPLE 6**

$\psi = \{(x, |x|) \mid x \in \mathbb{R}\}$. This is the *absolute value function*. Recall that **EXAMPLE 7** absolute value is defined in pieces, that is,

$$\text{if } x \geq 0, \ |x| = x$$

but

$$\text{if } x < 0, \ |x| = -x.$$

For example, some ordered pairs would be $(4, 4)$, $(-4, 4)$, $(2.79, 2.79)$, $(-\frac{3}{4}, \frac{3}{4})$. In other words, the absolute value function performs a magic trick that makes every number positive, or more mundanely, associates with each number an intuitive notion of its size. $D_\psi = \mathbb{R}$ and $R_\psi = [0, \infty)$.

EXAMPLE 8   For each $n \in \mathbb{N}$, let $\phi_n = \{(x, y) \mid y = x^n\}$. These are called the *power functions*. Thus

$$\phi_2 = \{(x, y) \mid y = x^2\}, \qquad \phi_5 = \{(x, y) \mid y = x^5\},$$

etc. Also, $D_{\phi_n} = \mathbb{R}$ for each $n \in \mathbb{N}$ and

$$R_{\phi_n} = \begin{cases} \mathbb{R} & \text{if } n \text{ is odd} \\ [0, \infty) & \text{if } n \text{ is even.} \end{cases}$$

Thus, if n is odd (say $n = 3$), the output of the function can be either positive or negative $((-2)^3 = -8, (2)^3 = 8)$, but if n is even (for example $n = 2$), the "output" can be only positive $[(-2)^2 = 4, (2)^2 = 4]$.

EXAMPLE 9   The *greatest integer* function is defined by $g = \{(x, n) \mid x \in \mathbb{R} \text{ and } n \text{ is the largest integer for which } n \leq x\}$. A more common notation for this is $\lfloor x \rfloor$, so that $\lfloor 2\frac{1}{2} \rfloor = 2$; $\lfloor \pi \rfloor = 3$; $\lfloor 7 \rfloor = 7$, and $\lfloor -\frac{3}{2} \rfloor = -2$. (This function is closely related to the postage function of Problem 1(v) at the end of the chapter.) Then, $D_g = \mathbb{R}$, while $R_y = \mathbb{Z}$.†

EXAMPLE 10   Any computer program can be viewed as a function whose domain is the set of all possible data sets (each data set being one domain element), and whose range is the set of all resulting outputs from the computer. A specific example follows.

1.2e.   A Flowchart Example

Let us look at a sample flowchart to compute values of a particular function, both to strengthen our understanding of the concepts of function, domain, and range, and to keep flowcharting fresh in our minds. The function with which we will work is a pay-check function for a sales department of Springspeed's Discount Store. Each sales person is paid weekly on the basis of $100 plus a comission of 10% of his gross sales that week. For simplicity, let us assume that the withholding tax, social security, and other deductions standardly total 19.5% of each person's gross pay. The flowchart is written so that the payroll department provides weekly sales totals, one at a time, and the flowchart provides the corresponding net pays, one at a time. The flowchart in Figure 1.2 gives a

---

† This function occurs in both Basic and Fortran for positiye numbers. In Basic, the library function INT is just the greatest integer function, while in Fortran any change of mode from real to integer (by the library function IFIX or by an assignment such as I=X) is a use of the greatest integer function.

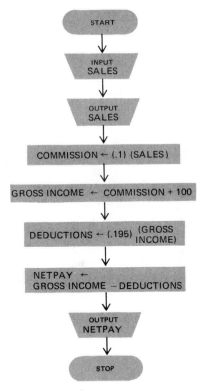

FIGURE 1-2

more precise definition of the function and may clarify your understanding that the domain of our paycheck function is the set of all individual weekly sales totals, while the range is the set of all possible net pays.

1.3. GRAPHS

1.3a. Cartesian Coordinate System

The seemingly simple step we are about to take from viewing relations algebraically to viewing them geometrically is far more profound than it may seem. Historically, from the death of Archimedes (212 B.C.) until the early 1600s, very little progress was made in the development of mathematics. Then, in 1619, René Descartes, in a series of three dreams, found the key to what is now called analytic geometry. He published these ideas as part of the *Method*† in 1637. With the influx of these new ideas, mathematics took on much of the character it still has today.

What is analytic geometry? We shall discuss this subject in some depth in Chapter 6, but basically it is a simple marriage of algebra and

---

† The full title, would you believe, is *A Discourse on the Method of Rightly Conducting the Reason and Seeking Truth in the Sciences, Further, the Dioptric, Meteors, and Geometry, Essays in This Method.*

geometry—this marriage of algebra and geometry takes place under the auspices of the ordered pair. We have seen the ordered pair in its algebraic role with functions and relations; let us now refresh our memories of the geometric interpretation of the ordered pair as a point in the plane.

First, one takes two perpendicular lines in a plane, one horizontal and vertical. We label each with a scale, as in Figure 1.3. The horizontal line is called the x-axis and the vertical line is called the y-axis.† Together they are called the coordinate axes.

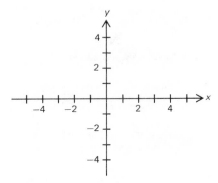

FIGURE 1-3

These axes divide the plane into four regions, called quadrants, which are standardly referred to by the Roman numerals shown in Figure 1.4.‡

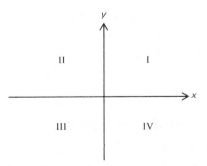

FIGURE 1-4

Within this framework, each point in the entire plane can be uniquely labeled by an ordered pair of numbers (a, b), where a is the horizontal distance from the point to the y-axis and b is the distance from the point to the x-axis. It is sometimes more convenient to think of a as the number on the x-axis directly below (or above) the point, and b as the number on

† In problems in which the letters x and y are not used, one may refer to the axes using the appropriate letters such as the s and m axes in Figure 1.9 on p. 55.
‡ Formal definitions of the quadrants are asked for in Problem 55.

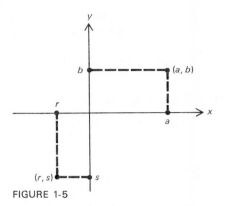

FIGURE 1-5

the *y*-axis to the left (or right) of the point, as indicated in Figure 1.5. (Do you see that this is equivalent to the first description of *a* and *b*? If not, look at the rectangle formed and think about it.)

For the point (*a*, *b*), the number *a* is called the *x*-coordinate, or *abscissa* of the point; and *b* the *y*-coordinate or *ordinate* of the point. The whole system of axes and labeling is called the *Cartesian coordinate system*.†

DEFINITION 1.7
Abscissa, Ordinate, and Cartesian Coordinate System

It is this coordinate system which makes possible the marriage of algebra and geometry to which we now turn our attention. Let us begin with any relation whose domain and range are composed of real numbers. As you will remember, such relations are often represented by some algebraic expression. The set of ordered pairs comprising the relation can be viewed geometrically as the set of points in the plane whose coordinate pairs are the same as the pairs comprising the relation. The result is a kind of photograph of the relation, called the *graph* of the relation. One important convention to keep in mind is that no matter what letters are used in a particular problem, the horizontal axis (the *x*-coordinate or abscissa) is usually used for the domain (i.e., first) element, while the vertical axis (*y*-coordinate or ordinate) is usually used for the range (i.e., second) element.

DEFINITION 1.8
Graph

Analytic geometry which we shall discuss at greater length later, takes the fullest advantage of the relationship between algebra and geometry. Even without a digression into analytic geometry, we can see that simply graphing a relation can be of value in several ways. Graphs of relations can help one to understand how the relations behave and what properties they enjoy. Domains and ranges are more clearly evident. One can see important patterns at a glance that might be very hard to see if one looked only at the set of ordered pairs or at an equation. Often one can more quickly describe a relation to someone else by its graph than by a more formal definition. In short, graphing can aid both understanding and communication, and so is a skill to cultivate from the start.

The marriage of relations and their graphs through the medium of

_____

† Other coordinate systems in the plane are possible, and, in fact, others are often used. Can you invent any?

ordered pairs is so close that one often bypasses the technical difference and refers to a graph as if it were the relation itself. As in earlier circumstances, we shall agree to such license only because we shall feel free to be precise when any ambiguity or misconception arises.

1.3b. Graphing Techniques

GRAPHING STRATEGIES

The only tool we have at present to construct graphs is that of actually plotting some of the points on the graph and looking for a pattern. A convenient way to handle this is with the use of a table of values. Let us proceed by example.

EXAMPLE 11

We shall go into great detail in this rather simple example so that you will have a model for future problems. The problem is to graph $\{(x, y) \mid y = 3x + 2\}$.

1. We begin by drawing a table for values:

x						
y						

2. Next, we choose a value of x to use to compute the first ordered pair. Suppose we choose $x = 2$. Then, using $x = 2$ in $y = 3x + 2$, we get $y = 3(2) + 2 = 6 + 2 = 8$. Thus our first ordered pair is $(2, 8)$:

x	2					
y	8					

3. Now we choose several more points. For $x = 0$, $y = 3(0) + 2 = 2$, and we have $(0, 2)$. For $x = -2$, $y = 3(-2) + 2 = -6 + 2 = -4$, and we have $(-2, -4)$. Hence our table of values becomes:

x	2	0	-2			
y	8	2	-4			

4. At this point, we should sketch our Cartesian coordinate system, establishing what looks to be a reasonable scale from the few points computed so far. (There will be cases in which you may want one unit on the axes to equal 10, 100, or even 1000. There may even be times when you want different scales on each axis, e.g., 1 unit = 20 on the x-axis and 1 unit = 100 on the y-axis.) Any scale you choose is fine if it shows the behavior of the graph to its best advantage, and as long as you *clearly* indicate the scale used on the axes.

5. Next, we plot the points we have so far and check to see if we need still more (Figure 1.6). In our example, the points seem to lie on a straight line.

6. Now we compute a few more points to check ourselves. For $x = 1$, $y = 3(1) + 2 = 5$, and we have $(1, 5)$. For $x = -1$, $y = 3(-1) + 2 = -1$, and we have $(-1, -1)$.

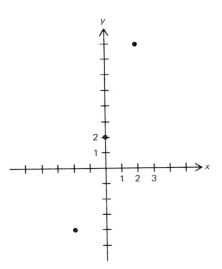

FIGURE 1-6

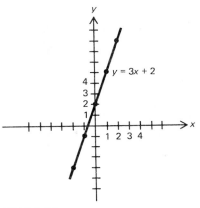

x	y
2	8
0	2
−2	−4
1	5
−1	−1

FIGURE 1-7

7. If we feel fairly confident at this point, we draw a smooth curve† through the points we have so far. If uncertain of our results, we return to repeat steps 5 and 6 again and again until we do feel confident. (See the final graph in Figure 1.7.)

GRAPHING STRATEGIES

To graph the function $f = \{(x, y) \,|\, y = x^2 - 1\}$, we might start out by constructing Table 1.1. This would yield the points plotted in Figure 1.8a. Although some pattern is evident, we should try some additional, more carefully selected values, as suggested in Table 1.2. These yield the points indicated in Figure 1.8b, which are enough to suggest the resulting graph clearly. We may draw the full line, since the domain of f is $\mathbb{R}$; i.e., there are no values of x that do not have corresponding value y.

EXAMPLE 12

Just how many points one should plot depends on which points are chosen and one's ability to see patterns clearly. One should plot enough points to be sure his graph is correct.

To graph the relation $g = \{(s, m) \,|\, m^2 = s\}$, we might start by trying to complete the table

EXAMPLE 13

s	−2	−1	0	1	2
m					

only to find that there are no values of m corresponding to $s = -2$ or $s = -1$, and that for $s = 2$ we may not know offhand what the correct value of m is. This already suggests two considerations. The first is to

---

† Mathematicians use the word *curve* for any continuous graph, i.e., any graph that has no holes, or jumps in it, even if it is straight or jagged.

TABLE 1.1

x	y
−3	8
−2	3
0	−1
1	0
3	8

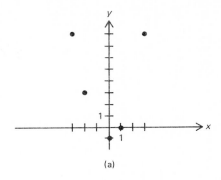

(a)

TABLE 1.2

x	y
−3	8
−2	3
0	−1
1	0
3	8
2	3
−1	0
$-\frac{1}{2}$	$-\frac{3}{4}$
$\frac{1}{2}$	$-\frac{3}{4}$

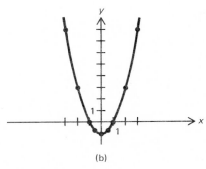

(b)

FIGURE 1-8

check the domain of the relation. There is no sense in attempting to tabulate values that are not in the domain. In this example, the domain is [0, ∞). Second, there is reason to choose domain values to tabulate, for which you can compute the corresponding range values without too much effort. This should not be done to excess, however. If a region of the graph is unclear and only messy computations will clarify, compute these values.

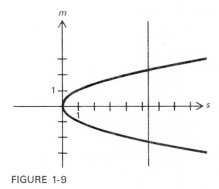

FIGURE 1-9

54   *RELATIONS AND FUNCTIONS*

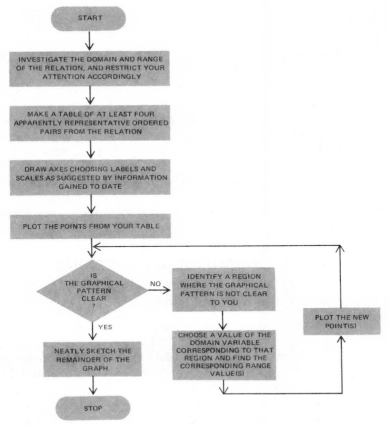

FIGURE 1-10

In the case at hand, it would be wise to pick s to be a perfect square as often as possible.

The alert reader may have noticed that when $a = 1$, there are two possible values of m: both $+1$ and -1. Thus g is not a function.

Digressing for a moment we note that a graphical version of the restriction on a function is that there be at most one point of the graph above or below each point on the x-axis. Equivalently, each vertical line may intersect the graph of the function in at most one point. Figure 1.9 exhibits the graph of a relation that is not a function. Note that while some vertical lines may intersect the graph in only one point (here there is only one, the m-axis), the fact that *some* lines intersect more than once is sufficient to disqualify the relation as a function.

NOTE

After regrouping our forces and having applied more calculations, we might end up with the table of Figure 1.9, and consequently the graph of Figure 1.9 is the graph of Example 13.

s	0	1	4	9	$\frac{1}{4}$
m	0	± 1	± 2	± 3	$\pm\frac{1}{2}$

To gain some sense of perspective in this matter of graphing, let us note that while patience in plotting points can produce a graph that is accurate to the eye, precise information about such things as maximum or minimum values and details of curvature cannot be found in this way. One needs the more powerful tools of calculus to grapple with most of these issues. Furthermore, extensive point plotting may be inappropriate, as when only a sketch is needed, or unnecessary if one observes special properties. We shall therefore be on the lookout for new strategies, short cuts, and other aids to graphing as we proceed. Nevertheless, the plotting of points will remain an important tool when no others are available. The flowchart of Figure 1.10 gives a summary of the procedure we have developed above for such graphing.

1.4. WAYS OF COMBINING FUNCTIONS

1.4a. Functional Composition

DEFINITION 1.9
Functional Composition

There are many ways in which functions may be combined. One of the most common is demonstrated in the following example. Let us view a certain woodworker as a function. Each time she gets a piece of wood she makes something from it. (A long plank becomes a bookcase; a certain size piece of plywood becomes a stool; and a wood scrap is fashioned into a toy boat.) The domain of the function is a set whose elements are pieces of wood; the range is a set of objects constructed. The woman's son enters the picture next, as a second function. He paints each object his mother makes. This second function has as domain the range of the first function (the set of unpainted objects) and as range a set of painted objects. The efforts of mother and son may be viewed separately, as we have just done, or may be grouped together to form a single function whose domain is the domain of the first component function (i.e., the set of pieces of wood) and whose range is the range of the second component function (i.e., the set of finished objects). See Figure 1.11.

Such a combination of functions is called *functional composition*. More formally if f and g are functions and $R_g \subseteq D_f$, then $f \circ g$ denotes the composite function $\{(x, z) \mid x \in D_g, z \in R_f$ and there is some y so that $(x, y) \in g$ and $(y, z) \in f\}$. Looking at the composite function "element-wise" we write $(f \circ g)(x) = f(g(x))$.

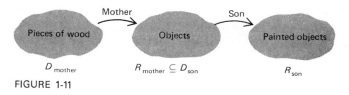

D_{mother} $R_{\text{mother}} \subseteq D_{\text{son}}$ R_{son}

FIGURE 1-11

Composite functions occur so frequently in mathematics that one **FUNCTIONAL COMPOSITION** often fails to see them. For example, the function h specified by $h(x) = \sqrt{1-x}$ can be viewed as the composite $f \circ g$, where g is specified by $g(x) = 1-x$ and $f(x) = \sqrt{x}$. Given a value of x, one first applies g to get **EXAMPLE 14** $1-x$, then applies f to the result to get $\sqrt{1-x}$. Symbolically, for $g(x) = 1-x$, $f(x) = \sqrt{x}$,

$$
\begin{aligned}
h(x) = (f \circ g)(x) = f(g(x)) \quad & \text{(Read this "}f\text{ circle }g\text{ of }x\text{ equals }f\text{ of }g\text{ of }x\text{")} \\
= f(1-x) \quad & \text{(Replacing }g(x)\text{ by its value)} \\
= \sqrt{1-x} \quad & \text{(Substituting the value in parentheses,} \\
& (1-x)\text{ for }x\text{ in }f(x)\text{).}
\end{aligned}
$$

The order in which the functions are composed is important. In the **NOTE** previous example,

$$
\begin{aligned}
(g \circ f)(x) = g(f(x)) = g(\sqrt{x}) \quad & \text{(Replacing }f(x)\text{ by its value, }\sqrt{x}\text{)} \\
= 1 - \sqrt{x} \quad & \text{(Substituting }\sqrt{x}\text{ for }x\text{ in }g(x) = 1-x\text{)}
\end{aligned}
$$

which is quite different from $(f \circ g)(x)$. Notice carefully that the function to the right of the little circle is done first, not the function one sees first when reading.

We can also compose functions for specific values of the argument. **EXAMPLE 15** Thus if $h(x) = \sqrt{x} + 2$ and $k(x) = (x-1)/x^2$, we can find $(h \circ k)(2)$ as follows:

$$
\begin{aligned}
(h \circ k)(2) &= h(k(2)) \\
&= h\left(\frac{2-1}{2^2}\right) \quad \left(\text{Since } k(2) = \frac{2-1}{2^2}\right) \\
&= h(\tfrac{1}{4}) \\
&= \sqrt{\tfrac{1}{4}} + 2 \quad \text{(Using the rule for } h) \\
&= \tfrac{1}{2} + 2 = 2\tfrac{1}{2}
\end{aligned}
$$

$(f \circ g)$ (initial data) - final output

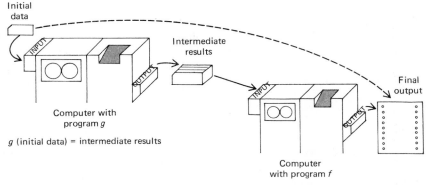

FIGURE 1-12

FUNCTIONAL COMPOSITION

In the analogy between functions and computer programs that was introduced in Section 1.2b, the composition of two functions could be viewed as two such programs for which the output of one computer becomes the input for the other. See Figure 1.12.

1.4b. Arithmetic Operations with Functions

ARITHMETIC OPERATIONS WITH FUNCTIONS

Later in this book, and increasingly in later courses you may take, it will be convenient to view some functions as being built up from simpler ones by processes of addition, subtraction, multiplication, and division as well as by composition. Such operations will probably seem very natural to you, but it is valuable to establish agreement on precisely what we shall mean by such operations with functions.

EXAMPLE 16

For example, the function k specified by $k(x) = 3x + \lfloor x \rfloor$ can easily be thought of as the sum of the two functions $d + t$, where $d(x) = 3x$ and $t(x) = \lfloor x \rfloor$, while the function b given by $b(x) = (x^2 - 3)(2x + 1)$ can be viewed as $w \cdot p$, the product of $w(x) = x^2 - 3$ and $p(x) = 2x + 1$.

DEFINITION 1.10
Addition, Subtraction, Multiplication, and Division of Functions

More formally, if f and g are real-valued functions having a common domain, we define the functions $f + g$, $f - g$, $f \cdot g$, and f/g by the rules

$$(f + g)(x) = f(x) + g(x)$$
$$(f - g)(x) = f(x) - g(x)$$
$$(f \cdot g)(x) = f(x) \cdot g(x)$$
$$\left(\frac{f}{g}\right)(x) = f(x)/g(x) \text{ provided } g(x) \neq 0.\dagger$$

In other words, these operations with *functions* are accomplished by performing the same-named, real-number operations with the *values of the two functions*.‡

EXAMPLE 17

Thus if $f(x) = x^2 - 1$ and $g(x) = 3x$,

$$(f + g)(x) = f(x) + g(x) = (x^2 - 1) + 3x = x^2 + 3x - 1$$
$$(f - g)(x) = f(x) - g(x) = (x^2 - 1) - 3x = x^2 - 3x - 1$$
$$(f \cdot g)(x) = f(x) \cdot g(x) = (x^2 - 1) \cdot (3x) = 3x^3 - 3x$$
$$\frac{f}{g}(x) = \frac{f(x)}{g(x)} = \frac{x^2 - 1}{3x} = \frac{x}{3} - \frac{1}{3x}, \text{ provided } x \neq 0.$$

The difference between composition and the four arithmetic operations with functions can be illustrated by the diagrams of Figure 1.13. Note that, in composition, the functions act one after the other, while with the arithmetic operations they act in tandem.

EXAMPLE 18

A particular function may often be viewed as being constructed from

---

† The provision that $g(x) \neq 0$ for f/g may require deleting some elements from the common domain of f and g so as to avoid division by zero.

‡ It is worth pointing out that the use of the symbols $+$, $-$, $\cdot$, and $/$ is motivated by the use of these familiar operations with the values $f(x)$ and $g(x)$ but are really new operations. After all f and g are *sets* and addition (etc.) of sets has not been defined before.

$$x \xrightarrow{\ f\ } f(x) \xrightarrow{\ g\ } g(f(x))$$
$$g \circ f$$

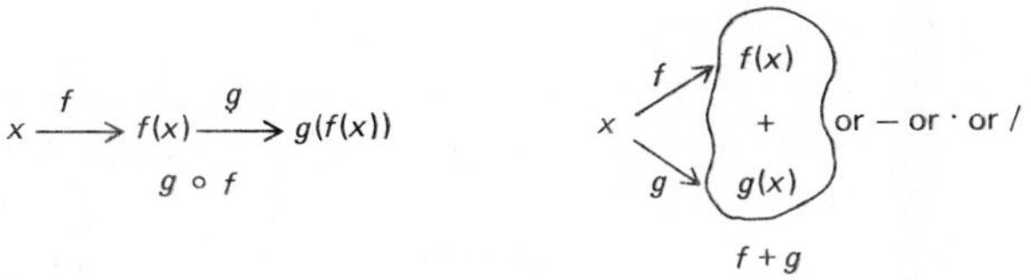

(a)

or $-$ or $\cdot$ or $/$

$f + g$

(b)

FIGURE 1-13

simpler functions in any one of several ways. If $h(x) = x^4 - x^2$, h may be viewed as

$a - b$ (by choosing $a(x) = x^4$ and $b(x) = x^2$), or as

$a \cdot b$ (by choosing $a(x) = x^2$ and $b(x) = x^2 - 1$ [$(a \cdot b)(x) = a(x) \cdot b(x)$ $= x^2(x^2 - 1) = x^4 - x^2$]), or as

$a \circ b$ (by choosing $a(x) = x^2 - x$ and $b(x) = x^2$ [$(a \circ b)(x) = a(b(x))$ $= a(x^2) = (x^2)^2 - (x^2) = x^4 - x^2$]).

We may also use exponents with functions to indicate repeated multiplications, as is done with numbers. Thus f^2 is the function $f \cdot f$, while $f^4 = f \cdot f \cdot f \cdot f$. When considering particular elements of such a function, we may write $y = f^2(x)$ or $w = f^4(x)$, keeping the exponent as part of the function name, as well as $y = (f(x))^2$ or $w = (f(x))^4$.

But beware—$f(x^4)$ is entirely different from $f^4(x)$. For example, if $f(x) = 3x + 1$, then $f(x^4) = 3x^4 + 1$ but $f^4(x) = 81x^4 + 108x^3 + 54x^2 + 12x + 1$. **NOTE**

Similarly, we write $4 \cdot f$ to mean $f + f + f + f$ or $-3g$ to mean $0 - (g + g + g)$, where 0 denotes the zero function. (See p. 47.)

1.4c. Algebra of Functions

Since functions are sets, we could also combine functions by using set operations such as union and complementation. Although perfectly legal, these operations are of at most peripheral interest to us in dealing with the elementary functions and so will not be pursued here. **ALGEBRA OF FUNCTIONS**

Nevertheless, now that we have some operations which may be applied to real-valued functions (i.e., functions with range $\subseteq \mathbb{R}$), it is appropriate to ask about the resulting algebra of such functions. Not only is it appropriate, but it is also important if we are going to use these operations with any degree of confidence and ease. We shall carefully ask and answer a few questions, then state some other results, the proofs of which are left as problems.

What does it mean to say $f = g$ for two functions f and g? **QUESTION 1**

Since f and g are sets, we appeal to the definition of set equality. Thus two functions are *equal* if they contain precisely the same ordered pairs. In actually checking for equality of functions, we check that $f(x) = g(x)$ for all x in the domain. **DEFINITION 1.11**
Equality of Functions

1.4. *WAYS OF COMBINING FUNCTIONS* 59

QUESTION 2   Is addition of functions commutative?

If f and g are any two functions with the same domain and with ranges subsets of $\mathbb{R}$, then we need to ask whether $(f+g)(x)$ is the same as $(g+f)(x)$ for every x in the common domain of f and g. But, for each x,

$$
\begin{aligned}
(f+g)(x) &= f(x) + g(x) && \text{by definition of addition of functions} \\
&= g(x) + f(x) && \text{by commutativity of ``}+\text{'' in } \mathbb{R} \\
&= (g+f)(x) && \text{by definition of addition of functions.}
\end{aligned}
$$

so we may conclude that addition of real-valued functions *is* commutative.

QUESTION 3   Is the multiplication of functions associative?

If f, g, and h have the same domain, and ranges $\subseteq \mathbb{R}$, we are asking whether $f \cdot (g \cdot h) = (f \cdot g) \cdot h$, i.e., whether $(f \cdot (g \cdot h))(x) = ((f \cdot g) \cdot h)(x)$ for all x. Again the algebra of real numbers provides our answer:

$$
\begin{aligned}
(f \cdot (g \cdot h))(x) &= f(x) \cdot [(g \cdot h)(x)] = f(x) \cdot [g(x) \cdot h(x)] && \text{by definition} \\
&&& \text{of multiplication of functions, used twice} \\
&= (f(x) \cdot g(x)) \cdot h(x) && \text{by associativity of ``}\cdot\text{'' in } \mathbb{R} \\
&= ((f \cdot g)(x)) \cdot h(x) = ((f \cdot g) \cdot h)(x) && \text{by the defini-} \\
&&& \text{tion of multiplication of functions, used twice.}
\end{aligned}
$$

Therefore we have proved that the multiplication of real-valued functions is associative.

QUESTION 4   Is functional composition commutative?

Rephrased, this question asks whether $(f \circ g)(x) = (g \circ f)(x)$ for all x. Here we cannot just use the definition of functional composition and the algebra of real numbers, since the definition does not make use of such algebra. Instead, we might try a few examples to see if we can find a way to prove the result in general or else find a counterexample (i.e., an example which would show that composition does not always commute).

In fact, Example 14 is sufficient to show that for at least some functions f and g, there are values of x for which $(f \circ g)(x) \ne (g \circ f)(x)$.

This provides a counterexample † and the answer to Question 4 is no; functional composition is not commutative. Other results include:

1. The addition of real-valued functions is associative.
2. The multiplication of real-valued functions is commutative.
3. The multiplication of real-valued functions distributes over the addition of real-valued functions.
4. Functional composition is associative.

As a final word in this section, we note that while functions and relations do not need to have domains and/or ranges subsets of $\mathbb{R}$, our major consideration after this chapter will be real-valued functions of a

---

† Most results in mathematics are expressed in absolute terms: If these hypotheses hold, then this conclusion must be true. To prove such a statement is true is often hard because one must be sure that it holds in *all* cases. But if a claim like this is made, which is actually not always true (such as "If two functions can be composed, then the composition is commutative"), it is easy to prove the falsity of the claim by exhibiting a case in which the hypotheses are true but the conclusion is false, such as we have just done. An example used to disprove a claim is called a counterexample. Be sure you see that while an example can "unprove" a claim, examples cannot be used to prove a claim.

real variable (i.e., $F : \mathbb{R} \rightarrow \mathbb{R}$). For this reason, we shall stop specifying "real-valued" for functions, letting this fact be understood from context.

1.4d. Functions and Their Arguments

There are other properties of functions than just their behavior under the operations $+$, $-$, $\cdot$, $\div$, and $\circ$. One class of properties is concerned with the effect on functional values of changes in the function's argument. For example, a toy manufacturer might decide to make some children's balls. The surface area of the ball is a function of the radius, specifically $S(r) = 4\pi r^2$. The manufacturer might wish to know the effect of doubling the radius on the amount of material needed to make the ball (i.e., on the surface area). Many people seem to feel that doubling the radius would double the amount of material needed, i.e., $S(2r) = 2S(r)$. Actually, $S(2r) = 16\pi r^2$, while $2(S(r)) = 8\pi r^2$, so that the intuited relationship does not hold.

More generally, one must be careful when dealing with expressions such as $f(x + y)$, $f(x \cdot y)$, $f(1/x)$, $f(x^3)$, and so on. Usually, $f(x + y) \neq f(x) + f(y)$, $f(x \cdot y) \neq f(x) \cdot f(y)$, $f(1/x) \neq 1/f(x)$, or $f(x^3) \neq (f(x))^3$, as one might be tempted to claim. Indeed, *no* statements of this type can be made that would apply to *all* functions (see Problem 14). As we look at each group of elementary functions, however, we shall look for those relationships of this nature, which do exist within the group.

1.5. INVERSES OF FUNCTIONS

1.5a. Inverses

In the last section, we purposely overlooked an operation that we could perform on any function. We could have reversed each ordered pair in the function. If $f = \{(1, 1), (2, 4), (3, 9), (4, 16)\}$, this new "operation" would produce $\{(1, 1), (4, 2), (9, 3), (16, 4)\}$. Such a reversal is natural on at least two levels. One is the level of innate curiosity. Second is the pragmatic recognition that from time to time we each reverse the direction of many associations we use in day to day living. For example, a professor, instead of assigning grades to her students, might wish to associate with each grade the names of those students who earned that grade so that she could encourage the A and B students to major in her subject. Or instead of associating each item on a menu with its price, one might well think first of a price and associate the price with the appropriate items on the menu.

So while the pragmatist sees value in this operation, the experimentalist is intrigued that the professor in the example above no longer has a function. Since $(A, \text{Matilda})$, $(A, \text{Archibald})$, and $(A, \text{Pygmalion})$ are all in her new set of ordered pairs. But the mathematician is not disturbed in the slightest. There are several paths he can take in reaction. One extreme is to say "Functions are not the only things of value in the universe. Relations really are useful. Where would we be without a relation such as those associating one address or phone number with several different people?" and let it go at that. The other is to say "A function is called 'nice' if, when the ordered pairs comprising the function are reversed, the resulting set of ordered pairs is also a function" and at this extreme mathematicians would deign to study only "nice" functions.

Within these extremes we shall steer a middle course. Although recognizing the usefulness of relations, we are still interested in keeping life as simple as it was when we first began to restrict our attention to functions, and so we shall hope that these new relations with reversed ordered pairs will be functions as often as possible.

Let us turn again to the computer program analogy for functions. The reversal of ordered pairs corresponds to starting with the output and ending up with the corresponding input.† We would be trying to undo the work done by the program.

The new relation resulting from such a reversal is an object worthy

DEFINITION 1.12

Inverse of f

of our study; we begin with a formal definition. The *inverse* of a function *f* is the set of ordered pairs that result from reversing each ordered pair in *f*, and is denoted by f^{-1}.

More succinctly, we may write, this definition as

$$f^{-1} = \{(y, x) \mid (x, y) \in f\}.$$

The notation f^{-1}, which is the one most commonly used for the inverse of *f*, should *not* be thought of as *f* to the minus one power, but just as a new notation meaning *only* the inverse of *f*.

Since f^{-1} is formed by reversing the ordered pairs of *f*, it follows immediately that $D_{f^{-1}} = R_f$ and $R_{f^{-1}} = D_f$.

NOTE

For those accustomed to using the word inverse in mathematical settings to mean "the undoing of an operation" such as -5 as the additive inverse of 5 since $5 + (-5) = 0$ (the additive identity element), or $\frac{1}{3}$ as the multiplicative inverse of 3 since $3 \cdot \frac{1}{3} = 1$ (the multiplicative identity element), the inverse of a function is the "functional composition inverse" of *f* because whenever f^{-1} is also a function, $f^{-1} \circ f = i$ (the identity function).‡

We have seen in the professor-grade example that the inverse of a function is a relation but not always a function. Nevertheless, the notation $f^{-1}(y)$ is borrowed from functional notation to denote the set of all the original domain elements *x* for which $f(x) = y$, whether or not f^{-1} is a function, i.e., $f^{-1}(y) = \{x \in D_f \mid f(x) = y\}$.

Returning to our example of the function $c : \mathbb{N} \to \mathbb{N}$ that counts the number of letters in the word for each number, we can write $c^{-1}(3) = \{1, 2, 6, 10\}$, since $c(1) = c(2) = c(6) = c(10) = 3$ and there is no other $n \in \mathbb{N}$ so that $c(n) = 3$. Also, our function diagram (Figure 1.14) can aid visualization as long as we keep in mind that f^{-1} need not be a function.

† This does *not* mean to reverse the order of statements in the program! It simply will not work. If you don't believe it, try an example.
‡ See Example 3.

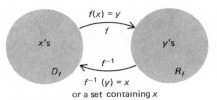

$f(x) = y$

$f^{-1}(y) = x$
or a set containing x

FIGURE 1-14

Let us also note in passing that we could just as well talk of inverses of relations and use the same notations developed above for functions; indeed, we shall feel free to do so. We have chosen, however, to put our emphasis on inverses of functions because our main concern in later chapters will be with functions.

Concerning vocabulary, one should note that the expression "inverse function" is ordinarily (but not universally) reserved for an inverse that *is* a function and should not be confused with the more general term "inverse of a function," which may or may not be a function.†

1.5b.   Restricting the Domain

We have already mentioned several times that f^{-1} need not be a function. In many circumstances, it is useful to restrict a function somehow so that its inverse is a function. The procedure is often quite simple. Suppose we have

$$f = \{(x, y) \mid y = x^2\}. \quad [\text{Figure 1.15 (left)}]$$

Its inverse is

$$f^{-1} = \{(y, x) \mid x = \pm\sqrt{y}\}. \quad [\text{Figure 1.15 (right)}]$$

Clearly, f^{-1} is not a function. (See Figure 1.15.)

But while f^{-1} is not a function, it can be easily carved into two pieces which are functions, as suggested by Figure 1.16. But rather than perform such surgery of f^{-1}, most mathematicians consider it more appropriate

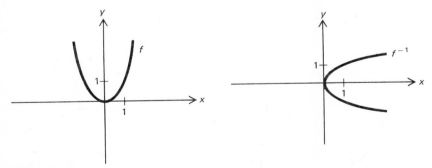

FIGURE 1-15

† Some mathematicians use the term *converse relation* where we have used inverse of a relation, while retaining the term *inverse function* where we have used inverse function, thereby reducing the danger of confusing the two concepts. We have chosen not to do so, primarily because it is not a widely accepted convention.

RESTRICTING DOMAINS to partition the domain of f and come up with several functions whose union is f and whose inverses *are* functions.

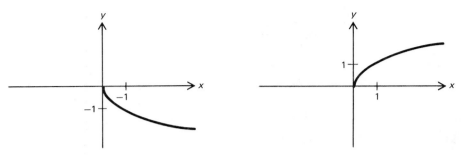

FIGURE 1-16

EXAMPLE 19   In our example, we break the domain of f into two pieces: $(-\infty, 0]$ and $[0, +\infty)$, and consider

$$f_1: \{(x, y) \,|\, y = x^2, x \in (-\infty, 0]\}$$

and

$$f_2: \{(x, y) \,|\, y = x^2, x \in [0, \infty)\}.$$

Do you see clearly that $f = f_1 \cup f_2$? (See Figure 1.17.) Then f_1^{-1} and f_2^{-1} are what we originally thought of as sensible inverse *functions* for f—their graphs are those of Figure 1.16.

EXAMPLE 20   Consider another example: if $g = \{(x, y) \,|\, y = |x+1|\}$, then $g^{-1} = \{(y, x) \,|\, x = \pm y - 1\}$. (See Figure 1.18.) Most people would choose to define

$$g_1 = \{(x, y) \,|\, y = |x+1|, x \in (-\infty, -1]\}$$

and

$$g_2 = \{(x, y) \,|\, y = |x+1|, x \in [-1, \infty)\}.$$

The graphs of g_1^{-1} and g_2^{-1} are shown in Figure 1.19. Clearly, g_1^{-1} and g_2^{-1} are both functions.

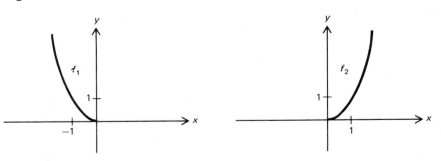

FIGURE 1-17

64   *RELATIONS AND FUNCTIONS*

RESTRICTING
DOMAINS

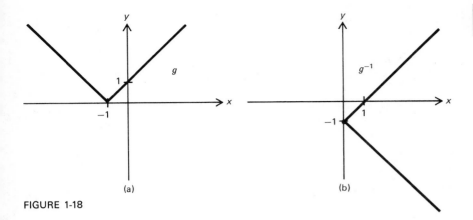

FIGURE 1-18

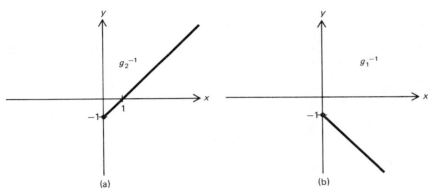

FIGURE 1-19

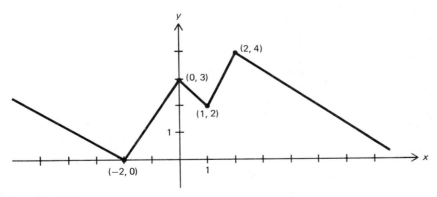

FIGURE 1-20

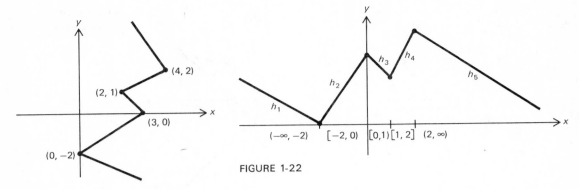

FIGURE 1-22

FIGURE 1-21

While g could be carved into more pieces, each of whose inverse would be a function, the choice made here is preferred because it uses the fewest number of pieces possible. Last, let us look briefly at the graph of a rather complicated function, which we shall not describe algebraically: Figure 1.20. Then h^{-1} would be as shown in Figure 1.21. We might partition the domain of the original function as in Figure 1.22 and then obtain the inverses shown in Figure 1.23. There are no hard and fast rules we can give you for subdividing the domain of a function to get inverses that are functions. Sometimes, the way of restricting the domain is obvious,

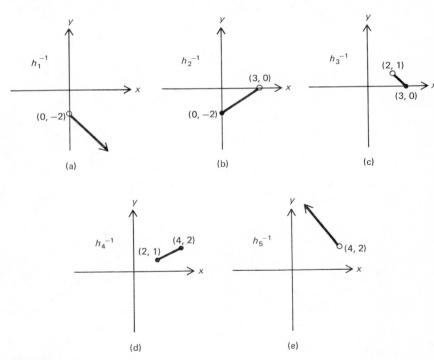

FIGURE 1-23

66   *RELATIONS AND FUNCTIONS*

sometimes not; often, there are several equally correct and useful partitions —you will have to use your own judgment.

1.5c. Graphing Inverses

Because our concern is largely with functions with domain and range subsets of the reals, and we have already noted how helpful graphs are in studying functions, a natural question to raise is "What can we tell about the graph of f^{-1} if we know the graph of f?" Figure 1.24 shows several typical points (·) and the points that result when the coordinates are interchanged ($\times$). Closer examination reveals a pattern. In particular, consider the single pair of points in Figure 1.25. Using plane geometry, one can see that by drawing the graph of $\{(x, y)\,|\,y = x\}$ and the lines PQ, PO, QO, two congruent triangles PRO and QRO are formed.† Therefore, angles PRO and QRO are right angles, and the length of the side PR equals the length of the side RQ.

GRAPHING INVERSES

The result of interchanging coordinates is therefore equivalent to moving the point perpendicularly through the line $y = x$ to the location on the other side of that line, as far from the line as the original point was. This would be true no matter which side of the line the original point was on or which quadrant contained the original point. (Do *you* believe this? See Problem 85.)

NOTE

† For a formal geometric proof, one needs two additional points T and S as in the following diagram. One proof would be:

First, prove $\triangle OQS \cong \triangle OTP$,

$OS \cong OT$	since each has length b
$QS \cong TP$	since each has length a
$\angle OSQ \cong \angle OTP$	since each is a right angle by construction
$\triangle OSQ \cong \triangle OTP$	by S.A.S.

Then, prove $\triangle OPR \cong \triangle OPR$.

$OP \cong OQ$	since they are correspondingly parts of congruent triangles
$OR \cong OR$	by identity
$\angle TOR \cong \angle ROS$	by construction
$\angle TOP \cong \angle QOS$	since they are corresponding parts of congruent triangles
$\angle POR \cong \angle ROQ$	differences of congruent angles are congruent
$\triangle OPR \cong \triangle OQR$	by S.A.S.

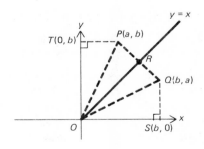

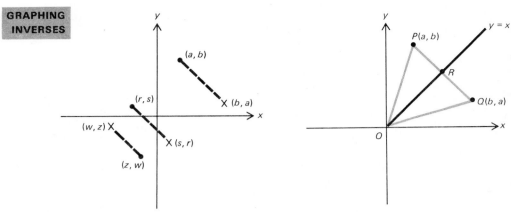

FIGURE 1-24   FIGURE 1-25

The new point, the one produced by reversing the elements of the ordered pair, is called the reflection of the first point through the graph $y = x$. The graph of f^{-1} is then the reflection, point by point, of the graph of f through the graph of $y = x$. Figure 1.26 gives an example of the graphs of f and f^{-1}. You should study it carefully to see how the reflection idea is applied.

Some people find another technique helpful in graphing inverses. Given the graph of f, one first "flips" the graph around the y-axis (or reflects it in the y-axis), then rotates the plane $90°$ clockwise, and finally relabels the axes appropriately. This is shown in the following sketches:

If you use this technique, you should convince yourself that it is equivalent to the point reflections just discussed.

1.5d. Finding Inverses Using an Algebraic Rule

If a function is specified by an algebraic rule, it is sometimes possible to express the inverse in a neat way by an algebraic rule.

EXAMPLE 21   For example, if $f = \{(x, y) \,|\, y = 3x - 5\}$, it would be nice to be able to write $f^{-1} = \{(y, x) \,|\, x = \text{some rule depending on } y\}$. We know that $f^{-1} = \{(y, x) \,|\, y = 3x - 5\}$, so we are really looking for a way to rewrite $y = 3x - 5$ in a form $x = [\text{expression involving } y]$. For this example, this is not hard. Solving the equation for x, we have

$$x = \frac{y + 5}{3}$$

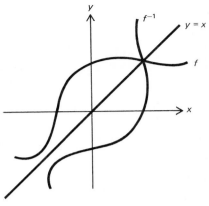

FIGURE 1-26

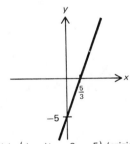

(a) $\{(x, y) \mid y = 3x - 5\}$ (original function).

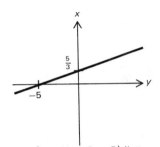

(b) $\{(y, x) \mid y = 3x - 5\}$ (letters interchanged in ordered pair only).

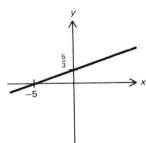

(c) $\{(x, y) \mid x = 3y - 5\}$ (letters interchanged in rule only).

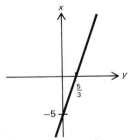

(d) $\{(y, x) \mid x = 3y - 5\}$ (letters interchanged in both ordered pair and rule).

FIGURE 1-27

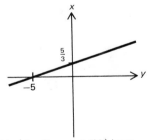

(e) $\{(y, x) \mid x = (y + 5)3\}$ (same as (b) but with rule solved for range variable).

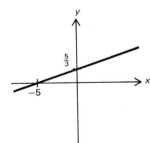

(f) $\{(x, y) \mid y = (x + 5)/3\}$ (same as (c) with rule solved for range variable, and same as (e) with x and y in traditional roles).

ALGEBRAIC RULES FOR INVERSES

so

$$f^{-1} = \left\{ (y, x) \mid x = \frac{y+5}{3} \right\}.$$

EXAMPLE 22

In general, one wants to manipulate the equation so as to isolate the variable that stands for the elements originally in the domain of f. However, this is not always easy.

If $f = \{(x, y) \mid y = x^2 + 2x + 1\}$, we wish to solve $y = x^2 + 2x + 1$ for x.

$$y = x^2 + 2x + 1$$
$$y = (x + 1)^2$$

and thus $\quad \pm\sqrt{y} = x + 1 \quad$ or $\quad \pm\sqrt{y} - 1 = x$

So, $\quad f^{-1} = \{(y, x) \mid x = \sqrt{y} - 1 \quad$ or $\quad x = -\sqrt{y} - 1\}$

Now let us think of graphing inverses from such algebraic specifications. In Example 21 above, it would be easy to become confused by the roles played by x and y in $\{(y, x) \mid x = (y + 5)/3\}$. Many people find it easier to rewrite f^{-1} with different letters, such as $\{(a, b) \mid b = (a + 5)/3\}$, and then realize that a corresponds to the x-axis (domain) and b with the y-axis (range), and graph f^{-1} accordingly. Others are willing to translate back to x and y, but now in the traditional roles with x representing the argument and y the value: $\{(x, y) \mid y = (x + 5)/3\}$. A careful examination of Figure 1.27 may help you understand such manipulations, and Problem 19 gives you the opportunity to experiment.†

EXAMPLE 23

If $f = \{(x, y) \mid y = \sqrt{x + 1}\}$, then $f^{-1} = \{(y, x) \mid y = \sqrt{x + 1}\}$. If $y = \sqrt{x + 1}$, then $y^2 = x + 1$ or $x = y^2 - 1$ but with $y \geq 0$. We have $f^{-1} = \{(y, x) \mid x = y^2 - 1, y \geq 0\}$. Here we had to be careful not to introduce more for the inverse than allowed by the original function; hence the need for y to be greater than or equal to 0.

---

† If you are still confused, after trying Problem 19, here is another way of looking at the changes of form with which we are concerned. The most likely source of confusion in comparing

$$f = \{(x, y) \mid y = 3x - 5\} \text{ (a)}$$
$$f^{-1} = \left\{ (y, x) \mid x = \frac{y+5}{3} \right\} \text{ (b)}$$
$$f^{-1} = \left\{ (x, y) \mid y = \frac{x+5}{3} \right\} \text{ (c)}$$

is in the uncertainty of whether or when the x's are the same objects and the **y's are the same objects**. It is helpful just not to expect any such sameness and merely read each set as a **separate entity**, thinking of arguments and values rather than x's and y's. If you want x and y in **traditional roles for graphing f^{-1}**, though, and the translation suggested in the body of the text is not helpful, try going from

$$f^{-1} = \left\{ (y, x) \mid x = \frac{y+5}{3} \right\}$$

to

$$f^{-1} = \left\{ (a, b) \mid b = \frac{a+5}{3} \right\}$$

to

$$f^{-1} = \left\{ (\hat{x}, \hat{y}) \mid \hat{y} = \frac{\hat{x}+5}{3} \right\}.$$

The use of $\wedge$ notation will maintain a feeling of unlikeness between the argument of f (namely x) and the argument of f^{-1} (namely $\hat{x}$) and similarly between y and $\hat{y}$, while retaining the familiar type of letter for each role.

SUMMARY OF IMPORTANT CONCEPTS AND NOTATION

EQUALITY OF ORDERED PAIRS: (Definition 1.1) Two ordered pairs (a, b) and (s, t) are equal if and only if $a = s$ and $b = t$.

RELATION: (Definition 1.2) A relation is any set of ordered pairs.

DOMAIN: (Definition 1.3) If T is a relation, then the set of all the first elements in the ordered pairs that make up T is called the domain of T (denoted D_T).

RANGE: (Definition 1.3) If T is a relation, then the set of all the second elements in the ordered pairs that make up T is called the range of T (denoted R_T).

FUNCTION (Definition 1.4a) A function is a set of ordered pairs no two of which have the same first element. Stated another way, (Definition 1.4b) if some rule assigns to each element x of one set X a single element y from a second set Y, then the set of ordered pairs (x, y) is called a function.

CODOMAIN: (Definition 1.5) If f is a function and $R_f \subseteq C$, then C may be called the codomain of f. $f : X \rightarrow Y$ denotes that f is a function with domain X and codomain Y.

ARGUMENT: (Definition 1.6) An argument of a function is any element of the domain of the function.

VALUE: (Definition 1.6) A value of a function is any element of the range of the function.

QUADRANTS: (Section 1.3a) The quadrants are the four regions into which the axes divide the plane. (See Figure 1.4 and Problem 55.)

ABSCISSA: (Definition 1.7) The abscissa (or x-coordinate) of the point (a, b) is the number a.

ORDINATE: (Definition 1.7) The ordinate (or y-coordinate) of the point (a, b) is the number b.

CARTESIAN COORDINATE SYSTEM: (Definition 1.7) If two perpendicular lines each have a number scale so that the point of intersection is labeled 0 on both lines, then these lines, together with the associated labeling of each point in the plane, are called a Cartesian Coordinate System.

GRAPH: (Definition 1.8) The graph of a relation is the set of points in the plane that have the same coordinate pairs as the pairs comprising the relation.

GRAPHING STRATEGIES: (Section 1.3b and Figure 1.10)

FUNCTIONAL COMPOSITION: (Definition 1.9) If f and g are functions and $R_g \subseteq D_f$, then the composite function $f \circ g$ is $\{(x, z) \mid z = f[g(x)]\}$.

ADDITION, SUBTRACTION, MULTIPLICATION, AND DIVISION OF FUNCTIONS: (Definition 1.10) If f and g are real-valued functions having the same domain, the function $f + g$ is defined by the rule:

$$(f + g)(x) = f(x) + g(x)$$

Similarly, $f - g$, $f \cdot g$, and f/g are defined by the rules:

$$(f - g)(x) = f(x) - g(x)$$
$$(f \cdot g)(x) = f(x) \cdot g(x)$$

and

$$\frac{f}{g}(x) = \frac{f(x)}{g(x)} \text{ provided that } g(x) \neq 0.$$

ALGEBRA OF FUNCTIONS: (Section 1.4c)

EQUALITY OF FUNCTIONS: (Definition 1.11) Two functions are equal if they contain precisely the same ordered pairs.

INVERSE OF A FUNCTION (OR RELATION) f: (Definition 1.12) The inverse of f, denoted f^{-1}, is the set of ordered pairs that results from reversing the elements of each ordered pair of f.

RESTRICTING DOMAINS: (Section 1.5b)

GRAPHING INVERSES: (Section 1.5c)

ALGEBRAIC RULES FOR INVERSES: (Section 1.5d)

PRACTICE EXERCISES

Section 1.2b

1. Find the domain and range for each of the following relations.
   (a) $\{(w, x), (c, z), (2, 9), (3, 4)\}$
   (b) $\{(2, 4), (3, 9), (4, 16), (5, 25)\}$
   (c) $\{(a, b), (q, t), (m, t), (e, z)\}$
   (d) $\{(B, \text{Bach}), (B, \text{Beethoven}), (B, \text{Brahms})\}$
   (e) $\{(\text{Bach}, B), (\text{Brahms}, B), (\text{Beethoven}, B)\}$
   (f) $\{(1, a), (b, 9), (t, t), (\sqrt{2}, 5), (0, z)\}$
   (g) $\{(1, 2), (7, 5), (5, 1), (2, 7)\}$
   (h) $\{(1, 1), (5, 5), (2, 2), (7, 7)\}$
   (i) $\{(a, b), (z, m), (g, f), (a, t), (a, a)\}$
   (j) $\{(x, y) \mid y = x + 1\}$
   (k) $\{(x, y) \mid y = 2x + 7\}$
   (l) $\{(\alpha, \beta) \mid \beta = -3 - \alpha\}$
   (m) $\{(s, t) \mid t = s^3\}$
   (n) $\{(y, x) \mid x = 3y - 12\}$
   (o) $\{(j, s) \mid j = s^2\}$
   (p) $\{(\psi, \sigma) \mid \sigma = 3 - \sqrt{\psi}\}$
   (q) $\{(a, b) \mid b = \sqrt{a + 1}\}$
   (r) $\{(m, n) \mid n^2 + m^2 \geq -2nm\}$
   (s) $\{(a, b) \mid ab = 1\}$
   (t) $\{(\alpha, \beta) \mid \alpha^2 + \beta^2 = 25\}$
   (u) $\{(w, k) \mid k = w + 1/w\}$
   (v) $\{(x, y) \mid y$ is the current U.S. postage for a letter weighing x ounces$\}$
   (w) $\{(A, B) \mid A \subseteq \mathbb{R}, B \subseteq \mathbb{R}, \text{ and } A \subseteq B\}$
   (x) $\{(c, d) \mid c < d\}$
   (y) $\{(e, f) \mid e \neq f, e, f \in \mathbb{R}\}$
   (z) $\{(m, n) \mid m$ is an even divisor of n, n and $m \in \mathbb{N}\}$.

Section 1.2c

2. Which of the relations in Problem 1 are functions? Verify that each of the others is not a function by exhibiting for each, two different ordered pairs with the same first element.

3. If the following are functions specified by $\sigma(y) = y/(y-3)$, $\phi(a) = a^3 - 2a + 7$, $f(x) = x^2$, $g(z) = 3z + 2$, and $h(x) = (1/x) - 3$, find the following:

(a) $f(4)$	(b) $\phi(-5)$	(c) $\phi(6)$
(d) $h(0)$	(e) $f(-4)$	(f) $\phi(1/2)$
(g) $\sigma(1)$	(h) $\sigma(4)$	(i) $\phi(0)$
(j) $\sigma(3)$	(k) $\phi(-2)$	(l) $h(-1)$
(m) $g(25)$	(n) $\phi(100)$	(o) $\phi(-17)$
(p) $\sigma(-83)$	(q) $\sigma(-25\frac{1}{2})$	(r) $g(-7.21)$
(s) $f(2.2)$	(t) $g(-8\frac{1}{2})$	(u) $\sigma(5.3)$
(v) $h(\frac{1}{9})$	(w) $f(-4.7)$	(x) $g(2000)$
(y) $h(2500)$	(z) $f(674)$	

4. Given the following functions, find and simplify each of the following expressions. $\sigma(y) = y/(y-3)$, $\phi(a) = a^3 - 2a + 7$, $f(x) = x^2$, $g(z) = 3z + 2$, $h(x) = (1/x) - 3$.

(a) $\sigma(x^2)$	(b) $\phi(x+3)$	(c) $f(x+2)$
(d) $\sigma(5x)$	(e) $\phi(2x-3)$	(f) $\phi(5x)$
(g) $f(2x)$	(h) $2 \cdot f(x)$	(i) $g(a \cdot b)$
(j) $g(a) \cdot g(b)$	(k) $f(a+b)$	(l) $f(a) + f(b)$
(m) $h(1/a)$	(n) $1/h(a)$	(o) $f(g(3))$
(p) $g(f(3))$	(q) $g(x-y)$	(r) $g(x) - g(y)$

Section 1.2d

5. Find or simplify each of the following.

(a) $p_3(5)$	(b) $\lfloor 2.3 \rfloor$	(c) $\phi_3(4)$
(d) $p_{-1/2}(10)$	(e) $\lfloor -6.2 \rfloor$	(f) $\phi_5(2)$
(g) $p_0(12)$	(h) $\lfloor \sqrt{2} \rfloor$	(i) $\phi_3(p_{10}(1))$
(j) $\psi(\phi_3(-3))$	(k) $\phi_{\lfloor 3 \rfloor}(5)$	(l) $(\lfloor -\sqrt{2} \rfloor)(p_5(\sqrt{2}))$
(m) $p_{-2}(2x+5)$	(n) $\phi_2(x+w)$	(o) $\phi_4(2x-3)$
(p) $p_0(x^3 - 5x)$		

Section 1.3b

6. Graph each of the following relations.
   (a) $a = \{(x, y) \mid y = 2x - 5\}$
   (b) $b = \{(r, s) \mid s = 5 - r^2\}$
   (c) $c = \{(a, b) \mid a + b = -1\}$
   (d) $d = \{(w, z) \mid w^2 + z^2 = 25\}$
   (e) $e = \{(x, y) \mid y^2 = 5 - x\}$
   (f) $f = \{(\alpha, \beta) \mid \beta = \alpha^2\}$
   (g) $g = \{(\sigma, \delta) \mid \delta = \sigma^4\}$
   (h) $h = \{(x, y) \mid y = (x-3)(x-2)\}$

(i) $i = \{\alpha, \beta) \,|\, \beta = \alpha^3\}$

(j) $j = \{(d, e) \,|\, e = d^3 - d\}$

(k) $k = \{(f, g) \,|\, g = f - f^3\}$

(l) $l = \{(m, p) \,|\, 16m^2 + 9p^2 = 25\}$

(m) $m = \{(z, w) \,|\, w = \sqrt{z + 1}\}$

(n) $n = \{(r, \psi) \,|\, \psi = \sqrt{r}\}$

(o) $o = \{(t, b) \,|\, b = |t|\}$

(p) $p = \{(\square, \triangle) \,|\, \square = 5\}$

(q) $q = \{(A, B) \,|\, B \leq 3A + 1\}$

(r) $r = \{(c, w) \,|\, \sqrt{w} = \sqrt{c}\}$

(s) $s = \{(\triangle, \square) \,|\, \square = \triangle + 2\}$

(t) $t = \{(\zeta, \eta) \,|\, \eta = \lfloor \zeta \rfloor\}$ (See Example 9)

(u) $u = \{(\beta, \alpha) \,|\, \alpha = (\beta + 1)/(\beta - 2)\}$

(v) $v = \{(x, y) \,|\, y = x - \lfloor x \rfloor\}$ (See Example 9)

(w) $w = \{(s, t) \,|\, |t| = |s|\}$

(x) $x = \{(m, n) \,|\, m \leq n\}$

(y) $y = \left\{ (w, t) \,|\, t = \dfrac{3 + w^2}{w + 2} \right\}$

(z) $z = \{(x, y) \,|\, y = 3t \text{ and } x = t^2 \text{ for some } t \in \mathbb{R}\}$

(aa) $aa = \{(x, y) \,|\, y = |t| \text{ and } x = 1/t \text{ for some } t \in \mathbb{R}\}$

(bb) $bb = \{(f, g) \,|\, g = |f^2 - 4|\}$

(cc) $cc = \{(s, b) \,|\, 2 \leq b \leq 5\}$

(dd) $dd = \{(k, m) \,|\, m = k(k + 1)(k - 2)\}$

Section 1.4a

7. Given the following functions, find and/or simplify each of the given expressions: $f(x) = \dfrac{x + 1}{x - 1}$; $g(x) = \sqrt{x - 3}$; $h(x) = 1/x$; $p(x) = 4$; $k(x) = 5 - 2x$; $m(x) = x^2 - 3x$; and $r(x) = 1/(3x^2)$.

(a) $(m \circ g)(5)$

(b) $(g \circ m)(5)$

(c) $(h \circ f)(-3)$

(d) $(p \circ r)(3)$

(e) $(f \circ m)(-2)$

(f) $(f \circ g)(x)$

(g) $(g \circ h)(x)$

(h) $(f \circ k)(x)$

(i) $(g \circ f)(x)$

(j) $(f \circ h)(x)$

(k) $(r \circ m)(x)$

(l) $(h \circ f)(x)$

(m) $(f \circ f)(x)$

(n) $(h \circ h)(x)$

(o) $(p \circ m)(x)$

(p) $(m \circ h)(x)$

(q) $(f \circ (g \circ k))(0)$

(r) $(k \circ (f \circ h))(2)$

(s) $((g \circ p) \circ r)(2)$

(t) $(f \circ (g \circ h))(x)$

(u) $(g \circ (r \circ r))(x)$

(v) $((k \circ k) \circ k)(x)$

(w) $(k \circ (f \circ h))(x)$

(x) $(r \circ (r \circ r))(x)$

(y) $[(f \circ g) \circ (f \circ k)](x)$

(z) $[(r \circ m) \circ (h \circ k)](x)$

Section 1.4b

8. Combine the following functions as indicated below: $f(x) = \dfrac{x+1}{x-1}$;

$g(x) = \sqrt{x-3}$; $h(x) = \dfrac{1}{x}$; $k(x) = 5 - 2x$; $m(x) = x^2 - 3x$; $r(x) = \dfrac{1}{3x^2}$;

and $p(x) = 4$.

  (a) $\left(\dfrac{f}{k}\right)(5)$ (b) $(r \cdot m)(2)$ (c) $(f - m)(0)$

  (d) $(k + m)(-2)$ (e) $(r \cdot h)(10)$ (f) $\left(\dfrac{f}{r}\right)(2)$

  (g) $(p + f)(3)$ (h) $(m - k)(17)$ (i) $(m \cdot h)(x)$

  (j) $(m + h)(x)$ (k) $\left(\dfrac{m}{h}\right)(x)$ (l) $(m \cdot k)(x)$

 (m) $(f + k)(x)$ (n) $(g + k)(x)$ (o) $(k + p)(x)$

  (p) $(p + m)(x)$ (q) $\left(\dfrac{h}{p}\right)(x)$ (r) $(f \cdot (g + h))(4)$

  (s) $m^2(3)$ (t) $r^3(-2)$ (u) $(g^2 + r^2)(7)$

  (v) $(3f - 2k)(x)$ (w) $(h \cdot (f + k))(x)$ (x) $h^4(x)$

  (y) $(m \cdot h)^2(x)$ (z) $(f + k)^2(x)$

9. Using the same functions as in Problem 8, find

  (a) $[g \circ (m + k)](x)$ (b) $[g \circ (m \cdot h)](x)$

  (c) $[r + (h \circ m)](x)$ (d) $[h \circ (f - m)](x)$

  (e) $(3f \circ 2k)(x)$ (f) $(p \circ (h \cdot p))(x)$

  (g) $\left(\dfrac{f}{m} \circ h\right)(x)$ (h) $\left(\dfrac{f \circ 2g}{r}\right)(x)$

  (i) $((h + p) \circ (h - p))(x)$ (j) $[(f \circ f) - (m \circ m)](x)$

  (k) $f^3(x)$ (l) $f(x^3)$

 (m) $k(3x)$ (n) $(3k)(x)$

  (o) $r(\tfrac{1}{2})$ (p) $\left(\dfrac{1}{r}\right)(2)$

  (q) $m(x + w)$ (r) $m(x) + m(w)$

  (s) $p(x^2)$ (t) $p^2(x)$

10. Decompose each of the following functions into the product of two functions.

  (a) $d(x) = (x - 2)(x + 1)$ (b) $e(x) = x^2 - 16$

  (c) $q(x) = 3x - 6$ (d) $t(x) = x^3 - 4x$

  (e) $a(x) = x^2 - 7x + 12$

11. Decompose each of the following functions into the composition of two functions.

  (a) $p(x) = x^2 - 16$ (b) $o(x) = x^2 + 6x + 9$

  (c) $i(x) = 1/(x + 2)$ (d) $n(x) = \sqrt{x - 3}$

  (e) $t(x) = 1/\sqrt{2 - x}$

12. Decompose each of the following functions into the sum, product,

difference, quotient and/or composition of other functions in as many ways as you can.

(a) $s(x) = x^2 - 16$
(b) $h(y) = 7/\sqrt{76 - y}$
(c) $a(w) = w^3 - 2w$
(d) $r(t) = 1/(t^2 - 2t + 1)$
(e) $o(z) = \sqrt{z^2 - 7}$
(f) $n(x) = x^4 + 4x^2 + 4$

Section 1.4c

13. Find values for, or simplify, each pair of expressions and then indicate what algebraic property of functions is being illustrated or disproved by each such pair. Let $w(x) = \dfrac{3x}{2 - x^2}$; $p(x) = 4 + x$; $s(x) = \dfrac{1}{x} + 1$; and $t(x) = x^2 + 3x$.

(a) $[w \circ (p \circ s)](\frac{1}{2})$
$\quad [(w \circ p) \circ s](\frac{1}{2})$
(b) $(t \circ p)(-3)$
$\quad (p \circ t)(-3)$
(c) $[p \cdot (s + t)](4)$
$\quad [p \cdot s + p \cdot t](4)$
(d) $[t \circ (s \cdot p)](-2)$
$\quad [(t \circ s) \cdot (t \circ p)](-2)$
(e) $[(p \cdot s) \cdot t](x)$
$\quad [p \cdot (s \cdot t)](x)$
(f) $[s \circ (p + t)](x)$
$\quad [(s \circ p) + (s \circ t)](x)$
(g) $(w + t)(x)$
$\quad (t + w)(x)$
(h) $[(p + s) + t](x)$
$\quad [p + (s + t)](x)$
(i) $(w \circ s)(x)$
$\quad (s \circ w)(x)$
(j) $[s + (t \circ p)](x)$
$\quad [(s + t) \circ (s + p)](x)$

Section 1.4d

14. To see that formulas such as $f(x + y) = f(x) + f(y)$ do hold for some functions and not for others, find a function from among those listed for which each of the following formulas hold. (i) $f(x) = 2^x$; (ii) $f(x) = 2x$; (iii) $f(x) = x^2$; and (iv) $f(x) = 1/x$.

(a) $f(x + z) \neq f(x) + f(z)$
(b) $f(x + z) = f(x) + f(z)$
(c) $f(x \cdot w) \neq f(x)f(w)$
(d) $f(x \cdot w) = f(x)f(w)$
(e) $f(cx) \neq cf(x)$
(f) $f(cx) = cf(x)$
(g) $f(1/x) \neq 1/f(x)$
(h) $f(1/x) = 1/f(x)$
(i) $f(x + t) \neq f(x)f(t)$
(j) $f(x + t) = f(x)f(t)$

Section 1.5a

15. Exhibit the inverse of each of the following functions:

(a) $\{(1, 2), (3, 4), (5, 6), (7, 8), \cdots\}$
(b) $\{(1, a), (2, b), (3, c), (4, d), \cdots, (26, z)\}$
(c) $\{(1, 2), (3, 4), (4, 3), (2, 1)\}$
(d) $\{(1, 0), (2, 0), (3, 0), (4, 0), \cdots\}$
(e) $\{(-1, 1), (-2, 2), (-3, 3), \cdots, (-99, 99)\}$
(f) $\{(\cdot, *), (+, +), (-, -), (\div, /)\}$
(g) $\{(x, y) \mid y = 3x + 2\}$
(h) $\{(s, t) \mid t = s\}$

(i) $\{(k, w) \mid k$ is a senator from the state of $w\}$
(j) $\{(\alpha, \beta) \mid \beta = \alpha^2\}$

16. Given the function specified by the following rules, find each of the indicated objects: $m(x) = 5x - 7$; $a(x) = |x + 1|$; $t(x) = x^2 - 3$; and $h(x) = \lfloor x - 2 \rfloor$.

   (a) $m^{-1}(0)$ (b) $t^{-1}(0)$ (c) $a^{-1}(0)$
   (d) $m^{-1}(6)$ (e) $h^{-1}(2)$ (f) $m^{-1}(-120)$
   (g) $t^{-1}(-2)$ (h) $t^{-1}(13)$ (i) $a^{-1}(5)$
   (j) $a^{-1}(\sqrt{2})$ (k) $h^{-1}(-20)$ (l) $a^{-1}(-1)$
   (m) $h^{-1}(6)$ (n) $m^{-1}(\frac{1}{2})$ (o) $t^{-1}(-5)$
   (p) $h^{-1}(\frac{1}{2})$

Section 1.5b

17. For each of the functions graphed below, list the sets into which the domain should be subdivided in order that the inverse over each such subset be a function. You should list as few sets as possible to accomplish this.

(a)
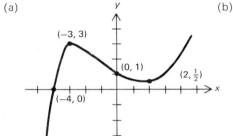

(b)
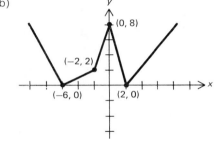

(c)
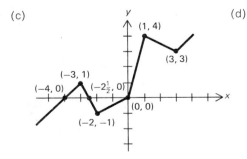

(d)
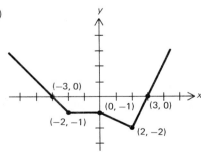

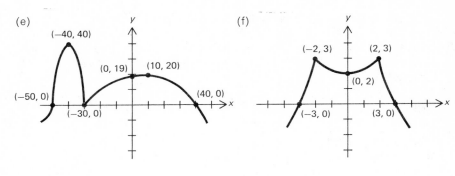

(e)

(-40, 40)
(0, 19)
(10, 20)
(-50, 0)
(-30, 0)
(40, 0)

(f)

(-2, 3)
(2, 3)
(0, 2)
(-3, 0)
(3, 0)

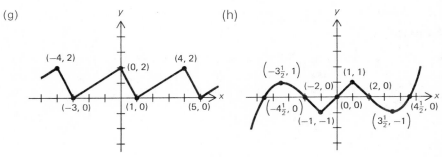

(g)

(-4, 2)
(0, 2)
(4, 2)
(-3, 0)
(1, 0)
(5, 0)

(h)

$\left(-3\frac{1}{2}, 1\right)$
(1, 1)
(-2, 0)
(2, 0)
$\left(-4\frac{1}{2}, 0\right)$
(0, 0)
$\left(4\frac{1}{2}, 0\right)$
(-1, -1)
$\left(3\frac{1}{2}, -1\right)$

Section 1.5c

18. Copy the graph of f in each problem below and on the same axes sketch the graph of f^{-1}.

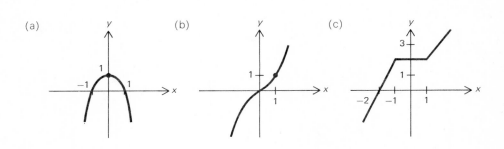

(a)

(b)

(c)

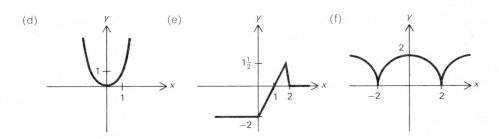

(d)

(e)

(f)

(g)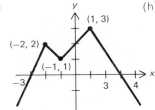

(h)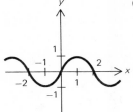

(i)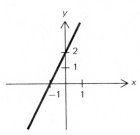

(j)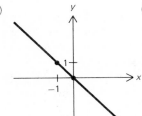

(k)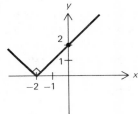

(l)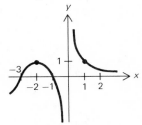

Section 1.5d

19. If $f=\{(x, y)\,|\,y=3-2x\}$ examine each of the following and decide which are representations of f, which are f^{-1} and which are neither. Then graph each by carefully plotting points and compare with your initial identifications.

(a) $\{(x, y)\,|\,x=3-2y\}$

(b) $\{(y, x)\,|\,x=3-2y\}$

(c) $\{(x, y)\,|\,y=3-2x\}$

(d) $\{(y, x)\,|\,y=3-2x\}$

(e) $\left\{(x, y)\,\middle|\,x=\dfrac{3-y}{2}\right\}$

(f) $\left\{(y, x)\,\middle|\,x=\dfrac{3-y}{2}\right\}$

(g) $\left\{(x, y)\,\middle|\,y=\dfrac{3-x}{2}\right\}$

(h) $\left\{(y, x)\,\middle|\,y=\dfrac{3-x}{2}\right\}$

20. For each of the following relations, find the inverse and express it in as neat a form as you can (e.g., have any algebraic rule(s) solved for the range variable if possible).

(a) $f=\{(r, s)\,|\,s=16-3r\}$

(b) $g=\{(x, y)\,|\,y=-\tfrac{1}{2}x+2\}$

(c) $h=\{(a, b)\,|\,b=2-a^2\}$

(d) $k=\{(w, z)\,|\,z=w^2-6w+9\}$

(e) $\phi=\{(s, t)\,|\,t=\sqrt{s}+1\}$

(f) $\sigma=\{(x, y)\,|\,y=x^3\}$

(g) $\psi=\{(m, k)\,|\,k=12+5m\}$

(h) $m=\{(i, j)\,|\,i, j\in\mathbb{Z}, j=(-1)^i\}$

(i) $p=\{(f, e)\,|\,e=\sqrt{f}-2\}$

(j) $q=\{(x, y)\,|\,y=1\}$

(k) $A=\{(s, t)\,|\,t=|s|\}$

(l) $T = \{(a, b) \mid b = 2 - |a|\}$
(m) $r = \{(y, r) \mid r = 4y^2 + 12y + 9\}$
(n) $j = \{(q, h) \mid h = \sqrt{9 - q^2}\}$
(o) $s = \{(\alpha, \beta) \mid \beta = |\alpha^2 - 4|\}$
(p) $t = \{(x, y) \mid y = 2^x\}$
(q) $u = \{(a, b) \mid b = a - |a|\}$
(r) $v = \{(x, y) \mid y \in \{3, 4, 5\}, x \in \{x \mid 1 \le x \le 10, x \in \mathbb{N}\}\}$

PROBLEMS

21. Consider the relation that assigns to each living person the name of his father and the maiden name of his mother. Determine whether this relation is a function or not, explain why or why not, and specify its domain and range. Write the relation in set form.

22. Let g be the relation that associates with each country its gross national product for last year. Is g a function? (Justify your answer.) What is its domain?

23. Let $n = \{(r, s) \mid r$ is a state in the USA and s is a U.S. senator from $r\}$ and $m = \{(r, s) \mid s$ is a state in the USA and r is a U.S. senator from $s\}$.
    Determine if each of n and m are functions (explaining why or why not) and find the domain and range of each.

24. Let d be the relation that associates with the states in the United States their Republican senator(s). To the best of your knowledge, state the domain and range and determine if d is a function.

25. Describe the activity of the following persons in terms of a concern with functions and/or relations. That is, explain how such a person's job *can* be viewed as a search for relations or as a use of preexisting relations, even though the person is unlikely to think of his job in such terms. Give examples of such relations.
    (a) A political scientist
    (b) A historian
    (c) A psychologist
    (d) A lawyer
    (e) A mechanic

26. Set up functions to find:
    (a) The surface area S of a cube if one side is x

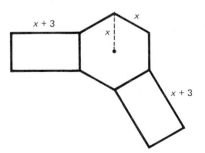

(b) The surface area S of a rectangular prism if its width is w, length twice the width w, and height half the width

(c) The length of fence needed to enclose the area shown above

(d) The area of the figure above

(e) The volume of a cylinder with a hemispherical top

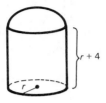

(f) The cost of building a box if the material for the top and bottom cost $3 a square foot and the sides cost $4 a square foot with the length ℓ, width twice the length, and height 3 feet more than the length (all given in feet)

(g) The volume of concrete needed to fill the mold for a square column of height h and side $h^2/20$

(h) The length of the edges of a cube given its volume V

(i) The dimensions of a rectangular prism with length twice the height, height h, and width three times the height given the surface area

(j) The height of a cylinder with radius 5, given its volume

(k) The area of the figure below

(l) The length around the outside of the figure below

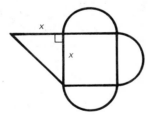

27. A corral is to be built in the shape indicated in the diagram below, where the length is twice the width. The total cost of fence needed is a function of the width of the corral.

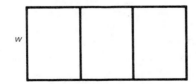

(a) If the fence costs $3.50 per meter, give the algebraic rule that defines this function.

(b) If the fence costs $5/meter for the outside walls and $3/meter for the interior walls, give the algebraic rule that defines this function.

28. A rancher has a herd of 1000 cattle, which he is preparing for market. The cattle are now worth 50 cents/pound and weigh an average of 450 lb. For each day he keeps them, the average weight increases by 2 lb, the market value drops 1 cent per day and it costs him 75 cents per day to keep each steer. Thus the value of the herd is a function of the number of days the herd is kept. Give the algebraic rule that defines this function.

29. A carpenter wishes to make a toy chest with a square end. If he uses wood costing 50 cents/sq ft for the bottom, wood at 75 cents/sq ft for the sides, and wood at 90 cents/sq ft for the top, and he wishes to spend $10 for the wood, give the algebraic rule that specifies the final volume as a function of the length of a side of the square ends.

30. After many years of research, horticulturalist Marvin Springspeed has developed a cubical apple (for efficient shipping). If the tree provides 5 cubic centimeters (cc) of substance to each apple each day over the period of growth of the apple, give the algebraic rule that specifies the length of edge of an apple as a function of how long it has been on the tree.

31. A 100 meter tower is to be supported by four guy wires, each fastened in the ground at the same distance from the base of the tower. Express algebraically the length of wire needed as a function of the distance from the base of the pole to one of the points where the wire is fastened to the ground.

32. A winch is pulling Hengel Agehog's yacht toward a pier. The winch is 12 ft above the water surface, and the rope is tied to the yacht 5 ft above the surface. (a) Give the algebraic rule specifying the length of rope yet to be pulled in as a function of the distance the boat is from the pier. (b) Comment on the domain and range of this function.

33. A printed page is to contain 30 sq in. of print and have margins of 1 in. on each side and the bottom, and $1\frac{1}{2}$ in. on top. Express the area of the whole page as a function of the length of a line of type.

34. The Springspeed Bus Company will charter a bus for $3 per person if 40 people ride. If the company subtracts 5 cents from every fare for each person in excess of 40 (i.e., if 45 ride, each of the 45 pays $2.75), express the total revenue as a function of the number of people riding (assume that there are at least 40 people).

35. An open box is to be made by cutting congruent squares from the corners of a rectangular piece of metal, and folding up the sides so produced. Express the volume as a function of the length of the edge of the squares if the original rectangles are 25 cm × 35 cm.

36. Calvin Butterball has a 16-ft ladder leaning against his house. It begins to slide down. Express the height of the top of the ladder as a function of the distance of the foot of the ladder from the house. (Assume flat ground and a vertical house wall.)

37. Maynard Basil Holywright is driving home from work one winter night. He passes directly under a street light and notices his car's shadow lengthening ahead of him. If the top of the car is 5 ft from the road and the light is 30 ft above the road:
   (a) Express the distance of the tip of the shadow from the light as a function of the distance of the car from the point directly under the light.
   (b) Express the length of the shadow as a function of the distance of the car from the point under the light.
   (c) Express the distance from the tip of the shadow to the point under the light as a function of the distance of the car from the point under the light.

38. A newly built reservoir has the shape of a triangular prism whose end is an isosceles triangle of base 750 m and depth 50 m. The length of the reservoir is 1 km. Think of the reservoir filling and express
   (a) The volume as a function of the depth of the water
   (b) The depth as a function of the volume.

39. Oceanographer Hengel Agehog needs to get some lunch. She is on the beach. One half mile inland is a road, parallel to the beach; 10 miles down the road is a restaurant. Her motorcycle will go 30 mph in the sand and 70 mph on the road. Express the time it will take her to reach the restaurant as a function of the distance she travels on the road. (Assume that the entire $\frac{1}{2} \times 10$ mile rectangle is sand on which she can drive.)

40. Criticize (i.e., give both the good and bad features of) the following definition: A function is a correspondence between two sets of numbers so that for each element in either set there is precisely one element corresponding to it in the other set.

41. Let $g = \{(x, y) \,|\, x$ is a student from the USA and y is the hometown of $x\}$ and $h = \{(y, z) \,|\, y$ is a town in the USA and z is the state in which y is located$\}$. Consider the set $A = \{(x, z) \,|\,$ there is some town y in the USA so that $(x, y) \in g$ and $(y, z) \in h\}$?
   (a) Is A a relation? (Explain.)
   (b) Is A a function? (Explain.)
   (c) Redefine A without reference to g and h.
   (d) What is the domain of A?
   (e) What is the range of A?

42. Since the domain of a function can be any set, it could be a set of ordered pairs, or ordered triples, or ordered n-tuples for any $n \in \mathbb{N}$. Such functions are called functions of several variables. Examples include the area of a rectangle as a function of its length and width (here the domain would consist of ordered pairs), and a student's grade as a function of homework scores, test scores, and the final exam (here the domain would consist of n-tuples).
   (a) Verify as clearly as you can that each of these examples actually is a function.
   (b) Find three examples of functions of several variables other than

the above two. State the domain and range for each as clearly and completely as you can. (Hint: study the problems of the preliminary chapter.)

43. Since the range of a function can be any set, it could be a set of ordered pairs, or ordered triples, or ordered n-tuples for any $n \in \mathbb{N}$. Such functions are often called vector-valued functions. Examples include the set of divisors of a positive integer n (in increasing order) as a function of n, and the season's list of scores as a function whose domain is a set of professional football teams.
    (a) Verify as clearly as you can that each of these examples actually is a function.
    (b) Find examples of two relations that can be reinterpreted as vector-valued functions other than the two given here. Clearly show that they are functions in the new setting.

44. Problems 42 and 43 can be combined to yield vector-valued functions of several variables. Find an example of such a function; show that it is a function; and state its domain and range as clearly and completely as you can.

In Problems 45–50, a function is described. Write a flowchart to compute values for the given function.

45. Hengel Agehog's weekly pay is a function of the number of hours she works that week. She is paid at $3.50 per hour, earns time-and-one-half for any time over 40 hr, has 21% of her gross pay deducted for taxes and insurance, and pays $10 per week into Payroll Savings.

46. Though not recorded in the published account of his voyages, Gulliver learned that in Lilliput, postage was a function of the height of the sender. The rate was 30 scails for each centimeter or fraction of a centimeter for first class handling.

47. The Fibonacci function F has domain $\mathbb{N}$ and is defined by $F(1) = 1$, $F(2) = 1$, $F(k) = F(k-1) + F(k-2)$ for $k > 2$.

48. The golden ratio can be approximated by ratios of "adjacent" values of the Fibonacci function of Problem 47. Specifically, let the approximating function be

$$A(n) = \frac{F(n)}{F(n-1)}.$$

49. Generalized Fibonacci functions G have domain $\mathbb{N}$, $G(1)$, and $G(2)$ specified in any way one wishes and $G(n) = G(n-1) + G(n-2)$ for $n > 2$. Your flowchart should handle all of these conditions by having values of $G(1)$ and $G(2)$ be among those items to be inputted.

50. One's electric bill is a function of the number of kilowatt hours of electricity used per month. Consolidated Utilities Company charges for electric power according to the following schedule: 7 cents per kilowatt hour for the first 500 kwh, 5 cents per kilowatt hour for the next 500 kwh, and 3 cents per kilowatt hour for any beyond that.

51. Algorithms can also describe relations that are not functions. Write

a flowchart that will compute values for the relation which associates with each positive integer the positive integers less than the given one. The input should be one domain element at a time; the output should include the domain element and all corresponding range elements.

52. Write a flowchart to compute values of the relation that associates with each positive integer the positive integers which are even divisors of the integer. The input should be one domain element at a time; the output should include the domain element and all corresponding range elements.

53. Write a flowchart to compute the function of several variables that associates with the interest rate, principal, and number of years, the interest accumulated over those years if interest is compounded annually.

54. Relations and functions can be, and often are, developed through the vehicle of Cartesian products (see Problem 36, in the preliminary chapter). A relation is defined to be any subset of a Cartesian product, and functions are relations in which no two ordered pairs have the same first element.

   (a) Show that a relation by our definition is also a relation by this new definition. (Hint: Find two sets whose Cartesian product contains the given relation.)

   (b) Show that a relation by the new definition is a relation by our definition. [(a) and (b) together show that the two definitions are equivalent.]

   (c) Show that r and $D_r \times R_r$ need not be the same.

55. Give a definition of each quadrant that uses restrictions on the coordinates of the points found in that quadrant.

56. Write a flowchart to compute and output a table of values for a function. It should be written so that the table will really be useful to you. Questions to consider in constructing this should include: how flexible is the algorithm; does it allow different procedures for different problems; of how much value do you expect it to be for you?

57. Functions that consist of only finitely many ordered pairs of numbers have graphs consisting of only finitely many points. No smooth line may legitimately be drawn through these points so that a different kind of graph is often resorted to, such as pie graphs, bar graphs, pictographs, broken line graphs, and so on. Draw a graph (i) of this type, (ii) in the coordinate plane, for each of the following functions:

   (a) (13, 890), (14, 910), (15, 905), (16, 895), (17, 915) (hypothetical Dow Jones stock average as function of date for a hypothetical week.)

   (b) (1972, 285), (1973, 300), (1974, 320), (1975, 290) (size of graduating classes at Springspeed College).

   (c) (1, 3), (2, 1½), (3, 2⅓), (4, 2) (average number of cavities for leading brands of toothpaste identified by number).

58. If one refers to Problem 43, one could view the relation $\{(x, y) \mid y^2 = x\}$ as a function whose range consists of ordered pairs: $\{(x, (y, z)) \mid y =$

$+\sqrt{x}$, $z = -\sqrt{x}$}. One can then graph the function-version by using a three-dimensional coordinate system.

    (a) Draw, as best you can, the graph of this "complete square root function."

    (b) Compare the graph of (a) with the graph of the relation $\{(x, y) \mid y^2 = x\}$.

    (c) Construct a statement about the graphical restriction on functions with ranges consisting of ordered pairs which corresponds to the statement that the graph of an "ordinary" function has at most one point on each vertical line.

59. Clearly explain why $R_g \subseteq D_f$ must hold in order to define $f \circ g$.

60. If f, g, and h are the functions specified by $f(x) = x^2 - 2$, $g(x) = \sqrt{x}$, and $h(x) = 1/x$, find the domain and range for each of the following functions.

    (a) $f \cdot h$ (b) $f \circ h$ (c) $g \circ f$

    (d) $g \circ h$ (e) $g + h$ (f) $h \circ f$

    (g) $f \cdot g$ (h) $f + h$ (i) $h \circ h$

    (j) f/h

61. If f and g are two functions each specified by an equation, give

    (a) An interpretation of $f \cap g$ in terms of sets.

    (b) An interpretation of $f \cap g$ in terms of high school algebra, where f and g are specified by equations.

    (c) An interpretation of $f \cap g$ in graphical terms.

62. Prove that if f and g are functions, $f \cup g$ need not be a function.

63. Prove that if f and g are functions, $f \cap g$ is also a function.

64. Prove result 1 at the end of Section 1.4c.

65. Prove result 2 at the end of Section 1.4c.

66. Prove result 3 at the end of Section 1.4c.

67. Prove result 4 at the end of Section 1.4c.

68. Does functional composition distribute over the addition of real-valued functions? Prove or cite a counterexample.

69. Does functional composition distribute over the multiplication of real-valued functions? Prove or cite a counterexample.

70. Find a specific function that behaves for the addition of real-valued functions as zero does for the addition of numbers.

71. Find a specific function that behaves for multiplication of real-valued functions as 1 does for multiplication of numbers.

72. Find a specific function that is an identity for functional composition; i.e., find a function g so that $f \circ g = f$ for all functions f.

73. In Section 1.4, it has been tacitly assumed that if f and g are functions, then so are $f \circ g$, $f + g$, $f - g$, $f \cdot g$, and f/g. To remedy this, assume f and g are functions for which the operation in question is defined.

    (a) Prove $f \circ g$ is a function.   (b) Prove $f + g$ is a function.

    (c) Prove $f \cdot g$ is a function.   (d) Prove $f - g$ is a function.

    (e) Prove f/g is a function, provided that $g(x)$ is never 0.

74. Using assignment instructions such as $z \leftarrow f(x)$ but not allowing the use of $+$, $-$, $\cdot$, $\div$, or "$\circ$" with functions names alone, write a

flowchart for the evaluation of each of the following:

(a) $(f+g))(x)$ (b) $(f \cdot g)(x)$ (c) $(f \circ g)(x)$

(d) $(f/g)(x)$ (e) $(f-g)(x)$ (f) $(g \circ f)(x)$

75. Which of the inverses in Practice Exercise 20 are inverse functions?

76. Draw the graphs of f and f^{-1} from Example 23 on one set of axes. Also draw the graph that would have resulted had we not stipulated that $y < 0$.

77. Could there be a function that had no inverse at all? Explain.

78. Show by example that the inverse of a relation which is not a function may be a function.

79. Prove that $(f^{-1})^{-1} = f$ (i.e., that the inverse of the inverse is the function itself). This result paves the way to deriving the graph of f from the f^{-1}. It can be interpreted as saying "If one applies the reflection technique already devised to the graph of f^{-1}, one will get the graph of f."

80. Discuss the inverse of the greatest integer function with reference to the concerns of Section 1.5b.

81. Find an expression for $(f \cdot g)^{-1}$ in terms of f^{-1} and g^{-1} or explain why no such expression exists. (Assume that f^{-1} and g^{-1} are functions, too.)

82. Find an expression for $(f \circ g)^{-1}$ in terms of f^{-1} and g^{-1} or explain why no such expression exists. (Hint: Try to construct a function ψ so that $\psi \circ (f \circ g) = $ the identity function.) (Assume that f^{-1} and g^{-1} are functions, too.)

83. A secret code can be viewed as a function that associates with letters other letters or symbols. The inverse of such a function is the key to the code. Write the function and its inverse for the code in which

The quick brown fox jumped over the lazy dogs

becomes

Yjr wiovl ntpem gpc ki,]rf pbrt yjr ;sxu fphd/

84. Consider the following six functions: $b(x) = x$; $u(x) = 1/x$;

$$e(x) = \frac{1}{1-x}; \ t(x) = 1 - x; \ a(x) = \frac{x-1}{x}; \text{ and } y(x) = \frac{x}{x-1}.$$

(a) Complete the following table for functional composition (the given entry is $e \circ t$)

$\circ$	b	e	a	u	t	y
b						
e					u	
u						
t						
y						

(b) Use the results of part (a) to help you find the inverse of each of the six functions.

(This set of functions under the operation of functional composition is an example of what mathematicians call a noncommutative group. The study of groups occurs in the branch of mathematics called abstract algebra.)

85. Redo the analysis associated with Figure 1.25 but with P being in quadrant IV.

86. In connection with the algebra of functions, it is possible to define a kind of relation, known as a partial ordering, on a set of real-valued functions with the same domain. One says $f \lessdot g$ (or (f, g) is an element of the relation) and says that f is subordinate to g, provided $f(x) \le g(x)$ for all x in their domain.
   (a) Verify that this relation is *reflexive*, i.e., $f \lessdot f$ for all functions f.
   (b) Verify that this relation is *antisymmetric*, ie., if $f \lessdot g$ and $g \lessdot f$, then $f = g$.
   (c) Verify that this relation is *transitive*, i.e., if $f \lessdot g$ and $g \lessdot h$, then $f \lessdot h$.
   (d) Show that $\lessdot$ is not the same kind of relation as $\le$ is in $\mathbb{R}$, in that one can have functions f and g so that $f \not\lessdot g$ and $g \not\lessdot f$, i.e., f and g are not related by $\lessdot$. (Recall that if a and b are any two real numbers, either $a \le b$ or $b \le a$. This order relation is called a total ordering, or a linear ordering.)
   (e) Show by example that order is important (i.e., that $f \lessdot g$ does not mean that $g \lessdot f$; one says the relation is not symmetric).

87. If $f(x) = x^2 + 4$ with domain $[-1, 2]$,
   (a) What is the range of f?
   (b) What subset of the domain of f will produce a function $\hat{f}$ whose inverse $\hat{f}^{-1}$ would be a function with the property that $R_{\hat{f}} = R_f$?

88. Produce a convincing argument (proof) that the flip-rotate-relable technique mentioned in the footnote at the end of Section 1.5c is equivalent to the point reflection technique described in the text proper.

Problems 89–91 deal with another pair of properties that some functions with domains and ranges subsets of $\mathbb{R}$ have: A function f is said to be odd if $f(-x) = -f(x)$ for every $x \in D_f$. A function f is said to be even if $f(-x) = f(x)$ for every $x \in D_f$.

89. The recognition of oddness or evenness can be a graphing aid, in that each refers to a kind of symmetry. To see this, graph each of the following and note what is called symmetry in the y-axis for the even functions, symmetry in the origin for the odd functions.
   (a) $n(x) = x^3$ (odd)
   (b) $u(t) = 1/t$ (odd)
   (c) $m(w) = w^2$ (even)
   (d) $b(z) = |z|$ (even)
   (e) $e(y) = y$ (odd)
   (f) $r(y) = y^3 - y$ (odd)
   (g) $t(a) = a^2 + 2$ (even)
   (h) $h(x) = 1/x^2$ (even)
   (i) $e(m) = m^5$ (odd)
   (j) $o(w) = \sqrt[3]{w}$ (odd)
   (k) $r(t) = t^3$ (even)
   (l) $y(x) = |x^3 - x|$ (even)

Thus, if one determines the graph of f for $x \in [0, \infty)$, one can, by symmetry, sketch the graph for $x \in (-\infty, 0]$, provided that f is odd or even.

90. Most functions are neither odd nor even. Examine each of the following and by use of the definitions determine which are odd, which even, and which neither.

(a) $p(x) = \dfrac{x+1}{x-1}$

(b) $i(y) = 3y^3 - 5y$

(c) $h(z) = \lfloor z \rfloor$

(d) $e(w) = 3w + 5$

(e) $q(t) = \dfrac{1}{t^3 + t}$

(f) $r(m) = |m^2 + 2m|$

(g) $f(a) = \sqrt{a+2}$

(h) $s(p) = \dfrac{|p|}{p}$

(i) $g(i) = (i - 2)^2$

(j) $e(r) = r^3 + 2$

(k) $b(s) = 6 - s^2$

(l) $r(v) = v + \dfrac{1}{v}$

(m) $f(p) = \left| v + \dfrac{1}{v} \right|$

91. Let f and g represent even functions, and h and k represent odd functions. Then, again using the definitions, determine whether each of the following is odd, even, or neither. Your work should provide proof.

(a) $f \circ h$

(b) $f \circ g$

(c) $h \circ k$

(d) $h \circ f$

(e) $f + g$

(f) $f + h$

(g) $f \cdot g$

(h) $f \cdot h$

(i) $f - g$

(j) $f - h$

(k) f/g

(l) f/h

The following problems (92–95) are concerned with another property that a function may have that relates to the issue of inverse functions: A function is said to be one-to-one (or injective) if no two ordered pairs have the same second element.

92. Using the definitions of function and one-to-one, prove that the inverse of a one-to-one function is necessarily a function.

93. In reference to the previous problem, prove that the inverse of a one-to-one function is also one-to-one.

94. Give a geometric interpretation of the definition of one-to-one for functions whose domains and range are subsets of $\mathbb{R}$.

95. To determine whether a function is one-to-one or not, one typically asks "Suppose $f(x) = f(z)$?" and attempts to show that x must then be the same as z. Success in this attempt establishes that f is one-to-one. (Do you see why?) Determine which of the following functions are injective. For those that are, supply your proof. For those that are not, exhibit two ordered pairs with different first elements but the same second element.

(a) $p(x) = x^2$ (b) $o(t) = t^3$ (c) $w(m) = |m|$

(d) $e(w) = 3w - 2$ (e) $s(r) = 4 - 2r$ (f) $s(v) = 1/v$

(g) $e(a) = 1/a^2$ (h) $r(g) = \sqrt{g}$ (i) $i(x) = \lfloor x \rfloor$

(j) $e(y) = \sqrt[3]{y}$ (k) $s(z) = z^2 + 2z$

CHAPTER 2

POLYNOMIAL FUNCTIONS

2.1. INTRODUCTION

In Chapter 1, we established a base of common vocabulary and embarked on an effort to understand functions as objects of study in the way mathematicians view them. We move now to the study of particular types of functions that are important both to mathematics as an intellectual discipline and to the use of mathematics in the "real world." Within this study, we shall continue to ask questions that bear on general philosophical understandings as well as questions related to particular facts.

It may help your overall understanding if you see a few basic questions as common to our study of each type of elementary function. First, of course, is "What kind of functions are we looking at—how are they defined or characterized?" This is a matter of establishing a frame of reference for our explorations.

The second is "How can we find values of these functions for certain arguments?" This is a natural question to ask, and will be easily answered in some cases and not so easily in others.

The third is "How can we find values of f^{-1}?" The most common version of this is to ask for $f^{-1}(0)$. Why the concern for $f^{-1}(0)$? First, we

DEFINITION 2.1
Zeros and $f^{-1}(0)$

need to recognize that $f^{-1}(0)$ is the set of numbers, x, for which $f(x) = 0$. Each of these values of x is called a *zero* for the function f. If f describes some physical phenomenon such as the motion of a (hopefully) peaceful missile, the zeros could be the time of launch and the time of impact. Or f could be a function called a derivative (one of the major objects of study in the calculus) whose zeros provide precise information about maximum and minimum values of a related function (such as a cost function, revenue function, efficiency function, etc). So in applying mathematics to the solution of real problems of various types, the ability to find zeros, i.e., to find $f^{-1}(0)$, is vital. Furthermore, the search for zeros includes all the activities you are accustomed to thinking of as solving equations. And last, because zeros have been found to be of frequent practical importance, it is common practice in graphing functions to insist on precise location of the zeros (i.e., x-intercepts) whenever possible; therefore, to draw authoritative graphs, one needs to be able to find zeros.

The fourth is "What properties do these functions have and how do they relate with one another?" This is a more general question and has an algebraic flavor. In part, this is a question of abstract mathematical nature, but it also has the practical aspect of concern with manipulations that may aid the search for zeros or the finding of values for functions, very much like the attention to factoring in algebra is, on the one hand, asking seemingly silly questions of how to write the same thing different ways and, on the other hand, is vital to solving equations and being able to use algebra as a tool.

The fifth is "What do these functions look like?" or "What are their graphs?" This question, as noted in Chapter 1, is included because graphs provide another path to seeing and understanding functions.

We start the study of the elementary functions by investigating what we call polynomial functions because they are in some senses the most natural and provide a bridge between algebra and the study of functions. They are natural because they may be used to represent many phenomena and may often be used to approximate more complicated functions when only approximate answers are needed. Furthermore, they are easy to find values for, unlike many of the other elementary functions; this helps to explain the popularity of polynomial functions among users of mathematics. Among polynomial functions, our first concern will be two special types: *linear* and *quadratic*.

2.2. TWO SIMPLE TYPES OF FUNCTIONS

2.2a. Linear Functions

LINEAR FUNCTIONS

One of the most familiar functions used to describe phenomena is the linear function. If nails cost 50 cents per pound, then the weight of nails one gets is a linear function of how much he pays for the nails ($w(d) = 2d$ or $w = d/0.50$). If a cook wishes to make $\frac{2}{3}$ of a recipe, the new amount of each ingredient needed is a linear function of the amount called for in the recipe ($N(A) = (\frac{2}{3})A$). Centigrade temperature is a linear function of

Fahrenheit temperature $(C(F) = \frac{5}{9}(F - 32))$. If a politician knows that he will get 100,000 votes with no campaigning and will get 100 additional votes for each second of TV time he uses, then the number of votes he can get is a linear function of the number of seconds of TV time he uses $(V(t) = 100{,}000 + 100t)$, and so on.

In formal language a function f is said to be *linear* if for each $x \in D_f$, $f(x) = ax + b$, where a and b are fixed real numbers. Such functions are called linear because their graphs are straight lines (or subsets of straight lines if D_f is a proper subset of $\mathbb{R}$). Although it would be esthetically pleasing to verify that graphs of linear functions are straight lines, such a proof would entail a longer journey into the realm of analytic geometry than is appropriate in this chapter (see Chapter 6).

DEFINITION 2.2
Linear Function

The four examples cited above satisfy the definition of linear function, as we now show. If $w(d)$ is the weight of nails (in pounds) purchased by d dollars, then $w(d) = 2d$ (so, in the form $y = ax + b$, $a = 2, b = 0$) defines the linear function w. If $n(A)$ is the new amount of an ingredient when the recipe calls for an amount A, $n(A) = \frac{2}{3}A$ (so $a = \frac{2}{3}$, $b = 0$). If $C(F)$ is the centigrade temperature corresponding to a Fahrenheit temperature F, $C(F) = \frac{5}{9}(F - 32)$ (so $a = \frac{5}{9}$, $b = -\frac{160}{9}$). If $v(t)$ is the number of votes received, and if t seconds of TV time are used, $v(t) = 100t + 100{,}000$ (here $a = 100$, $b = 100{,}000$).

EXAMPLE 1

Graphs of linear functions are easy to draw because one needs only two points to uniquely determine a straight line. Thus one need only plot two points, and one of these can be read directly from the equation if it is in the form $y = ax + b$, namely, the point $(0, b)$. The value b is called the *y-intercept* of the graph, since it is the point where the line crosses the y-axis. More generally, the y-intercept of a function f is the value $f(0)$.

DEFINITION 2.3
y-intercept

Let us draw the graph of the function f specified by $f(x) = \frac{1}{2}x + 3$. Choosing $x = 2$ quickly produces the point $(2, 4)$. Plotting this point together with the y-intercept $(0, 3)$, and drawing the graph, one gets the graph of Figure 2.1.

EXAMPLE 2

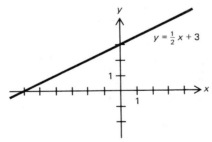

FIGURE 2-1

In the case when $a = 0$ in the rule for a linear function, i.e., $f(x) = b$, we have what we have previously called a constant function (Section 1.2d). Its graph is a horizontal straight line. Look, for example, at $f(x) = 3$. Notice that the table of values looks like

EXAMPLE 3

x	0	1	5
y	3	3	3

etc., resulting in the graph of Figure 2.2.

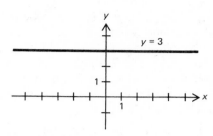

FIGURE 2-2

Some mathematicians choose to isolate constant functions and not call them linear. Although this can occasionally be a useful categorization, we shall include constant functions with linear functions because their graphs are straight lines.

As mentioned in Section 2.1, one of the questions often asked about functions is "What are their sets of zeros?" For linear functions, the answer is easily found. Algebraically, this translates into the question "What are the solutions of $ax + b = 0$?" The use of algebra assures us that if $a \neq 0$, then $-b/a$ is the only zero. If $a = 0$, either there are no zeros at all if $b \neq 0$ (e.g., if $f(x) = 0x + 3$, the equation $0x + 3 = 0$ has no solutions) or else every real number is a zero if $b = 0$ (for then $f(x) = 0x + 0 = 0$ for all values of x; i.e., f is the zero function).†

A graphical analysis of the question reinforces our algebraic solution and aids further understanding. Recall from Section 2.1 that zeros correspond to intersections of the graph with the x-axis. Also recall from geometry that two lines intersect in exactly one point unless they are parallel or coincident. Thus we should expect linear functions to have exactly one zero unless their graphs are parallel to, or coincident with, the x-axis. The only straight lines that are parallel to the x-axis are those with equation $y = c$ for some constant c. Thus we are considering the constant functions $f(x) = c$, such as $f(x) = \sqrt{2}$, which has a graph parallel to the x-axis through $(0, \sqrt{2})$. If $c \neq 0$, the line is truly parallel and has no x-intercept, while if $c = 0$, the line is the x-axis. Thus our algebraic conclusions are all accounted for and reaffirmed.

† Notice that the solution of an equation such as $3x + 5 = 3 - 2x$ straightforwardly transforms into the equation $5x + 2 = 0$. The solution of $5x + 2 = 0$ (i.e., $x = -\frac{2}{5}$) is the zero of the linear function $f(x) = 5x + 2$. More generally, solving any equation can be restated as finding zeros for a (related) function.

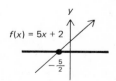

2.2b.   Quadratic Functions

Quadratic functions are also common. These are functions f that may be specified by a rule $f(x) = ax^2 + bx + c$ for each $x \in D_f$, where a, b, and c are fixed real numbers and $a \neq 0$. The condition that $a \neq 0$ is to exclude functions that are really linear functions, for if $a = 0$, $f(x) = ax^2 + bx + c$ becomes $f(x) = bx + c$.

DEFINITION 2.4
**Quadratic
Functions**

Examples of quadratic functions include:

EXAMPLE 4

a. The area of a circle is a quadratic function of the radius of the circle ($A(r) = \pi r^2$, so that $a = \pi$, $b = 0$, and $c = 0$).

b. The height of an object dropped off a 100-ft high cliff (ignoring air resistance) is a quadratic function of the time since it was dropped [$h(t) = -16t^2 + 100$, where $h(t)$ is the distance of the object from the ground, $a = -16$, $b = 0$, $c = 100$, and $D_h = [0, 2\frac{1}{2}]$ (why?)].

c. The kinetic energy of a car is a quadratic function of its velocity [$E(v) = (m/2)v^2$ if m is the mass of the car in slugs, $a = m/2, b = 0$, and $c = 0$].

d. The area of a rectangular field of perimeter 300 m is a quadratic function of the length L of one side ($A(L) = L \cdot (300 - L) = -L^2 + 300 \cdot L$, where $a = -1$, $b = 300$, and $c = 0$).

e. The receipts of a manufacturer for yearly sales of widgets is a quadratic function of the number sold if the price decreases from a high of $10 each, by 5 cents for each 100 sold: $\left(R(n) = n \cdot \left(1000 - 5 \cdot \left(\dfrac{n}{100} \right) \right) = -\dfrac{n^2}{20}\right.$ $+ 1000n$, where $a = -\frac{1}{20}$, $b = 1000$, and $c = 0 \Big)$.

Let us turn first to the question of finding zeros. For quadratic functions, finding all the zeros is harder than with linear functions. While you may already be familiar with the answer, we shall take some time to derive it in order to demonstrate one of the frequently used techniques of mathematics, that of observing a pattern and trying to generalize it.
Our goal is to solve the equation

$$ax^2 + bx + c = 0. \qquad (1)$$

Not seeing a key to this, one might try a number of different specific examples. In so doing, he would notice that equations such as $x^2 - 25 = 0$ are easily solved by rewriting the equation as $x^2 = 25$, taking square roots of both sides and arriving at the set of solutions $\{5, -5\}$. In a similar fashion, the left side of $x^2 - 6x + 9 = 0$ is a trinomial square, so that the equation can be rewritten as $(x - 3)^2 = 0$ to yield the solution set $\{3\}$. One thinks "It would be nice if all quadratic equations could be rewritten as squares!" and then seeks to turn wishful thinking into reality. The equation $x^2 - 4x - 5 = 0$ can be rewritten as $x^2 - 4x + 4 = 9$, transformed to $(x - 2)^2 = 9$, then solved by taking square roots of both sides to get $x - 2 = \pm 3$. The final result is the solution set $\{-1, 5\}$. Now see the complete solution below:

$$x^2 - 4x - 5 = 0$$
$$x^2 - 4x = 5$$
$$x^2 - 4x + 4 = 5 + 4 = 9 \qquad \textit{(continued)}$$

$$(x-2)^2 = 9$$
$$(x-2) = \pm 3$$
$$x = \pm 3 + 2 = \begin{cases} \nearrow 3 + 2 = 5 \\ \searrow -3 + 2 = -1. \end{cases}$$

yielding

$$\{-1, +5\}.$$

Taking courage in hand one attempts to do this for equation (1) as it stands. Being more comfortable with $1 \cdot x^2$ instead of ax^2, we first divide by a to get

$$x^2 + \frac{b}{a}x + \frac{c}{a} = 0. \qquad (2)$$

For the left side to be a trinomial square, the constant term would have to be $(\frac{1}{2} \cdot b/a)^2$ (since $(x + \Box)^2 = x^2 + 2\Box x + \Box^2$ shows the constant term to be the square of half the coefficient of the x-term).† Thus we should add $-c/a$ to both sides of the equation and then add $(b/2a)^2$ to both sides, yielding the equation

$$x^2 + \frac{b}{a}x + \left(\frac{b}{2a}\right)^2 = \left(\frac{b}{2a}\right)^2 - \frac{c}{a}. \qquad (3)$$

This, in turn, transforms into

$$\left(x + \frac{b}{2a}\right)^2 = \frac{b^2 - 4ac}{4a^2}. \qquad (4)$$

(Be sure you understand the algebra involved.)

We may proceed provided that $\dfrac{b^2 - 4ac}{4a^2} \geq 0$. Otherwise, we would be attempting to take the square root of a negative number which is not possible in $\mathbb{R}$. This equation is equivalent to saying that $b^2 - 4ac \geq 0$, because $4a^2$ must be positive (why?). But if $b^2 - 4ac \geq 0$, we have

$$x + \frac{b}{2a} = \pm\sqrt{\frac{b^2 - 4ac}{4a^2}} = \pm\frac{\sqrt{b^2 - 4ac}}{2a} \qquad (5)$$

and finally

$$x = \pm\frac{\sqrt{b^2 - 4ac}}{2a} - \frac{b}{2a}$$

$$x = -\frac{b \pm \sqrt{b^2 - 4ac}}{2a}. \qquad (6)$$

FORMULA 2.1
Quadratic
Formula

This equation is often called the *Quadratic Formula*.

Our conclusions, then, are that if $b^2 - 4ac \geq 0$, the quadratic function f specified by $f(x) = ax^2 + bx + c$ has

$$f^{-1}(0) = \left\{\frac{-b + \sqrt{b^2 - 4ac}}{2a}, \frac{-b - \sqrt{b^2 - 4ac}}{2a}\right\}$$

† See "completing the square" in Appendix A, Section A-18.

while if $b^2 - 4ac < 0$, $f^{-1}(0) = \emptyset$. The latter holds not because we have not been clever enough to find such zeros but because Equation (4), which is equivalent to equation (1), cannot be true for any real value of x when $b^2 - 4ac < 0$. In general, the failure to find zeros does not necessarily mean that they do not exist. One must find conclusive evidence, as we have just done for $b^2 - 4ac < 0$, that none could possibly exist.

Note also that if $b^2 - 4ac = 0$, then the two zeros are really the same. Such a zero is called a *repeated zero* or a *zero of multiplicity two*.†

As some readers may already know, when $b^2 - 4ac < 0$, one can find zeros for the quadratic function $f(x) = ax^2 + bx + c$, provided that one is willing to extend the domain of f to a number system larger than $\mathbb{R}$. This number system is called the complex number system. We defer consideration of this situation to Chapter 7 on complex numbers.

Since the number $b^2 - 4ac$ has prominence in the formula, it is given a special name: the *discriminant* of the function.

DEFINITION 2.5
Discriminant

Find all the real zeros of f if $f(x) = x^2 - 4x - 6$. In this example, $a = 1$, $b = -4$, and $c = -6$. Employing Equation (6), we compute:

EXAMPLE 5

$$x = \frac{-b \pm \sqrt{b^2 - 4ac}}{2a}$$

$$x = \frac{-(-4) \pm \sqrt{(-4)^2 - 4(1)(-6)}}{2 \cdot 1}$$

$$x = \frac{4 \pm \sqrt{16 + 24}}{2}$$

$$x = \frac{4 \pm \sqrt{40}}{2} = \frac{4 \pm \sqrt{4 \cdot 10}}{2} = \frac{4 \pm 2\sqrt{10}}{2} = 2 \pm \sqrt{10}.$$

So, $x = 2 + \sqrt{10}$ and $x = 2 - \sqrt{10}$ are the two solutions. Notice that the discriminant is positive here ($+40$), and we found two distinct real zeros. This is in harmony with our observation earlier in this section.

We choose to leave these numbers in the form $2 + \sqrt{10}$ and $2 - \sqrt{10}$ as exact zeros, although if this were being used in a real-life problem, one would probably wish approximate values, which can be gotten by approximating $\sqrt{10}$ (often using a table of values of square roots). As long as there are no clear reasons to do otherwise, however, answers should be left in exact form, not approximated.

Find all the real zeros of g if $g(x) = 3x^2 + 6x + 3$. This time Equation (6) yields

EXAMPLE 6

$$x = \frac{-6 \pm \sqrt{6^2 - 4 \cdot 3 \cdot 3}}{2 \cdot 3}$$

$$= \frac{-6 \pm \sqrt{36 - 36}}{6}$$

† The concept of multiplicity of a zero is dealt with more generally in Section 2.4b.

$$= \frac{-6 \pm \sqrt{0}}{6}$$

i.e.,

$$x = -1 \pm 0 = -1.$$

The discriminant $= 0$, and we find a single real zero of multiplicity two.

EXAMPLE 7   Find the zeros of $r(x) = x^2 - x - 6$. For this example, $a = 1$, $b = -1$, and $c = -6$, and we could proceed as in the previous examples. However, an observant student might have noticed that $r(x)$ is easily factorable to $r(x) = (x - 3)(x + 2)$ and recall from his algebra courses that because a product of real numbers is zero only if one or more of the factors is zero, one need only solve $x - 3 = 0$ and $x + 2 = 0$ to find the zeros in this instance, thus saving a lot of computation. (Try this problem using the quadratic formula and see how much work it would be.) It is a good idea to check quadratics for possible easy factorizations unless your factoring is notoriously bad (in which case you should see Appendix A, Sections A-10–A-12). The value of the quadratic formula is not that one always wants to use it but that it is always available, always applicable, and always correct.

Now let us consider graphs of quadratic functions. As it turns out, they are always a special kind of curve, a parabola (or part of a parabola if $D_f \subseteq \mathbb{R}$). The basic shape of parabolas is suggested by several sketches in Figure 2.3. Reflectors of flashlights, mirrors for telescopes, radar antennas, and many other objects have parabolic cross sections, due to some interesting properties the curves possess. But these properties, along with any proof that all quadratic graphs actually are parabolas, is deferred to your study of analytic geometry and the calculus.†

If we accept that graphs of quadratic functions are always parabolas, there are several graphing aids which we can establish to reduce the job of plotting points. Let us look at a preliminary outline of these short cuts in the form of a list of the corresponding questions a mathematician might ask if confronted with our situation for the first time. "Can I easily determine whether a particular graph will open up or down?"‡ "Can I find the exact point that is the high or low point of the graph?," "Are the intercepts easily found?"

The last question is the most easily answered. The quadratic formula gives us the x-intercepts *if* there are any, and the constant term (c in the general form) is the y-intercept, since $f(0) = a \cdot 0^2 + b \cdot 0 + c = c$.

EXAMPLE 8   As an example, if $\sigma(x) = 2x^2 + 4x - 6$, σ has zeros 1 and -3, so has x-intercepts at 1 and -3. The y-intercept in this example is -6.

The other two questions are answerable together by returning to the algebra used in deriving the quadratic formula. Writing the value $f(x)$ in

---

† Actually, we could define parabola in terms of graphs of quadratic equations but choose not to do so because what we are actually claiming is that all quadratic functions have similarly shaped graphs (whatever they are called).

‡ As a general curve, parabolas may open up, down, sideways, or in any direction. But (without proof) as the graph of a quadratic function, it must open either up or down.

general form $ax^2 + bx + c$, we may complete the square as before but without attempting to solve for zero, with the result:

$$f(x) = a\left[x^2 + \frac{b}{a}x + \left(\frac{b}{2a}\right)^2 + \frac{c}{a} - \left(\frac{b}{2a}\right)^2\right]$$

(See Equations (3) and (4))

$$= a\left[\left(x + \frac{b}{2a}\right)^2 + \frac{4ac - b^2}{4a^2}\right]$$

$$= a\left(x + \frac{b}{2a}\right)^2 + \frac{4ac - b^2}{4a}. \tag{7}$$

The only term in the last line that depends on x (and thus affects a change in the value of y) is the first one, $a(x + b/2a)^2$. From this, we can find that the parabola will open upward if $a > 0$ and downward if $a < 0$. The explanation is that since $(x + b/2a)^2 \geq 0$ for all x, if $a > 0$, then the first term can only increase in the positive (up) direction; if $a < 0$, it can only increase in the negative (down) direction.

This same analysis of Equation (7) above allows us to determine the high or low point precisely. This high or low point is called the vertex of the parabola. If $a > 0$, so that the curve opens upward, then the vertex will occur when the first term of Equation (7) has its smallest value. Because this term is always ≥ 0, its smallest value will be zero. If we solve

$$a\left(x + \frac{b}{2a}\right)^2 = 0,$$

we find that the solution is $x = -b/2a$. The corresponding y value can then be derived from Equation (7), since for $x = -b/2a$,

$$f(x) = a \cdot 0^2 + \frac{4ac - b^2}{4a} = \frac{4ac - b^2}{4a}.$$

In other words, the vertex is at

$$\left(-\frac{b}{2a}, \frac{4ac - b^2}{4a}\right).$$

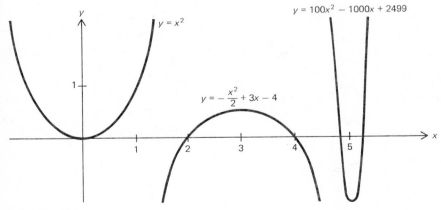

FIGURE 2-3

In the case of $a < 0$, we would be expecting the vertex to be a high point. The term $a(x + b/2a)^2$, being ≤ 0, will attain its largest value when it actually equals zero, and a quick run through reasoning similar to that of the first case should convince you that the vertex is again at

$$\left(-\frac{b}{2a}, \frac{4ac - b^2}{4a}\right).$$

Thus this formula for the vertex is valid in both cases.

EXAMPLE 9
These techniques enable us to find that $f(x) = 3x^2 - 7x + 2$ opens up (since $3 > 0$) and has vertex at

$$\left(-\left(\frac{-7}{2 \cdot 3}\right), \frac{4 \cdot 3 \cdot 2 - (-7)^2}{4 \cdot 3}\right) = \left(\frac{7}{6}, \frac{-25}{12}\right)$$

and $y(x) = -2x^2 + x - 5$ opens down (since $-2 < 0$) with vertex at

$$\left(-\left(\frac{-1}{(2)(-2)}\right), \frac{4(-2)(-5) - (1)^2}{4(-2)}\right) = \left(\frac{1}{4}, \frac{-39}{8}\right).$$

and $y(x) = 9x^2 + 18x + 5$ opens up with vertex at $(-1, -4)$.

Putting these techniques together with the intercept information previously discovered, we can now graph quadratic functions in the manner of the following example.

EXAMPLE 10
Draw the graph of f if $f(x) = 3x^2 - 4x + 1$. By the analysis just completed, since $3 > 0$, the parabola opens upward; the minimum point is

$$\left(-b/2a, \frac{-b^2 + 4ac}{4a}\right) \text{ or } \left(\frac{-(-4)}{2(3)}, \frac{-(-4)^2 + 4 \cdot 3 \cdot 1}{4 \cdot 3}\right)$$

or $(\frac{4}{6}, -\frac{4}{12})$ or $(\frac{2}{3}, -\frac{1}{3})$. By using the quadratic formula, the zeros of f are at $x = \frac{1}{3}$ and $x = 1$; the y-intercept is $+1$. This yields Figure 2.4(a). Plotting a few more points for precision we have the graph of Figure 2.4(b).

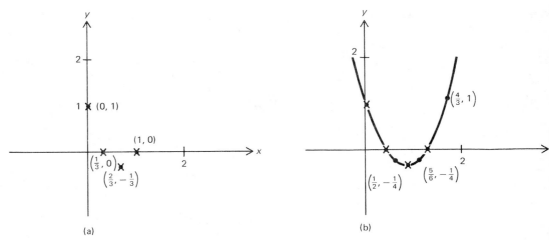

(a)   (b)

FIGURE 2-4

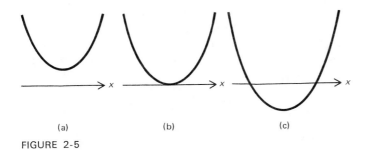

(a)   (b)   (c)

FIGURE 2-5

As was the case with linear functions, our graphical knowledge of
quadratic functions can enhance an understanding of the algebraic conclusions about zeros. It seems intuitively quite clear that there are three
possible spatial relationships between a horizontal line and a parabola
(whether opening up or down). These are suggested by Figure 2.5: no
intersection, intersection in one point, or intersection in two points.

Actually, the derivation of the quadratic formula is a proof that these
actually are the only three possibilities. When the horizontal line is the x-axis,
these points correspond directly to cases of no real zeros, one repeated real
zero, and two distinct real zeros. Thus the fact that a function such as
$\psi(x) = 6x^2 + 5x + 5$ has no real zeros means that the graph lies entirely
above or below (here above) the x-axis, while $f(x) = 9x^2 + 6x + 1$
(having a repeated real zero) "bounces off" the x-axis at one point, the
graph never going all the way through the axis, and $g(x) = 2x^2 + 7x - 1$
"lives" on both sides of the x-axis with two separate crossing points.

As was true for linear functions, quadratic functions and their graphs
can be more completely understood and appreciated when the tools of
analytic geometry are applied to them. (See Chapter 6.)

2.3. OTHER POLYNOMIAL FUNCTIONS

2.3a. General Polynomials

Having studied linear and quadratic functions in the last section, a
whole family of functions may have been suggested to you. After looking
at the rules $4x + 5$ and $7x^2 + 2x - 1$, one might think of other functions like
$\frac{4}{3}x^3 + 3x^2 - 2x + 7$ and $8x^4 + \sqrt{7}x^3 + 9x^2 - x + 7$, and so on. In more
general language, after considering $ax + b$ and $ax^2 + bx + c$, one might
think of looking at $ax^3 + bx^2 + cx + d$ or $ax^4 + bx^3 + cx^2 + dx + e$. Such
functions have practical use, both in their own right and in their ability to
closely approximate other functions that are more difficult to work with.
Rather than tackle each one individually, we shall try to investigate such
functions all at once. These functions (including linear and quadratic
functions) are called polynomial functions. Formally, a function f is a

DEFINITION 2.6
Polynomial
Function

DEFINITION 2.7
Degree
DEFINITION 2.8
Lead Coefficient

polynomial function if it can be specified by a rule of the form

$$f(x) = a_n x^n + a_{n-1} x^{n-1} + \cdots + a_1 x + a_0 \qquad (8)$$

for some nonnegative integer n, where $a_0, a_1, \cdots, a_n$ are real number constants. Such a function is said to be of *degree n* or of *n*th degree if $a_n \neq 0$, i.e., if there really is a term involving x^n and no term with x to a higher power than n, and a_n is called the *lead coefficient*.

For example, $f(x) = 3x^5 + 2x - \sqrt{3}$ is a polynomial of degree 5 with lead coefficient 3, having $a_5 = 3$, $a_4 = 0$, $a_3 = 0$, $a_2 = 0$, $a_1 = 2$, $a_0 = -\sqrt{3}$, and $m(x) = 0x^6 + 2x^2 - \frac{7}{3}x^4 + 3$ is a polynomial of degree 4 with lead coefficient $-\frac{7}{3}$ where $a_4 = -\frac{7}{3}$, $a_3 = 0$, $a_2 = 2$, $a_1 = 0$, $a_0 = 3$.

Let us turn first to the problem of finding zeros for polynomial functions. It would be desirable to have a formula to produce these zeros that would require our only knowing the coefficients (i.e., $a_0, a_1, \cdots, a_n$) and the operations of algebra (namely, addition, subtraction, multiplication, division, and extraction of roots). The search for such formulas has occupied the efforts of many mathematicians over the course of centuries. We have already derived such formulas for polynomials of degree 1 and 2, and more complicated formulas were developed for polynomials of degree 3 and 4 (see Appendix E). The case of fifth degree polynomials remained one of the great unsolved problems of mathematics until the first third of the 19th century. It was finally proved by Neils Abel and, independently at about the same time, by Evariste Galois† that it was in fact impossible to construct a formula that would find all the zeros of every fifth degree polynomial if one knew all the coefficients and used the operations of algebra. This simultaneously dashed all hopes for general formulas for polynomials of degree higher than 5.‡

Since we will not be able to rely on formulas for polynomials of degree greater than 4, and since the formulas which do exist for third and fourth degree polynomials are complicated, we find it advisable to change our tactics. In so doing we will hope to develop sufficiently powerful tools that our omission of the third and fourth degree formulas will not be regretted at a later date. Our new approach will be more along the line of "divide and conquer." Instead of trying to find the zeros all at once, we will worry about finding one at a time, hoping then to be able to simplify the remaining problem. This project will be our major objective in this section and the following one.

2.3b. Rational Zero Theorem

One large class of polynomials is easy prey for one zero-detecting technique. However, the route of discovery for this tool is not one that readily suggests itself to an observer.

---

† See E. T. Bell, *Men of Mathematics* (New York: Simon and Schuster, 1937) for biographies of these two interesting mathematicians.
‡ Results of Section 2.4 can show that such polynomials have fifth degree factors; therefore, to find their zeros, one would need to be able to find the zeros of fifth degree polynomials.

This brings to light another of the mathematician's procedures, but one that seldom shows up in textbooks or formal presentations. It is a form of trial and error. The investigator, baffled by a large question, tries to solve simpler but related problems hoping to extend the results later and eventually find an answer to the main question.

In this process, many questions are asked that prove to be of no help, and the route toward an answer may end up being quite circuitous. Once established, a result is examined and often a shorter but less obvious path is found. In the interests of efficiency and elegance, the original route with all its blind alleys is forgotten. The student, however, while he may be able to follow the reasoning, fails to see a motivation behind the final form of the derivation or proof and feels that he must have missed something ,when in fact the only motivation left is the verification of the result. In such cases, the student should seek to understand why the result is of any value, why one would want to know the result, and in so doing learn what questions a mathematician asks, for in some senses this is a more important aspect of mathematical discovery than the proving or disproving of conjectured results. One needs to know what problem to solve before he can solve it.

To return to the problem of finding zeros, one might say to himself "All right, if I don't see a way to find just any real zero, is there perhaps a way to find a rational zero (if the function has one) ? After all, many of the zeros I have run into in previous problems are rational numbers." †

Such a zero is one that can be written as t/b (short for top/bottom), where t and b are integers and the fraction t/b is in lowest terms.

If t/b were a zero for a polynomial function p of degree n, it would be true that

$$a_n\left(\frac{t}{b}\right)^n + a_{n-1}\left(\frac{t}{b}\right)^{n-1} + \cdots$$
$$+ a_1\left(\frac{t}{b}\right) + a_0 = 0 \quad (9)$$

by substituting the value of the zero t/b for x in the general form of a polynomial. We next decide to restrict our attention further by requiring that all the coefficients be integers. Then we clear fractions by multiplying both sides of (9) by b^n. This process yields

From this point on, we shall carry a specific polynomial and zero through the proof, to clarify the ideas we are using. If we use $p(x) = 16x^3 + 16x^2 - 9x - 9$ as our polynomial, we have :

If $\frac{3}{4}$ were a zero for our polynomial function p of degree 3, it would be true that

$$16(\tfrac{3}{4})^3 + 16(\tfrac{3}{4})^2$$
$$- (9\tfrac{3}{4}) - 9 = 0 \quad (9)$$

by substituting the value of the zero $\frac{3}{4}$ for x in the polynomial $p(x)$ above. We next decide to restrict our attention further by requiring that all the coefficients be integers. Then we clear fractions by multiplying both sides of (9) by 4^3. This process yields

† Remember that the rational numbers $\mathbb{Q}$ are numbers that can be written as fractions: 3, $\frac{2}{3}$, $-\frac{27}{3}$, etc.

$a_n t^n + a_{n-1}t^{n-1}b + a_{n-2}t^{n-2}b^2$

$+ \cdots + a_1 t b^{n-1} + a_0 b^n = 0$ (10)

A seemingly whimsical transferral of the last term to the right side of the equation would allow us to write the equation as

$t(a_n t^{n-1} + a_{n-1}t^{n-2}b + \cdots$

$+ a_1 b^{n-1}) = -a_0 b^n$ (11)

or as

$(a_n t^{n-1} + \cdots + a_1 b^{n-1})$

$$= \frac{-a_0 b^n}{t} \quad (12)$$

Next we note that the left side is an integer. (This is true because we restricted $\{a_n, a_{n-1}, \cdots, a_2, a_1\}$ to integers. By definition, b and t are integers and therefore t or b to any power is an integer. Thus each term is an integer, since all the products $a_n t^{n-1}$, $a_{n-1}t^{n-2}b$, $\cdots$, $a_2 t^1 b^{n-2}$, $a_1 b^{n-1}$ are products of integers. Finally, the sum

$(a_n t^{n-1} + \cdots + a_1 b^{n-1})$

is the sum of integers and, therefore, an integer itself.) This would mean that the right side would have to be an integer, too; hence t must actually be a factor of $-a_0 b^n$ (i.e., divide $-a_0 b^n$ evenly). Since t and b have no common factors, t and b^n also have none, so tt must be a factor of a_0. In other words the numerator (t) of any rational zero (t/b) for a polynomial function with integer coefficients must be an even divisor of the constant term of the polynomial (a_0 above).

In a similar way, moving the first term of the left side of (12) to the right side leads the way to finding that b must be a factor of a_n; i.e., the denominator (b) of any rational zero (t/b) for a poly-

$16 \cdot 3^3 + 16 \cdot 3^2 \cdot 4$

$- 9 \cdot 3 \cdot 4^2 - 9 \cdot 4^3 = 0$ (10)

A seemingly whimsical transferral of the last term to the right side of the equation would allow us to write the equation as

$3(16 \cdot 3^2 + 16 \cdot 3 \cdot 4$

$- 9 \cdot 4^2) = 9 \cdot 4^3$ (11)

or as

$16 \cdot 3^2 + 16 \cdot 3 \cdot 4 - 9 \cdot 4^2$

$$= \frac{9 \cdot 4^3}{3} \quad (12)$$

Next we note that the left side is an integer. (This is true because we restricted $\{16, 16, -9, -9\}$ to integers. By definition, 4 and 3 are integers and therefore 3 or 4 to any power is an integer. Thus each term is an integer, since all the products $16 \cdot 3^2$, $16 \cdot 3 \cdot 4$, $-9 \cdot 4^2$ are products of integers. Finally, the sum $(16 \cdot 3^2 + 16 \cdot 3 \cdot 4 - 9 \cdot 4^2)$ is the sum of integers and, therefore, an integer itself.) This would mean that the right side would have to be an integer, too; hence 3 must actually be a factor of $9 \cdot 4^3$ (i.e., divide $9 \cdot 4^3$ evenly). Since $\frac{3}{4}$ was in lowest terms (i.e., 3 and 4 have no common factors), 3 is not a factor of 4^3, and so it must be a factor of 9. In other words, the numerator (3) of any rational zero ($\frac{3}{4}$) for a polynomial function with integer coefficients must be an even divisor of the constant term of the polynomial (9 above).

In a similar way, moving the first term of the left side of (12) to the right side leads the way to finding that 4 must be a factor of 16; i.e., the denominator (4) of any rational zero ($\frac{3}{4}$) for a polynomial function with integer co-

nomial function with integer co-efficients must be an even divisor of the lead coefficient (a_n above).

efficients must be an even divisor of the lead coefficient (16 above).

A good exercise at this point would be to try this second half of the proof either with the first half in hand to help you understand the process and reasoning there, or without the book at all (after some studying) to be sure you have mastered the proof.

These two facts may seem to be obscure or of little value on first glance, but at this point, any help toward finding zeros for higher degree polynomials is appreciated. That these results are a help is easier to see when we state them more succinctly in the following theorem.

The Rational Zero Theorem states that if the polynomial function specified by $p(x) = a_n x^n + \cdots + a_0$ has all integer coefficients, then all the rational zeros for p are contained in the set of all rational numbers whose numerators are divisors of a_0 and whose denominators are divisors of a_n.

THEOREM 2.1
Rational Zero Theorem

Notice that the derivation above constitutes a proof of this theorem.

Now let us look at an example to see the value of this result by finding all the rational zeros of p if $p(x) = 4x^3 + 24x^2 - x - 6$. Since $a_n = 4$ and $a_0 = -6$, the rational zeros (if, indeed, there are any) must be members of the set

EXAMPLE 11

$$\{\tfrac{1}{1}, \tfrac{1}{2}, \tfrac{1}{4}, \tfrac{2}{1}, \tfrac{2}{2}, \tfrac{2}{4}, \tfrac{3}{1}, \tfrac{3}{2}, \tfrac{3}{4}, \tfrac{6}{1}, \tfrac{6}{2}, \tfrac{6}{4}, -\tfrac{1}{1}, -\tfrac{1}{2}, -\tfrac{1}{4}, -\tfrac{2}{1}, -\tfrac{2}{2}, -\tfrac{2}{4}, -\tfrac{3}{1}, -\tfrac{3}{2},$$
$$-\tfrac{3}{4}, -\tfrac{6}{1}, -\tfrac{6}{2}, -\tfrac{6}{4}\}.$$

```
DØ YØU NEED INSTRUCTIØNS ØN THE USE ØF THIS PRØGRAM?        N
DEGREE ØF PØLYNØMIAL?        3.
CØEFFICIENTS ØF PØLYNØMIAL?50.
25.
-43.
-18.
THE PØLYNØMIAL BEING INVESTIGATED IS ØF  3TH DEGREE
WITH INTEGER CØEFFICIENTS (FRØM HIGHEST DEGREE TØ
LØWEST):
        50        25       -43       -18
A LIST ØF THE PØTENTIAL RATIØNAL ZERØS (IN DECIMAL
FØRM) IS:
    1.00000     .50000     .20000     .10000     .04000     .02000
    2.00000     .40000     .03000    3.00000    1.50000     .60000
     .30000     .12000     .06000    6.00000    1.20000     .24000
    9.00000    4.50000    1.80000     .90000     .36000     .18000
   18.00000    3.60000     .72000   -1.00000    -.50000    -.20000
    -.10000    -.04000    -.02000   -2.00000    -.40000    -.03000
   -3.00000   -1.50000    -.60000    -.30000    -.12000    -.06000
   -6.00000   -1.20000    -.24000   -9.00000   -4.50000   -1.80000
    -.90000    -.36000    -.18000  -18.00000   -3.60000    -.72000

DØ YØU WANT TØ CHECK THE LIST ØF PØTENTIAL ZERØS
TØ SEE IF ANY ØF THEM ARE ACTUALLY ZERØS?       YES


THE ACTUAL RATIØNAL ZERØS ARE LISTED BELØW ALØNG
WITH VALUES WHICH MAY BE ZERØS, BUT WHICH DØ NØT (QUITE)
APPEAR SØ DUE TØ RØUND-ØFF ERRØR,

   ACTUAL ZERØS         PØSSIBLE ZERØS-ASSØCIATED FUNCTIØNAL VALUE


        X                    X                  F(X)

                         .9000000          -.6821210E-12
   -1.0000000                                 0.0
                        -.4000000          -.1136368E-12
```

FIGURE 2.6

Removing repetitions and abbreviating, we have the set

$$\{\pm1, \pm\tfrac{1}{2}, \pm\tfrac{1}{4}, \pm2, \pm3, \pm\tfrac{3}{2}, \pm\tfrac{3}{4}, \pm6\}$$

as containing all the rational zeros of p.

While this may not seem to be such a bargain, it is more than a little reassuring to know without a doubt that this set has captured all possible rational zeros. Let us hope the pleasure of our discovery will not be damped by discovering that p has no rational zeros.

At this point, we would have to substitute each of the 16 numbers into the formula for p to check if it is a zero—the theorem does *not* say that they all are. We shall soon develop a way to shorten this job, motivated at least in part by the tedium of this procedure. By the way, the zeros in this example are, in fact, $\tfrac{1}{2}$, $-\tfrac{1}{2}$, and -6.

Program **RATZER** of Appendix C has been written to find rational zeros in the manner just described for polynomials of degree ≤10, and does so with a minimum of frills. Figures 2.6 and 2.7 exhibit some outputs from the program that give further examples of the results achievable by use of the rational zero theorem above in the effort to find zeros for polynomials. Interpretation of the outputs from this program will be facilitated by the table of decimal forms of some common rational numbers found in Appendix D, Table 5.

```
DØ YØU WANT TØ STUDY ANØTHER PØLYNØMIAL?        YES
DEGREE ØF PØLYNØMIAL?      5.
CØEFFICIENTS ØF PØLYNØMIAL?21.
29.
24.
24.
3.
-5.
THE PØLYNØMIAL BEING INVESTIGATED IS ØF  5TH DEGREE
WITH INTEGER CØEFFICIENTS (FRØM HIGHEST DEGREE TØ
LØWEST):
      21       29       24       24        3       -5
A LIST ØF THE PØTENTIAL RATIØNAL ZERØS (IN DECIMAL
FØRM) IS:
    1.00000      .33333      .14286      .04762    5.00000    1.66667
     .71429      .23810    -1.00000     -.33333     -.14286    -.04762
   -5.00000    -1.66667     -.71429     -.23810
DØ YØU WANT TØ CHECK THE LIST ØF PØTENTIAL ZERØS
TØ SEE IF ANY ØF THEM ARE ACTUALLY ZERØS?       YES


THE ACTUAL RATIØNAL ZERØS ARE LISTED BELØW ALØNG
WITH VALUES WHICH MAY BE ZERØS, BUT WHICH DØ NØT (QUITE)
APPEAR SØ DUE TØ RØUND-ØFF ERRØR,

   ACTUAL ZERØS              PØSSIBLE ZERØS-ASSØCIATED FUNCTIØNAL VALUE


       X                          X                  F(X)

                              .3333333            -.8526513E-13
   -1.0000000                                          0.0
                             -.7142857            -.8526513E-13
```

FIGURE 2.7

2.4. SOME GENERAL RESULTS ABOUT ZEROS

2.4a. Factor Theorem

Using the rational zero theorem, we find that the function f specified by $f(x) = x^3 + x^2 - 2x - 2$ has only one actual rational zero, namely, -1. It would be nice if the knowledge of this zero would enable us to produce a related problem that would give us clues to find any more zeros f might possess or be sure there were no more. Instead of a direct attack on the question "How can I simplify this problem?" we find that a seemingly unrelated observation about quadratic functions will provide the key to the first useful result.

Here another facet of the doing of mathematics, as distinct from the learning of mathematics, comes to light. Many of the most significant parts of mathematics have come into being as offshoots of the attempt to answer questions that seemed to bear no relation to the discovery made. This is not an uncommon occurrence in any discipline.

Look next at Figure 2.8, which is a page of output from a program to solve quadratic equations. It is not unreasonable to think that one's eye might be caught by a number of examples in which each of the zeros is a factor of the constant term. A further analysis of this apparent coincidence could lead one to discover that each time the rule for a function is factored, each factor exhibits a zero, for example,

1. $x^2 + 5x + 4$ has zeros -1 and -4, and in factored form is $(x + 4)(x + 1) = (x - (-4))(x - (-1))$.
2. $2x^2 + 2x - 144$ has zeros 8 and -9, and in factored form is $2(x + 9)(x - 8) = 2(x - (-9))(x - (8))$.
3. $3x^2 - 23x + 30$ has zeros 6 and $1\frac{2}{3}$, and in factored form is $(x - 6)(3x - 5) = (x - 6)(3)(x - (5/3))$.

One might conjecture as a result of the above data that if z is a zero for a quadratic function f, then $(x - z)$ is a factor for $f(x)$.[†] The validity of this conjecture can be checked by using the quadratic formula. The quadratic formula gives all the zeros for a quadratic function $f(x) = ax^2 + bx + c$; therefore, our conjecture would say that both

$$\left[x - \left(\frac{-b + \sqrt{b^2 - 4ac}}{2a} \right) \right] \quad \text{and} \quad \left[x - \left(\frac{-b - \sqrt{b^2 - 4ac}}{2a} \right) \right]$$

are factors of $f(x)$. If we multiply these together, we would get

$$x^2 - \left[x \left(\frac{-b - \sqrt{b^2 - 4ac}}{2a} \right) \right] - \left[x \left(\frac{-b + \sqrt{b^2 - 4ac}}{2a} \right) \right]$$
$$+ \left[\left(\frac{-b + \sqrt{b^2 - 4ac}}{2a} \right) \left(\frac{-b - \sqrt{b^2 - 4ac}}{2a} \right) \right].$$

After a good bit of algebraic manipulation (try it yourself), we get $x^2 + (b/a)x + c/a$. It is not hard to see that $f(x)$ is just the product $a(x^2 + (b/a)x + c/a)$. That is, $f(x) = (x - z_1)(x - z_2) \cdot a$, where z_1 and z_2 are the zeros of f,—i.e., $z_1 = (-b + \sqrt{b^2 - 4ac})/2a$, $z_2 = (-b - \sqrt{b^2 - 4ac})/2a$.

---

† While this is admittedly not the most efficient way to see this, such roundabout routes are realistic in "real life" mathematics, and we believe you should see samples of such.

This is of marginal value for quadratic functions since we already know how to find the zeros for them, but it provides the basis for further conjecture.

By reanalyzing a known result, a path toward a new result is found. The new conjecture is that if z is a zero for any polynomial function p, then $(x - z)$ is a factor for $p(x)$. To attempt to prove that this is true, we need some way to determine whether or not $(x - z)$ is a factor of $p(x)$. A criterion is provided by the algebra of real numbers.

When one learns how to divide algebraic expressions, it is common procedure to note that if $p(x)$ and $d(x)$ are polynomials, and $d(x)$ has degree less than or equal to the degree of $p(x)$, then there are two other polynomials $q(x)$ with degree $\leq p(x)$ and $r(x)$ with degree $< d(x)$ with the

THE ZEROS OF	1.0000 X**2	+	-10.0000 X	+	-24.0000	ARE	12.0000 AND	-2.0000
THE ZEROS OF	-4.5000 X**2	+	9.7500 X	+	3.2000	ARE	-0.2895 AND	2.4562
THE ZEROS OF	3.0000 X**2	+	-23.0000 X	+	30.0000	ARE	6.0000 AND	1.6667
THE ZEROS OF	1.0000 X**2	+	4.3000 X	+	-3.5000	ARE	00.7000 AND	-5.0000
THE ZEROS OF	-2.0000 X**2	+	3.0000 X	+	4.0000	ARE	-0.8508 AND	2.3508
THE ZEROS OF	15.0000 X**2	+	-11.0000 X	+	-14.0000	ARE	1.4000 AND	-0.6667
THE ZEROS OF	14.0000 X**2	+	29.0000 X	+	-70.0000	ARE	1.4286 AND	-3.5000
THE ZEROS OF	1.0000 X**2	+	-6.0000 X	+	-16.0000	ARE	8.0000 AND	-2.0000
THE ZEROS OF	3.0000 X**2	+	5.0000 X	+	2.0000	ARE	-0.6667 AND	-1.0000
THE ZEROS OF	1.0000 X**2	+	-4.0000 X	+	-5.0000	ARE	5.0000 AND	-1.0000
THE ZEROS OF	2.3000 X**2	+	-7.4000 X	+	1.0500	ARE	3.0686 AND	00.1488
THE ZEROS OF	1.0000 X**2	+	5.0000 X	+	4.0000	ARE	-1.0000 AND	-4.0000
THE ZEROS OF	1.0000 X**2	+	4.0000 X	+	4.0000	ARE	-2.0000 AND	-2.0000
THE ZEROS OF	3.1000 X**2	+	-6.3000 X	+	1.2500	ARE	1.8094 AND	00.2228
THE ZEROS OF	7.0000 X**2	+	12.0000 X	+	2.0000	ARE	-0.1871 AND	-1.5272
THE ZEROS OF	1.0000 X**2	+	0.0000 X	+	-25.0000	ARE	5.0000 AND	-5.0000
THE ZEROS OF	00.0020 X**2	+	00.0750 X	+	-0.0150	ARE	00.1989 AND	-37.6989
THE ZEROS OF	4.0000 X**2	+	10.0000 X	+	3.0000	ARE	-0.3486 AND	-2.1514
THE ZEROS OF	1.0000 X**2	+	14.0000 X	+	49.0000	ARE	-7.0000 AND	-7.0000

FIGURE 2.8

property that

$$p(x) = q(x) \cdot d(x) + r(x). \tag{13}$$

This is a way of saying that a polynomial $p(x)$ (the dividend) can always be divided by another polynomial $d(x)$ (the divisor), provided $d(x)$ is of no higher degree than $p(x)$, with a resulting quotient polynomial $q(x)$ and remainder polynomial $r(x)$. Furthermore, the degree of $r(x)$ is less than the degree of $d(x)$.

For example, if

$$p(x) = 4x^4 + 2x^3 - 4x^2 + 6x - 1$$

and

$$d(x) = 2x^2 - 4x + 1$$

then

$$
\begin{array}{r}
2x^2 + 5x + 7 \\
2x^2 - 4x + 1 \overline{) 4x^4 + 2x^3 - 4x^2 + 6x - 1} \\
4x^4 - 8x^3 + 2x^2 \\
\hline
10x^3 - 6x^2 + 6x \\
10x^3 - 20x^2 + 5x \\
\hline
14x^2 + x - 1 \\
14x^2 - 28x + 7 \\
\hline
29x - 8
\end{array}
$$

and

$$q(x) = 2x^2 + 5x + 7$$
$$r(x) = 29x - 8$$

and we may write $p(x) = q(x) \cdot d(x) + r(x)$ as

$$(4x^4 + 2x^3 - 4x^2 + 6x - 1) = (2x^2 - 4x + 1)(2x^2 + 5x + 7) + (29x - 8).$$

Remember, now, that we would like to show that $(x - z)$ is a factor of $p(x)$. Given Equation (13) and the restrictions on it mentioned above, we could show that $(x - z)$ was a factor of $p(x)$ by showing that dividing $p(x)$ by $(x - z)$ resulted in $r(x)$'s being zero.

To clarify this, suppose we have $x^3 + 2x^2 - 5x + 2$ with zero $z = 1$, and want to show that $(x - 1)$ is a factor. We perform the following division

$$
\begin{array}{r}
x^2 + 3x - 2 \\
x - 1 \overline{) x^3 + 2x^2 - 5x + 2.} \\
x^3 - x^2 \\
\hline
3x^2 - 5x \\
3x^2 - 3x \\
\hline
-2x + 2 \\
-2x + 2 \\
\hline
\end{array}
$$

and get $x^3 + 2x^2 - 5x + 2 = (x - 1)(x^2 + 3x - 2)$, where

$$p(x) = x^3 + 2x^2 - 5x + 2$$
$$d(x) = x - 1$$

$$q(x) = x^2 + 3x - 2$$

and

$$r(x) = 0$$

as we suggested above. To accomplish the proof of $(x - z)$'s being a factor of $p(x)$, we know that there are some polynomials $q(x)$ and $r(x)$ so that

$$p(x) = q(x)(x - z) + r(x) \qquad (14)$$

for all x, not just for z, where $(x - z)$ is the $d(x)$ of Equation (13). Since the degree of $r(x)$ must be less than the degree of $(x - z)$ by the restrictions set above, it must have degree zero [$(x - z)$ is of first degree and the only degree less than first is zero]. That is, $r(x)$ must in fact be a constant. We can determine the value of that constant by substituting z in Equation (14):

$$p(z) = q(z)(z - z) + r(z)$$
$$0 = q(z) \cdot 0 + r(z)$$
$$0 = r(z).$$

THEOREM 2.2
Factor Theorem

So if z is a zero of p, then $p(x) = q(x)(x - z)$ so that $(x - z)$ is indeed a factor of $p(x)$ and our conjecture has proved to be correct. That is, if z is a zero for a polynomial function p, then $(x - z)$ is a factor of $p(x)$. This result is usually called the *Factor Theorem*.

THEOREM 2.3
Converse of the
Factor Theorem

The Factor Theorem should not be confused with its converse.† The Factor Theorem starts with a zero and claims that something is a factor. The *converse* would say that if $(x - z)$ is a factor of $p(x)$, then z is a zero for p. (Do you see the difference? If not, read both again.) This result is also true (and useful). The proof is short this time. To say that $(x - z)$ is a factor of $p(x)$ means that there is a polynomial $q(x)$ so that $p(x) = q(x)(x - z)$. Thus, substituting z into this equation, we see that

$$p(z) = q(z)(z - z)$$
$$= q(z) \cdot 0$$
$$= 0,$$

and therefore, z is indeed a zero for p.

Now let us review our position and see what value the factor theorem and its converse have in our search for zeros. We started this discussion with the idea that it would help us simplify the problem of finding all the zeros of a polynomial once we had found one zero. If we wish to find the zeros of p and know that α is one zero, the Factor Theorem assures us that there is a polynomial q so that $p(x) = (x - \alpha)(q(x))$.

EXAMPLE 12

Let us look at a specific case. Suppose we have $p(x) = x^3 - 4x^2 + x + 6$ and we know that 2 is a zero of $p(x)$. The Factor Theorem tells us

---

† Many theorems have the basic form "If P is true, then Q is true," where P and Q stand for some statement. The *converse* of such a statement is "If Q is true, then P is true." As an example, the converse of "If x is a horse, then x is a four-legged animal" is "If x is a four-legged animal, then x is a horse." Here it is clear that the converse is not the same as the original statement. As is also suggested by this example, the truth of the converse is a separate matter from the truth of the original statement. Be careful not to mistake a statement and its converse when only one is known to be true.

that

$$p(x) = x^3 - 4x^2 + x + 6 = (x - 2)(\text{some } q(x)),$$

and long division tells us that $q(x) = x^2 - 2x - 3$. Thus we may write

$$x^3 - 4x^2 + x + 6 = (x - 2)(x^2 - 2x - 3).$$

The simplification we have been seeking appears when we see that we now only need to find the zeros of $q(x)$. Let us see that this really *is* true.

First if β is a zero for q, then

$$\begin{aligned} p(\beta) &= (\beta - \alpha)(q(\beta)) \\ &= (\beta - \alpha) \cdot 0, \end{aligned}$$

since $q(\beta) = 0$ if β is a zero of q, and then

$$p(\beta) = 0,$$

so that β is a zero for p. Next if β is a zero for p and $\beta \neq \alpha$, then $p(\beta) = 0 = (\beta - \alpha)q(\beta)$; so, solving for q, $q(\beta) = 0/(\alpha - \beta) = 0$. Therefore, β is a zero for q.

"But," you ask, "is this really of any value?" The answer is yes. As an example, consider again the function cited at the start of this section, $f(x) = x^3 + x^2 - 2x - 2$. Knowing that -1 is a zero, we may divide $p(x)$ by $(x + 1)$ to find the $q(x)$ we know must exist; if f has any other zeros, they must also be zeros for q, and any zero for q is a zero for f. In this case, q is $x^2 - 2$ (i.e., $x^3 + x^2 - 2x - 2 = (x + 1)(x^2 - 2)$). The zeros of the polynomial q are easily found by inspection or the quadratic formula to be $\sqrt{2}$ and $-\sqrt{2}$.

In addition to this, another way to find zeros is suggested, which stems from the converse of the Factor Theorem. If the rule for a polynomial function can be factored, then each factor of the form $(x - \alpha)$ exhibits a zero for the polynomial. Thus a polynomial such as $t(x) = x^3 - 2x^2 - 15x$ can be factored as $x(x - 5)(x + 3)$, which yields the zeros $0, +5,$ and -3 (x is $x - 0$) for t by inspection.

Looking at the relationship between factors and zeros from still another viewpoint, we recall the property of real numbers that a product is zero if and only if at least one of the factors in the product is zero. Thus if the rule for a polynomial function can be factored in *any* form, the problem of finding zeros for the original polynomial reduces to finding the zeros for each factor, whether or not the factors are of the form $(x - \alpha)$. We leave a complete proof of this assertion to the problems (Problem 44) although the argument is quite similar to that used above for $p(x)$ expressed as $(x - \alpha)(q(x))$.

In summary, then, if we know that $f(x) = x^3 - 2x^2 - 5x + 6$ has a zero $x = 1$, then we know that we may write $f(x) = (x - 1)(\text{some } q(x))$, or in this case,

$$f(x) = (x - 1)(x^2 - x - 6)$$

and can find more zeros for f by finding the zeros of $x^2 - x - 6$. If we have $g(x) = x^4 - 7x^2 + 12$, which we can write as

$$g(x) = (x^2 - 4)(x^2 - 3) \qquad \text{or} \qquad g(x) = (x + 2)(x - 2)(x^2 - 3),$$

then we know that the zeros of g are $x = -2, x = 2, x = \sqrt{3},$ and $x = -\sqrt{3}$.

2.4b. Fundamental Theorem of Algebra

We have explored the close and useful relationship between zeros and factors but so far we have no way of knowing if or when all of the zeros of a polynomial function have been found. We have seen that linear functions have at most one zero and quadratic functions have at most two zeros. A natural conjecture would be that an nth degree polynomial function has at the most n zeros. Let us suppose for the moment that p is a polynomial function of degree n and that p has at least $n + 1$ zeros. Let us call these zeros $z_1, z_2, \cdots, z_{n+1}$. Making use of the Factor Theorem, we can conclude that each of the $n + 1$ expressions $(x - z_i)$ is a factor of $p(x)$, and hence

$$(x - z_1)(x - z_2) \cdot \cdots \cdot (x - z_{n+1})$$

is a factor of $p(x)$. But

$$(x - z_1)(x - z_2) \cdot \cdots \cdot (x - z_{n+1}) = x^{n+1} + \text{(terms of lower degree in } x),$$

which is of degree $n + 1$. Since p is of degree n, it is not possible for p to have a term of degree $n + 1$; hence p could not have had $n + 1$ zeros.

Thus, if one has found n zeros for a polynomial function of degree n, he can be sure he has found all possible zeros. However not all polynomial functions have n distinct real zeros; (for example $m(x) = x^2 + 4x + 4$), so this result is not wholly satisfying. Another step toward a complete answer is afforded with the observation that whereas $p(x) = x^2 - 8x + 16$ has only one zero (that being 4), the factor $(x - 4)$ occurs twice in a complete factorization of $p(x)$. Returning to our recently discovered results about factors and zeros, it seems worthwhile making note of such repetitions. For example, if one had found only two different zeros (α and β) for a polynomial t of degree 5, but could also verify that the factors $(x - \alpha)$ and $(x - \beta)$ occurred two and three times, respectively, in a factorization of t, then by the argument just used to show that an nth degree polynomial has at most n zeros, he could be sure the zeros α and β accounted for all of t's zeros. We now have reason to make a definition. A number α is said to be a *zero of multiplicity* m for a polynomial p, provided that $(x - \alpha)^m$ is a factor of $p(x)$ but $(x - \alpha)^{m+1}$ is not. A stronger version of our previous result would be that a polynomial of degree n has at most n zeros, including multiplicities. The earlier proof is still valid, since no restriction was made that the z_i's all had to be different.

DEFINITION 2.9
Zero of
Multiplicity m

Though this gives us a better answer, the best answer to the question "How many zeros does a polynomial function have?" is beyond our present knowledge to develop rigorously or prove. Nevertheless, we shall state the result because it is one of the most important theorems in mathematics.

THEOREM 2.4
Fundamental
Theorem of Algebra

The Fundamental Theorem of Algebra states that every polynomial function p of degree n has exactly n zeros, provided that the domain of p is the set of complex numbers, not just $\mathbb{R}$, and provided a zero of multiplicity m is counted m times.

The complex number system is a set of numbers that contains $\mathbb{R}$ as a subset, somewhat as $\mathbb{R}$ contains $\mathbb{Z}$ as a subset. We defer further consideration of complex zeros until Chapter 7, however, and leave this result with the assurance that as long as we restrict our attention to $\mathbb{R}$, we will have to be satisfied with the result established just prior to the Fundamental Theorem of Algebra as the answer to the question of how many zeros we can expect to find for a polynomial function, namely, that a polynomial of degree n has at most n real zeros including multiplicities. Thus

$$f(x) = x^{20} - 3x^{12} - x^5$$

might have as many as 20 real zeros;

$$g(x) = 2x^4 - 3x^3 + 5$$

might have as many as 4 real zeros,

$$h(x) = (x - 1)^2(x - 3)$$

has exactly 3 real zeros, and

$$j(x) = (x - 3)^3(x + 2)^2$$

has exactly 5 real zeros.

At this point, let us take a quick look back and another one ahead. We have discovered several ways for finding zeros for general polynomials but we have not found a way to find all the (real) zeros, and among the ways we *have* found there is some computational tedium.

NOTE

There are two directions in which to turn to relieve the tedium. One is to leave computation to computer programs; the other is to find some shortcuts. The former approach is fine, but the latter should not be ignored both because computers are not always handy and because machine computation can be improved (made faster and more accurate) by the use of the same shortcuts as human beings would use, so we look into shortcuts in the next section. As for our failure (and indeed that of mathematics) to find analytic techniques capable of tracking down all the real zeros of polynomials, we shall turn to algorithms for finding approximate values for zeros, although we shall venture only a short distance on this rather open-ended path.

2.5. SYNTHETIC DIVISION

2.5a. The Primary Algorithm

In this section, we shift gears from an essentially theoretical approach to one that is down-to-earth and mechanical. We have developed two needs which will be served by this section—that of easier computation of values of a polynomial function (for use with the Rational Zero Theorem) and easier division of polynomials by expressions of the form $x - \alpha$ (for use with the Factor Theorem). We shall deal with the latter first.

The reader should already be familiar with the long division of polynomial expressions† and so be able to find the results of such divisions as

$$x - 3 \overline{)\, x^3 + 2x^2 - 5x - 2} \qquad \text{and} \qquad x + 6 \overline{)\, x^5 - 36x^3 + 16x^2 - 576}.$$

† If not, see Appendix A, Section A-9.

However, the ordinary process of long division includes a lot of wasted motion, especially when the divisor is of the form $x - \alpha$. [Note that $(x + \alpha)$ is of this form if one rewrites it as $(x - (-\alpha))$.] The powers of x really serve the function of place holders—it is the coefficients that "do the work." Look, for example, at the first problem above worked out:

$$
\begin{array}{r}
x^2 +\ 5x + 10 \\
x - 3\overline{)\ x^3 + 2x^2 -\ 5x -\ 2} \\
\underline{x^3 - 3x^2 } \\
5x^2 -\ 5x \\
\underline{5x^2 - 15x } \\
10x -\ 2 \\
\underline{10x - 30} \\
28
\end{array}
$$

We could just as easily have left out all the x's, *provided* that we used columns to distinguish among the coefficients. Thus we would have the following:

$$
\begin{array}{r}
1 \quad\ \ 5 \quad\ \ 10 \\
1 - 3\overline{)\ 1 \quad\ \ 2 \quad -5 \quad -2} \\
\underline{1 \quad -3 } \\
5 \quad -5 \\
\underline{5 \quad -15 } \\
10 \quad -2 \\
\underline{10 \quad -30} \\
28
\end{array}
$$

This can be distilled even further. Of the four numbers in each rectangular array such as

$$
\begin{array}{cc}
\hline
10 & -2 \\
10 & -30 \\
\hline
\end{array}
$$

or, in general,

$$
\begin{array}{cc}
\hline
a & b \\
c & d \\
\hline
\end{array}
$$

only two are really important: a because it is used to determine the next term in the quotient, and d because it determines the value of a in the next cycle. The numbers b and c are dead weight, in that c is always the same as a, and b is merely copied from the dividend. (Look at the above example again and be sure you understand the work so far.)

Furthermore, the 1 in the divisor plays a minor role, since for the class of problems we are considering, it is always 1; and in the scheme we are developing, it enters each cycle in the question, "What do I need to multiply 1 by to get a?" Since the answer is always "a," it is not necessary to write down the 1.

Deleting the superfluous information we have mentioned above, we would come up with the following scheme:

```
              1     5    10
   -3) 1      2    -5    -2
             -3
            ─────
              5
                   -15
                  ─────
                    10
                         -30
                        ─────
                          28
```

Next we add arrows to indicate the source of the coefficients in the quotient. Look at the original problem to be sure you see this. Each rectangle contains a subtraction that needs to be done to start the next cycle:

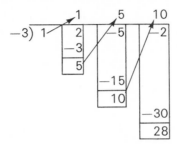

A new scheme is then suggested that will save space, avoid the need for rectangles and arrows, and reduce the occurrence of errors caused by confusing columns. We raise each subtraction problem so hey will all occur on the same level, as follows:

```
              1      5     10
   -3) 1      2     -5     -2
             -3    -15    -30
            ────  ─────  ─────
              5     10     28
```

Next we change each subtraction to addition. Rather than change signs one by one as we go, it is more efficient to change the divisor -3 to $+3$. In other words, the scheme becomes:

```
                                    line:
           1     5    10      ←  a
   3) 1    2    -5    -2      ←  b
           3    15    30      ←  c
         ────  ────  ────
           5    10    28      ←  d
```

Compare this with that just above and be sure you see that they are equivalent. Then, as a final distillation, notice that with the exception of the lead coefficient, the quotient's coefficients appear on the bottom line (line d) as well as on the top line (line a). If we rewrite the lead coefficient of the dividend (line b) on the bottom line (line d), we could omit the entire top line, with the result that the bottom line exhibits both the quotient and remainder:

$$\begin{array}{r}
\text{dividend} \\
\text{divisor} \rightarrow 3)\ \ 1 \quad 2 \quad -5 \quad -2 \\
\quad\quad\quad\quad 3 \quad 15 \quad 30 \\
\hline
1 \quad 5 \quad 10 \mid 28 \\
\end{array}$$

quotient　　　　remainder

Since the top line has been deleted, the ordinary division symbol is out of place and so has been replaced with a symbol to isolate the working part of the divisor. Also, lines have been inserted that provide a base for the additions and serve to isolate the quotient and remainder from the rest of the problem and from each other. The algorithm that results from all these efficiencies is called synthetic division.

DEFINITION 2.10
Synthetic Division

Notice that all the powers of x from the degree of the dividend through zero must be provided with a column even if not actually present in the dividend (i.e., even if the coefficient is zero). For example, $x^5 - 4x^3 + 5$ divided by $x - 3$ would be set up for synthetic division thus:

$$3)\ 1 \quad +0 \quad -4 \quad +0 \quad +0 \quad +5,$$

representing $x^5 + 0x^4 - 4x^3 + 0x^2 + 0x^1 + 5$ divided by $x - 3$. Also, the power of x associated with a coefficient in the quotient is one less than the power of x associated with the coefficient of the dividend directly above it. That is, in

$$+3)\ 1 \ +0 \ -4 \ +0 \ +0 \ \ +5 \leftarrow \text{this line represents}$$
$$\quad\quad +3 \ +9 \ +15 \ +45 \ +135 \quad\quad x^5 - 4x^3 + 5 \div x - 3.$$
$$\hline$$
$$1 \ +3 \ +5 \ +15 \ +45 \mid 140 \leftarrow \text{this line represents}$$
$$x^4 + 3x^3 + 5x^2 + 15x + 45$$
$$\text{with remainder } 140$$

Of the two leading 1's, the top one represents $1 \cdot x^5$ and the bottom one $1 \cdot x^4$. Similarly, the 3 in the bottom line is the coefficient of x^3, while the 0 above it corresponds to x^4, etc. If this is still not clear to you, go back over the development so far to see where each number in the initial framework for synthetic division originated, and once more to see how each of the multiplications and additions corresponds to similar procedures in the long division algorithm. In any event, it would be valuable for you to take a polynomial and divisor different from any that we have used so far in this section and carry through the development of synthetic division for yourself.

We review the mechanism of this simplified division algorithm by flowcharting it; see Figure 2.9. While the flowchart may look a little complicated, it is really quite a simple algorithm. As we mentioned above, this algorithm is called synthetic division.

EXAMPLE 13

Let us use the algorithm for the second example cited at the start of this section. We have $x^5 - 36x^3 + 16x^2 - 576$, divided by $x + 6$. If you understand the general polynomial form, you can write $\alpha = -6$, $n = 5$,

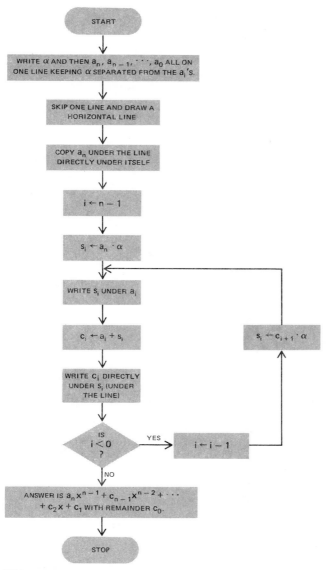

FIGURE 2-9. Synthitic division algorithm.

$a_5 = 1$, $a_4 = 0$, $a_3 = -36$, $a_2 = 16$, $a_1 = 0$, and $a_0 = -576$, giving

$$
\begin{array}{r|rrrrrr}
-6) & 1 & 0 & -36 & 16 & 0 & -576 \\
    &   & -6 & 36 & 0 & -96 & 576 \\
\hline
    & 1 & -6 & 0 & 16 & -96 & 0
\end{array}
$$

If you find the quantity of symbols in the flowchart a hinderance to your understanding, the following view of Example 13 may help:

initial problem: $\dfrac{x^5 + 36x^3 + 16x^2 - 576}{x - 6}$;

dividend: $x^5 - 36x^3 + 16x^2 - 576 = x^5 + 0x^4 - 36x^3 + 16x^2 + 0x - 576$

(note the need to represent all powers of x less than the degree of the dividend);

divisor: $x - 6$.

The usual long division form is transformed as follows:

$$x - 6\overline{)\, x^5 + 0x^4 - 36x^3 + 16x^2 + 0x - 576}$$

(note sign $\quad\downarrow\quad\downarrow\quad\downarrow\qquad\downarrow\qquad\downarrow\quad\downarrow$
change) $\rightarrow$ 6) $1 +0 \quad -36 \quad +16 \quad +0 \quad -576$

though one ordinarily does not actually write the long division form first. Finally the pattern of addition and multiplications produces

$$\begin{array}{r|rrrrr}
6) & 1 & +0 & -36 & +16 & +0 & -576 \\
 & & -6 & 36 & 0 & -96 & +576 \\
\hline
 & 1 & -6 & 0 & 16 & -96 & 0
\end{array}$$

in which the bottom row (1 −6 0 16 −96 0) is interpreted as the quotient $1x^4 - 6x^3 + 0x^2 + 16x - 96$ with remainder 0.

2.5b. Evaluation of Polynomials

As a by-product of the synthetic division algorithm, we have an efficient way of evaluating polynomials for any particular argument we wish. Here is a case of getting more than we asked for when we get a result. Problem 45 asserts that $p(x) = q(x)(x - \alpha) + p(\alpha)$ for any real α. Interpreting the equation $p(x) = q(x)(x - \alpha) + p(\alpha)$ from our current standpoint, we can see that the value $p(\alpha)$ is just the remainder in the division of $p(x)$ by $(x - \alpha)$, i.e., $p(x)$ divided by $(x - \alpha) = q(x)$ with remainder $p(\alpha)$. Look this over again to be sure you understand it.

If, for example, $p(x) = x^5 - 5x^4 - 10x + 6$, and we wish to compute $p(6)$. We could compute $6^5 - 5 \cdot 6^4 - 10 \cdot 6 + 6 = 7776 - 5 \cdot 1296 - 10 \cdot 6 + 6 = 7776 - 6480 - 60 + 6 = 1242$, or perform the synthetic division that follows:

$$\begin{array}{r|rrrrr}
6) & 1 & -5 & 0 & 0 & -10 & 6 \\
 & & 6 & 6 & 36 & 216 & 1236 \\
\hline
 & 1 & 1 & 6 & 36 & 206 & 1242
\end{array}$$

Either computation yields $p(6) = 1242$, but most people find the synthetic division evaluation at least as quick and less likely to produce errors once they master it. Furthermore, from the point of view of computer programs, the synthetic division method of evaluation polynomials is more efficient in terms of time used, and the value so computed is likely to contain less error, due to the machine's finite arithmetic (see Section B2-5c, of Appendix B).

Recapitulating, the division of $p(x)$ by $(x - \alpha)$ (whether by synthetic division or not) will inform one whether or not α is a zero of p, and if not, will at least tell what the functional value $p(\alpha)$ is; each of these pieces of information is contained in the remainder term. If the remainder is zero (i.e.,

$p(\alpha) = 0$ so that α is a zero) the quotient polynomial is then immediately available for the next stage in the search for zeros.

Thus

EXAMPLE 14

$$-7) \ \underline{\begin{array}{cccc} 1 & 2 & -3 & -1 \\ & -7 & 35 & -224 \end{array}} $$
$$ 1 \quad -5 \quad 32 \quad |-225$$

tells one that (a) $x + 7$ is *not* a factor of $p(x) = x^3 + 2x^2 - 3x - 1$; (b) -7 is *not* a zero of $p(x)$; but (c) $p(-7) = -225$.

Similarly,

EXAMPLE 15

$$5) \ \underline{\begin{array}{cccc} 1 & -4 & -4 & -5 \\ & 5 & 5 & 5 \end{array}} $$
$$ 1 \quad 1 \quad 1 \quad | \ 0$$

tells one that (a) $x - 5$ *is* a factor of $g(x) = x^3 - 4x^2 - 4x - 5$; (b) 5 is a zero of $g(x)$; and (c) any remaining zeros of $x^3 - 4x^2 - 4x - 5$ are the zeros of $x^2 + x + 1$.

From another direction, if we want to check to see if $x + 7$ is a factor of $x^3 + 2x^2 - 3x - 1$, we perform the synthetic division as follows:

EXAMPLE 16

$$(\text{change sign}) \rightarrow -7) \ \underline{\begin{array}{cccc} 1 & 2 & -3 & -1 \\ & -7 & 35 & -224 \end{array}} $$
$$\phantom{(\text{change sign}) \rightarrow -7) } 1 \quad -5 \quad 32 \quad |-225$$

i.e., $x + 7$ is not a factor of $x^3 + 2x^2 - 3x - 1$.

Or, if we want to evaluate $f(x) = x^2 - 2x + 1$ at $x = -7$, we write

EXAMPLE 17

$$(\text{no sign change}) \rightarrow -7) \ \underline{\begin{array}{ccc} 1 & -2 & 1 \\ & -7 & 63 \end{array}} $$
$$\phantom{(\text{no sign change}) \rightarrow -7) } 1 \quad -9 \quad |64$$

which gives us the same result as $f(-7) = (-7^2) - 2 \cdot (-7) + 1 = 64$.

To see if α is a zero of $p(x)$ is the same as asking if $p(\alpha) = 0$, so that one uses α (not $-\alpha$) in the synthetic division.

A word of caution is necessary. Remember that the algorithm as we have developed it is based on division by $x - \alpha$; therefore, an expression like $x + 7$ has $\alpha = -7$, not $\alpha = 7$, and an addition is performed in each cycle unlike regular division, which used subtraction.†

2.6. GRAPHING

It is regrettably true that without making use of the calculus, there are few analytic shortcuts to graphing polynomial functions of degree greater than 2. Nevertheless, there are some valuable, primarily qualitative aids accessible to us. Synthetic division can be viewed as one shortcut, in that

---

† Some authors base their algorithm on division by $x + \alpha$ (in which case $x - 7$ has $\alpha = -7$), and end up with subtractions instead of additions. See Problem 52.

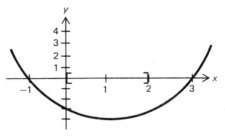

FIGURE 2-10

it can simplify the job of evaluating the function for plotting points. Remember that $p(\alpha)$ is the same as the remainder when using synthetic division with "divisor" α.

The y-intercept is easily found—it is just the constant term a_0. That this is so becomes obvious when one substitutes $x = 0$ into the general polynomial form:

$$p(0) = a_n 0^n + a_{n-1} 0^{n-1} + \cdots + a_1 0^1 + a_0 = 0 + 0 + \cdots + 0 + a_0 = a_0.$$

Finding and plotting as many zeros as possible can be of value in two respects. First, carefully plotted zeros make the graph more authoritative. Second, the multiplicities of the zeros dictate the appearance of the graph near the zeros. Let us consider this latter point in some detail.

If a zero z is of odd multiplicity, then $p(x) = (x - z)^n q(x)$, where n is odd and $q(z) \neq 0$ (i.e., $q(x)$ does not also have z as a zero). If one stays near enough to z to avoid other zeros of p, then $q(x)$ will be of one sign (since if it changed sign, it would have to go through zero).† Before seeing how this is useful, let us consider an example with which we shall parallel the above and the following discussion. Suppose that we have $p(x) = (x - 1)^3 (x^2 - 2x - 3) = (x - 1)^3 (x - 3)(x + 1)$. We see that in addition to the zeros 3 and -1, p has a zero of odd multiplicity (multiplicity 3), namely, $x = 1$. If we stay near 1, in this case keeping $x \in [0, 2]$ will qualify as "near," we can see that $q(x)$ (in this case $x^2 - 2x - 3$) is always of one sign. (See Figure 2.10, where it is graphically clear that $x^2 - 2x - 3$ is in fact always negative in $[0, 2]$.) If we further restrict x to be very close to z, and $x > z$, then $(x - z) > 0$ so that $(x - z)^n > 0$. However, if x is close to z but $x < z$, then $(x - z) < 0$, and since n is odd $(x - z)^n < 0$. Since the sign of $p(x)$ is dictated by the sign of $(x - z)^n$ and $q(x)$, and we know that one and only one of these factors changes sign as x passes through z, $p(x)$ must change sign as x passes through z. The actual signs depend on the sign of $q(z)$. In our example, if $x > z$, i.e., if $x > 1$, then $(x - 1) > 0$ and $(x - 1)^3 > 0$; as one specific case $x = 2 > 1$, then

$$(2 - 1) = 1 > 0 \qquad \text{and} \qquad (2 - 1)^3 = 1^3 = 1 > 0.$$

If $x < z$, i.e., if $x < 1$, then $(x - 1) < 0$ and $(x - 1)^3 < 0$; and as one specific

---

† Since the proof of this is a topic in the calculus, we shall borrow it without proof, noting that it is a reasonable and sensible conjecture to make, as well as being true.

case $x = \frac{1}{2} < 1$, then

$$(\tfrac{1}{2} - 1) = -\tfrac{1}{2} < 0 \qquad \text{and} \qquad (\tfrac{1}{2} - 1)^3 = (-\tfrac{1}{2})^3 = -\tfrac{1}{8} < 0.$$

Thus, for our example, $p(x)$ must change sign at $x = 1$. Having observed above that $q(x)$ is always negative between 0 and 2, then for $x < 1$, since $(x - 1)^3 < 0$, $(x - 1)^3 q(x) > 0$, since we are multiplying a negative number by a negative. For $x > 1$, since $(x - 1)^3 > 0$, $(x - 1)^3 q(x) < 0$, since we have a negative number times a positive. Figure 2.11 exhibits the graph of our $p(x)$ by the graphing program that is found in Appendix C.

If z were of even multiplicity, then n is even, whence $(x - z)^n > 0$ for all x near z, regardless of which side of z it is on so that there is no sign change. See if you can follow through a similar argument for $p'(x) = (x - 1)^2(x^2 - 2x - 3)$, which is just like our previous example except that p' has $(x - 1)^2$ and p has $(x - 1)^3$. Figure 2.12 exhibits the graph of this function, again using the graphing program.

The flatness of the graph near the zero is also dependent upon the multiplicity of the zero, for if x is very close to z, then $(x - z)$ is very close to zero; also, the larger n is, the smaller $(x - z)^n$ will be in absolute value,

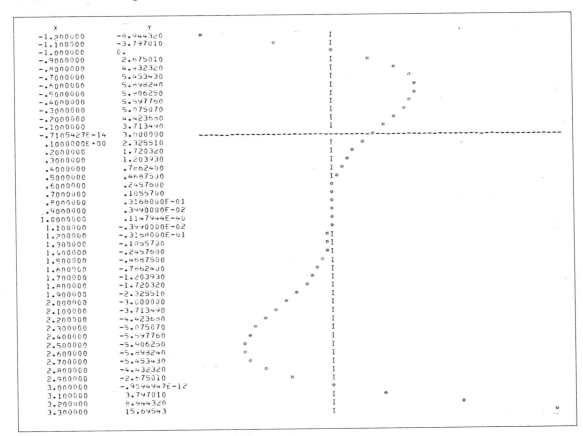

FIGURE 2.11. Computer graph of $p(x) = (x - 1)^3(x^2 - 2x - 3)$.

and the flatter the graph will be near z. Remember $(\frac{1}{4})^2 > (\frac{1}{4})^3$, i.e., $\frac{1}{16} > \frac{1}{64}$. Look also at the graph of Figure 2.13, which shows that, for each value of x between -1 and 1, the higher the power n, the smaller the absolute value of x^n. Figures 2.11 and 2.12 also illustrate the effect of multiplicity of a zero on the flatness of the graph near the zero. (See also Problem 42b of Chapter 5.)

It is important to note that the maximum (or minimum) value of a polynomial function, which occurs between two real zeros, need not be—indeed most likely is not—located at the midpoint of the interval; therefore, care is needed in determining such points as accurately as possible, i.e., if we have zeros $x = 1$ and $x = 3$, the maximum or minimum point of the graph will probably not occur at $x = 2$. (See Problem 12.)

Finally, we observe, albeit not too rigorously, that the highest degree term of a polynomial dominates the other terms when x becomes large enough in absolute value with the effect that in a larger scale picture, the graph of an nth degree polynomial looks essentially like $a_n x^n$. This is illustrated in the graphs of Figures 2.14 and 2.15, which are products of the

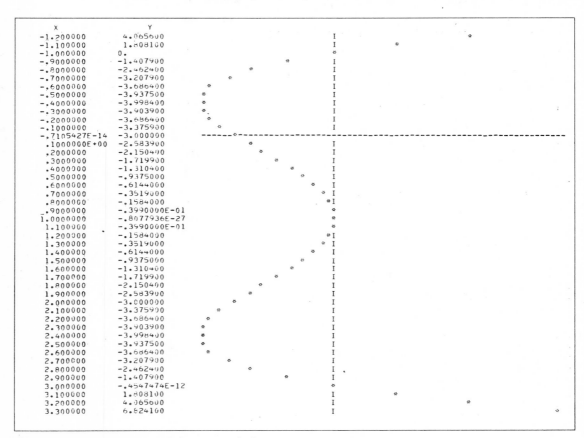

FIGURE 2.12. Computer graph of $p^1(x) = (x - 1)^2(x^2 - 2x - 3)$.

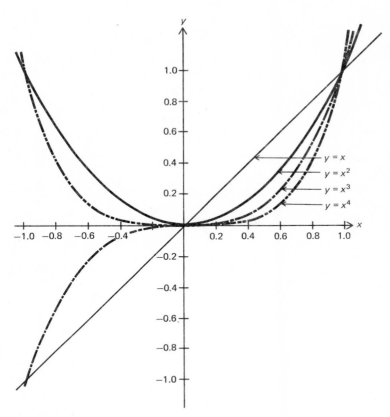

FIGURE 2-13

previously mentioned graphing program. Be sure you note the scope of both x and y values in each of those graphs to see the way in which they view the same graph from increasing distance. (Problem 46 of Chapter 5 discusses this phenomenon more formally.) Since the graph of $a_n x^n$ has the general appearance of one of the curves in Figure 2.16 when n is even, and one of the curves in Figure 2.17 when n is odd (exact shape dictated by the exact values of a_n and n), knowing the sign of a_n and parity† of n informs one whether the graph eventually rises to unbounded heights or falls to limitless depths as one follows x in either direction on its axis.

In summary, here is a list of procedures to follow in graphing:

1. Find the y-intercept.
2. Find as many real zeros as possible, including multiplicities.
3. Note the degree and sign of lead term of the polynomial.
4. Try to combine the above facts in a rough sketch.
5. Plot points where the sketch suggests that information is needed.

† That is, its "oddness" or "evenness."

```
        X               Y
    -2.200000      -7.096320          *                      I
    -2.100000      -2.936010                *                I
    -2.000000       0.                              *
    -1.900000       1.434010                        I           *
    -1.800000       3.064320                        I              *
    -1.700000       3.566430                        I                *
    -1.600000       3.594240                        I                *
    -1.500000       3.281250                        I              *
    -1.400000       2.741760                        I          *
    -1.300000       2.072070                        I       *
    -1.200000       1.351680                        I    *
    -1.100000        .6444900                       I *
    -1.000000       0.                              *
     -.9000000      -.5454900                    *  I
     -.8000000      -.9676800                   *   I
     -.7000000     -1.253070                  *     I
     -.6000000     -1.397760                 *      I
     -.5000000     -1.406250                 *      I
     -.4000000     -1.290240                  *     I
     -.3000000     -1.067430                   *    I
     -.2000000      -.7603200                    *  I
     -.1000000      -.3950100                       *I
    -.1421085E-13  -.5684342E-13  ------------------*----------------------------
     .1000005E+00   .3950100                        I *
     .2000000       .7603200                        I   *
     .3000000      1.067430                         I    *
     .4000000      1.290240                         I     *
     .5000000      1.406250                         I      *
     .6000000      1.397760                         I      *
     .7000000      1.253070                         I    *
     .8000000       .9676800                        I  *
     .9000000       .5454900                        I *
    1.000000       0.                               *
    1.100000       -.6444900                     *  I
    1.200000      -1.351680                    *     I
    1.300000      -2.072070                  *       I
    1.400000      -2.741760                *         I
    1.500000      -3.281250             *            I
    1.600000      -3.594240            *             I
    1.700000      -3.566430            *             I
    1.800000      -3.064320              *           I
    1.900000      -1.934010                 *        I
    2.000000       -.6821210E-12                 *
    2.100000       2.936010                        I      *
    2.200000       7.096320                        I             *
    2.300000      12.72343                         I                        *
```

FIGURE 2.14. Computer graph of $f(x) = (x^2 - 1)(x^2 - 4)x$.

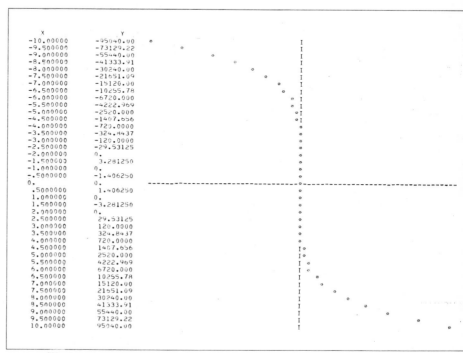

FIGURE 2.15. Computer graph of $f(x) = (x^2 - 1) \cdot (x^2 - 4) \cdot x$.

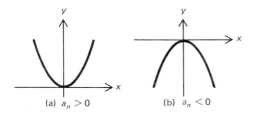

(a) $a_n > 0$ (b) $a_n < 0$

FIGURE 2-16

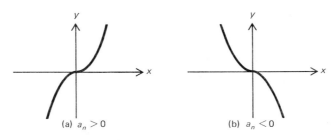

(a) $a_n > 0$ (b) $a_n < 0$

FIGURE 2-17

2.7. APPROXIMATE ZEROS

Motivated initially by the impossibility of finding all zeros of all poly-nomial functions by exact methods, we turn briefly to the consideration of ways to generate approximate values for zeros.

Indeed, in the world of "real-life applications," the reliance on methods of approximation is growing steadily. This is true even where exact solutions might be obtainable. There are several reasons for this. For one, good ap-proximations are often all that are needed in the practical setting. For a second, the increasing availability of computers makes feasible the use of algorithms that would be very tedious and time-consuming by hand. And for a third, the standard use of a program that uses an approximation al-gorithm would often be cheaper to use than human skills and/or a more elaborate program employing a variety of exact analytic tools.

The search for algorithms to generate approximate zeros goes on in the branch of mathematics called numerical analysis. While many of the known algorithms are based on mathematical concepts and tools that are beyond the scope of this book, there are several effective algorithms whose "workings" we can understand quite readily with minimal reference to higher mathematics. We consider one such algorithm in this section, and then provide opportunities for you to explore this subject further in Problems 56 and 57 of this chapter, and later in Problems 38 and 39 of Chapter 6.

2.7a. Successive Halving

The algorithm we shall consider is called the Method of Successive Halving. To understand how it works, one needs to know that if p is a polynomial, and $l < r$ (l for left, r for right), $p(l)$ and $p(r)$ are of opposite

SUCCESSIVE
HALVING

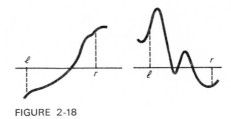

FIGURE 2-18

sign, then p has at least one zero in the open interval (l, r).† See Figure 2.18.

If one bisects the interval, then there is a zero for p either at the midpoint or in one of the two subintervals formed. See Figure 2.19. Since the midpoint of $[l, r]$ is the average of the values l and r, it has the value $\dfrac{l+r}{2}$.

$\left(\text{For example if } l = 4 \text{ and } r = 6, \text{ the midpoint would be } 5 : \dfrac{4+6}{2}\right).$ ‡ Re-

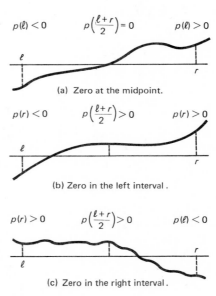

$p(l) < 0 \qquad p\left(\dfrac{l+r}{2}\right) = 0 \qquad p(l) > 0$

(a) Zero at the midpoint.

$p(r) < 0 \qquad p\left(\dfrac{l+r}{2}\right) > 0 \qquad p(r) > 0$

(b) Zero in the left interval.

$p(r) > 0 \qquad p\left(\dfrac{l+r}{2}\right) > 0 \qquad p(l) < 0$

(c) Zero in the right interval.

FIGURE 2-19

stating the above, either $p\left(\dfrac{l+r}{2}\right) = 0$ or p has a zero in $\left(l, \dfrac{l+r}{2}\right)$ or $\left(\dfrac{l+r}{2}, r\right)$.

† This is a special case of a result proved in the calculus, which states that if f is a function that is continuous on a closed interval $[l, r]$, and $f(l)$ and $f(r)$ are of opposite sign, then f has at least one zero in (l, r). An intuitive explanation of this result is that if the graph of f can be drawn without lifting the pencil from the paper, and the graph begins and ends on different sides of the x-axis, then it must cross the axis at least once in between.

‡ See Section 6.2a for further derivation of this.

If we could tell which, we could repeat this procedure over and over, getting shorter and shorter intervals that contain a zero for p. Eventually, a zero would be reached or the interval would be short enough so that either endpoint would be an acceptable approximation. The problem remains of determining, at each stage, which of the two subintervals contains a zero for p. If $p\left(\dfrac{l+r}{2}\right) \neq 0$, then it is either positive or negative, and so opposite in sign from $p(l)$ or $p(r)$. See Figure 2.17 again. If opposite from $p(l)$, then there is a zero in $\left(l, \dfrac{l+r}{2}\right)$. If opposite from $p(r)$ then there is a zero in $\left(\dfrac{l+r}{2}, r\right)$.

EXAMPLE 18

The flowchart of Figure 2.20 exhibits the algorithm at this level of development, and our tracing an example with it should be illuminating. Let us consider

$$f(x) = 2x^3 + 7x^2 - 10x - 35$$

with initial values of l and r being 1 and 4, respectively. We first compute $Y1 = f(1) = -36$ and $Y2 = f(4) = 165$ (note that f does change sign from l to r as required). Next M is assigned the value $(1 + 4)/2 = 2\frac{1}{2}$ and $y = f(2\frac{1}{2}) = 15$. Since $15 \neq 0$, we ask if 15 has the same sign as -36. The answer is no; therefore, r becomes $2\frac{1}{2}$, $Y2$ becomes 15, while l remains 1 and $Y1$ remains -36. Thus if we were to tabulate our results to date we would have

l	M	r	$Y1$	Y	$Y2$
1	2.5	4	-36	15	165
1		2.5	-36		15

We are ready now for the next cycle, since $2.5 - 1 = 1.5 \not< .00001$. Now M becomes 1.75 $[= (2.5 + 1)/2]$; Y becomes -20.34375; and $Y \neq 0$. Also, Y does have the same sign as $Y1$ so that l is changed to 1.75 and $Y1$ to -20.34375. The table now becomes

l	M	r	$Y1$	Y	$Y2$
1	2.5	4	-36	15	165
1	1.75	2.5	-36	$-20.34 \cdots$	15
1.75		2.5	$-20.34 \cdots$		15

Cycling again, we find $M = 2.125$, $Y = -5.449 \cdots$ so that l is changed again in this cycle. It may also help to look at this example graphically. Figure 2.21 exhibits several cycles, with the graph of f sketched in only schematically.

Using the stopping criterion " Is $r - l < .0001$?" of the flowchart, the algorithm continues through the computation of 15 values of M. Figure 2.22 exhibits the output of a program to do successive halving which was run with this particular example, so that you may see the step by step progress stated above. (Notice, though, that the program follows a slightly different flowchart, in that the output is done every cycle and includes all

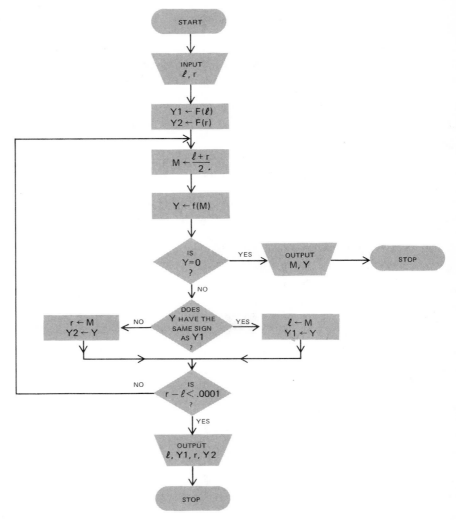

FIGURE 2-20

three values, not just the two best choices. It is worth noticing that while the last value of M is the best approximation for the zero in this case, had the algorithm stopped with cycle 10, 11, or 12, for example, the best approximation would have been r, i.e., there is no guarantee that M is always better than both l and r.

By now, the eager programmer may be asking whether it is legitimate to ask "Does Y have the same sign as $Y1$?" or if this requires too much human interpretation. An efficient way of asking this question is "Is $Y \cdot Y1 > 0$?," which certainly is at a machine-implementable level. (Do you see why this is a valid modification of the question?)

There are, of course, shortcomings in this algorithm. For one, it takes

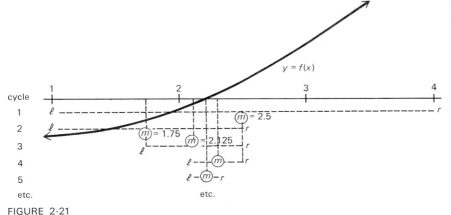

FIGURE 2-21

many more cycles to get good values than some other more sophisticated algorithms. Most of the latter methods rely on concepts from calculus and so are not considered at this stage. Some, however, concern material we have covered here, and you are encouraged (in Problem 57) to devise other algorithms for this task. A second shortcoming is that by requiring $f(l)$ and $f(r)$ to be of opposite sign, this algorithm is incapable of finding zeros of even multiplicity (try it on $y = (3x - 1)^2$ with $l = 0$ and $r = 1$). Perhaps your efforts with Problem 57 will produce an algorithm without this shortcoming. As mentioned earlier, the problem of finding algorithms to approximate zeros for all kinds of functions is one of the important (and by no means completed) activities in the branch of mathematics called numerical analysis. The major emphasis of this section should remain on the reasons for finding approximate zeros, rather than on the specific algorithm, and second, on the fact that you have already learned enough mathematics to invent techniques of merit by yourself.

CYCLE	L	M	R	F (L)	F (M)	F (R)
1	1.00000000	2.50000000	4.00000000	-36.00000000	15.00000000	165.00000000
2	1.00000000	1.75000000	2.50000000	-36.00000000	-20.34375000	15.00000000
3	1.75000000	2.12500000	2.50000000	-20.34375000	-5.44921875	15.00000000
4	2.12500000	2.31250000	2.50000000	-5.44921875	4.04150391	15.00000000
5	2.12500000	2.21875000	2.31250000	-5.44921875	-0.88238525	4.04150391
6	2.21875000	2.26562500	2.31250000	-0.88238525	1.53430939	4.04150391
7	2.21875000	2.24218750	2.26562500	-0.88238525	0.31472683	1.53430939
8	2.21875000	2.23046875	2.24218750	-0.88238525	-0.28662837	0.31472683
9	2.23046875	2.23632813	2.24218750	-0.28662837	0.01334824	0.31472683
10	2.23046875	2.23339844	2.23632813	-0.28662837	-0.13681516	0.01334824
11	2.23339844	2.23486328	2.23632813	-0.13681516	-0.06177726	0.01334824
12	2.23486328	2.23559570	2.23632813	-0.06177726	-0.02422546	0.01334824
13	2.23559570	2.23596191	2.23632813	-0.02422546	-0.00544135	0.01334824
14	2.23596191	2.23614502	2.23632813	-0.00544135	0.00395276	0.01334824
15	2.23596191	2.23605347	2.23614502	-0.00544135	-0.00074447	0.00395276

TOLERANCE TEST MET. ALGORITHM STOPPED

FIGURE 2.22

2.7. APPROXIMATE ZEROS   129

2.7b. Relationship to General Results about Zeros

As a final word in this section, once an approximate zero has been found, the problem can be reduced as discussed in Section 2.4 to get an approximate $q(x)$ whose zeros could be investigated. One needs to be careful, however, that in getting approximations from approximations, one does not generate so much error as to seriously contaminate the results. (The interested student is directed to a numerical analysis book for further investigation.)

2.8. SOME APPLICATIONS

APPLICATIONS

Aside from mention of a few examples of linear and quadratic functions, we have not encountered any applications of polynomial functions. In this section, we shall work a few examples to give some feeling of the use of polynomials both in real life and in mathematical contexts. However, we wish to emphasize that pure mathematics as a formal discipline is concerned with pattern structure, general rules, and relationships. Such are often inspired by practical problems, and once the machinery has been developed it is made available to all who might find it useful. While we do not intend to assign applied mathematics to a second-class role, we recognize that most students at this level see mathematics *only* as applications and we hope to help you see that there is much more to mathematics than solving problems.

We also wish to make clear that the examples in this chapter, indeed the whole book, are not intended to be fully representative of what you may encounter yourself. Indeed, we encourage you to speak with people active in other disciplines of interest to you, and learn of other kinds of applications. But now to our examples.

EXAMPLE 19

Sociologist Melody Dampfire observes that for a certain well-run job training program, the number of people taken off welfare roles seems to be a linear function of the money spent on training. In particular, during a month in which $50,000 was spent, 10 persons were removed from the welfare rolls, while a month in which $62,000 was spent, 13 persons were removed from the welfare rolls.

(a) Find the rule for this function; (b) determine how much would need to be spent in a month to get 20 persons removed from welfare rolls; (c) interpret the value $f(0)$, where f is the function being studied; and (d) determine the amount needed to remove one person from welfare according to this model.

a. Since a linear function is to be used, we would be dealing with $p(d) = a \cdot d + b$, where p represents the number of people taken off welfare as a result of spending d dollars, and a and b are unknown constants we need to find. Knowing that (50,000, 10) and (62,000, 13) are in the function, we have the pair of equations

$$10 = a \cdot 50000 + b$$
$$13 = a \cdot 62000 + b,$$

which may be solved simultaneously to determine a and b. These are
$a = \frac{1}{4000}$, $b = -2\frac{1}{2}$. Therefore, the function is

$$p(d) = \left(\frac{1}{4000} \cdot d\right) - (2\tfrac{1}{2}).$$

b. This is now a matter of a simple equation to solve (and corresponds to a use of the inverse of the function):

$$20 = \frac{1}{4000} \cdot d - 2\tfrac{1}{2}$$

$$22\tfrac{1}{2} = \frac{d}{4000}$$

$$\$90{,}000 = d.$$

c. $f(0)$ (or in our new notation $p(0)$) $= -2\frac{1}{2}$. This means that if no money were spent in a month, the welfare roles would increase by $2\frac{1}{2}$ in that month. It also emphasizes that the function is only an approximation of reality, since $\frac{1}{2}$ person is not realistic. (Indeed, a linear function may have been an incorrect choice.)

d. Here we are viewing numbers of people as the independent variable, which suggests an appeal to the inverse of our function. The algebraic rule for the inverse is easily found in this case and is in fact:

$$d = d(p) = 4000p + 10{,}000.$$

From this, it is not complicated to see that a change of 1 in the value of p (say, from p_0 to $p_0 + 1$) causes a change in the required dollar amount of

$$d(p_0 + 1) - d(p_0) = 4000(p_0 + 1) + 10{,}000 - (4000p_0 + 10{,}000)$$
$$= 4000p_0 + 14{,}000 - 4000p_0 - 10{,}000$$
$$= 4000$$

so that the cost of removing one more person from the welfare rolls (the marginal cost) is $4000. An a by-product of having found the inverse, we can also easily see that an expenditure of $10,000 is required just to keep a constant welfare population. This could have the economic implication that even if the number of welfare recipients were reduced to zero, a continued expense might be needed to maintain this situation.

Business tycoon Calvin Butterball produces widgets and believes that **EXAMPLE 20** market conditions are accurately described by the function of Example 4e, p. 95. How many widgets should he produce to maximize his receipts?

Referring again to Example 4e, you will note that the function is $R(n) = -(n^2/20) + 1000n$. This is a quadratic function whose parabolic graph opens downward. Hence the largest receipts will correspond to the vertex of the parabola (which is the point on the graph with the largest R coordinate). Using the formula for the vertex of such a parabola, we find the point to be

$$\left(-\frac{b}{2a}, \frac{4ac - b^2}{4a}\right) \qquad a = -\tfrac{1}{20}, b = 1000, c = 0$$

$$\left(\frac{-1000}{2(-\frac{1}{20})}, \frac{4(-\frac{1}{20})(0) - (1000)^2}{4(-\frac{1}{20})}\right)$$

$$\left(\frac{1000}{\frac{1}{10}}, \frac{0 - 1,000,000}{-\frac{1}{5}}\right)$$

$$(10,000, 5,000,000).$$

So, he should produce 10,000 widgets, with a resulting maximum revenue of $5 million.

EXAMPLE 21   Where do the graphs of $y = x^3 + 2x + 3$ and $y = -2x^2 + 5x + 9$ intersect?

This should remind you of solving simultaneous equations to find points of intersection. Indeed, we wish to solve the two equations

$$y = x^3 + 2x + 3$$
$$y = -2x^2 + 5x + 9$$

simultaneously; thereby we find the points (x, y), which are common to both functions. Because these are nonlinear equations, the usual elimination procedure will not work. We may use a form of substitution, however, and set the two expressions for y equal to each other.

$$x^3 + 2x + 3 = -2x^2 + 5x + 9.$$

This is equivalent to

$$x^3 + 2x^2 - 3x - 6 = 0,$$

which brings us back to finding zeros of polynomial functions. As you can by now determine, the solution set is $\{-2, \sqrt{3}, -\sqrt{3}\}$. Then, returning to the original question, the points of intersection are found by substituting these three values of x into one of the two original functions to get the three points: $(-2, -9)$, $(\sqrt{3}, 3 + 5\sqrt{3})$, $(-\sqrt{3}, 3 - 5\sqrt{3})$.

EXAMPLE 22   An object projected upward from the surface of an airless body such as the moon, acted on only by gravity, has its altitude, velocity, and acceleration described by three related functions. Suppose that on one such body the functions for an astronaut, who has just jumped up, are:

$$a(t) = -5 \qquad \text{(acceleration)}$$
$$v(t) = -5t + 12 \qquad \text{(velocity)}$$
$$h(t) = -\frac{5t^2}{2} + 12t \qquad \text{(altitude)},$$

where t is time in seconds; h is in feet; v is in feet per second (ft/sec); and a is in feet per second per second (ft/sec²).

The derivation of such functions is one of the jobs of the calculus. Meanwhile we can proceed to ask and answer the following questions:

a. What is the maximum height she reaches in her jump?
b. How long does it take to reach that altitude?
c. When does she land on the ground again?
d. How fast is she going when she lands?

a. Using the function h (a quadratic) and the vertex formula as in
Example 18, we find the highest point of the jump to be described by

$$\left(\frac{-12}{2(-\frac{5}{2})}, \ \frac{12(-\frac{5}{2})(0)-12^2}{4(-\frac{5}{2})}\right)$$

$$\left(\frac{12}{5}, \ \frac{144}{20}\right)$$

$$\left(2\frac{2}{5}, \ 7\frac{1}{5}\right)$$

which tells us that the maximum height is $7\frac{1}{5}$ ft, and that the height is reached in $2\frac{2}{5}$ sec.

b. We just answered that while solving part a.

c. This is translated into finding the zeros of h. They are easily found to be at $t=0$ and $t=4\frac{4}{5}$. The $t=0$ case corresponds to when she jumped, while $t=4\frac{4}{5}$ tells us she landed $4\frac{4}{5}$ seconds later.

d. Her velocity at "landing" is given by

$$V\left(4\frac{4}{5}\right) = -5\cdot\frac{24}{5}+12 = -12,$$

which means that her velocity was 12 ft/sec toward the surface.

2.9. A PROJECT

In studying the polynomial functions, we have had an introduction to how mathematics is done. We take this opportunity to involve the reader more actively by taking a brief look at some other functions, called algebraic functions, which although not always classed as elementary functions, are closely enough related to warrant our attention in this chapter. In particular, you will be asked to carry out some limited investigations of these functions yourself with only a moderate amount of guidance.

As a general class of functions, the algebraic functions are ones that can be expressed by a rule which involves a finite number of extractions of roots, additions, subtractions, multiplications, and divisions of the argument of the function and constants.† Examples of such rules include the following:

$$y = 37x^{3000} - \frac{3}{x^{10}} + {}^{13}\sqrt{x^3} - 1$$

$$t = \frac{3x^4 - 4x + 2}{5x^2 - 2x + 1}$$

$$s = \frac{\sqrt[3]{x}}{3x^5 - 5x^2 + 2} + \frac{3x}{13}$$

† If the rule is $f(x) = $ (expression in x), one has an *explicit* algebraic function. Other algebraic functions (called implicit algebraic functions) do not permit solving the rule for the function value as in the example $f = \left\{(x, y) \mid \sqrt{xy} - x^2y^5 = 3\left(\frac{x}{y}\right) + \sqrt[3]{y}\right\}.$

and

$$w = \left(100x^{100} + 99x^{99} + \cdots + x + 0 - \frac{1}{x} - \frac{2}{x^2} - \cdots \right.$$
$$\left. - \frac{99}{x^{99}} - \frac{100}{x^{100}} \right) \sqrt[100]{x} \, \sqrt[99]{x} \cdots \sqrt[2]{x} \cdot x.$$

You should quickly see that the set of polynomial functions is a subset of the set of algebraic functions, since each is defined by a rule employing only a finite number of additions, subtractions, multiplications, and divisions.

One particularly noteworthy subset of the algebraic functions is the set of rational functions. A function is called a rational function if its rule can be expressed as the quotient of two polynomial functions. We begin our investigation with these functions.

Phase 1. "How does one find the zeros of rational functions?"

a. To gain some insight, first try to solve the equations

$$\frac{x^2 - x - 6}{2x^2 + 1} = 0 \quad \text{and} \quad \frac{x^2 - x - 6}{3x^3 + 4x - 1} = 0.$$

b. Then formulate a conjecture for the method of finding the zeros in the general case, i.e., for $p(x)/q(x)$, where p and q are both polynomials.

c. Next try to prove your conjecture.

d. Carefully find and check the zeros of

$$\frac{6x^3 + 5x^2 - 2x + 1}{2x^2 - x - 3}.$$

Do you need to revise your efforts in parts (b) and (c) in the light of this example?

Phase 2. "Since we may not divide by zero, what happens near the zeros of the denominator?"

a. Again, some examples should provide fertile ground for initial conjectures. Try sketching graphs of the following functions near the zeros of the denominator:

$$\frac{1}{x - 2}, \quad \frac{1}{(x - 2)^2}, \quad \frac{1}{(x - 2)^3}, \quad \frac{x^2 - 1}{x - 2}, \quad \frac{x^2 - 9}{x - 2}, \quad \frac{x^2 - 1}{(x - 2)^2}.$$

Approach 1: Do this by hand computation of values.

Approach 2: Do this using an existing program that computes functional values and/or graphs.

Approach 3: Write a program to assist in working these examples and then use it.

b. Again, formulate a conjecture of the effect on the graph of $p(x)/q(x)$ of having x near a zero of q. Be as precise and clear in your statement(s) as you can.

c. Provide as thorough a justification of your conjecture(s) as you can. (While formal proof is probably beyond your reach now, you should be able to argue quite convincingly without reference to specific examples, much as was done in Section 2.6.)

d. Reconsider your claims in the light of the example $\dfrac{x^2 - 4}{x - 2}$. (Be careful.)

Phase 3. What happens as x becomes very large positive and very large negative?

   a. A variety of behavior is possible, as evidenced by the following examples:

$$\frac{x}{x+2}, \quad \frac{x^2}{x+2}, \quad \frac{x}{x^2+2}, \quad \frac{2x^2-x-3}{2x-x^2}.$$

      Approach 1: Examine the behavior (graph?) of each for values of x that are large positive and values that are large negative.

      Approach 2: As in phase 2, use an existing program.

      Approach 3: As in phase 2, write and use a program.

   b. Formulate a conjecture that details conditions under which the various qualitative graphical phenomena you encountered in part (a) may occur. Your conjectures should be aimed at providing qualitative graphical aids that would not require point plotting.

   c. Explain why your conjectures seem plausible without depending solely on reference to the examples of part (a). (Again, proof in the formal sense would become too involved.) You may find it helpful to consider algebraic manipulations to express such functions in other forms (such as doing implied divisions, breaking fractions into separate pieces, factorizations, etc.)

Phase 4. In phases 1–3 you have learned several features of rational functions. Here these features are put to use in sketching graphs of such functions, graphs whose behavior is as attractive as it is unexpected to the casual plotter of points. With this expectation, sketch graphs of each of the following rational functions.

1. $b(x) = \dfrac{x+2}{x-1}$

2. $e(y) = \dfrac{y-3}{(y+2)(y-1)}$

3. $t(z) = \dfrac{3z^2 - 5z - 2}{z^2 + 1}$

4. $h(w) = \dfrac{w^2 - 9}{2w^2 + 4w - 30}$

5. $a(t) = \dfrac{2t^2 - 8t + 6}{t^2 + t}$

6. $n(a) = \dfrac{a^3 - 3a^2 - 4a}{a^2 + 2a - 8}$

7. $n(b) = \dfrac{b^3 - 3b^2 - 4b}{b^2 + 1}$

8. $e(m) = \dfrac{m^2 + 3}{2m^2 + 1}$

9. $b(s) = \dfrac{s+3}{2s^3 - s^2 - 12s + 6}$

10. $o(t) = \dfrac{3t+1}{4-t}$

11. $n(x) = \dfrac{x^4 - 7x^2 + 12}{x^3 - 6x^2 + 11x - 6}$

12. $n(v) = \dfrac{v^2 - 3v}{v^3 + v^2 - 2v}$

13. $i(h) = \dfrac{h^2 + 4h + 4}{2h^3 + 6h^2 - h - 3}$

14. $e(x) = \dfrac{2x^3 - x^2 - 12x + 6}{x+3}$

Phases 2 and 3 bring one into proximity with a concept called limit, which is the foundation upon which the calculus is built. While the technical details of limits are quite demanding, the underlying ideas are quite straightforward; attention to these two phases can help provide a stronger basis for understanding the nature of the limit concept.

Also, the concept of the asymptote, which is associated with some of the behavior in phases 2 and 3 is introduced now. An asymptote to a graph is a line that the graph approaches but does not meet. For example, the graph of $y = 1/x$ has the x- and y-axes as asymptotes. A more precise definition of asymptote requires some notions from calculus that so we shall content ourselves for now with this less formal definition. Note that asymptotes may be vertical, horizontal, or oblique (neither horizontal nor vertical), thereby accounting for all the possible straight lines in the plane.

We conclude our introduction to algebraic functions by considering the root functions.

ALGEBRAIC
FUNCTIONS

Phase 5. Consider the root functions R_n defined by $R_n(x) = x^{1/n}$, where $n \in \mathbb{N}$.

    a. Verify that the root function R_n is related to the power functions $\phi_n(x) = x^n$ by means of inverses of relations. Discuss the issue of function versus relation wherever it appears as an issue. Also discuss domains and ranges.

    b. Using your observations from part (a), sketch graphs of $R_1, R_2, R_3, R_4, R_5,$ and R_6.

Phase 6. Let $T_r(x) = x + r$, where $r \in \mathbb{R}$.

    a. Study the composite functions $T_r \circ R_n$ first by graphing some examples of your choice, then by formally describing the effect(s) T_r has on the graph of R_n. Also, specify similarities and differences in domains and/or ranges of R_n and $T_r \circ R_n$.

    b. Study the composite functions $R_n \circ T_r$ first by graphing some examples of your choice, then by formally describing the effect(s) T_r has on the graph of R_n. Also, specify similarities and differences in domains and/or ranges of R_n and $R_n \circ T_r$.

NOTE   You will see questions similar to those of phase 6 reappearing in Chapter 3 (Section 8) and Chapter 6 (Section 3), as well as in occasional problems.

Phase 7. Now put together what you have learned in phases 1 through 6, both in content and methodology, to sketch graphs of the following algebraic functions.

1. $a(x) = \sqrt{x - 5} + 2$

2. $s(w) = \sqrt{\dfrac{w + 3}{w - 2}}$

3. $y(t) = \sqrt[3]{\dfrac{t + 3}{t - 2}}$

4. $m(p) = \sqrt[3]{p - 5} + 2$

5. $p(y) = \sqrt{y^2 - y - 6}$

6. $t(a) = \dfrac{\sqrt{a^2 + 4a + 4}}{a^2 + 7a + 12}$

7. $o(e) = \dfrac{e^2 + 7e + 12}{\sqrt{e^2 + 4e + 4}}$

8. $t(i) = i + \dfrac{1}{i}$

9. $i(o) = o^2 + 1/o$ 10. $c(u) = u^2 - \dfrac{1}{u}$

2.10. SOME THINGS NOT DONE

There are many other topics and investigations that involve polynomial functions—we mention just a few.

In gathering data in the hopes of discovering a pattern, one may wish to use a polynomial function to describe the pattern; often a straight line seems appropriate. The choice of the polynomial may be suggested by the nature of the phenomena being investigated, by the apparent shape assumed by the data when graphed, or by the desire to use a polynomial in place of a more esoteric function. The choice of a polynomial may be, in fact, inappropriate, but one can still ask for the polynomial of a certain degree that best *fits* his data. There are various meanings for the word *best*, but in any case the question we leave unanswered is "How does one generate such a polynomial?"

We have also skirted the problem of inverses of polynomial functions in general. To the extent that we can graph p, we can graph p^{-1}, but there remains the problem that in order to subdivide the domain of p to get smaller functions whose inverses are functions, we would need to be able to find the high and low points on the graphs precisely. Except for quadratic functions, we have not been able to do this. We have, without knowing it, dealt with the problem of finding $p^{-1}(b)$ because finding $p^{-1}(b)$ is equivalent to solving the equation $p(x) = b$, which in turn is equivalent to the equation $p(x) - b = 0$. But $p(x) - b$ is merely a new polynomial, say q, and so $p^{-1}(b) = q^{-1}(0)$, and we have studied the finding of zeros for polynomials.

There is also a technique for locating certain intervals that must contain the zeros of polynomials, called Descartes' Rule of Signs, which we have not developed because we believe that it is of marginal importance. The interested student may wish to ask his instructor for a reference in which to find this rule.

What about relationships among the functions and formulas for such terms as $p(x + w)$? While not beyond our abilities to answer, these questions prove to be of less value, and so we omit them.

And perhaps you have other questions in mind as well. If so, we encourage you to ask your instructor for guidance or perspective on your questions.

SUMMARY OF IMPORTANT CONCEPTS AND NOTATION

ZEROS: (Definition 2.1) The set of values of x for which $f(x) = 0$ is called the set of zeros of $f(x)$.

$f^{-1}(0)$: (Definition 2.1) $f^{-1}(0)$ is the set of zeros of f.

LINEAR FUNCTION: (Definition 2.2) A function f is said to be linear if for each $x \in D_f$, $f(x) = ax + b$, where a and b are fixed real numbers.

Y-INTERCEPT: (Definition 2.3) The y-intercept of a function is the value $f(0)$.

QUADRATIC FUNCTIONS: (Definition 2.4) A quadratic function is one that may be specified by a rule $f(x) = ax^2 + bx + c$ for each $x \in D_f$, where a, b, and c are fixed real numbers and $a \neq 0$.

QUADRATIC FORMULA: (Formula 2.1) $x = \dfrac{-b \pm \sqrt{b^2 - 4ac}}{2a}$

DISCRIMINANT: (Definition 2.5) The discriminant is the number $b^2 - 4ac$ from the quadratic formula.

POLYNOMIAL FUNCTION: (Definition 2.6) A function f is a polynomial function if it can be specified by a rule of the form

$$f(x) = a_n x^n + a_{n-1} x^{n-1} + \cdots + a_1 x + a_0$$

for some nonnegative integer n, where $a_0, a_1, \cdots, a_n$ are real number constants.

DEGREE: (Definition 2.7) A polynomial f is said to be of degree n if it can be specified by a rule $f(x) = a_n x^n + \cdots + a_0$, where $a_n \neq 0$.

LEAD COEFFICIENT: (Definition 2.8) The lead coefficient for a polynomial in the form $a_n x^n + \cdots + a_1 x + a_0$ is a_n.

RATIONAL ZERO THEOREM: (Theorem 2.1) If a polynomial function specified by $p(x) = a_n x^n + \cdots + a_0$ has all integer coefficients, then all the rational zeros for p are contained in the set of all rational numbers whose numerators are divisors of a_0 and whose denominators are divisors of a_n.

FACTOR THEOREM (Theorem 2.2) If z is a zero for a polynomial function p, then $(x - z)$ is a factor of $p(x)$.

CONVERSE OF THE FACTOR THEOREM: (Theorem 2.3) If $(x - z)$ is a factor of $p(x)$, then z is a zero for p.

ZERO OF MULTIPLICITY M: (Definition 2.9) A number q is said to be a zero of multiplicity m for a polynomial p, provided that $(x - q)^m$ is a factor of $p(x)$ but $(x - q)^{m+1}$ is not.

FUNDAMENTAL THEOREM OF ALGEBRA: (Theorem 2.4) Every polynomial function p of degree n has exactly n zeros, provided that the domain of p is the set of complex numbers, not just $\mathbb{R}$, and provided a zero of multiplicity m is counted m times.

SYNTHETIC DIVISION: (Definition 2.10, Sections 2.5a and 2.5b, and Figure 2.9.)

BEHAVIOR NEAR ZEROS: (Section 2.6)

BEHAVIOR AWAY FROM ZEROS: (Section 2.6)

SUCCESSIVE HALVING: (Section 2.7a and Figure 2.20)

PRACTICE EXERCISES

Section 2.2a

1. Sketch the graph of the following. Indicate the y-intercept in each case, and find all real zeros for each.

    (a) $t(x) = 4x - 5$ (b) $o(t) = 2t + 5$

(c) $p(z) = 3z$

(d) $o(y) = 5$

(e) $s(x) = x$

(f) $o(r) = 3r$

(g) $g(x) = -4x + 7$

(h) $y(t) = \frac{3}{4}$

(i) $i(x) = \frac{1}{2}$

(j) $s(x) = 3x - 5$

(k) $g(y) = -4y - 7$

(l) $e(m) = -4$

(m) $o(z) = 4z + 7$

(n) $m(x) = 7x$

(o) $e(t) = 12 - 7t$

(p) $t(w) = 3 + 8w$

(q) $r(a) = a + 1$

(r) $y(p) = 2p - 1$

Section 2.2b

2. Find all the real zeros of the functions specified by the following rules.

(a) $M(a) = a^2 - a - 1$

(b) $a(t) = t^2 - t + 1$

(c) $t(w) = 2w^2 + 3w + 4$

(d) $h(s) = 4s^2 - 20s + 25$

(e) $e(y) = -3y^2 - 4y + 2$

(f) $m(z) = z^2 + 6z + 9$

(g) $a(x) = 2 + 3x^2 - x$

(h) $t(v) = -2v^2 - 3v + 5$

(i) $i(x) = x^2 + 1$

(j) $c(t) = 4t^2 - 12t + 9$

(k) $s(n) = 3n + 1 - n^2$

(l) $i(x) = 3x^2 + 2x - 1$

(m) $3x^2 + 4x - 5 = s(x)$

(n) $2x^2 - 3x - 4 = c(x)$

(o) $r(x) = -x^2 + 3x + 5$

(p) $e(x) = -2x^2 + 10x + 21$

(q) $a(m) = 4m^2 - 5m + 3$

(r) $t(b) = 4b^2 + 4b + 1$

(s) $i(j) = j^2 - 15j - 1$

(t) $v(w) = 2w^2 + 3$

(u) $e(t) = 4 - 20t + 25t^2$

3. Sketch the graph of each function in Problem 2.

Section 2.3a

4. Each of the following rules specifies a function. For those which are polynomials, give the degree of the function. For those which are not polynomials, explain why not.

(a) $i(x) = 37x^{10} - 4x^7 + 2$

(b) $n(z) = z^2 - 4z + 3/z$

(c) $t(s) = \sqrt{s^2 + 1}$

(d) $e(w) = 35$

(e) $g(a) = 2a^2 + 7 - a$

(f) $r(t) = (t + 2)(t - 3)(t^2 + 5)$

(g) $a(y) = (3y + 2)/5$

(h) $s(x) = x^5 + x^2 + x$

(i) $d(k) = |k^2 + 2k - 3|$

(j) $e(j) = \dfrac{j^3 - 2j^2 + j - 2}{j^2 + 1}$

(k) $r(x) = \dfrac{x^2 - x - 6}{x - 3}$

(l) $i(t) = \dfrac{t^2 - t - 7}{t - 3}$

(m) $v(b) = b - 17 + 3b^5$

(n) $a(m) = \lfloor m^2 - 2m + 1 \rfloor$
(o) $t(y) = (y + 1)(y^2 + 5) + (y^3 - 2)$
(p) $i(w) = \sqrt{2}w^3 + 2w^2 - 5$
(q) $v(z) = -89z^{100} + 35z^{37} - \frac{5}{2}z + 2$
(r) $e(x) = 3x - 5$

Section 2.3b

5. Find all the rational zeros for the functions specified by the following rules.
(a) $n(x) = x^3 - 2x^2 - 5x + 6$
(b) $e(t) = t^3 - 6t^2 + t - 6$
(c) $w(s) = s^4 - 3s^2 + 2$
(d) $t(a) = a^4 + 5a^2 + 4$
(e) $o(y) = 10y^3 + 41y^2 + 2y - 8$
(f) $n(w) = 9w^3 - 54w^2 + 47w - 10$
(g) $s(x) = 2x^4 - 3x + 1$
(h) $e(x) = 2x^3 - x^2 + 2x - 1$
(i) $i(x) = 3x^4 + x - 17$
(j) $b(x) = 2x^3 + 5x^2 + x - 2$
(k) $n(x) = 3x^3 + x^2 + x - 2$
(l) $i(x) = x^4 + 2x^3 - 13x^2 - 38x - 24$
(m) $z(t) = 2t^4 + 9t^3 - 7t^2 - 9t + 5$

Section 2.4a

6. By using your knowledge of factoring and by the grace of the converse of the factor theorem, find as many zeros as you can for the following functions.
(a) $z(x) = x^3 - 5x^2 + 6x$
(b) $e(y) = y^4 - 16$
(c) $n(w) = (w - 1)(w - 2)(w - 3)(w - 4)$
(d) $o(m) = m^4 - 6m^2 - 17$
(e) $a(s) = s^3 - 36s$
(f) $b(t) = 5t^3 + 20t^2 - 60t$
(g) $e(z) = 4 - 2z - 4z^2 + 2z^3$
(h) $s(x) = x^3 - 2x^2 - 3x + 6$

7. Factor each of the following polynomials as completely as possible:
(a) $v(x) = x^2 - 5x + 2$
(b) $o(x) = 2x^4 + 3x^3 - x^2 - 3x - 1$ if -1 and $-\frac{1}{2}$ are zeros.
(c) $n(x) = 3x^4 + 3x^3 - 50x^2 + 47x - 15$ if 3 and -5 are zeros.
(d) $n(x) = x^3 - 7x - 6$ if -1, -2, $+3$ are zeros.
(e) $e(x) = x^4 + 2x^3 - 7x^2 - 20x - 12$ if -2 and $+3$ are zeros
(f) $u(x) = x^4 - 7x^2 - 6x$ if $+1$ is a zero.
(g) $m(x) = x^3 + 9x^2 + 26x + 24$ if -3 is a zero.
(h) $a(x) = 2x^3 - 2x^2 - 10x - 6$ if 3 is a zero.
(i) $n(x) = x^2 + 3x - 2$
(j) $n(x) = 3x^4 - 3x^3 - 39x^2 + 75x - 36$ if 3 and 1 are zeros.

Section 2.4b

8. Using the Factor Theorem and any other tools developed so far, find all the (real) zeros for each of the functions specified below.
   (a) $g(x) = x^3 - 3x^2 - 3x + 9$, and $\sqrt{3}$ is one zero.
   (b) $a(y) = x^3 + \sqrt{6}x^2 - 3x - 3\sqrt{6}$, and $\sqrt{3}$ is one zero.
   (c) $u(x) = x^3 - 2x^2 - 5x + 10$, and 2 is a zero.
   (d) $s(w) = 4w^4 - 8w^3 - 141w^2 + 2w + 35$ and 7, and $\frac{1}{2}$ are zeros.
   (e) $s(t) = t^4 - 8t^3 + 24t^2 - 32t + 16$, and 2 is a zero.
   (f) $r(x) = x^3 - 5x^2 + 8x - 4$, and 2 is a zero of multiplicity 2.
   (g) $i(x) = x^4 - 2x^3 - 3x^2 + 8x - 4$, and 1 is a zero of multiplicity 2.
   (h) $e(x) = x^4 - 2x^2 + 1$ if -1 and 1 are each zeros of multiplicity 2.
   (i) $m(a) = 2a^4 + 7a^3 + 9a^2 + 5a + 1$ if $-\frac{1}{2}$ and -1 are the only real zeros, of unknown multiplicities.
   (j) $a(r) = 12r^4 + 17r^2 - 5$ if $\pm\frac{1}{2}$ are zeros.
   (k) $n(s) = 3s^5 + s^4 - 3s^3 - s^2 - 18s - 6$ if $-\frac{1}{3}$ is one real zero.
   (l) $n(z) = (z^2 + 3z - 2)(z^2 - 4z + 3)$.

Section 2.5a

9. Do the following divisions using synthetic division:
   (a) $(3x^3 - 7x^2 + 4x - 2)/(x - 6)$
   (b) $(x^3 - 9x^2 + 26x - 24)/(x - 2)$
   (c) $(x^5 - 2x^3 + x - 1)/(x + 2)$
   (d) $(x^4 + x^2 + x + 1)/(x + 1)$
   (e) $(x^4 + 1)/(x + 1)$
   (f) $(x^3 - 8)/(x - 2)$
   (g) $(x^3 + 2x^2 + 2x + 4)/(x - 3)$
   (h) $(x^4 - 21x^2 - 100)/(x + 5)$
   (i) $(x^3 + 27)/(x + 3)$
   (j) $(2x^3 - 17x^2 + 47x - 42)/(x - 3)$
   (k) $(x^4 - 2x^3 - 71x^2 + 72x + 1260)/(x + 6)$
   (l) $(x^4 - 2x^3 - 71x^2 + 72x + 1260)/(x - 7)$

Section 2.5b

10. Evaluate the following functions for the given arguments using synthetic division.
   (a) $a(x) = 3x^3 + 2x^2 + x + 1$ for $x = 4$
   (b) $r(s) = s^4 - 2s^3 - 71s^2 + 72s + 1260$ for $s = 4$
   (c) $c(t) = t^4 - 8t^3 + 24t^2 - 32t + 16$ for $t = 2$
   (d) $h(w) = w^4 - 8w^3 + 24w^2 - 32w + 16$ for $w = 3$
   (e) $i(z) = z^5 - 25z^3 - 27z^2 + 675$ for $z = 4$
   (f) $m(y) = y^5 - 25y^3 - 27y^2 + 675$ for $y = 5$
   (g) $e(b) = b^5 - 25b^3 - 27b^2 + 675$ for $b = 6$
   (h) $d(v) = v^4 - 20v^2 + 64$ for $v = 3$
   (i) $e(b) = b^4 - 20b^2 + 64$ for $b = -4$
   (j) $s(u) = u^5 - 2u^3 + u - 1$ for $u = -2$

11. Each of the following rules specifies a function for which the given set is a subset of the set of all zeros for the function. Use synthetic division to produce a simpler function whose zeros are the zeros not yet known for the original function.
   (a) $m(w) = w^3 + w^2 - 4w + 6, \{-3\}$
   (b) $a(t) = t^3 + 3t^2 - 25t - 75, \{5\}$
   (c) $g(z) = z^4 + 3z^3 - 29z^2 - 75z + 100, \{-5, 1\}$
   (d) $i(x) = x^3 - 2x^2 - 5x + 10, \{2\}$
   (e) $c(m) = m^4 - 8m^3 + 24m^2 - 32m + 16, \{2\}$

Section 2.6

12. Draw precise graphs of the following functions over the indicated intervals.
   (a) $s(p) = p^3 - p^2$ on $[0, 1]$ (b) $e(f) = 2f^2 - f^3$ on $[0, 2]$
   (c) $t(i) = i^4 + i^3$ on $[-1, 0]$

13. Sketch graphs of the following functions, plotting only those points available by inspection of the rule.
   (a) $d(x) = (x - 1)^3(x + 2)(x - 3)$
   (b) $a(x) = (x - 1)(x + 2)^3(x - 3)^2$
   (c) $v(x) = (x - 1)(x + 2)(x - 3)$
   (d) $i(x) = (x - 1)^2(x + 2)^2(x - 3)$
   (e) $d(x) = (x - 1)(x + 2)^2(x - 3)^3$
   (f) $s(x) = (2x - 1)^2(x + 2)^2$
   (g) $c(x) = (3 - x)(x + 2)x^2$
   (h) $o(x) = (7 - 2x)(1 - 2x)^2(3 + 2x)$
   (i) $t(x) = (x + 1)^2(x + 2)^2(x + 3)^2$
   (j) $t(x) = x^5(x + 3)^3(x - 1)^2(x^2 - 1)$

Section 2.7

14. Notice that if $p(x) = x^2 - a$, the zeros of p are $\pm\sqrt{a}$. Thus one can approximate $\sqrt{a}$ by approximating the zeros of $x^2 - a$. Use successive halving through five cycles (not using a program) to find approximate values for

   (a) $\sqrt{2}$ (b) $\sqrt{3}$ (c) $\sqrt{5}$ (d) $\sqrt{\frac{1}{2}}$

   Let $l = 0, r = a + 1$ as starting values (i.e., stop at the fifth midpoint).

PROBLEMS

15. Prove that the parabola $y = ax^2 + bx + c$ can intersect the straight line $y = d$ in at the most two points.

16. For each of the functions specified by the given rules, find the set of all values for k that will result in the function's having at least one real zero.
   (a) $a(x) = 2x^2 + 4x + k$ (b) $b(y) = 3y^2 + ky + 1$
   (c) $c(z) = kz^2 - 2z + 7$ (d) $d(x) = 3x^2 - 2x + k$

(e) $e(z) = 5z^2 - kz + 2$ (f) $f(z) = kz^2 + kz - 4$
(g) $g(x) = kx^2 + kx + k$ (h) $h(x) = kx^2 + kx - k$

17. (a) If f is a quadratic function specified by $f(x) = x^2 + bx + c$, and f has two distinct real zeros r_1 and r_2, prove that (i) $r_1 + r_2 = -b$; (ii) $r_1 \cdot r_2 = c$.
    (b) What are the analogous results for $ax^2 + bx + c = g(x)$?

18. Using Problem 17, prove that if the quadratic function $g(x) = ax^2 + bx + c$ has real zeros z_1 and z_2, then the vertex occurs at $x = (z_1 + z_2)/2$, i.e., midway between the zeros.

19. Write out in detail the explanation for the formula to give the vertex of the parabola for $f(x) = ax^2 + bx + c = 0$ when $a < 0$.

20. (a) Why is the domain of function h of Example 4b just $[0, 2\frac{1}{2}]$?
    (b) Why must $4a^2$ be positive (i.e., why not either negative or zero) on p. 96.

21. For machine computation, Equation (6) occasionally is unsuitable because of the roundoff error of machine arithmetic. In most such cases, another form of this equation is adequate:

$$\frac{2c}{-b \pm \sqrt{b^2 - 4ac}}.$$

   (a) Prove that these two numbers really are zeros of Equation (1).
   (b) Derive these zeros by completing the square in another way than was done in Section 2.2b. [Hint: Divide by c and complete the square on the "right side" of the trinomial.]

22. Write a flowchart to find the zeros of any linear function, with the coefficients as input.

23. Write a flowchart to find all the real zeros for any quadratic function, with the coefficients as input. What does your program do if there is only one or no real zeros or if the function is really linear ($a = 0$)?

Problems 24–30 deal with a particular kind of linear function that is described in Problem 24.

24. A quantity q is said to vary directly with (or be directly proportional to) a quantity p, provided that q is a linear function of p with rule $q(p) = k \cdot p$ for some constant k (i.e., the y-intercept is zero). The constant k is called the constant of proportionality. Write the functional representation of the following relationships.
    (a) The pay a plumber earns varies directly with the number of hours he works.
    (b) The area of a circle varies directly with the square of its radius.
    (c) If an object travels at a constant speed, the distance traveled varies directly with the time traveled.

25. Prove that if x varies directly with y, and y varies directly with z, then x varies directly with z.

26. The amount of fertilizer needed for a field is directly proportional to the area of the field. If Calvin Butterball needs 300 lb of fertilizer to cover an area 10×1000 ft, how much will he need to cover a field that is 90×550 ft?

27. The surface area of a sphere varies directly with the square of the radius of the sphere. If a sphere of radius 3 in. had a surface area of 36π sq in., how large a radius is needed to have an area of 256π sq in.

28. The amount of the chemical skwart produced in a certain reaction varies directly with the amount of the chemical oiluol used in the reaction. If 36 g of skwart are produced when 28 g of oiluol are used,
    (a) What is the constant of proportionality?
    (b) How much oiluol is needed to produce 100 g of skwart?

29. Prove that if x varies directly with y, then y varies directly with x.

30. Prove that a linear function is additive if and only if it defines a direct variation. (A function f is said to be additive provided that $f(x+y) = f(x)+f(y)$ for all x and y in D_f.)

31. Discuss the inverse of
    (a) Linear functions   (b) Quadratic functions
    (Suggestion: Consider such topics as the specification of the inverse, whether the inverse is a function—sometimes, always, never?—and graphs.)

32. For each of the following claims, provide a proof or a counterexample.
    (a) The composition of two linear functions is always linear.
    (b) The composition of two quadratic functions is always quadratic.
    (c) Linear functions are always injective. (See Problems 92–95 of Chapter 1.)
    (d) Quadratic functions are never injective. (See Problems 92–95 of Chapter 1.)
    (e) Linear functions are never odd. (See Problems 89–91 of Chapter 1.)
    (f) Quadratic functions are always even. (See Problems 89–91 of Chapter 1.)

33. Write a flowchart that inputs two ordered pairs of numbers (c, d) and (r, s) and outputs the coefficients m and b of the equation $y = mx + b$ of the line determined by the points (c, d) and (r, s) in the plane. Be sure your flowchart handles all cases.

34. Write a flowchart that inputs three ordered pairs (x_1, y_1), (x_2, y_2) and (x_3, y_3) and outputs the coefficients A, B, and C of the equation $y = Ax^2 + Bx + C$ for the parabola determined by these three points. Be sure your flowchart is able to deal with any combination of three points.

35. We have used the word intersection in connection with sets and in connection with two (or more) graphs. Precisely and concisely explain why these two uses of the word are compatible.

36. Use the graphing program in Appendix C and/or hand graphing to test the effect of changing each coefficient in the defining rules for linear and quadratic functions. After gathering data,
    (a) Formulate as clearly as you can just what the effect is (conjecture a theorem if you will)
    (b) See if you can explain, justify, or prove your assertion more formally. In particular, you might try the following sets of examples.

(The asterisk indicates that these may help you decide if your conjecture is really independent of the other coefficient(s).)

(i) $f(x) = 2x - 10$
$f(x) = 2x - 5$
$f(x) = 2x - 1$
$f(x) = 2x - \frac{1}{2}$
$f(x) = 2x + 0$
$f(x) = 2x + 1$
$f(x) = 2x + 5$
*$f(x) = -x + 3$
*$f(x) = -x + 8$
*$f(x) = -x - 2$

(ii) $g(x) = -4x + 1$
$g(x) = -2x + 1$
$g(x) = -x + 1$
$g(x) = 0x + 1$
$g(x) = x/2 + 1$
$g(x) = 3x + 1$
$g(x) = 4x + 1$
*$g(x) = -2x + 5$
*$g(x) = -3x + 5$
*$g(x) = -x/2 + 5$

(iii) $q(x) = x^2 + x - 5$
$q(x) = x^2 + x - 3$
$q(x) = x^2 + x - 1$
$q(x) = x^2 + x - \frac{1}{3}$
$q(x) = x^2 + x + 0$
$q(x) = x^2 + x + \frac{1}{2}$
$q(x) = x^2 + x + 1$
$q(x) = x^2 + x + 2$
$q(x) = x^2 + x + 5$
*$q(x) = 3x^2 - 2x + 1$
*$q(x) = 3x^2 - 2x - 5$
*$q(x) = 3x^2 - 2x + 0$

(iv) $p(x) = x^2 + x - 5$
$p(x) = x^2 + 2x - 5$
$p(x) = x^2 + 5x - 5$
$p(x) = x^2 + x/2 - 5$
$p(x) = x^2 + x/3 - 5$
$p(x) = x^2 - 2x - 5$
$p(x) = x^2 - 3x - 5$
$p(x) = x^2 - x - 5$
$p(x) = x^2 - 5x - 5$
*$p(x) = 3x^2 - 2x + 1$
*$p(x) = 3x^2 - 3x + 1$
*$p(x) = 3x^2 - 0x + 1$

(v) $r(x) = x^2 + x - 5$
$r(x) = x^2/2 + x - 5$
$r(x) = -5x^2 + x - 5$
$r(x) = -6x^2 + x - 5$
$r(x) = 2x^2 + x - 5$
$r(x) = 5x^2 + x^2 - 5$
$r(x) = 3x^2 + x - 5$
$r(x) = -x^2/3 + x - 5$
$r(x) = -2x^2 + x - 5$
*$r(x) = 3x^2 - 2x + 1$
*$r(x) = 2x^2 - 2x + 1$
*$r(x) = 0x^2 - 2x + 1$

37. Prove the second half of the Rational Zero Theorem as suggested on p. 105.

38. Dissatisfied with being restricted to integer coefficients in the Rational Zero Theorem, Marvin Springspeed wants to broaden the theorem. Show him a way to apply the Rational Zero Theorem to polynomials with rational coefficients; i.e., carefully write how to do so. (Hint: If it is not obvious to you, try some simple examples for clues, e.g., $f(x) = x^2/4 + x + 1$ or $g(x) = x^2/2 - x/2 - 3$.)

39. A more restrictive but also useful relative of the Rational Zero Theorem results if one adds the requirement that the lead coefficient be 1. State this theorem in as neat a form as you can.

40. For each of the following claims, produce a proof or a counterexample. In each, suppose that p is any polynomial function of degree n, and q is any polynomial function of degree m.
    (a) $p \cdot q$ is always of degree $m + n$.
    (b) If $n = m$, then the degree of $p + q$ is always n.
    (c) $p \circ q$ is always of degree $m + n$.
41. Write a flowchart that will input the degree and coefficients for a polynomial, and then one (or more) arguments for the polynomial; and that will output the argument(s) and corresponding value(s) of the function (do not use synthetic division).
42. Write a flowchart that will input the degree and integer coefficients of a polynomial of degree ≤ 15. Then complete the following.
    (a) Output the set of rational numbers that contains all the rational zeros (allowing repetitions).
    (b) Repeat (a) but with no repetitions in the list.
    (c) Output just the rational zeros.
43. Is the Factor Theorem valid for linear functions? Explain carefully.
44. Prove that if $p = qr$, where p, q, and r are all polynomial functions, then the set of zeros of p is the union of the set of zeros for q and r. Thus any factorization of a polynomial can simplify the problem of finding zeros.
45. A result that is closely related to the Factor Theorem is the Remainder Theorem. This asserts that if p is a polynomial function and α is any real number, the remainder $r(x)$ left after division of $p(x)$ by $(x - \alpha)$ is a constant, and in fact is the number $p(\alpha)$. This is, $p(x) = (x - \alpha)q(x) + p(\alpha)$ for some $q(x)$. Prove this theorem.
46. Find an example to show that a polynomial of degree n may have fewer than n real zeros.
47. Prove that two functions with the same set of zeros (including multiplicities) need not be identical.
48. Devise a way to use synthetic division to find the quotient $(x^4 - 7x^2 + 120)/(x^2 - x - 6)$.

Problems 49 and 50 investigate another method of evaluating polynomial functions.

49. Another method of evaluating polynomial functions that is more efficient for machine use than the ordinary brute force way is called Horner's Method. It is built on the observation that

$$a_n x^n + a_{n-1} x^{n-1} + \cdots + a_1 x + a_0$$

can be written as

$$((\cdots [(a_n x + a_{n-1})x + a_{n-2}]x + a_{n-3} \cdots)x + a_1)x + a_0.$$

Compare the number of multiplications and divisions needed by each of these methods.
50. Write a flowchart to evaluate $p(\alpha)$ using Horner's Method, with input of $\alpha, n, a_n, a_{n-1}, \cdots, a_0$. Compare this with the flowchart of Problem 53.

51. Find all the real zeros of each of the following functions.
    (a) $t(u) = u^4 + 3u^3 - 24u^2 - 28u + 48$
    (b) $h(x) = x^3 + 3x^2 - 13x - 15$
    (c) $i(w) = w^4 + 7w^2 - 144$
    (d) $n(t) = 6t^3 + 25t^2 + 3t - 4$
    (e) $k(y) = 10y^3 - 57y^2 - 19y + 6$
    (f) $c(w) = 6w^4 + 21w^3 + 27w^2 + 15w + 3$
    (g) $l(a) = 2a^4 + a^3 - 11a^2 - 5a + 5$
    (h) $e(x) = x^5 - 10x^4 + 40x^3 - 80x^2 + 80x - 32$
    (i) $a(x) = x^4 - 2x^3 - 11x^2 + 12x + 36$
    (j) $r(b) = 4b^4 + 12b^3 - 13b^2 - 3b + 3$
    (k) $l(m) = m^4 - 8m^3 + 24m^2 - 32m + 16$
    (l) $y(t) = t^4 + 3t^2 - 18$
52. Derive an algorithm for synthetic division based on division by $x + \alpha$ (from the existing algorithm or from scratch) and check it on several examples.
53. The synthetic division algorithm is an efficient way for computer evaluation of the values of polynomial functions. Here the positions of the coefficients are indicated by subscript instead of by column. Modify the flowchart of Figure 2.9 so that it is suitable for computer use, with input of α, n, a_n, a_{n-1}, $\cdots$, a_0 and output of $p(\alpha)$. (This is really quite a simple flowchart.)
54. Draw graphs of the following polynomial functions.
    (a) $h(x) = x^3 + 6x^2 + 12x + 8$
    (b) $o(w) = w^3 + 4w^2 + 4w$
    (c) $m(x) = -x^3 - x^2 + 2x$
    (d) $e(z) = z^3 - 2z^2 + 2z - 4$
    (e) $o(y) = y^4 + 4y^3 + 6y^2 + 4y + 1$
    (f) $m(t) = t^4 - 5t^2 + 4$
    (g) $o(p) = -p^4 - 4p^3 + 6p^2 + 4p - 5$
    (h) $r(s) = s^4 - s^3 - s^2 - s - 2$
    (i) $p(y) = y^4 - 3y^3$
    (j) $h(x) = x^3 - 5x^2 - x + 5$
    (k) $i(b) = b^4 + 5b^2 + 4$
    (l) $s(t) = t^3 + t^2 + t + 1$
    (m) $m(a) = a^4 - 8a^3 + 24a^2 - 32a + 16$
55. Write a flowchart to give a table of values for a polynomial function (to use as an aid in graphing). Synthetic division should be used to compute values; the input should include the degree, coefficients, endpoints, and step size (see Problem 53).
56. Since two points determine a line, a variation on the method of successive halving is formed by taking in each cycle, not the midpoint of the interval $[l, r]$ but the point in $[l, r]$ where the line through $(l, f(l))$, $(r, f(r))$ intersects the x-axis. Write a flowchart for this using a particular polynomial function. This method is called the Method of False Position. (Hint: Use similar triangles.)
57. Suggest another method for approximating zeros of polynomials.

Write as complete a flowchart as you can, and explain any techniques you need in order to complete the flowchart.

58. Explain in more detail, the validity of replacing "Does Y have the same sign as $Y1$?" with "Is $Y \cdot Y1 > 0$?" on p. 128.

59. Examine the effect of composing polynomial functions with non-polynomial functions. Specifically, sketch graphs of each of the following sets of functions:
(a) If $f(x) = |x|$ and $g(x) = x^2 - 2x$, graph $f, g, f \circ g$, and $g \circ f$.
(b) If $f(x) = \lfloor x \rfloor$ and $g(x) = x^2 - 2x$, graph $f, g, f \circ g$ and $g \circ f$.
(c) If $f(x) = 1/x$ and $g(x) = x^2 - 2x$, graph $f, g, f \circ g$, and $g \circ f$.

60. In algebra, you learned to solve a few fairly simple types of inequalities. The tools developed in this chapter enable one to solve a much wider variety of inequalities, namely, polynomial inequalities, provided that one believes that no polynomial can change its sign without going through zero. This is a true assertion but requires learning a substantial amount of calculus to prove. Use this idea to help solve the following inequalities (it may help to think graphically, too).
(a) $x^3 - 2x^2 - x + 2 > 0$ (b) $x^3 - 2x^2 - x + 2 \leq 0$
(c) $x^4 - 5x^2 + 4 < 0$ (d) $x^4 - 5x^2 + 4 \geq 0$
(e) $x^4 - 2x^3 - 3x^2 + 4x + 4 \geq 0$ (f) $x^3 - 3x^2 - x + 8 < 5$
(g) $x^4 + 3x^3 - 3x^2 - 11x > 6$ (h) $2x^3 - 3x^2 + 4x + 4 \leq 7$

61. The Springspeed Bus Company will charter a bus for $3 per person for 40 persons but will reduce every fare by 5 cents for every rider in excess of 40 (thus 42 persons ride at $2.90 each). What number of people will produce the largest gross receipts for the company? (Trial and error will not be given any credit.)

62. Maynard Basil Holywright has purchased 200 ft of fence with which he wishes to make a divided rectangular enclosure, as sketched. What dimensions will provide him with the largest total area?

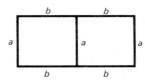

63. If an object thrown into the air has its altitude described by $a(t) = -16t^2 + 48t + 160$, where t is time measured in seconds:
(a) What is the maximum altitude achieved?
(b) When is the maximum altitude achieved?
(c) When does the object hit the ground?

64. The Leonia Drug Company is testing a new pain killer. They find that the only effect of any consequence to drivers is a reduction in visual perception, and that such perception is measured by $v(d) = -2d^3 - d^2 + 7d + 9$, where d is the dose of drug in milligrams and v is perception measured in p-units. If driving is safe only when v is at least 3 p-units, what doses of the drug will be safe for drivers to take?

65. Dampfire Research Associates has done some efficiency studies for

the Butterball Manufacturing Company and has found that when a
worker learns a new job on an assembly line, his job efficiency
increases at first as he becomes more familiar with the procedures,
but then decreases as boredom sets in. If, in particular, efficiency is
described by $e(w) = 36w - 4w^2 + 108$, where w is time in weeks,
(a) How long does it take for his efficiency to reach the zero level?
(b) When is he at his peak efficiency?

66. Among the products of the Butterball Manufacturing Company is
quality reproductions of great paintings. The company finds that for
most paintings, there is an initial cost of \$30,000 to set up production,
and a subsequent cost of \$3.50 in materials and labor to produce each
copy (shipping, etc., included).
(a) Write a rule describing the total cost of this operation as a function
of the number of copies produced. What kind of function is it?
(b) If the company expects to sell these items so that the wholesale
price will be 30% above cost, find the rule for the total revenue for
the operation (again as a function of the number produced). What
kind of function is this?
(c) Determine the minimum number that must be produced if the
wholesale price is to be no more than \$20 and the company is to
at least break even.

67. What is the smallest vertical distance between the parabola $y = x^2 + 4$
and the straight line $y = 2x - 3$?

68. Solve the following sets of simultaneous (nonlinear) equations.
(a) $8x^3 + 20x^2 - 4 = y$ (b) $y + 2x^2 - x = 3$
    $4x^3 - 10x^2 - 6x = 2y$ $4y - 8x = 5$
(c) $y^2 - 3x + 2y = 1$
    $4x - 2y^2 = -6$

69. A civil engineer Hengel Agehog is concerned with bridge safety. She
finds that due to a variety of factors, one index of stability is given by
$s(l) = 24l^3 - 38l^2 + 19l - 3$, where l is the length in miles. If s must
be positive for the bridge to be safe, what lengths of bridge are
possible?

70. Recreation director Marvin Springspeed is designing a swimming
facility, which will have a rectangular pool and a circular pool, the
diameter of the circular pool to equal the width of the rectangular pool.
If the combined perimeters are to be 200 m, what is the largest possible
area (total) for the pools?

71. What is the smallest possible value for the sum of a number (not
necessarily $\in \mathbb{Z}$) and its square?

72. If the sum of two numbers is to be 25, how large can their product be?
How small?

73. Holywright's Appliance Store sells their best electric dryer for \$290
and a comparable gas dryer for \$330. If the cost of electricity is
2.1 cents per load and the cost of gas is 1.58 cents per load, determine
the two functions, describing the overall cost of operation for each dryer.
Then use these functions to discuss briefly the economics involved
in choosing between the two models.

CHAPTER 3

CIRCULAR FUNCTIONS

3.1. INTRODUCTION

You have probably heard of trigonometric functions—perhaps you have even studied them. Today, however, mathematicians more frequently refer to these functions as the circular functions.

Why? Originally the trigonometric functions were seen primarily in limited practical settings, useful in solving problems of indirect measurement ("How far away is that star?" or "How wide is that aligator-infested swamp?"), which involved the use of similar triangles (trigonometry means triangle measurement). In time, however, these functions achieved more independent importance both in the physical sciences and mathematics so that their true nature as functions can be more fully grasped if one approaches them through the medium of a circle instead of a triangle.

In short, those who have previously met these functions as trigonometric functions should open their minds to a new approach, while those who have not yet had the pleasure of making their acquaintance need not feel disadvantaged since their role in trigonometry will be developed later in the chapter.

3.2. THE BASIC FUNCTIONS

3.2a. The Unit Circle and P(t)

DEFINITION 3.1
Unit Circle

Consider the circle of radius 1, which has its center at the origin (Figure 3.1). This circle is called the *unit circle*. Further, let us take special note of the point (1, 0), which is the intersection of the circle with the positive x-axis. One can imagine the circle to be a race track and the point (1, 0) to be the starting line. Next picture an infinitesimally small racing car that drives around the track in a counterclockwise direction. One could define a function that associated with the distance the car had traveled from the starting line, the point on the track where the car was then located. For example, in Figure 3.2, we would associate the length of arc AB with point B, the length of arc AC with point C, etc.

DEFINITION 3.2
P(t)

In more formal fashion, we define the function P with domain ℝ by P(t) = the point on the unit circle that is the terminal point of an arc t, units long, whose initial point is (1, 0). The positive direction for measuring the arc is counterclockwise. In this definition, it is important to recall that an arc may have a length greater than the circumference of the circle, i.e., the race car may make more than one trip around the track. Also, negative arc lengths mean that the arc is measured in the reverse (clockwise) direction (i.e., the car might go around the track backward).

Since the circumference of the unit circle is 2π,† several of the points

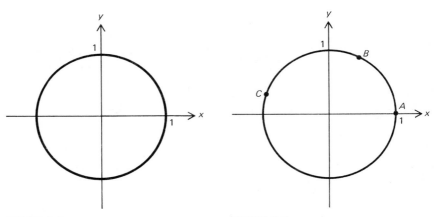

FIGURE 3-1   FIGURE 3-2

on the circle are readily identified as being associated with certain multiples of π through the function P just defined.

EXAMPLE 1

Figure 3.3(a) exhibits several such points. The distance around the circle from $P(0)$ to $P(\pi)$ is one-half the distance from $P(0)$ back to $P(2\pi)$ ($\frac{1}{2} \cdot 2\pi = \pi$). Thus $P(\pi)$ is one-half the way around the circle.

---

† The circumference of any circle is $C = 2\pi r$ or πD, and in this case, $r = 1$ so that $C = 2\pi \cdot 1$ or $\pi \cdot 2 = 2\pi$.

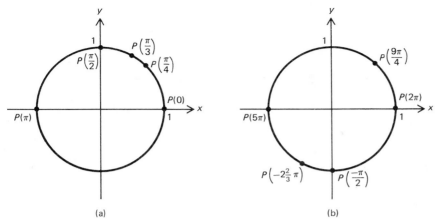

(a)　　　　　　　　　　　　　(b)

FIGURE 3-3

Similarly, $P(\pi/2)$ is $\frac{1}{4}$ of the way around the circle ($\frac{1}{4} \cdot 2\pi = \pi/2$); $P(\pi/4)$ is $\frac{1}{8}$ of the way around the circle ($\frac{1}{8} \cdot 2\pi = \pi/4$); $P(\pi/3)$ is $\frac{1}{6}$ of the way around the circle ($\frac{1}{6} \cdot 2\pi = \pi/3$), etc. Figure 3.3(b) exhibits $P(t)$ for several values of t that are negative and several values of t greater than 2π. You should verify the positions of these points in your own mind. Note that $9\pi/4 = 2\pi + \pi/4$; $5\pi = 2\pi + 2\pi + \pi$ (i.e., $2\frac{1}{2}$ revolutions); $-\pi/2$ is $\frac{1}{4}$ of a "backward revolution"; and $-2\frac{2}{3}\pi = -2\pi + -2\pi/3$.

To reiterate, $P(t)$ is the point on the unit circle corresponding to a distance of t units traveled along the circle, where the positive direction is counterclockwise.

The function P is sometimes called the winding function, since it can be thought of as winding a number line† around and around the circle, with the origin of the line at the point $(1, 0)$ on the circle. The point $P(t)$ is then the point of the circle "touching" the point t on the number line.

It might be helpful here to think of the circle as a spool of some type and the number line as a tape measure being wrapped around it. Then, for example, $P(3)$ is the point on the edge of the spool where the "3" of the tape touches the spool. See Figure 3.4.

3.2b. Sine and Cosine

The next step is to define two more functions, let us call them α and β. **SINE AND COSINE**
These functions have the set of points making up the unit circle as domain. Then α associates with each point on the circle its x-coordinate and β associates with each point on the circle its y-coordinate. Thus if Q is a

---

† A number line is just a straight line with units marked on it. Each point on the line corresponds to some real number, and vice versa.

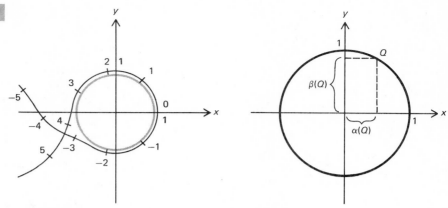

FIGURE 3-4   FIGURE 3-5

point on the unit circle,

$$\alpha(Q) = \text{the } x\text{-coordinate of } Q$$
$$\beta(Q) = \text{the } y\text{-coordinate of } Q.$$

See Figure 3.5.

Next we compose each of α and β with P to generate two new functions. Let us begin with α. You will remember that P consists of ordered pairs:

(arc length along the circle, point on the circle),

and we saw above that α consists of ordered pairs:

(point on circle, x-coordinate of point on circle).

The composition we shall perform is

$$(\alpha \circ P)(t) = \alpha(P(t)),$$

where t is an arc length. Recalling that P acts first, we get

$\alpha(P(t)) = \alpha(\text{the point on the unit circle } t \text{ units from } (1, 0))$
$= \text{the } x\text{-coordinate of the point on the unit circle } t \text{ units}$
  from $(1, 0)$.

Thus $\alpha \circ P$ produces x-coordinates of points on the unit circle. The domain of this new function is the set of arc lengths (which are real numbers); therefore, the domain is $\mathbb{R}$. The range, the set of x-coordinates, must be $[-1, 1]$. A similar argument works for $\beta \circ P$, and it would solidify your understanding if you tried to carry through a similar development yourself, either in your head or on paper.

The details in the above development are often hard to digest immediately. If you find this the case, perhaps a reference to Section 1.4a on functional composition, followed by a rereading of this subsection, would help.

The composite functions $\alpha \circ P$ and $\beta \circ P$ have special names, *cosine* and *sine*, much as we gave other special functions names in Section 1.2d. We shall see that it is, in fact, these composite functions and some of their siblings that will command our attention in this chapter. The names are

abbreviated cos and sin, respectively; a formal definition of each is then
given by the rules below:

$$\sin t = (\beta \circ P)(t)$$
$$\cos t = (\alpha \circ P)(t)$$

To reiterate, sine and cosine of t are, respectively, the y and x coordinates of the point $P(t)$ on the unit circle. They are best understood initially
as composite functions, as just presented. However, we shall quickly
view them as functions in their own right, just as one views $f(x) = x^2 + 3$
as a function in its own right instead of as the composition $h \circ g$ of the
simpler functions $g(x) = x^2$ and $h(x) = x + 3$.

Let us find a few values for sine and cosine (see Figure 3.6).

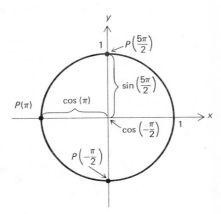

FIGURE 3-6

EXAMPLE 2

$\cos(\pi) = \alpha(P(t)) = x$-coordinate of the point on the circle π units
  from $(1, 0) = -1$.
$\cos(-\pi/2) = \alpha(P(-\pi/2)) = x$-coordinate of the point on the
  circle $-\pi/2$ units from $(1, 0) = 0$.
$\sin 5\pi/2 = \beta(P(5\pi/2)) = y$-coordinate of the point on the circle
  $5\pi/2 = 2\pi + \pi/2$ units from $(1, 0) = 1$.

It is also worth reemphasizing that sine and cosine have domain $\mathbb{R}$.
One speaks of $\sin(5)$ or $\cos(1.3)$ just as one does $p(5)$ or $f(1.3)$.

In dealing with literal arguments such as t and x, we often omit
parentheses and write $\sin t$ or $\cos x$ (instead of $\sin(t)$ and $\cos(x)$), when
no confusion is likely.

Finally, because of the convenient location of $P(t)$ for t being various
multiples and fractions of π, one frequently finds it useful to consider
$\sin t$ and $\cos t$ for these multiples of π even in preference to integer
and rational arguments. Also, one usually writes $\sin \pi$, not $\sin 3.14$; and
$\cos 2\pi$, not $\cos 6.28$, since 3.14 and 6.28 are only approximate values for
π and 2π and we wish to keep our statements precise.

3.3. EVALUATING THE BASIC CIRCULAR FUNCTIONS

3.3a. Computing Values

It is not hard to see that finding values for sine and cosine is not a simple task. This is in contrast to polynomial functions, which required only a finite number of additions, subtractions, multiplications, and divisions, which were specified in the algebraic rule for the function. Although as we noted in the previous section, a few values are easy to find, sin 1 or even cos $\pi/12$ are quite another matter. What we need is a simple computational rule that would generate the coordinates of $P(t)$ for any value of t we wished. Unfortunately, there are no such simple formulas.

A few more values than were found in Section 3.2b can be derived without undue fuss; they are values which occur frequently enough in applications to warrant such special attention. We look at them in the hopes of finding a clue as to how to proceed for other values.

The point $P(\pi/4)$ $(\pi/4 = \frac{1}{8} \cdot 2\pi)$ is the point of intersection of the unit circle and the line with equation $y = x$. This can be seen, since the line with equation $y = x$ bisects the arc in the first quadrant, and $P(\pi/2)$ is the point $(0, 1)$. See Figure 3.7. One way to find the x and y coordinates of $P(\pi/4)$ (i.e., to evaluate cos $\pi/4$ and sin $\pi/4$) is to begin with the equation of the circle. It is $x^2 + y^2 = 1$. Since $P(\pi/4)$ is on the line $y = x$ and on the circle $x^2 + y^2 = 1$, we need only find the coordinates of the intersection of these two curves. Thus we must solve

$$x^2 + y^2 = 1 \tag{1}$$

$$y = x \tag{2}$$

simultaneously. By substituting Equation (2) into Equation (1), we get

$$x^2 + x^2 = 1$$
$$2x^2 = 1$$
$$x^2 = \tfrac{1}{2}$$

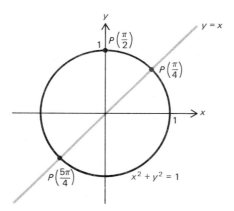

FIGURE 3-7

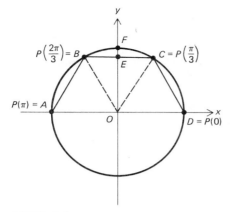

FIGURE 3-8

and

$$x = \pm\sqrt{\tfrac{1}{2}} = \pm\frac{\sqrt{1}}{\sqrt{2}} \cdot \frac{\sqrt{2}}{\sqrt{2}} = \pm\frac{\sqrt{2}}{2}.$$

The $+\sqrt{2}/2$ yields the point in the first quadrant, whereas the value $-\sqrt{2}/2$ yields the point of intersection, which occurs in the third quadrant. In other words, we have found that $\sin \pi/4 = \cos \pi/4 = \sqrt{\tfrac{1}{2}}$, or as it is usually written, $\sqrt{2}/2$, and in addition $\sin 5\pi/4 = \cos 5\pi/4 = -\sqrt{2}/2$.

This information can readily be used to generate the values for $\sin 3\pi/4$, $\cos 3\pi/4$, $\sin 7\pi/4$, $\cos 7\pi/4$, and indeed, $\sin t$ and $\cos t$ for any value of t for which $P(t)$ is coincident with $P(\pi/4)$, $P(3\pi/4)$, $P(5\pi/4)$, or $P(7\pi/4)$, but we defer this perfectly legitimate investigation for a few minutes, in the interests of finding other values less closely related to those we have just found.

Unfortunately, we do not know anything about the equations of other lines through particular points of interest, so it appears that we shall have to try a new strategy.

An essentially trial and error search for a new strategy might result in finding a geometric approach. For example, in the diagram of Figure 3.8, the triangles BOC and OCD are both found to be equilateral. Angle $BOC =$ angle $COD =$ angle AOB, since the arcs subtended are equal; then, since the measure of AOD is $180°$, each of the three is $60°$. Because OD, OC, OB, and OA are all radii, they are equal, and the three triangles are isosceles and congruent to one another. But the base angles are equal, and therefore are $60°$ also, and the triangles are, in fact, equilateral. Thus the length of BC is 1. Since arcs CF and FB have length $\pi/6$, both $\angle COF$ and $\angle BOF$ are $30°$ making line OF the perpendicular bisector of BC so that the length of EC is $\tfrac{1}{2}$. This is just the x-coordinate of $P(\pi/3)$; therefore, $\cos \pi/3 = \tfrac{1}{2}$. Since OC has length 1 and ED has length $\tfrac{1}{2}$, the Pythagorean Theorem may be applied to triangle OEC with the resulting observation that

$$OE = \sqrt{1^2 - (\tfrac{1}{2})^2} = \frac{\sqrt{3}}{2}.$$

3.3. *EVALUATING THE BASIC CIRCULAR FUNCTIONS* 157

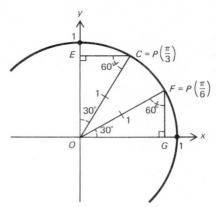

FIGURE 3-9

Hence, sin $\pi/3 = \sqrt{3}/2$.

Then, as shown in Figure 3.9, the congruence of triangles OEC and OGF can be established to derives in $\pi/6 = \frac{1}{2}$ and cos $\pi/6 = \sqrt{3}/2$. (Again, there are related values that can be obtained from these values, but we continue to defer such study.)

While it is evident that the industrious geometer could produce other values for sine and cosine, we have exhausted most of the useful supply of information here. This is primarily because other more general methods have been discovered for finding values for sine and cosine. Were such methods not in existence, then the more values we could compute using geometry, the better off we would be. The values tabulated below, which we have found so far, are among the most widely used values and the reader is strongly advised to learn them now.

	0	$\dfrac{\pi}{6}$	$\dfrac{\pi}{4}$	$\dfrac{\pi}{3}$	$\dfrac{\pi}{2}$
sin t	0	$\dfrac{1}{2}$	$\dfrac{\sqrt{2}}{2}$	$\dfrac{\sqrt{3}}{2}$	1
cos t	1	$\dfrac{\sqrt{3}}{2}$	$\dfrac{\sqrt{2}}{2}$	$\dfrac{1}{2}$	0

Two useful approaches remain for the evaluation of sin t and cos t. One provides the direct computation of (approximate) values, while the other enables one to find sin t and cos t for "related values" of t. Let us consider the former first.

NOTE The growth of any one area of mathematics does not occur in isolation from the growth in other areas, yet in learning mathematics one often tries to learn a substantial amount in one area without depending on the knowledge of some other area that had grown with the area being studied. We found fairly frequent reference to the calculus in our study of poly-

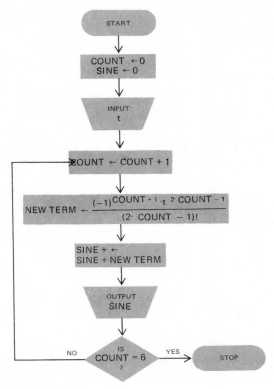

FIGURE 3-10

nomial functions, for example. It is perhaps unfortunate for our present discussion that the calculus is also tied up with the circular functions, because the most useful means of direct computation of values comes from the calculus.

Here, then, we only state the results and make a few observations. The formulas we need are

$$\sin t = t - \frac{t^3}{3!} + \frac{t^5}{5!} - \frac{t^7}{7!} + \cdots + \frac{(-1)^{n+1}t^{2n-1}}{(2n-1)!} + \cdots$$

$$\cos t = 1 - \frac{t^2}{2!} + \frac{t^4}{4!} - \frac{t^6}{6!} + \cdots + \frac{(-1)^{n+1}t^{2n-2}}{(2n-2)!} + \cdots$$

where each sum continues infinitely, the nth term being given by the general expression indicated in each formula.†

While exact values result only when meaning is given to the infinite sums (this is discussed in more detail in Section 5.3), useful approximate values are turned out quite painlessly, using only a few of the terms.

For example, the flowchart of Figure 3.10 is designed to find approximate values of sin t by using the first six terms in the sum. We shall trace

EXAMPLE 3

† The factorial function $N!$ is described on pp. 16–17.

3.3. *EVALUATING THE BASIC CIRCULAR FUNCTIONS* 159

this flowchart to approximate sin 2 in order to see more clearly how these formulas behave.

The first time through the loop having input 2 for t, COUNT is assigned the value 1 (we are taking care of the first term now) so NEWTERM is assigned the value

$$\frac{(-1)^2 2^1}{1!} = \frac{1 \cdot 2}{1} = 2,$$

whence sine assumes the value $0 + 2 = 2$, clearly a bad approximation, since sine can never be larger than 1. Because $1 \neq 6$, we loop back, COUNT becomes 2, and NEWTERM is then assigned the value

$$\frac{(-1)^3 \cdot 2^3}{3!} = \frac{-1 \cdot 8}{3 \cdot 2 \cdot 1} = -\frac{4}{3}.$$

Sine therefore becomes $2 - \frac{4}{3} = \frac{2}{3} = .666\overline{6}$, which is at least a possible value for sine. The next run through the loop sees

$$\text{NEWTERM} = \frac{(-1)^4 2^5}{5!} = \frac{32}{5 \cdot 4 \cdot 3 \cdot 2 \cdot 1} = \frac{4}{15}$$

and $\text{SINE} = \frac{2}{3} + \frac{4}{15} = \frac{14}{15} = 0.933\overline{3}$, while the fourth pass yields NEWTERM $= -\frac{8}{315}$ and $\text{SINE} = \frac{286}{315} \approx .907936$.

The fifth pass produces $\text{SINE} = 20624/22675 = .907347 \cdots$ and the sixth and final pass yields $\text{SINE} = .909296 \cdots$. The actual value of sin 2 is $.9092974 \cdots$ so that, in this case, six terms has done a reasonably good job.

Figure 3.11 exhibits the results of a short computer program to carry out similar computations for several other values of t. Problems 85–87 ask you to write some other flowcharts related to these formulas, but the results in Figure 3.11 should indicate that there is a problem deciding how many terms one needs to take in order to be assured of a good approximation. We leave such considerations to a course in numerical analysis, however, and note only that the library functions available in most programming languages make use of this approach to finding values of sine and cosine, and include sophisticated features for determining when an approximation is good enough.

3.3b. Related Values for t Greater than 2π and Periodicity

The other approach, that of finding related values, is within our grasp and is also of value. We consider one aspect of it here and another in Section 3.5. Some readers will have noticed already that $P(t)$ is the same as $P(t + 2\pi)$. This is because $P(t + 2\pi)$ is the point 2π units further around the unit circle than $P(t)$, and since the circumference of the circle is exactly 2π, the addition of 2π to the argument merely takes one once around the circle. [Remember, e.g., that $P(\pi/4) = P(9\pi/4) = P(2\pi + \pi/4)$.]

One can extend this reasoning to $P(t + k2\pi)$ where k is any integer, positive or negative, since any number of complete circuits around the

```
NUMBER
  OF          SIN  2.00            COS  2.00
TERMS

    1        2.000000000          1.000000000
    2         .6666666667          .3333333333
    3         .9333333333          .4666666667
    4         .9679365079          .4539682540
    5         .9893474427          .4546737213
    6         .9892961360          .4546480680


NUMBER
  OF          SIN -3.00            COS -3.00
TERMS

    1       -3.000000000          1.000000000
    2        1.500000000          -.5000000000
    3        -.5250000000          .1750000000
    4        -.9107142857E-01      .3035714286E-01
    5        -.1453125000          .4843750000E-01
    6        -.1408745942          .4695819805E-01
    7        -.1411306272          .4704354240E-01
    8        -.1411195543          .4703988479E-01
    9        -.1411200174          .4704000580E-01
   10        -.1411200079          .4704000262E-01


NUMBER
  OF          SIN 10.00            COS 10.00
TERMS

    1       10.00000000           1.000000000
    2     -156.6666667          -15.66666667
    3      676.6666667           67.66666667
    4    -1307.460317          -130.7460317
    5     1448.271605           144.8271605
    6    -1056.939234          -105.6939234
    7      548.9651501           54.89651501
    8     -215.7512231          -21.57512231
    9       65.39450233           6.539450233
   10      -16.81185014          -1.681185014
   11        2.761090926          .2761090926
   12       -1.107079245         -.1107079245
   13        -.4623842162        -.4623842162E-01
   14        -.5542211149        -.5542211149E-01
   15        -.5429111520        -.5429111520E-01
   16        -.5441272770        -.5441272770E-01
   17        -.5440121137        -.5440121137E-01
   18        -.5440217912        -.5440217912E-01
   19        -.5440210647        -.5440210647E-01
   20        -.5440211137        -.5440211137E-01
```

FIGURE 3.11

circle, in either direction, lands one where one started.† For this reason, the computation of values for sin t and cos t is trivial for values of t outside the interval $[0, 2\pi]$, provided that one knows sin t and cos t for all the values of t in $[0, 2\pi]$. Thus,

EXAMPLE 4

$$P\left(\frac{7\pi}{3}\right) = P\left(\frac{\pi}{3} + 2\pi\right) = P\left(\frac{\pi}{3} + (1)2\pi\right) = P\left(\frac{\pi}{3}\right)$$

$$P\left(\frac{-11\pi}{3}\right) = P\left(\frac{\pi}{3} - 4\pi\right) = P\left(\frac{\pi}{3} + (-2)2\pi\right) = P\left(\frac{\pi}{3}\right)$$

$$P\left(\frac{61\pi}{3}\right) = P\left(\frac{\pi}{3} + 20\pi\right) = P\left(\frac{\pi}{3} + 10 \cdot 2\pi\right) = P\left(\frac{\pi}{3}\right) \text{ etc.}$$

† This can be more formally proved using mathematical induction (see Problem 30 of Chapter 5), but meanwhile we trust the logic of the argument just given.

If you are skeptical that $P\left(\dfrac{-11\pi}{3}\right)$ really is the same point as $P(\pi/3)$, trace the path around the circle yourself by counting out 11 " $-\pi/3$ " 's one after the other.

DEFINITION 3.4
Periodic Function

In general, a function f for which there is some number p with the property $f(x+p) = f(x)$ for all $x \in D_f$ is called a *periodic function* because of the repetitive nature of the values assumed by the function. Since $P(t)$ is periodic, $[P(t) = P(t+2\pi), p = 2\pi$ in this case], it is quite clear to see that sin t and cos t are periodic : as $P(t)$ courses around and around the unit circle, the coordinates of $P(t)$ cycle and recycle through their respective values (see also Problem 32). For example, knowing that $P(\pi/3) = (\frac{1}{2}, \sqrt{3}/2)$ (i.e., cos $\pi/3 = \frac{1}{2}$, sin $\pi/3 = \sqrt{3}/2$), we can conclude that $P(7\pi/3) = (\frac{1}{2}, \sqrt{3}/2)$, and so sin $7\pi/3 = \sqrt{3}/2$ and cos $7\pi/3 = \frac{1}{2}$. Or in more general language, $(\cos(t+2\pi), \sin(t+2\pi)) = P(t+2\pi) = P(t) = (\cos t, \sin t)$, whence cos $(t+2\pi) = \cos t$ and $\sin(t+2\pi) = \sin t$, no matter what t is chosen.

DEFINITION 3.5
Period

If there is a smallest positive number p that satisfies the condition in the definition of periodic function, it is called the *period* of the function in question. Our point function P has period 2π, and a careful look at sine and cosine suggests that they, too, have period 2π. The only real question is whether or not there is a smaller positive number than 2π that will fill the role of p. Problem 33 provides an opportunity to prove that there is no such smaller number p for sine, or cosine, although a careful look at the circle should be quite convincing.

EXAMPLE 5

Some other examples of periodic functions include :
1. t, where $t(h)$ is the time of day h hours after 1 :00 AM January 1, 1900 ; $t(h) = t(h+24)$.
2. L, where $L(t)$ is the location (relative to the sun) of a comet in an elliptical orbit about the sun, $L(t) = L(t + \text{some specific time of return to the same point in orbit})$.
3. h, where $h(x) = x - \lfloor x \rfloor$ (see Example 9, p. 48), $h(x) = h(x+1)$.
4. p, where $p(r)$ is the vertical distance from the road of a point on a car tire after r revolutions of the wheel, $p(r) = p(r+1)$.

3.3c. Relating to Values of t Less than $\pi/2$

Returning to sine and cosine, one need consider only those values of t in $[0, 2\pi]$ in order to understand and be able to evaluate sine and cosine. But, in fact, one can reduce the domain still further if one is interested in finding values for sin t and cos t by referring to other known values. For any applications of the circular functions that require hand computations, this ability is of considerable value, and it will also help us to understand the behavior of the functions better.

Suppose that $P(t)$ is in the second quadrant (see Figure 3.12). We may construct the right triangle $P(t)SO$ shown by the solid lines in Figure 3.12 whose legs are sin t and cos t in length. We may then construct a right triangle in quadrant I, congruent to $P(t)SO$, as also shown in the

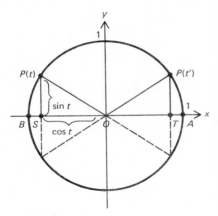

FIGURE 3-12

figure. The vertex of that triangle, which lies on the circle, is $P(t')$ for some real number t' in $(0, \pi/2)$. If we could easily find out what t' was, given t, and if we knew $\sin(t')$ and $\cos(t')$, then all that would be needed to get sin t and cos t is to adjust the sign of $\sin(t')$ and $\cos(t')$ appropriately. Furthermore, an analogous triangle exists if $P(t)$ is in quadrant III or IV, too, as shown by the dotted lines in Figure 3.12. The value of t' is all that is left to be found—and that is quite easy. It is a fact of plane geometry that equal central angles subtend equal arcs, and by the congruence of the triangles $P(t)SO$ and $P(t')TO$, and therefore the equality of $\angle P(t)OS$ and $\angle P(t')OT$, we deduce that the arc starting at $A = P(0) = (1, 0)$ and ending at $P(t')$ is the same length as the arc starting at $P(t)$ and ending at $(-1, 0) = B = P(\pi)$. Thus arc $\overset{\frown}{AP(t')} = \overset{\frown}{AB} - \overset{\frown}{AP(t)}$, or, in our more brief notation, $t' = \pi - t$. If t is in the third quadrant, t' would be $t - \pi$, while if t is in the fourth quadrant, $t' = 2\pi - t$. Be sure you understand each of these. It will be quite helpful later if you understand them rather than just memorize them.

In short, all one really needs to know is values for sin t and cos t for $t \in [0, \pi/2]$. All the remaining values can be deduced from these by adding or subtracting multiples of 2π and finding the number $t' \in [0, \pi/2]$, as described above. Thus tables of values for the circular functions rarely go beyond $t = \pi/2$.† Let us consider a few examples using Table 1, Appendix D.

Find sin 1.23. This can be read directly from the table as 0.9425.   **EXAMPLE 6**

Find cos 0.897. This can be approximated from Table 1, Appendix D,   **EXAMPLE 7**
using linear interpolation. (See Appendix A, Section A.21 if you are not sure about interpolation.) Since 0.897 is $\frac{7}{10}$ of the distance from 0.89 to 0.90, we assume cos 0.897 is $\frac{7}{10}$ of the distance from cos 0.89 to cos

---

† When we come to applying the circular functions to angles, we shall find tables of values constructed in an even more compact way. For the time being, though, one should use a table that lists values for t as a real number or in radians (we shall explain this terminology later, too) such as Table 1, Appendix D.

0.90. That is, cos $0.897 \approx \cos 0.89 + \frac{7}{10}(\cos 0.90 - \cos 0.89) \approx 0.6294$
$+ \frac{7}{10}(-0.0078) \approx 0.6239$.

EXAMPLE 8   Find sin 2.0. Then $P(2)$ is in the second quadrant (since $\pi/2 \approx 1.57 <$
$2 < \pi \approx 3.14$) so $t' = \pi - t \approx 3.14 - 2 = 1.14$ and so sin $2.0 \approx \sin 1.14 \approx$
0.9086. Note that the sign of sin 2 is the same as the sign of sin 1.14,
since the y-coordinates of $P(2)$ and $P(1.14)$ are of the same sign.

EXAMPLE 9   Find cos 4.0. Then $P(4)$ is in quadrant III, so $t' \approx 4 - 3.14 = 0.86$.
Cos $0.86 \approx 0.6524$ so that cos $4.0 \approx -0.6524$, since the x-coordinates
of P(4) and P(0.86) are opposite in sign.

EXAMPLE 10   Find sin 12.0. Since $12 > 2\pi$, we subtract as many multiples of 2π
as needed to reduce the problem to sin t for $t \in [0, 2\pi)$. Since $2\pi \approx 6.28$,
we need only subtract one such amount. In other words, sin $12.0 \approx$,
$\sin(12.0 - 6.28) = \sin 5.72$. Further, $P(5.72)$ is in quadrant IV (i.e.

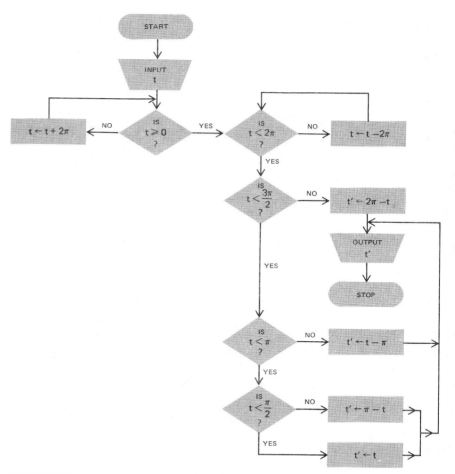

FIGURE 3-13

between $3\pi/2 \approx 4.71$ and $2\pi \approx 6.28$), so $t' = 2\pi - t \approx 6.28 - 5.72 = 0.56$ and so $\sin 12.0 \approx -\sin 0.56 \approx -0.5312$.

It should be emphasized that while we can, in theory find exact, values, the tabulated values are only approximate to start with, and the use of an approximation for π in finding t' introduces additional error.

As a final note, we see that by means of reference to the "first quadrant values" of t, the exact coordinates of $P(2\pi/3), P(3\pi/4), P(5\pi/6), P(7\pi/6), P(5\pi/4), P(4\pi/3), P(5\pi/3), P(7\pi/4),$ and $P(11\pi/6)$ are easily found. Therefore, one needs only to remember the exact values previously tabulated, and our promise to revisit these values has now been kept.

Thus to find the exact value of $\sin 7\pi/6$, we proceed as follows. Since $P(7\pi/6)$ is in the third quadrant, $t' = 7\pi/6 - \pi = \pi/6$ so that $\sin 7\pi/6 = -\sin \pi/6 = -\frac{1}{2}$.

EXAMPLE 11

Or, if we wanted to find $\cos 11\pi/4$, we would first subtract 2π : $\cos 11\pi/4 = \cos(11\pi/4 - 2\pi) = \cos 3\pi/4$. But $P(3\pi/4)$ is in the second quadrant, so $\cos 3\pi/4 = \cos(\pi - 3\pi/4) = \cos \pi/4$ except for sign. Since cosine is negative in the second quadrant, $\cos 3\pi/4 = \cos 11\pi/4 = -\sqrt{2}/2$.

EXAMPLE 12

The finding of values of t' may be conveniently summarized in the flowchart of Figure 3.13. Be sure you understand this procedure well enough so that you could rederive (not merely reproduce) this flowchart to help explain the process to a friend.

3.4. GRAPHS AND SOME OTHER CIRCULAR FUNCTIONS

3.4a. Graphs of Sine and Cosine

Now that we have some ways of finding values of $\sin t$ and $\cos t$, we are in a position to try to graph the functions, and we can profitably use the relationships developed in the last section. The reader is encouraged to get or make a full-sized piece of graph paper and construct the graph of cosine as we do the graph for sine below. Your active participation in this project will benefit your understanding.

GRAPHS OF SINE AND COSINE

While we could use the library functions with a graphing program (such as the one in Appendix C), there is much understanding of the functions to be gained by following through an analytic investigation, together with the satisfaction derived from seeing the graphs unfold before us. In addition, keep your eyes open to general strategies and the way previously learned information is used.

Before setting up axes, we should check the domain and range to settle on the placement of axes and which scales to use. The domain for sine is $\mathbb{R}$ and the range is $[-1, 1]$. (Do you remember why?) Furthermore, we have seen that the sine function is periodic with period 2π, and the repetitive nature of periodic functions suggests that the graph will repeat. Therefore, we pay particular attention to a length of one period and then

FIGURE 3-14

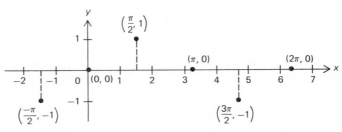

FIGURE 3-15

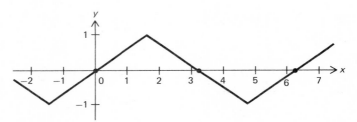

FIGURE 3-16

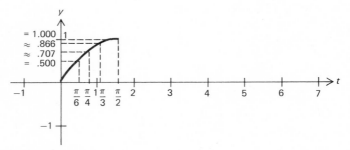

FIGURE 3-17

extend the graph. This translates graphically into including at least $[0, 2\pi]$ on the x-axis and $[-1, 1]$ on the y-axis, which results in Figure 3.14.

Using the first values we found in Section 3.2b, we have the situation depicted in Figure 3.15 for sine. From this information alone, one would have to be lucky to guess the right shape—Figure 3.16 exhibits the most obvious guess, which is wrong.

As soon as we recognize the need for more points, we should recall the apparent special value of the arguments in $(0, \pi/2)$, since other values can be referred back to these points, and then try to use the relationships of the last section. Such a plotting yields Figure 3.17 for sine.

Next observe that as t moves from $\pi/2$ to π, t' moves from $\pi/2$ back to 0, so a mirroring of the first part of the graph occurs for $t \in (\pi/2, \pi)$. This produces Figure 3.18 without more computation. (You might check a couple of values such as $3\pi/4$ to be sure.)

As $P(t)$ moves through the third quadrant, $P(t')$ moves back along the arc from $(1, 0)$ to $(0, 1)$, but the sign of sine is now negative. So the graph is the same shape as the first segment, but the values are negative. The result is shown in Figure 3.19. Finally, the period is completed as $P(t')$ moves along the arc from $(0, -1)$ to $(1, 0)$ yielding Figure 3.20. (Again you might check such values as $7\pi/4$, $7\pi/6$, or $5\pi/3$ to be sure we have not made an error.)

By virtue of the periodicity, this basic pattern is repeated over and over

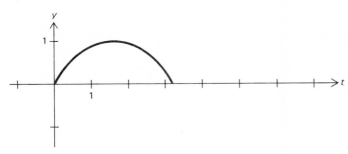

FIGURE 3-18

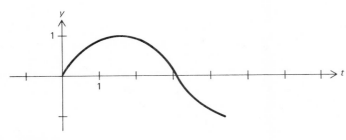

FIGURE 3-19

3.4. *GRAPHS AND SOME OTHER CIRCULAR FUNCTIONS* 167

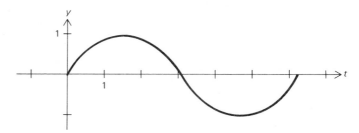

FIGURE 3-20

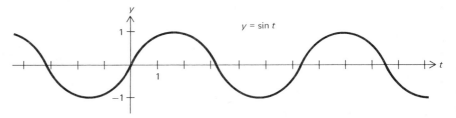

FIGURE 3-21

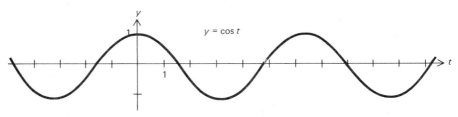

FIGURE 3-22

in each direction, finally to produce the graph of sine shown in Figure 3.21. Note that we needed to plot points only for $t \in [0, \pi/2]$.

The corresponding graph for cosine should look like Figure 3.22. (Is this the way your completed graph looks?)

Because of the importance of zeros for functions, the x-axis is often marked off in multiples of π rather than integers when the circular functions are being graphed. This produces graphs such as Figure 3.23 and 3.24.

The maximum and minimum points in each graph are easily found, unlike our experience with polynomials, because we know that the ordinates of all maximum and minimum points must be 1 and -1, respectively. A quick look at the unit circle establishes the appropriate values of the abscissa (x-values). For sine, maximums occur for those values of t for which $P(t) = (0, 1)$, namely, $\{t \mid t = \pi/2 + k2\pi$ for some $k \in \mathbb{Z}\}$, i.e.,

$$t \in \left\{ \cdots, \frac{\pi}{2} + (-2)2\pi, \frac{\pi}{2} + (-1)2\pi, \frac{\pi}{2} + (0)2\pi, \frac{\pi}{2} + (1)2\pi, \frac{\pi}{2} + (2)2\pi, \cdots \right\}$$

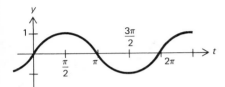

FIGURE 3-23

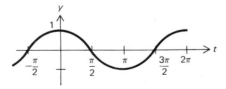

FIGURE 3-24

or

$$t \in \left\{ \cdots, \frac{\pi}{2} - 4\pi, \frac{\pi}{2} - 2\pi, \frac{\pi}{2}, \frac{\pi}{2} + 2\pi, \frac{\pi}{2} + 4\pi, \cdots \right\}$$

or

$$t \in \left\{ \cdots, \frac{-7\pi}{2}, \frac{-3\pi}{2}, \frac{\pi}{2}, \frac{5\pi}{2}, \frac{9\pi}{2}, \cdots \right\}.$$

Minimums occur for precisely those values of t for which $P(t) = (0, -1)$, i.e., $\{t \mid t = 3\pi/2 + k2\pi$ for some $k \in \mathbb{Z}\}$, i.e.,

$$t \in \left\{ \cdots, \frac{3\pi}{2} + (-2)2\pi, \frac{3\pi}{2} + (-1)2\pi, \right.$$

$$\left. \frac{3\pi}{2} + (0)2\pi, \frac{3\pi}{2} + (1)2\pi, \frac{3\pi}{2} + (2)2\pi, \cdots \right\}$$

or

$$t \in \left\{ \cdots, \frac{3\pi}{2} - 4\pi, \frac{3\pi}{2} - 2\pi, \frac{3\pi}{2}, \frac{3\pi}{2} + 2\pi, \frac{3\pi}{2} + 4\pi, \cdots \right\}$$

or

$$t \in \left\{ \cdots, \frac{-5\pi}{2}, \frac{-\pi}{2}, \frac{3\pi}{2}, \frac{7\pi}{2}, \frac{11\pi}{2}, \cdots \right\}.$$

Similarly, for cosine, the maximums occur for $t \in \{t \mid t = k2\pi$ for some $k \in \mathbb{Z}\}$ and minimums for $t \in \{t \mid t = \pi + k2\pi$ for some $k \in \mathbb{Z}\}$.

We have mentioned zeros; let us formalize them. The set of zeros for sine is $\{t \mid t = k\pi$ for some $k \in \mathbb{Z}\}$, i.e.,

$$t \in \{ \cdots, -3\pi, -2\pi, -\pi, 0, \pi, 2\pi, 3\pi, \cdots \}$$

and for cosine is

$$\{t \mid t = \pi/2 + k\pi \text{ for some } k \in \mathbb{Z}\},$$

i.e.,

$$t \in \left\{ \cdots, \frac{-5\pi}{2}, \frac{-3\pi}{2}, \frac{-\pi}{2}, \frac{\pi}{2}, \frac{3\pi}{2}, \frac{5\pi}{2}, \cdots \right\}$$

as can be seen by again looking at the unit circle.

By now you should have a reasonably good picture of the functions sine and cosine even though there are more properties to be investigated, and even though we have not yet looked into any applications. But before we turn to further investigations or applications, let us take some time to meet some siblings of sine and cosine.

3.4b. Definition and Evaluation of the Other Circular Functions

**THE OTHER CIRCU-
LAR FUNCTIONS**

The most important of the remaining circular functions is called the tangent function, abbreviated tan, and is defined by:

DEFINITION 3.6

Tangent

$$\tan t = \frac{\sin t}{\cos t}$$

or

$$\tan t = \frac{y}{x}$$

where $P(t)$ has coordinates (x, y). (It should be quite clear to you that these are identical definitions.)

Then there are three other circular functions, of lesser importance but not to be ignored: the secant, cosecant, and cotangent functions, which are denoted and defined, respectively, by

DEFINITION 3.7

**Secant, Cosecant,
Cotangent**

$$\sec t = \frac{1}{\cos t}$$

$$\csc t = \frac{1}{\sin t}$$

$$\cot t = \frac{1}{\tan t}.$$

(cot t is also denoted ctn t in some books, and is usefully recognized as $(\cos t)/(\sin t)$.)

All four of these functions can be viewed as being built from sine and cosine (and the constant function $g(x) = 1$) by the division of functions. Recall that it was important when dividing functions to exclude from the domain of the quotient function, any zeros of the function in the denominator. Thus tangent and secant have domain being all of $\mathbb{R}$ except $\{t \mid t = \pi/2 + k\pi$ for some $k \in \mathbb{Z}\}$, i.e., the zeros of cosine, while cosecant and cotangent have domain being all of $\mathbb{R}$ except $\{t \mid t = k\pi$ for some $k \in \mathbb{Z}\}$, i.e., the zeros of sine.

NOTE

While it might have been excusable to define the sine and cosine functions as we have in a formal way without any practical motivation, many readers may have little patience left for playing the juggling game that we seem to be playing now with these two functions. However, we shall soon see that the tangent function is a workhorse, and the other three can at least be well used to avoid the use of fractions in expressions that involve sine, cosine, and tangent.

A first nontaxing observation is that these four new functions are periodic with periods at least no larger than 2π.

For example,

EXAMPLE 13

$$\tan(t + 2\pi) = \frac{\sin(t + 2\pi)}{\cos(t + 2\pi)} = \frac{\sin t}{\cos t} = \tan t$$

(by virtue of the periodicity of sine and cosine). Most people's intuition would assert at this point that surely no positive number less than 2π could serve as the period for any of these functions, since no smaller number worked for either sine or cosine from which these are constructed. But ah, the fraility of intuitive logic. This argument is, in fact, not sound.

Indeed, one of the hard jobs in developing mathematical proficiency is learning of what a proof really consists and developing a sufficient grasp of logic to see when an argument really is valid. The difficulty is attested to by the fact that mathematicians not infrequently claim to have proved thus-and-such a result only to find out later that the result is false and their original logic was therefore faulty.

In this particular instance, let us suppose that no one pointed out the fallacy of the logic and let us see how and when we might extricate ourselves from the misconception.

Evaluations of all four functions depend on the values of sine and cosine; since we know how to find the values of sine and cosine, we can find the values of the other four. For example, the special values learned for sine and cosine produce the following table of values, which include the four new functions.

	0	$\frac{\pi}{6}$	$\frac{\pi}{4}$	$\frac{\pi}{3}$	$\frac{\pi}{2}$
$\sin t$	0	$1/2$	$\sqrt{2}/2$	$\sqrt{3}/2$	1
$\cos t$	1	$\sqrt{3}/2$	$\sqrt{2}/2$	$1/2$	0
$\tan t$	0	$\sqrt{3}/3$	1	$\sqrt{3}$	Undefined
$\sec t$	1	$2\sqrt{3}/3$	$\sqrt{2}$	2	Undefined
$\csc t$	Undefined	2	$\sqrt{2}$	$2\sqrt{3}/3$	1
$\cot t$	Undefined	$\sqrt{3}$	1	$\sqrt{3}/3$	0

The computations involved in arriving at the values in the above table are not difficult.

For example,

EXAMPLE 14

$$\tan \frac{\pi}{4} = \frac{\sin \pi/4}{\cos \pi/4} = \frac{\sqrt{2}/2}{\sqrt{2}/2} = \frac{\sqrt{2}}{2} \cdot \frac{2}{\sqrt{2}} = 1.$$

Also $\csc 0 = 1/\sin 0 = 1/0$ which is undefined, since it involves an attempt to divide by zero (see Appendix A, Section A.19 on division by zero if you are not sure of this). As a third example,

$$\sec \frac{\pi}{3} = \frac{1}{\cos \pi/3} = 1 \bigg/ \left(\frac{1}{2}\right) = 2.$$

NOTE

Memorization of the above table is strongly advised. While you could always derive the values in it if you knew the sine and cosine values, you will save yourself much time by learning this table now. There are a large number of facts in future sections which are best committed to memory, so do not let all the memorization "pile up." In general, memorizing facts as you go will not only speed up your homework, but will help cement the information in place as you make use of it.

As before, values of t not in $[0, \pi/2]$ can be related to a $t' \in [0, \pi/2]$. We shall use several examples to illustrate the techniques involved.

EXAMPLE 15

Find cot 5. $\cot 5 = \cos 5/\sin 5$, and since $P(5)$ is in the fourth quadrant, $\cot 5 \approx \cos(6.28 - 5)/-\sin(6.28 - 5) = \cos 1.28/-\sin 1.28 \approx$ $0.2867/-0.9580 \approx -0.2993$.

Obviously, computations such as 0.2867/0.9580 are not ones that we want to perform frequently. Fortunately, an obliging computer has already done this for us, so we need only look in Table 1, Appendix D, for values of tangent, cotangent, secant, and cosecant. In other words, the above example becomes $\cot 5 = \cos 5/\sin 5 \approx \cos 1.28/-\sin 1.28 = -\cot 1.28 \approx -0.2993$.

We still have a problem, though—that of sign. Let us see if we can find some general rules for dealing with the sign problem for $t \in [0, 2\pi]$. (Anything greater than 2π can easily be reduced back to the interval $[0, 2\pi]$.)

Let us begin by looking at Figure 3.25, which shows the signs of sine and cosine in all four quadrants (remember that the sine is the y-coordinate and the cosine is the x-coordinate):

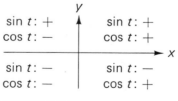

FIGURE 3.25

Since the other four functions are based on sine and cosine, we can find the signs of each of them, as shown in Figure 3.26.

The diagram in Figure 3.26 is much too complex to be of much help, so we list just the functions that are positive in each quadrant. See Figure 3.27.

$\begin{cases} \sin t: + \\ \cos t: - \\ \tan t = \sin t/\cos t: +/- \to - \end{cases}$
$\begin{cases} \csc t = 1/\sin t: \qquad 1/+ \to + \\ \sec t = 1/\cos t: \qquad 1/- \to - \\ \cot t = \cos t/\sin t: -/+ \to - \end{cases}$

$\begin{cases} \sin t: + \\ \cos t: + \\ \tan t = \sin t/\cos t: +/+ \to + \end{cases}$
$\begin{cases} \csc t = 1/\sin t: \qquad 1/+ \to + \\ \sec t = 1/\cos t: \qquad 1/+ \to + \\ \cot t = \cos t/\sin t: +/+ \to + \end{cases}$

$\begin{cases} \sin t: - \\ \cos t: - \\ \tan t = \sin t/\cos t: -/- \to + \end{cases}$
$\begin{cases} \csc t = 1/\sin t: \qquad 1/- \to - \\ \sec t = 1/\cos t: \qquad 1/- \to - \\ \cot t = \cos t/\sin t: -/- \to + \end{cases}$

$\begin{cases} \sin t: - \\ \cos t: + \\ \tan t = \sin t/\cos t: -/+ \to - \end{cases}$
$\begin{cases} \csc t = 1/\sin t: \qquad 1/- \to - \\ \sec t = 1/\cos t: \qquad 1/+ \to + \\ \cot t = \cos t/\sin t: +/- \to - \end{cases}$

FIGURE 3.26

$\begin{matrix} \sin t \\ \csc t \end{matrix}$ $\begin{cases} \sin t \\ \cos t \\ \tan t \\ \csc t \\ \sec t \\ \cot t \end{cases}$ i.e., all six functions

$\begin{matrix} \tan t \\ \cot t \end{matrix}$ $\begin{matrix} \cos t \\ \sec t \end{matrix}$

FIGURE 3.27

Notice that whenever a function is positive, its reciprocal function is also positive; therefore, all we really need to remember is that if one from each pair of functions in quadrants II–IV is positive, its reciprocal will also be positive. This gives us Figure 3.28.

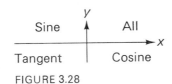

Sine | All
Tangent | Cosine

FIGURE 3.28

The following mnemonic (memory aid) might help you:

All Students Take Calculus.

Quadrant →	I	II	III	IV
	↑	↑	↑	↑
	all +	sin +	tan +	cos +.

If we add the rules for relating a number in quadrants II–IV back to the

3.4. *GRAPHS AND SOME OTHER CIRCULAR FUNCTIONS* 173

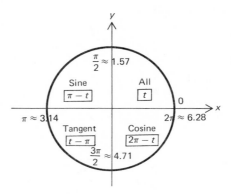

FIGURE 3-29

first quadrant, we have a summary diagram (Figure 3.29) that will help us find values of all six functions.

Now let us return to some examples and use the tools we have developed.

EXAMPLE 16   Find tan 4. This is a third quadrant value ($3.14 < 4 < 4.71$). A tangent in the third quadrant is positive, so $\tan 4 \approx +\tan(4 - 3.14) = \tan 0.86 \approx 1.162$ from Table 1, Appendix D.

EXAMPLE 17   Find sec 2.5. Here $P(2.5)$ is in the second quadrant and the secant is negative. Thus $\sec 2.5 \approx -\sec(3.14 - 2.5) = -\sec 0.64 \approx -1.247$ from Table 1, Appendix D.

EXAMPLE 18   Find csc $11\pi/6$. Here $P(11\pi/6)$ is in the fourth quadrant ($3\pi/2 < 11\pi/6 < 2\pi$) and the cosecant is negative; therefore, $\csc 11\pi/6 = -\csc(2\pi - 11\pi/6) = -\csc \pi/6 = -2$, since we know that $\csc \pi/6 = 2$.

EXAMPLE 19   Find tan 17. Since 17 is greater than $2\pi \approx 6.28$, the first step is to reduce tan 17 to the interval $[0, 2\pi]$: $\tan 17 \approx \tan(17 - 6.28) = \tan 10.72 = \tan(10.72 - 6.28) \approx \tan 4.44$. This puts us in the third quadrant, and tangent is positive there. Thus $\tan 17 \approx \tan 4.44 \approx \tan(4.44 - 3.14) = \tan 1.30 \approx 3.602$ from Table 1, Appendix D.

EXAMPLE 20   Find sec $13\pi/6$. Here $13\pi/6 > 2\pi$, so that $\sec 13\pi/6 = \sec(13\pi/6 - 2\pi) = \sec \pi/6$, and we know that $\sec \pi/6 = 2\sqrt{3}/3$.

3.4c.   Graphs of Other Circular Functions

Now we turn to the graphs of tangent, secant, cosecant, and cotangent. In particular, let us start with cosecant. We have already noted that the domain of cosecant is all of $\mathbb{R}$ except $\{k\pi \mid k \in \mathbb{Z}\}$ so that at each of these nonexistent abscissas, we draw a vertical dotted line as a reminder that the graph never intersects that line. Since cosecant $= 1/\text{sine}$, the range of cosecant $= \{y \mid y = 1/z \text{ for some } z \in [-1, 1]\} = (-\infty, -1] \cup [1, \infty)$. (Are you sure you see why?)

So far, then, we have reason to set up axes such as in Figure 3.30.

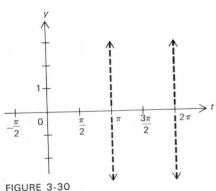

FIGURE 3-30

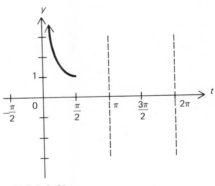

FIGURE 3-31

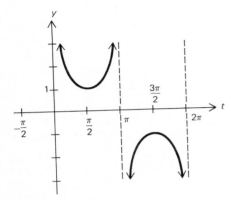

FIGURE 3-32

(We have chosen to mark the t-axis in multiples of π for accuracy.) A careful plotting of points for $t \in (0, \pi/2]$ yields Figure 3.31. As t gets closer to zero from the right, sine t gets closer to zero, so that $1/\sin t$ becomes very large. Indeed, there is no bound—$1/\sin t$ can be made as large as one pleases by choosing t sufficiently close to zero (i.e., sufficiently small). Thus the graph gets steeper and steeper, never crossing the y-axis. The quadrant-by-quadrant reference back to the first quadrant values produces Figure 3.32. (Verify this for yourself.)

From a glance at this graph, it is clear that the period of cosecant is, in fact, no less than 2π and so must actually be 2π; the extension of the graph by periodicity yields Figure 3.33 as the graph of cosecant. A similar analysis for secant produces Figure 3.34.

One might be tempted to think that each section of these graphs is a parabola, since it is a curve which opens upward or downward, and does so symmetrically. This is not the case at all—the curvature of *any* parabola is quite different from that of the sections of secant and cosecant. If you are skeptical, see Problem 35.

3.4. *GRAPHS AND SOME OTHER CIRCULAR FUNCTIONS* 175

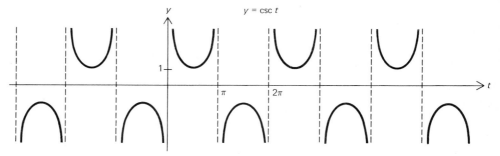

FIGURE 3-33

The graphs of tangent and cotangent are harder to obtain, since both sine and cosine come into play and the results of their interactions are harder to see. Let us consider the tangent first. It is undefined for each $x \in \{\pi/2 + k2\pi \mid k \in \mathbb{Z}\}$; so, we again put vertical dotted lines in the graph to indicate this fact. The range is less obvious and is perhaps best approached through an example.

Remembering the special values tabulated in the preceding subsection, we see that tan t increases in value from 0 at $t = 0$ through $1/\sqrt{3}(\approx 0.57)$ at $t = \pi/6$, 1 at $t = \pi/4$, and $\sqrt{3}(\approx 1.73)$ at $t = \pi/3$. Referring to the definition of tan t as sin t/cos t, we can see that as t moves from 0 to $\pi/4$, sin t is increasing in value and cos t is decreasing so that the fraction sin t/cos t is increasing, since both decreasing positive denominators and increasing positive numerators have the effect of making the fraction value larger positive. (Try a few examples to be sure you believe this.) Since the increase and decrease of sin t and cos t are smooth, we expect no missed values for the tangent between 0 and 1. As t proceeds from $\pi/4$ toward $\pi/2$, sin t continues to increase, while cos t continues to decrease so tan t continues to increase, and does so without any upper bound to its values. The range, therefore, is seen to include at least $[0, \infty)$.

Are there any negative values for tan t? There would be only if sin t and cos t had opposite signs. Because this happens for $t \in (\pi/2, \pi)$ (see Figure 3.29), there are at least some negative values. Since as t goes from $\pi/2$ to π, sin t goes through the same values but in reverse order as it did as t went

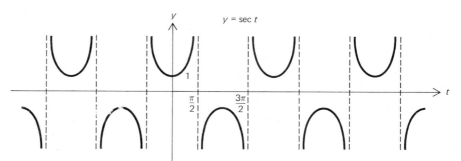

FIGURE 3-34

176   CIRCULAR FUNCTIONS

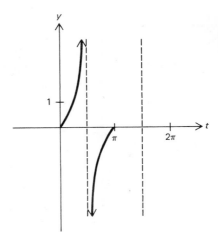

FIGURE 3-35

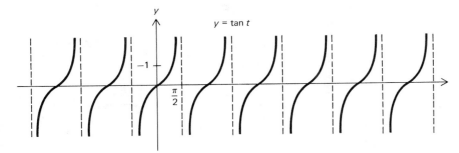

FIGURE 3-36

from 0 to $\pi/2$, and cos t goes through the negatives of the values it had between 0 and $\pi/2$ but in reverse order, tan t will go through the same values but with changed sign, so the range also includes $(-\infty, 0]$. Hence the range of tangent is, in fact, $\mathbb{R}$. Plotting points for $t \in [0, \pi/2)$ and again referring to those for $t \in (\pi/2, \pi]$, one has the portion of the graph of tangent exhibited in Figure 3.35.

As t goes from π to $3\pi/2$, sin t and cos t both repeat the values they had for $t \in (0, \pi/2)$, but this time they are both negative and this difference in sign from $(0, \pi/2)$ is erased when one looks at tan t. Then tan t follows exactly the same pattern in $(\pi, 3\pi/2)$ as it did in $(0, \pi/2)$. The same kind of argument relates tan t for $t \in (3\pi/2, 2\pi)$ to tan t for $t \in (\pi/2, \pi)$. Putting these analyses together with our knowledge that tangent is periodic, we get the graph of Figure 3.36.

It is evident from the graph we have just drawn that the period of tangent is π, not 2π. The graph has pointed out the error of logic mentioned earlier, and with this evidence we should seek an explanation. There are two routes one might follow: (1) check the original idea for error and (2) check the new evidence for error. Here we need only the former. With

3.4. *GRAPHS AND SOME OTHER CIRCULAR FUNCTIONS* 177

the suggestion that the period is π, we try π in the definition. Thus $\tan(t + \pi) = \sin(t + \pi)/\cos(t + \pi) = -\sin t/-\cos t = \sin t/\cos t = \tan t$. That $\sin(t + \pi) = -\sin t$ and $\cos(t + \pi) = -\cos t$ may not be initially obvious, but in search of an explanation we see that $P(t + \pi)$ is always on the opposite side of a diameter of the unit circle from $P(t)$. Thus the triangles with legs that are $\sin t$, $\cos t$, $\sin(t + \pi)$, and $\cos(t + \pi)$ can quickly be seen to be congruent. See Figure 3.37. The only possible difference between $\sin t$ and $\sin(t + \pi)$ or $\cos t$ and $\cos(t + \pi)$ is one of sign, and indeed the signs are opposite. We shall find analytic proofs of these two facts in the next section.

That no smaller number than π can be the period is evident because for no value of p less than π is $\tan(0 + p) = \tan 0$. That is to say, the zero of $\tan x$ does not repeat more often than every π units. For that matter, neither does any other value repeat for any p less than π, but it suffices to see this for just one value.

In a similar fashion, the graph of cotangent is found to be that given in Figure 3.38, and the period of cotangent is also π.

The only remaining questions to bring our knowledge of tangent, secant, cosecant, and cotangent up to date with our knowledge of sine

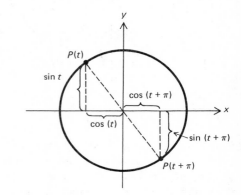

FIGURE 3-37

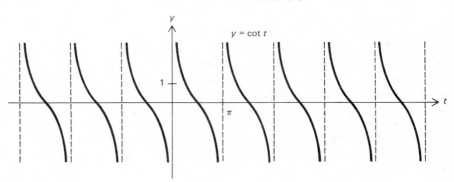

FIGURE 3-38

and cosine are to inquire about zeros and extreme values. Graphically, it is clear that neither secant nor cosecant have any zeros at all. This corresponds to the analytic observation that $\csc t = 1/\sin t = 0$ and $\sec t = 1/\cos t = 0$, can never be true. (An attempt to solve $1/\sin\ t = 0$, yields $\sin t \cdot 0 = 1$, and therefore $0 = 1$.)

GRAPHS OF THE OTHER CIRCULAR FUNCTIONS

$\tan t = 0$ precisely when $\sin t = 0$, since any fraction is zero when the numerator is zero, so the set of zeros for tangent is $\{k\pi \,|\, k \in \mathbb{Z}\}$, while for cotangent one gets the same set of zeros as for cosine, $\{\pi/2 + k\pi \,|\, k \in \mathbb{Z}\}$.

Questions of extreme values are left to the Problems.

3.5. RELATIONSHIPS AMONG THE FUNCTIONS

3.5a. Basic Identities

There are a sizable number of relationships among the circular functions which we shall arrive at in a variety of ways.

BASIC IDENTITIES

To begin with, tangent, secant, cosecant, and cotangent are all related to sine and cosine by definition.

Another valuable group of relationships is provided by the equation for the unit circle : $x^2 + y^2 = 1$. Since for any value of t, $(\cos t, \sin t)$ is a point on the circle (remembering that $\cos t$ is the x-coordinate of a point t on the circle and $\sin t$ is the y-coordinate), we deduce that for all values of t,

$$\cos^2 t + \sin^2 t = 1. \tag{1}$$

It is to be emphasized that this, as well as the definitions in the previous section and the other relationships to be developed in this section, is an *identity* not a conditional equation. That is, this is a statement that is always true, not one in which one is looking for a solution set of any kind. Any value of t will make these relationships true.

NOTE

As long as $\cot t \neq 0$, we may divide both sides of (1) by $\cos^2 t$ to get

$$\frac{\cos^2 t}{\cos^2 t} + \frac{\sin^2 t}{\cos^2 t} = \frac{1}{\cos^2 t}$$

or

$$1 + \tan^2 t = \sec^2 t. \tag{2}$$

Since one would not be concerned with $\tan t$ or $\sec t$ for those values of t for which $\cos t = 0$ (as tangent and secant are not defined for such values of t), Equation (2) is usually written without reference to this restriction.

Similarly, division of (1) by $\sin^2 t$ produces

$$\cot^2 t + 1 = \csc^2 t \tag{3}$$

as a valid identity throughout the domains of cotangent and cosecant.

A different kind of identity comes to light when one ponders such questions as whether $\sin(t + u)$ is readily expressed just in terms of $\sin t$ and $\sin u$. (See Section 1.4d.) The hope that such a formula might be $\sin(t + u) = \sin t + \sin u$ is quickly shattered by a counterexample such as $\sin(\pi/2 + \pi) = \sin\ 3\pi/2 = -1$, while $\sin\ \pi/2 + \sin\ \pi = 1 + 0 = 1$, and $1 \neq -1$.

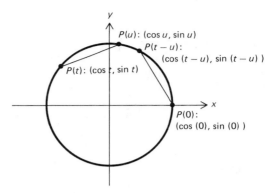

FIGURE 3-39

However, the question lingers as to whether or not some other not-too-complicated formula is valid for $\sin(t + u)$ in terms of $\sin t$ and $\sin u$. This question is reminiscent of Section 3.3 and the attempt to evaluate the circular functions for one argument in terms of their values for other arguments, i.e., the evaluation of functions in the second through fourth quadrants in terms of their values in the first quadrant.

As it turns out, the easiest such question to answer first is "What (if any) nice formula exists for $\cos(t - u)$?," although the derivation is not one that is likely to leap spontaneously into your mind. Rather it would be more likely to come by patient trial and error, so we will conveniently choose the right trial and omit the error.

Using Figure 3.39 as an aid to thinking and visualization, note that since the length of arc between $P(u)$ and $P(t)$ is the same as that between $P(0)$ and $P(t - u)$, the chords, $P(0)P(t - u)$, and $P(u)P(t)$ are the same length. Are you sure you see why this is true?

Using the distance formula† and recalling that sine and cosine give the coordinates of points on the circle, we have the length of

$$P(t - u)P(0) = \sqrt{(\cos(t - u) - \cos 0)^2 + (\sin(t - u) - \sin 0)^2}$$

and the length of

$$P(t)P(u) = \sqrt{(\cos t - \cos u)^2 + (\sin t - \sin u)^2}.$$

Since we know these arcs are equal, we have

$$\sqrt{[\cos(t - u) - \cos 0]^2 + [\sin(t - u) - \sin 0]^2}$$
$$= \sqrt{[\cos t - \cos u]^2 + [\sin t - \sin u]^2}.$$

Since $\cos 0 = 1$ and $\sin 0 = 0$, upon squaring both sides we get

$$[\cos(t - u) - 1]^2 + [\sin(t - u) - 0]^2 = [\cos t - \cos u]^2$$
$$+ [\sin t - \sin u]^2.$$

---

† Hopefully, the reader will recall from his study of algebra that the distance between any two points in the plane, $P(x, y)$ and $Q(a, b)$ is given by $\sqrt{(x - a)^2 + (y - b)^2}$. Those who do not remember or understand should consult Appendix A, Section A.20.

Expanding, we get

$\cos^2(t-u) - 2\cos(t-u) + 1 + \sin^2(t-u)$
$\qquad = \cos^2 t - 2\cos t \cos u + \cos^2 u + \sin^2 t - 2\sin t \sin u + \sin^2 u.$

Rearranging, we have

$\cos^2(t-u) + \sin^2(t-u) + 1 - 2\cos(t-u)$
$\qquad = \cos^2 t + \sin^2 t + \cos^2 u + \sin^2 u - 2[\cos t \cos u + \sin t \sin u].$

Using identity (1), this reduces to

$$1 + 1 - 2\cos(t-u) = 1 + 1 - 2[\cos t \cos u + \sin t \sin u],$$

since

$$\cos^2(t-u) + \sin^2(t-u) = 1, \cos^2 t + \sin^2 t = 1$$

and

$$\cos^2 u + \sin^2 u = 1.$$

Then we have by further simplification

$$\cos(t-u) = \cos t \cos u + \sin t \sin u. \tag{4}$$

We had hoped to get a formula for $\cos(t-u)$ in terms of $\cos t$ and $\cos u$; instead we get one in terms of $\cos t$, $\cos u$, $\sin t$, and $\sin u$. While perhaps a little disappointing, it is still a very valuable identity to keep in mind.

From Equation (4), one can derive a formula for $\cos(t+u)$. First, although one needs the insight to write $t + u$ as $t - (-u)$, for then

$$\cos(t+u) = \cos(t - (-u)) = \cos t \cos(-u) + \sin t \sin(-u.)$$

This could be simplified if we knew simplifications for $\cos(-u)$ and $\sin(-u)$. Using Equation (4), we find $\cos(-u) = \cos(0-u) = \cos 0 \cos u + \sin 0 \sin u$ or

$$\cos(-u) = \cos u. \tag{5}$$

In order to simplify $\sin(-u)$ we either need to refer back to the unit circle, or digress to another relationship. The latter turns out to be worth the effort.

Our trial and error mathematician may already have found, in using (4), that

$$\cos\left(\frac{\pi}{2} - t\right) = \cos\frac{\pi}{2}\cos t + \sin\frac{\pi}{2}\sin t = \sin t. \tag{6}$$

Since we have found $\sin t = \cos(\pi/2 - t)$, we see that

$$\sin(-t) = \cos\left(\frac{\pi}{2} - (-t)\right)$$

$$= \cos\left(\frac{\pi}{2} + t\right)$$

$$= \cos\left(t + \frac{\pi}{2}\right)$$

which we rewrite as

$$= \cos\left(t - \left(-\frac{\pi}{2}\right)\right)$$

using Equation (4),

$$= \cos t \cos\left(-\frac{\pi}{2}\right) + \sin t \sin\left(-\frac{\pi}{2}\right)$$

$$= 0 - \sin t.$$

Thus

$$\sin(-t) = -\sin t. \qquad (7)$$

Using Equations (5) and (7), we then have

$$\cos(t + u) = \cos(t - (-u)) = \cos t \cos(-u) + \sin t \sin(-u)$$
$$\cos(t + u) = \cos t \cos u - \sin t \sin u. \qquad (8)$$

Identity (6) also paves the way to analogous formulas for sine:

$$\sin(t + u) = \cos\left(\frac{\pi}{2} - (t + u)\right)$$

$$= \cos\left(\left(\frac{\pi}{2} - t\right) - u\right) \text{ by rewriting.}$$

By using Equation (4) we get

$$= \cos\left(\frac{\pi}{2} - t\right)\cos u + \sin\left(\frac{\pi}{2} - t\right)\sin u,$$

then by Equation (6)

$$= \sin t \cos u + \cos\left(\frac{\pi}{2} - \left(\frac{\pi}{2} - t\right)\right)\sin u$$

or

$$\sin(t + u) = \sin t \cos u + \cos t \sin u. \qquad (9)$$

Next,

$$\sin(t - u) = \sin(t + (-u))$$
$$= \sin t \cos(-u) + \cos t \sin(-u)$$

and using Equations (5) and (7) again,

$$\sin(t - u) = \sin t \cos u - \cos t \sin u. \qquad (10)$$

Special cases of Equations (8) and (9) occur when $t = u$. These special cases are

$$\sin 2t = 2 \sin t \cos t \qquad (11)$$
$$\cos 2t = \cos^2 t - \sin^2 t. \qquad (12)$$

Two other forms of Equation (12) are easily derived using Equation (1). These are

$$\cos 2t = \cos^2 t - \sin^2 t$$
$$= (1 - \sin^2 t) - \sin^2 t$$
$$\cos 2t = 1 - 2 \sin^2 t \qquad (13)$$

and

$$\cos 2t = \cos^2 t - \sin^2 t$$
$$= \cos^2 t - (1 - \cos^2 t)$$
$$\cos 2t = 2 \cos^2 t - 1 \qquad (14)$$

and these in turn easily lead to the last identities we shall deal with in this part of the section. Seeking formulas for $t/2$, one cleverly writes t as $2 \cdot t/2$ and uses Equations (13) and (14):

$$\cos t = \cos\left(2\left(\frac{t}{2}\right)\right) = 1 - 2 \sin^2 \frac{t}{2}.$$

Thus $\cos t = 1 - 2 \sin^2 t/2$ or $2 \sin^2 t/2 = 1 - \cos t$, and so

$$\sin^2 \frac{t}{2} = \frac{1 - \cos t}{2} \qquad \text{or} \qquad \sin \frac{t}{2} = \pm\sqrt{\frac{1 - \cos t}{2}}, \qquad (15)$$

the choice of signs dependent on the quadrant in which $P(t/2)$ lies. Similarly,

$$\cos t = \cos\left(2\left(\frac{t}{2}\right)\right) = 2 \cos^2 \frac{t}{2} - 1$$

or

$$\cos^2 \frac{t}{2} = \frac{\cos t + 1}{2} \qquad \text{or} \qquad \cos \frac{t}{2} = \pm\sqrt{\frac{\cos t + 1}{2}}, \qquad (16)$$

the choice of sign dependent on the quadrant in which $P(t/2)$ lies.

Although we have solved these for $\sin t/2$ and $\cos t/2$, the identities are more frequently useful in the forms

$$\sin^2 \frac{t}{2} = \frac{1 - \cos t}{2}$$

and

$$\cos^2 \frac{t}{2} = \frac{1 + \cos t}{2}.$$

Not only do these provide relationships involving halved arguments, but more clearly pave the way for an exchange of form from one involving exponents to one without exponents (or vice versa), and also provide single term replacements for expressions of the form $1 + \cos t$ and $1 - \cos t$ (and again vice versa).

Summarizing the identities developed to date, which exhibit a great deal of interdependence among the circular functions, we have the following list.

Definitional Identities

$\tan t = \sin t/\cos t.$
$\sec t = 1/\cos t.$
$\csc t = 1/\sin t.$
$\cot t = 1/\tan t = \cos t/\sin t.$

Fundamental Identities

$$\sin^2 t + \cos^2 t = 1.$$
$$1 + \tan^2 t = \sec^2 t.$$
$$\cot^2 t + 1 = \csc^2 t.$$

Addition Formulas

$$\cos(t - u) = \cos t \cos u + \sin t \sin u.$$
$$\cos(t + u) = \cos t \cos u - \sin t \sin u.$$
$$\sin(t + u) = \sin t \cos u + \cos t \sin u.$$
$$\sin(t - u) = \sin t \cos u - \cos t \sin u.$$

Double and Half-Argument Formulas

$$\sin 2t = 2 \sin t \cos t.$$
$$\cos 2t = \cos^2 t - \sin^2 t = 1 - 2 \sin^2 t = 2 \cos^2 t - 1.$$

$$\sin^2 \frac{t}{2} = \frac{1 - \cos t}{2}.$$

$$\cos^2 \frac{t}{2} = \frac{1 + \cos t}{2}.$$

Other Identities

$$\sin(-t) = -\sin t.$$
$$\cos(-t) = \cos t.$$
$$\cos(\pi/2 - t) = \sin t.$$

In addition to being aesthetically pleasing (it is nice to know how functions do interact), these identities are useful in simplifying complicated expressions involving the circular functions and in proving still other properties of these functions.

EXAMPLE 21   Simplify

$$\frac{\tan t + \cot t}{\sec t}.$$

Using definitions

$$\frac{\tan t + \cot t}{\sec t} = \frac{\sin t / \cos t + \cos t / \sin t}{1 / \cos t}$$

using algebra

$$= \frac{\sin^2 t + \cos^2 t}{\sin t \cos t} \cdot \frac{\cos t}{1}$$

using a fundamental identity and algebra

$$= \frac{1}{\sin t}$$

by definition

$$= \csc t.$$

Prove that $\sec\theta\csc\theta - 2\cos\theta\csc\theta = \tan\theta - \cot\theta$. Since this is not an equation to be solved, we cannot treat it as an equation. Rather, each side must be shown to be reducible to the same expression, or one side must be derived from the other, or the difference of the two sides must be shown to be equivalent to 0. For this example, we choose the first route. The reduction of both sides is often done in parallel to suggest what substitutions to make.

$$
\begin{array}{c|c}
\sec\theta\csc\theta - 2\cos\theta\csc\theta & \tan\theta - \cot\theta \\[2mm]
= \dfrac{1}{\cos\theta}\dfrac{1}{\sin\theta} - \dfrac{2\cos\theta}{\sin\theta} & = \dfrac{\sin\theta}{\cos\theta} - \dfrac{\cos\theta}{\sin\theta} \\[4mm]
= \dfrac{1 - 2\cos^2\theta}{\sin\theta\cos\theta} & = \dfrac{\sin^2\theta - \cos^2\theta}{\sin\theta\cos\theta} \\[4mm]
= \dfrac{-\cos 2\theta}{\sin\theta\cos\theta} & = \dfrac{-\cos 2\theta}{\sin\theta\cos\theta}
\end{array}
$$

Since the two sides are each equivalent to the same expression, their equality has been proved. Are you sure you see what identity was used on each side each step of the way?

EXAMPLE 23

Prove $\tan(t + \pi) = \tan t$.

$$
\begin{aligned}
\tan(t + \pi) &= \frac{\sin(t + \pi)}{\cos(t + \pi)} \\[2mm]
&= \frac{\sin t \cos\pi + \cos t \sin\pi}{\cos t \cos\pi - \sin t \sin\pi} \\[2mm]
&= \frac{\sin t(-1) + 0}{\cos t(-1) - 0} \\[2mm]
&= \frac{-\sin t}{-\cos t} \\[2mm]
&= \tan t.
\end{aligned}
$$

EXAMPLE 24

Other identities may require more insight to establish; e.g., one rather useful identity is

$$
\tan\frac{t}{2} = \frac{\sin t}{\cos t + 1}.
$$

To establish this, we first write

$$
\tan\frac{t}{2} = \frac{\sin t/2}{\cos t/2}
$$

by definition of tangent. Then motivated by the desire to come out with $\sin t$ in the numerator, we multiply the numerator and denominator by $2\cos t/2$ to make the numerator of the form $\sin 2\theta$.

$$
\frac{\sin t/2}{\cos t/2} = \frac{2\sin t/2 \cos t/2}{2\cos^2 t/2} = \frac{\sin t}{2\cos^2 t/2}.
$$

3.5.   *RELATIONSHIPS AMONG THE FUNCTIONS*   185

Be sure to note that this multiplication and division did not change any values, only the form, and that the division is legal (no division by zero) for all values of t for which $\tan t/2$ is defined; i.e., we have not reduced the domain of the functions we were considering initially. Next, to rid ourselves of the $\cos^2 t/2$ in the denominator we note that the half argument formula $\cos^2(t/2) = (1 + \cos t)/2$ provides us with the replacement $1 + \cos t$ for $2\cos^2(t/2)$, thus permitting us to write

$$\frac{\sin t}{2\cos^2 t/2} = \frac{\sin t}{\cos t + 1},$$

as originally claimed.

EXAMPLE 25

Prove that $\cos^4 x + \sin^4 x$ is identical to $1 - 2\sin^2 x \cos^2 x$. Initial attempts to reduce one side to the other or both to a common expression are likely to fail. More profitable here is to examine the difference of the two sides. That is, we exchange the problem

$$\cos^4 x + \sin^4 x = 1 - 2\sin^2 x \cos^2 x$$

for the equivalent problem

$$\cos^4 x + \sin^4 x - (1 - 2\sin^2 x \cos^2 x) = 0.$$

By regrouping, the left side readily reduces as follows

$$\begin{aligned}\cos^4 x + 2\sin^2 x \cos^2 x + \sin^4 x - 1 &= (\cos^2 x + \sin^2 x)^2 - 1 \\ &= 1^2 - 1 \\ &= 1 - 1 = 0\end{aligned}$$

which completes the proof.

3.5b. Other Identities

The following set of relationships, although less often used, are occasionally valuable. We develop them here so that you will have them as tools when you need them.

We begin with

$$\sin(t + u) = \sin t \cos u + \cos t \sin u$$

and

$$\sin(t - u) = \sin t \cos u - \cos t \sin u.$$

Adding these two identities, we get

$$\sin(t + u) + \sin(t - u) = 2\sin t \cos u$$

and thus

$$\sin t \cos u = \tfrac{1}{2}[\sin(t + u) + \sin(t - u)]. \tag{17}$$

By subtracting the same two identities, we get

$$\cos t \sin u = \tfrac{1}{2}[\sin(t + u) - \sin(t - u)]. \tag{18}$$

Similarly, we begin with

$$\cos(t + u) = \cos t \cos u - \sin t \sin u$$

and

$$\cos(t - u) = \cos t \cos u + \sin t \sin u.$$

Again, by adding, we get

$$\cos t \cos u = \tfrac{1}{2}[\cos(t + u) + \cos(t - u)] \qquad (19)$$

and by subtracting, we get

$$\sin t \sin u = -\tfrac{1}{2}[\cos(t + u) - \cos(t - u)]. \qquad (20)$$

Another set of identities, related to those above, is developed below. By letting

$$t = \frac{x + y}{2} \qquad \text{and} \qquad u = \frac{x - y}{2}$$

in Equation 17, we get

$$\sin \frac{x + y}{2} \cos \frac{x - y}{2} = \frac{1}{2}\left[\sin \frac{(x + y) + (x - y)}{2} + \sin \frac{(x + y) - (x - y)}{2}\right]$$

$$= \frac{1}{2}(\sin x + \sin y)$$

or

$$\sin x + \sin y = 2 \sin \frac{x + y}{2} \cos \frac{x - y}{2}. \qquad (21)$$

By similar substitutions, we get

$$\sin x - \sin y = 2 \cos \frac{x + y}{2} \sin \frac{x - y}{2} \qquad (22)$$

$$\cos x + \cos y = 2 \cos \frac{x + y}{2} \cos \frac{x - y}{2} \qquad (23)$$

$$\cos x - \cos y = -2 \sin \frac{x + y}{2} \sin \frac{x - y}{2}. \qquad (24)$$

You may have noted by now that Equations (17)–(20) allow us to conveniently write a product as a sum, and Equations (21)–(24) allow us to easily write a sum as a product (a kind of factorization).†

3.6. INVERSES OF THE CIRCULAR FUNCTIONS

The inverses (which we denote as $\sin^{-1}$, $\cos^{-1}$, etc.‡) of the circular functions are not so tame as we would like. Using the graphing device of Section 1.5c, we find the graphs of these inverses to be given by Figure 3.40. Not only do they fail to be functions, but they fail gloriously. For a particular x-value the set of $f^{-1}(x)$'s is either empty or contains infinitely many values. Although one occasionally wishes to deal with inverses of

† These eight identities were once called the Prosthaphaeretic Formulas, from the Greek meaning "to-and-from-taking" formulas—a rather descriptive if tongue-twisting name for them.
‡ Some other texts use arcsin for $\sin^{-1}$, arccos for $\cos^{-1}$, etc.

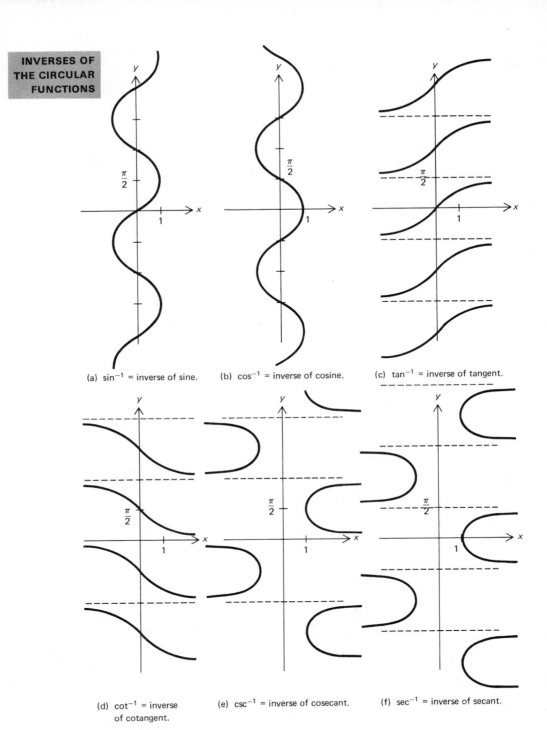

INVERSES OF THE CIRCULAR FUNCTIONS

(a) $\sin^{-1}$ = inverse of sine.

(b) $\cos^{-1}$ = inverse of cosine.

(c) $\tan^{-1}$ = inverse of tangent.

(d) $\cot^{-1}$ = inverse of cotangent.

(e) $\csc^{-1}$ = inverse of cosecant.

(f) $\sec^{-1}$ = inverse of secant.

FIGURE 3-40

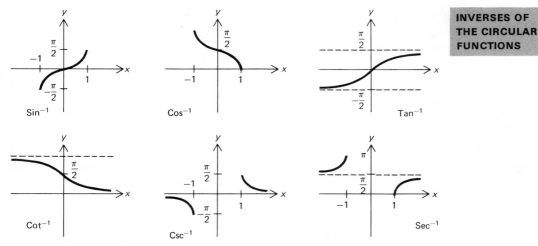

FIGURE 3-41

circular functions as relations, more frequently one wishes to know the principal value of the inverse. In other words, if one asks " For what value of t is $\sin t = \frac{1}{2}$? " he is likely to be interested in $\pi/6$, not $181\pi/6$ or even $5\pi/6$. For this reason, it is standard procedure to carve these relations down to create functions. This is done by restricting the domains of the parent circular functions in such a way as to be sure that each of the domains of the inverses is unchanged but the needed nonrepetition of range values occurs. (See Section 1.5b, where we first discussed the restriction of domains of functions.)

Looking at the graph of sine, a sensible (but not the only possible) restriction of the domain is to $[-\pi/2, \pi/2]$. The inverse function so created is denoted Sin^{-1} and can be defined by

$$\mathrm{Sin}^{-1} x = \text{the value of } t \in [-\pi/2, \pi/2] \text{ for which } \sin t = x.$$

This means that $\mathrm{Sin}^{-1}(-\sqrt{2}/2) = -\pi/4$ so that negative values of t gain some prestige thay may not previously have had in our eyes.

The standard restrictions for the other inverses are given in the following definitions : †

$\mathrm{Cos}^{-1} x = $ the value of $t \in [0, \pi]$ for which $\cos t = x$.
$\mathrm{Tan}^{-1} x = $ the value of $t \in (-\pi/2, \pi/2)$ for which $\tan t = x$.
$\mathrm{Sec}^{-1} x = $ the value of $t \in [0, \pi/2) \cup (\pi/2, \pi]$ for which $\sec t = x$.
$\mathrm{Csc}^{-1} x = $ the value of $t \in [-\pi/2, 0) \cup (0, \pi/2]$ for which $\csc t = x$.
$\mathrm{Cot}^{-1} x = $ the value of $t \in (0, \pi)$ for which $\cot t = x$.

The domains of both the inverses of the functions and the inverse functions are the ranges of the corresponding original circular functions, so that

$$D_{\cos^{-1}} = D_{\mathrm{Cos}^{-1}} = D_{\sin^{-1}} = D_{\mathrm{Sin}^{-1}} = [-1, 1]$$

† Again, there are alternate notations Arcsin, Arccos, etc.

$$D_{\tan^{-1}} = D_{\mathrm{Tan}^{-1}} = D_{\cot^{-1}} = D_{\mathrm{Cot}^{-1}} = \mathbb{R}$$
$$D_{\sec^{-1}} = D_{\mathrm{Sec}^{-1}} = D_{\csc^{-1}} = D_{\mathrm{Csc}^{-1}} = (-\infty, -1] \cup [1, \infty).$$

This information is summarized in Figure 3.41, which displays the graphs of these inverse functions.

We turn next to a few examples using these inverse functions and relations.

EXAMPLE 26   Find $\mathrm{Sin}^{-1} \frac{1}{2}$. A good way to handle problems involving inverse circular functions is to realize that the question being asked is: "What is the number (or angle†) whose sine is $\frac{1}{2}$?" We know from our special values that $\sin \pi/6 = \frac{1}{2}$. Since the definition of Sin^{-1} says that t must be in $[-\pi/2, \pi/2]$, and $\pi/6$ is in this interval, we have solved our problem. Therefore, $\mathrm{Sin}^{-1} \frac{1}{2} = \pi/6$.

EXAMPLE 27   Find $\cos^{-1} \frac{1}{2}$. Our question this time is: "What is the number or angle whose cosine is $\frac{1}{2}$?" Our immediate answer is $\pi/3$. But we are to find $\cos^{-1} \frac{1}{2}$, not $\mathrm{Cos}^{-1} \frac{1}{2}$. That means that $\pi/3$, $5\pi/3$, and $7\pi/3$ and many more numbers, positive and negative, are correct answers for our question. Thus, with some thought and reference back to the form used to write the zeros of various circular functions, we have

$$\cos^{-1} \frac{1}{2} = \left\{ \frac{\pi}{3} + 2n\pi \,|\, n \in \mathbb{Z} \right\} \qquad \text{and} \qquad \left\{ \frac{5\pi}{3} + 2n\pi \,|\, n \in \mathbb{Z} \right\},$$

or more properly using set notation,

$$= \left\{ \frac{\pi}{3} + 2n\pi \,|\, n \in \mathbb{Z} \right\} \cup \left\{ \frac{5\pi}{3} + 2n\pi \,|\, n \in \mathbb{Z} \right\}.$$

EXAMPLE 28   Find $\mathrm{Tan}^{-1}(-1.7)$. The question here is "What is the number or angle whose tangent is -1.7?" Using Table 1, Appendix D, we find that $\tan 1.04 \approx 1.704$. To get negative values for $\tan t$ and keep $t \in (-\pi/2, \pi/2)$, we need to choose $t = -1.04$, i.e., $\tan(-1.04) \approx -1.704$, so that $\mathrm{Tan}^{-1}(-1.7) \approx -1.04$.

EXAMPLE 29   Find $\mathrm{Sin}^{-1} \pi$. Careful—here's a deceptive problem. The domain of Sin^{-1} is $[-1, 1]$ and $\pi \notin [-1, 1]$ so that $\mathrm{Sin}^{-1} \pi$ is not defined. (Be careful that you keep a clear distinction in your mind between the circular functions and their inverses.)

3.7. APPLICATION TO ANGLES AND TRIANGLES

3.7a. Radian Measure

Suppose that (x, y) is a point on the unit circle. We have used the function P to associate such a point with various real numbers by saying that the point $(x, y) = P(t)$ if the arc of length t from $(1, 0)$ measured counter-clockwise ends at (x, y). By a different approach, we could associate the points on the unit circle with angles instead of real numbers.

---

† We shall soon be using angles so that it is a good idea to include this possibility in your thinking now for use later.

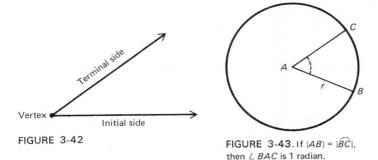

FIGURE 3-42

FIGURE 3-43. If $|AB| = |\overset{\frown}{BC}|$, then $\angle BAC$ is 1 radian.

Recall that an angle consists of two rays with a common vertex, one ray being called the initial side, the other the terminal side. See Figure 3.42. The measure of an angle is the amount of rotation needed to move the initial side to the terminal side, with positive measures being assigned to counterclockwise rotations, negative measures to clockwise rotations.

The units of angular measure, like all units of measure, are arbitrary. One could measure angles in the number of complete revolutions; therefore, a right angle would measure $\frac{1}{4}$ complete revolution. Although this unit is not frequently used in technical contexts, it is convenient in many less formal settings, and we find it useful here. Another more common system of units is that of degrees, minutes, and seconds. A degree is 1/360 of a complete revolution; a minute is 1/60 of a degree; and a second is 1/60 of a minute. The symbols for degrees, minutes, and seconds are °, ', ", so that 35°50'16" is read "35 degrees, 50 minutes, 16 seconds."

Another unit of angular measure that is used a good deal, particularly in mathematics, is the radian. A central angle of a circle has measure *1 radian* if it subtends an arc whose length is equal to the length of the radius of the circle. See Figure 3.43. This definition may seem ambiguous to the reader, in the sense that the use of different circles might permit angles of different sizes to be assigned the same measure. That this does not happen is seen by the following argument.

DEFINITION 3.8
Radian

Since the circumference of a circle of radius r is $2\pi r$ (see Figure 3.43 again), a central angle of one radian subtends an arc that is $r/2\pi r$ of the circumference. If $r/2\pi r$ seemed to you to come out of mid-air, consider this analogy. If it is 5 miles from Y to Z (Figure 3.44a) and you go 1 mile, you have gone one-fifth of the way; similarly, if you have gone x miles, you have gone $x/5$ths of the way. If you think of the circle of Figure 3.43 stretched out (Figure 3.44b), then a similar procedure produces $r/2\pi r$. Now $r/2\pi r = 1/2\pi$, which is totally independent of the radius and therefore of the size of the circle. In other words, 1 radian $= (1/2\pi)$ (one complete revolution). While this may seem to be a strange unit, it makes the connection between the circular functions and their application to angles and triangles especially close. The usual notation for radian is rad, so one might speak of an angle 3.5 rad.

Rotations of more than one complete revolution are permissible. We see that an angle whose measure is 361° consists of the same rays as one

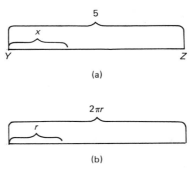

FIGURE 3-44

whose measure is 1°. This gives some reason to say that *two angles are equal* if their measures are equal rather than if they consist of the same rays. Under this convention, then, there is no loss of generality if all angles are viewed as having their vertices at the origin of the coordinate plane and their initial side coincident with the positive *x*-axis, and then to refer to angles by their measures. That is, we shall feel free to talk of "the angle 30°" instead of "the angle whose measure is 30°."

CIRCULAR FUNCTIONS OF ANGLES

Many people would be content at this juncture to let the domain of sine and cosine (and the other four functions as well) be either $\mathbb{R}$, as has been the case until now, or the set of angles, and so talk of sin $\pi/6$, sin $\pi/6$ rad, and sin 30° interchangeably. In fact, this is reasonable; therefore, we shall view the circular functions as functions of real numbers or of angles as suits our convenience.† With this increased freedom comes the need to keep context clearly in mind—sin 30 is a far cry from sin 30°.

In the angle setting, radians provide a valuable bridge: to find sin $x°$, if we can translate $x°$ to y rad, then sin $x° = \sin(y$ rad$) = \sin y$, which we

---

† A more complete and formal justification of the reasonableness of this dual usage for the circular functions would go as follows. We may associate each point (x, y) on the unit circle with those angles θ for which (x, y) is the intersection of the terminal side of the angle with the unit circle. This defines a function Q whose domain is the set of all angles (or all sizes of angles) and range the unit circle. Composing θ with the functions α and β of Section 3.2b we have the same ranges as we did for sine and cosine. To be technical, we might define the functions wine and cowine by

$$\text{win } \theta = (\beta \circ Q)(\theta) \qquad \text{and} \qquad \text{cow } \theta = (\alpha \circ Q)(\theta).$$

We would then have such facts as

$$\text{win } 30° = \text{win } \frac{\pi}{6} \text{ rad} = \sin \frac{\pi}{6} = \frac{1}{2}$$

$$\text{cow } 90° = \text{cow } \frac{\pi}{2} \text{ rad} = \cos \frac{\pi}{2} = 0$$

Indeed, the system of radian measure is defined so that

$$\text{win } (t \text{ rad}) = \sin t \qquad \text{and} \qquad \text{cow } (t \text{ rad}) = \cos t$$

for all real numbers t. This is because a central angle of t radians subtends an arc t units long, so that $P(t) = Q(t$ rad$)$. (See Problem 56.) So although there is a difference between the real number t and an angle of t radians, the fact that both are identified by the number t suggests that we overlook the difference between sine and wine, and use the same name for both. We would also do this for cosine and cowine. As mathematicians, we should be outraged at this brash suggestion, but as people and with a sense of historical conscience, we capitulate, for in reality, sine and cosine were originally conceived as functions of angles, and if one keeps context in mind, there need not be any confusion. Then, once we have allowed ourselves to speak of the sine of an angle, we let ourselves refer to the angle by its measure in any system. Thus $\sin(\frac{1}{4} \text{ revolution}) = \sin 90° = \sin(\pi/2 \text{ rad})$ are all ways of writing the sine of a right angle.

can evaluate by our earlier work. Thus translation between radians and degrees becomes an important technique.

To do this, we set up a ratio of degrees/radians and since 2π rad $= 360°$,

$$\frac{360°}{2\pi \text{ rad}} = \frac{x°}{y \text{ rad}}$$

provides an equation for finding degrees if radians are known, or for finding radians if degrees are known.

Find the number of degrees equivalent to 1 rad.   **EXAMPLE 30**

$$\frac{360}{2\pi} = \frac{x}{1}$$

$$2\pi x = 360$$

$$x = \frac{360}{2\pi} = \frac{180}{\pi} \approx 57°.$$

Find the number of radians equivalent to 25°.   **EXAMPLE 31**

$$\frac{360}{2\pi} = \frac{25}{y}$$

$$360 \cdot y = 25 \cdot 2\pi$$

$$y = \frac{50\pi}{360} = \frac{5\pi}{36} \text{ rad.}$$

Find the special values for sine and cosine as functions of angles in degrees.   **EXAMPLE 32**

Using the above equation, we find that

$$\frac{360}{2\pi} = \frac{x}{(\pi/2)}$$

$$2\pi x = 360 \left(\frac{\pi}{2}\right)$$

$$x = \frac{180\pi}{2\pi} = 90°$$

Therefore,

$$\frac{\pi}{2} \text{ rad} = 90°.$$

Similarly,

$$\frac{\pi}{3} \text{ rad} = 60°$$

$$\frac{\pi}{4} \text{ rad} = 45°$$

$$\frac{\pi}{6} \text{ rad} = 30°$$

$$0 \text{ rad} = 0°.$$

Thus we have the following table:

θ	0°	30°	45°	60°	90°
$\sin \theta$	0	$\dfrac{1}{2}$	$\dfrac{\sqrt{2}}{2}$	$\dfrac{\sqrt{3}}{2}$	1
$\cos \theta$	1	$\dfrac{\sqrt{3}}{2}$	$\dfrac{\sqrt{2}}{2}$	$\dfrac{1}{2}$	0

In practice, except for a few special angles such as those in the example above, the conversion from degrees to radians produces radian angles that are awkward to work with (as in Example 30). Hence one standardly finds tables for the circular functions with arguments tabulated by degrees instead of radians. (Such tables are usually called tables of the natural trigonometric functions.) The arrangement of these tables is more compact than that of Table 1, Appendix D, but we will look into triangles before explaining this.

3.7b. Solution of Right Triangles

Suppose that an engineer wants to build a bridge over a canyon. In order to design the bridge, he needs to know the width of the canyon but he cannot just go out and measure it directly. One way to find the distance (LP) is illustrated in Figure 3.45. A landmark L is chosen on one wall of the canyon and a point P on the opposing wall is marked. The engineer walks from P on a line perpendicular to the line PL and marks his stopping place as Q. He measures the length of PQ to be 35 ft and the angle LQP to be 60°. Recalling from plane geometry that similar triangles have sides proportional, he observes that if he constructs a triangle $P'Q'L'$ (similar to PQL) in the unit circle as shown in Figure 3.46,

$$\frac{\text{length of } QP}{\text{length of } Q'P'} = \frac{\text{length of } LP}{\text{length of } L'P'}.$$

Since $Q'P' = \cos \theta$ and $L'P' = \sin \theta$, length $QP/\cos \theta =$ length $LP/\sin \theta$.

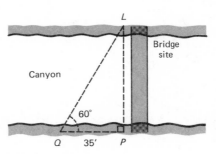

FIGURE 3-45

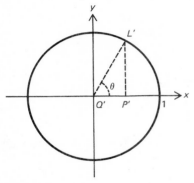

FIGURE 3-46

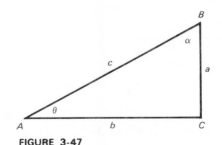

FIGURE 3-47

Since θ is known to be 60°, cos θ and sin θ are $\frac{1}{2}$ and $\sqrt{3}/2$, respectively (See Example 32.) Recall that QP is 35'. So the equation can be rewritten as $35/(\frac{1}{2}) = x/(\sqrt{3}/2)$ and $x = \sqrt{3} \cdot 35 \approx 60.62$ ft.

The point to be extracted from this is that any right triangle is similar to a right triangle with legs of length sin θ and cos θ, where θ is one of the acute angles in the triangle,† and thus the circular functions can be used to relate the parts of a right triangle to each other.

This brings us to the historical beginnings of trigonometry. The circular functions were originally just functions of acute angles. They were defined by ratios of sides of right triangles, which definitions can be reconstructed by the above techniques. We do this partially out of historical interest, since many of the uses of right angle trigonometry have been incorporated in computer programs and are no longer calculated by people. Enough individual uses, however, do remain in mathematics, engineering, and sciences to warrant their mention.

Let ABC be a right triangle as in Figure 3.47, with angle $BAC = \theta$, angle $ABC = \alpha$, angle $ACB = 90°$, the length of side $AC = b$, the length of side $BC = a$, and the length of side $AB = c$. Then

$$\frac{a}{c} = \frac{\sin \theta}{1} \quad \text{or} \quad \sin \theta = \frac{a}{c}$$

$$\frac{b}{c} = \frac{\cos \theta}{1} \quad \text{or} \quad \cos \theta = \frac{b}{c}$$

$$\frac{a}{b} = \frac{\sin \theta}{\cos \theta} \quad \text{or} \quad \tan \theta = \frac{a}{b}.$$

DEFINITION 3.9a
Triangle Definitions of Circular Functions

A commonly used alternative form results from noting that a is the length of the side opposite the angle θ, b the length of the side adjacent to angle θ, and c the length of the hypotenuse. Thus one has the colloquial

† Using the positive x-axis as one ray, copy one of the two acute angles θ with vertex at the origin and terminal side in the first quadrant. Drop a perpendicular from the intersection of the terminal side with the unit circle to the x-axis.

DEFINITION 3.9b
Colloquial
Definitions of
Circular Functions

definitions:

$$\sin \theta = \frac{\text{side opposite}}{\text{hypotenuse}}$$

$$\cos \theta = \frac{\text{side adjacent}}{\text{hypotenuse}}$$

$$\tan \theta = \frac{\text{side opposite}}{\text{side adjacent}}.$$

In this setting, a relationship between some of the functions appears that would not have seemed so useful before. Notice that using the notation of Figure 3.47, $\sin \alpha = b/c = \cos \theta$, $\cos \alpha = a/c = \sin \theta$, and θ and α are complementary angles. In other words, $\sin \alpha = \cos(90° - \alpha)$, while $\cos \alpha = \sin(90° - \alpha)$. Looking at the other functions, we see also that $\tan \alpha = \cot(90° - \alpha)$ and $\csc \alpha = \sec(90° - \alpha)$. Thus, if we know the values of the six functions for 3°, we also know them for 87° by an easy rearrangement. Or if we know them for 18°, we know them for 72°; so, we need only list each such set of numbers once if we label them wisely. Figure 3.48 shows how this can be done, and in fact how most tables of natural trigonometric functions are constructed.

For use with the values of θ listed at the left

θ	$\sin \theta$	$\cos \theta$	$\tan \theta$	$\cot \theta$	$\sec \theta$	$\csc \theta$	
0°	0	1	0	—	1	—	90°
1°	0.017	0.999	0.017	57.3	1	28.6	89°
2°	.	.	.	.	.	.	
	.	.	.	.	.	.	
44°	.	.	.	.	.	.	46°
45°	0.707	0.707	1	1	1.41	1.41	45°
	$\cos \theta$	$\sin \theta$	$\cot \theta$	$\tan \theta$	$\csc \theta$	$\sec \theta$	θ

For use with the values of θ at the right

FIGURE 3.48

The naming of the six functions now makes a little more sense than before, too. There are three functions—sine, tangent, and secant, and three cofunctions—cosine, cotangent, and cosecant, which are related to the three functions by the identities cited. The *cosine* of θ is also the sine of the *complement* of θ, and so on. Equation (6) of Section 3.5a and Problem 52 verify that these identities hold even when θ is an angle outside the values from 0° through 90°.

EXAMPLE 33   Let us consider some examples before proceeding. Solve the triangle of Figure 3.49. By *solve a triangle*, we mean find all the unknown parts: all the angles and sides not identified. For this triangle, we need to know c, and angles α and β. (We know that C is 90°.) We can find c using the

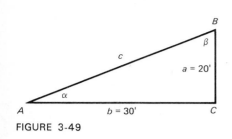

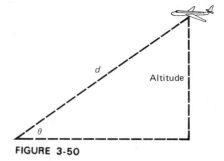

FIGURE 3-49

FIGURE 3-50

Pythagorean Theorem:

$$c^2 = a^2 + b^2$$
$$= 20^2 + 30^2$$
$$= 400 + 900$$
$$c^2 = 1300 \text{ and } c = \sqrt{1300} = 10\sqrt{13}.$$

Next we need to find angle α. For this, we can use the tangent function:

$$\tan \alpha = \tfrac{20}{30} \approx 0.6667.$$

Using Table 2, Appendix D, we see $\mathrm{Tan}^{-1}\, 0.6667 \approx 33°40'$. Thus $\alpha \approx 33°40'$. Since the sum of angles of any triangle is always $180°$, $\beta = 180° - (\alpha + C) \approx 180° - (33°40' + 90°) = 180° - 123°40'$. Thus $\beta \approx 56°60'$ and we have solved the triangle.

A computer-controlled airport needs to know the altitude of each **EXAMPLE 34** airplane in its vicinity. Radar is used to find the straight-line distance d to the airplane, and the angle of such a line with the ground is also monitored. See Figure 3.50. If the computer is fed the information that $d = 1000$ yd and $\theta = 25°$ for Flight No. 12π, what should the computer deduce is the altitude of Flight No. 12π, at that instant? Using the colloquial definition, we find that $\sin \theta = $ side opposite/hypotenuse, $\sin 25° = $ altitude/1000 yd. Table 2, Appendix D, tells us that $\sin 25° \approx 0.4226$ so that the altitude ≈ 422.6 yd or 1267.8 ft.

Sanitary sewer assessments are often based on the area of one's lot. **EXAMPLE 35** Find the area of the lot in Figure 3.51. First we divide the lot as in Figure

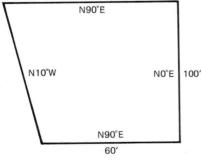

FIGURE 3-51

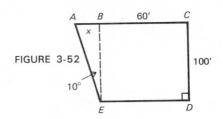

FIGURE 3-52

3.52. The area of the rectangle $BCDE$ is $60 \cdot 100 = 6000$ sq ft. The area of $\triangle ABE$ is $\frac{1}{2}$ (length EB) (length AB). The length of EB is 100', since $BCDE$ is a rectangle. Letting the length of AB be denoted by x, we have $\tan 10° = x/100$, and with the help of Table 2, $x = 17.63$, so that the triangular area is 881.5 sq. ft. Hence the total lot area is 6881.5 sq ft.

EXAMPLE 36 A ship using radar can tell that its distance from the coastal city of Seatown directly north of the ship is 28 miles. If there is no interference from sea currents or wind, at what angle away from due north should the ship head to reach Oceanview, which is 5 miles due east of Seatown? See Figure 3.53. Since $\tan \theta = $ side opposite/adjacent, $\tan \theta = 5/28 \approx 0.17858$. One may use Table 2, Appendix D, in reverse to find that $\theta \approx 10°$ to the nearest degree.

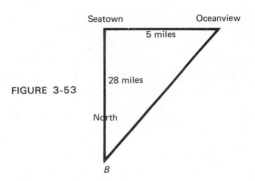

FIGURE 3-53

The solving of triangles is such a straightforward procedure that it quickly suggests the construction of a flowchart for the general method of solution. Ideally, we should like to be able to input any three pieces of data from a right triangle and have the algorithm generate all the other quantities and output everything. Actually, a fully general flowchart would need to test for various inconclusive cases and erroneous data (such as knowing only the three angles, or that the three sides have lengths 1, 2, and 3), and the amount of detail would obscure the simple individual patterns of solution we wish here to emphasize. Figure 3.54 therefore shows flowchart segments for two of the cases, and Problem 80 asks you to construct the remainder of a complete flowchart. The labeling of variables is according to the triangle in the figure.

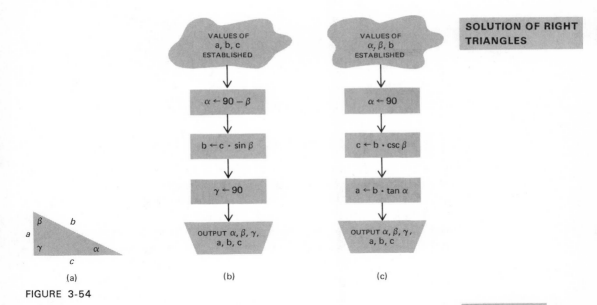

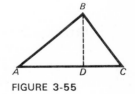

(a)

FIGURE 3-54

(b)

(c)

3.7c. Solution of Oblique Triangles

Certainly, there are many triangles of interest to the user of mathematics that are not right triangles. Such triangles are called *oblique* triangles. These may be solved by using a judiciously chosen altitude and solving two related right triangles. For example, in Figure 3.55, the triangle *ABC* is to be solved knowing the lengths of *AB* and *BC* and angle *A*. Dropping the altitude from *B* to *AC* (call it *BD*), we may find the length of *BD* using sine, the length of *AD* using cosine, then angle *DBC* using cosine, and length of *CD* by either sine or the Pythagorean Theorem. All parts of all three triangles are then known or found by simple addition or subtraction. (Now, did you really follow that logic without specific numbers. If not, read it again more slowly, or, better yet, take paper and pencil and try to do the problem using $AB = 6$, $BC = 4$, and angle $A = 30°$.)

The frequency with which oblique triangles occur suggests the search for a more efficient algorithm to use in solving such triangles. Indeed, there are two formulas that do provide more direct solutions. The possible motivations behind their discovery are not obvious, so we will see what they are and then briefly look back and try to see how they might have originally been derived.

We consider first the Law of Cosines. For no immediately apparent purpose, suppose that we place an arbitrary triangle, *ABC*, in the plane so that the vertex *B* is at the origin and the side *BC* lies on the positive *x*-axis. The point *C* will then have coordinates $(r, 0)$ where $r > 0$. While we may require the vertex *A* to be the upper half plane, since we are dealing with an arbitrary triangle, we may impose no other restriction on the coordinates (s, t) of *A* than that $t > 0$. Figure 3.56 displays several possible

DEFINITION 3.10
Solution of Oblique Triangles

FIGURE 3-55

LAW OF COSINES

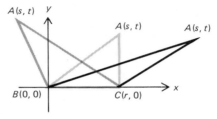

FIGURE 3-56

choices of (s, t), but you should go further to convince yourself that each and every triangle may be so situated and labeled.

If we now make the usual assignment of letters to sides, so that BC has length a, AC has length b, and AB has length c, we have a figure such as Figure 3.57. The distance formula may be ushered in now to tell us that indeed $b = \sqrt{(r - s)^2 + (t - 0)^2}$. However, by the labeling of the triangle (Figures 3.56 and 3.57) $r = a$ and $b = \sqrt{(a - s)^2 + (t - 0)^2}$ or $b^2 = s^2 + a^2 - 2sa + t^2$.

Wishing a formula in terms of a, b, and c only (we prefer not actually to have to introduce a coordinate system in every application), we note that $c = \sqrt{s^2 + t^2}$. This permits the following revision of the form:

$$b^2 = (s^2 + t^2) + a^2 - 2as$$
$$= c^2 + a^2 - 2as.$$

The s in the last term is the only obstacle left to overcome. But by dropping the altitude from A to BC, we find a right triangle in which $\cos B = s/c$, whence $s = c \cdot \cos B$. (Draw the modified diagram and convince yourself that it makes no difference whether $s > r$ or $\angle B > 90°$.) The net result is then that

FORMULA 3.1

Law of Cosines

$$b^2 = a^2 + c^2 - 2ac \cos B.$$

By successively placing each of the other two sides on the x-axis instead of BC, we get two other forms of this relationship:

$$c^2 = a^2 + b^2 - 2ab \cos C.$$
$$a^2 = b^2 + c^2 - 2bc \cos A.$$

These three formulas are the three forms of what is called the *Law of*

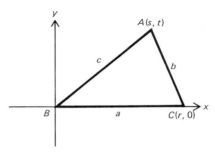

FIGURE 3-57

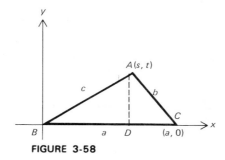

FIGURE 3-58

Cosines; as the derivation shows, they apply to all triangles, whether oblique or right.

As an example of the use of this law, suppose that we have a triangle **EXAMPLE 37** ABC to solve, in which we know that $a = 5$, $b = 6$, and $\angle C = 40°$. The second form of the Law of Cosines provides us with the further information that

$$c^2 = 25 + 36 - 2 \cdot 5 \cdot 6 \cos 40° \approx 61 - 60 \cdot (0.7660) = 61 - 45.96 = 15.04$$

or

$$c \approx \pm \sqrt{15.04} \approx \pm 3.9$$

from whence we choose $c \approx +3.9$. All that remains to be found are the two remaining angles, but we will find these more easily in a moment using the Law of Sines.

Before turning to the Law of Sines, however, look again at the Law of Cosines. Note its similarity with the familiar $a^2 = b^2 + c^2$ of the Pythagorean Theorem. Indeed, if $\angle A = 90°$, i.e., we have a right triangle with hypotenuse a, the Law of Cosines can be seen as a generalization of the Pythagorean Theorem. This in turn suggests the possibility that the origins of the Law of Cosines might have been in an effort to generalize this important classical theorem. On the other hand, it might also have come from idle "puttering" with the distance formula or from yet another motivation. As we noted in Section 2.3b, many proofs in mathematics are like this—the proof is straightforward but mysterious in its origins and motivations.

Meanwhile, if we continue a little further in this direction, we can uncover the second of our tools for oblique triangles. Let us return to the arbitrary triangle in the plane (see Figure 3.58), and let us include the altitude from A to BC, calling the foot of the altitude D. Having previously noted that $s = \cos B$, let us now note that using triangle BAD, we find $\sin B = t/c$, whence $t = c \sin B$. If we now "slide" the triangle into the position indicated in Figure 3.59, we could use triangle ADC to write $\sin C = t/b$ or $t = b \sin C$. Since the value of t would be unchanged by our "sliding" of the triangle, we may write

$$c \sin B = b \sin C,$$

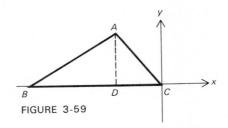

FIGURE 3-59

or, as it is more commonly written,

$$\frac{c}{\sin C} = \frac{b}{\sin B}.$$

Rotation of the triangle results in discovering that $c/\sin C = b/\sin B$ as well. Putting these together, we learn that in any oblique triangle ABC with sides labeled as usual,

FORMULA 3.2
Law of Sines

$$\frac{a}{\sin A} = \frac{b}{\sin B} = \frac{c}{\sin C}.$$

This is called the *Law of Sines*.

EXAMPLE 38
Now to complete the example begun after the Law of Cosines, we have the triangle of Figure 3.60. We may therefore write

$$\frac{6}{\sin B} = \frac{3.9}{\sin 40°} = \frac{5}{\sin A}$$

whence

$$\sin B \approx \frac{(6)(0.6428)}{3.9} \approx 0.9889 \text{ or } B \approx 81\tfrac{1}{2}°$$

and

$$\sin A \approx \frac{5(0.6428)}{3.9} \approx 0.8241 \text{ or } A \approx 55\tfrac{1}{2}°.$$

As a check, we find the sum of the angles of the triangle $= 177°$; there-

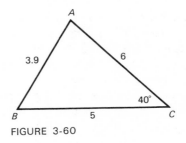

FIGURE 3-60

fore, our answers are not very accurate. The probable source of most of the error is our approximation of 3.9 for $\sqrt{15.04}$. (Indeed, using two more significant digits enables us to make up for this loss. This reminds us that one should not rely too heavily on quick hand computations, for had we found only $\angle B$ by the Law of Sines and then $\angle A$ as $180° - (\angle B + \angle C)$, we could have had significant error in $\angle A$.

Finally, we note no other obvious motivation for the Law of Sines and again assert that while disappointing, this is not unusual.

3.8. OTHER FUNCTIONS BASED ON THE CIRCULAR FUNCTIONS

In the application of the circular functions to the real world, one frequently needs a function which, though not exactly one of the six we have studied, is very closely related to them. Of primary concern to us will be functions whose rules are of the form

$$g(x) = a \cdot \text{cir}(bx + c)$$

where a, b, and c are constants, cir represents any of the six circular functions, and x is the variable. Our interest will center on techniques for graphing such functions, for in so doing we shall have a good chance to see their very close relationship to the six circular functions already discussed, and we shall also have a good chance to develop some general machinery for graphing.

Let us look in detail at $a \cdot \sin(bx + c)$, hoping to develop general patterns to shorten our work with the remaining five. As we have done upon occasion before this, we shall seek understanding through the analysis of subproblems. For example, let us attempt to determine the effect of the constant a. We might do this by graphing $y = \frac{1}{2}\sin x$, $y = \sin x$, $y = 2\sin x$, and $y = 3\sin x$ all on the same set of axes. Figure 3.61 exhibits the results of such an endeavor. It is apparent that the effect is to "stretch" the graph up and down or "squash" it flatter, depending on whether $a > 1$ or $a < 1$, at least as long as $a > 0$. It does not change the zeros or

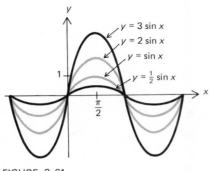

FIGURE 3-61

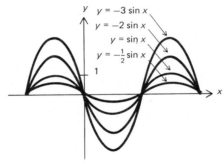

FIGURE 3-62

periodicity. If $a < 0$, the graph just reverses sides of the x-axis and gets "steeper" as $|a|$ gets larger, as shown in Figure 3.62. The vocabulary conventionally assigned to this is that $|a|$ is called the *amplitude* of the function or graph. That is, for $y = 3 \sin x$, the amplitude is $|3| = 3$; for $y = -4 \sin x$, the amplitude is $|-4| = 4$. Thinking in more general terms, as suggested by the specific examples, it becomes clear that the "flatness-steepness" analysis holds true in general, since multiplying $\sin x$ by a will yield maximum and minimum values of $a \cdot 1$ and $a(-1)$ (or $(-a)(+1)$ and $(-a)(-1)$), which means that the graph will be bounded by the lines $y = a$ and $y = -a$, producing what is called an amplitude envelope (Figure 3.63). Also, if $\sin x = 0$, $a \cdot \sin x = 0$ so that the zeros do in fact remain unchanged. Therefore, the larger $|a|$ is, the wider apart the bounds for the graph are and the steeper the graph will need to be in order to reach its maximum value between the zeros, which remain unchanged.

For cosine, a similar analysis holds and $|a|$ is again called the amplitude. For tangent and cotangent, however, since there are no maximum or minimum values, the changes are less striking. Zeros and values of x for which the function is undefined remain unchanged, but the individual points in the graph are shifted up or down, depending on the value of a. Problems 22–27 ask you to graph a few specific examples to see the patterns.

For secant and cosecant, while no maximum or minimum values exist in the strict sense, there are what are called relative maximum and minimum points. These are the tops of the lower loops and bottoms of the upper loops, respectively, and are relative in the sense that they are larger or

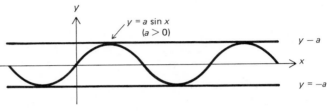

FIGURE 3-63

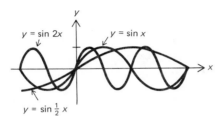

$y = \sin 2x$ $y = \sin x$

$y = \sin \frac{1}{2} x$

FIGURE 3-64

smaller than any other "nearby" values. If the constant a is greater than 1, positive values will be made larger positive and the relative minimums will be raised, while the negative values will be made larger negative and the relative maximums will be lowered. Similarly, if $a \in (0, 1)$, the loops will each be drawn in closer to the x-axis. In each case, the curvature changes somewhat.

If $a < 0$, the loops will change sides of the x-axis; their distance from the axis will increase as $|a|$ increases and decrease as $|a|$ decreases. Problems 24 and 25 ask you to graph a few specific examples to see this pattern more clearly, but your knowledge of the graphs of the circular functions, together with a verbal argument such as the above, should give you a reasonable, qualitative mental picture of what actually happens.

Since none of tangent, cotangent, secant, or cosecant have maximum or minimum values, the term amplitude is not used with a constant multiplier for these functions. The reader should be aware of the changes wrought by the multiplier even though the effect is not quite so dramatic as for sine and cosine.

Next we investigate the effect of b, and again we start with examples using sine. Figure 3.64 shows the graphs of $y = \sin bx$ for several positive values of b. It appears that as b gets closer to zero, the period of sin bx gets longer and longer. Can we give a more rigorous argument to back up this conjecture? What is the period of sine? It is the smallest positive number p so that $\sin(x + p) = \sin x$ for all $x \in D_{\sin}$. The period for sin bx ought to be the smallest number p so that $\sin b(x + p) = \sin bx$ for all $x \in D_{\sin bx}$. This might be clearer if we change notation for a moment: $\{(x, y) \mid y = \sin bx\}$ is a function related to sine, and might be called sine$_b$. The argument of sin$_b$ is x, and so the periodicity should be with respect to x, not bx. In other words, we do want to require $\sin b(x + p) = \sin bx$, not $\sin(bx + p) = \sin bx$ for all x. But then, since sin bx makes a direct use of sine, and we know the period of sine, we can make use of this information. Because one period for sine occurs from $x = 0$ to $x = 2\pi$, one period of sin bx should occur when the argument of sine, namely bx, goes from 0 to 2π. Now $bx = 0$ when $x = 0$ and $bx = 2\pi$ when $x = 2\pi/b$. This strongly suggests that p should have the value $2\pi/b$. That $2\pi/b$ is a legitimate candidate for the period is verified by $\sin b(x + 2\pi/b) = \sin(bx + 2\pi) = \sin bx$. That no smaller positive number usurps $2\pi/b$'s claim to be the period is verified by observing that if $0 < q < 2\pi/b$, then

3.8. *OTHER FUNCTIONS BASED ON THE CIRCULAR FUNCTIONS*   205

$bq < b \cdot 2\pi/b = 2\pi$. Since $\sin(bx + \alpha) \neq \sin bx$ for any positive $\alpha < 2\pi$, $\sin b(x + q)$ cannot be equal to $\sin bx$ for all $x \in D_{\sin bx}$.

If $b < 0$, one can transform the function to fit the first case by the use of the identity $\sin(-x) = -\sin x$. In other words if $b < 0$, let us call it $-\beta$ where $\beta > 0$. Then $\sin(-\beta x) = -\sin \beta x$, and $\sin \beta x$ fits the above analysis. We shall see what to do about the factor -1 shortly.

The above kind of analysis applies to any periodic function, as we now discuss. If f is periodic with period p, then if $b > 0$, $f(b(x + p/b)) = f(bx + p) = f(bx)$, while if $0 < q < p/b$, then $0 < bq < b \cdot p/b = p$ so that $f(b(x + q)) = f(bx + bq) \neq f(bx)$ for all $x \in D_f$. The effect of the b can be viewed as a kind of shrinking. The original function is told to look not at x itself but at a multiple of x so that the behavior of the new function occurs sooner (if $b > 1$) than that of the original function; so, in a sense, less space is needed for the same patterns to register. The reader is asked to apply the above reasoning and results to the remaining five circular functions in the Practice Exercises.

What about the constant c? By itself, it has the effect of shifting the graph to the right or left as the examples in Figure 3.65 indicate for sine. A general explanation of this is afforded by observing that for any function f, the related function $g = \{(x, y) \mid y = f(x + 3)\}$ is told that at x it is to be what f is 3 units to the right. In other words, the behavior of f occurs 3 units sooner, or the graph of f is shifted to the left 3 units. If 3 is replaced by c, the analysis shows that the graph is shifted c units, and the shift is to the right if $c < 0$ or to the left if $c > 0$.

COMBINED EFFECTS
OF a, b, AND c,
IN a·cir(bx+c)

How then are we to put these three individual effects together? Let us again look at sine and try to use reasoning before examples this time.

The effect of b and c are felt first in the sense that they affect the argument of sine and hence the position of the graph along the x-axis, whereas a merely multiplies the values sine produces. Is the period still $2\pi/b$? Because

$$\sin\left(b\left(x + \frac{2\pi}{b}\right) + c\right) = \sin(bx + 2\pi + c)$$
$$= \sin[(bx + c) + 2\pi] = \sin(bx + c),$$

the answer is yes. Is the shift still $|c|$ units? This is harder to answer because there is a question of which is being shifted: the original sine graph or the stretched or shrunken one? This is most easily dealt with

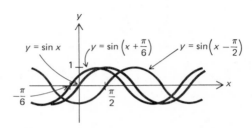

FIGURE 3-65

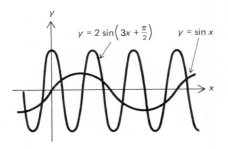

$y = 2 \sin\left(3x + \frac{\pi}{2}\right)$ $y = \sin x$

FIGURE 3-66

COMBINED EFFECTS
OF *a*, *b*, and *c*,
IN *a* . cir(*bx*+*c*)

by merely asking where a zero of the new function occurs. Once we have that information, the period dictates the spacing of zeros. An easy zero occurs when $bx + c = 0$ since $\sin 0 = 0$. The position of this zero dictates the start of a "standard period" for sine. But $bx + c = 0$ if $x = -c/b$, so that the shift is c/b units, and is still to the right if $c < 0$ and to the left if $c > 0$. The number $-c/b$ is often called the *phase shift* of the new function.

Sketch the graph of $2 \sin(3x + \pi/2)$. The period is $2\pi/3$, the beginning of a "standard period" occurs at $(-\pi/2)/3 = -\pi/6$, and the amplitude is 2. In other words, the graph is as sketched in Figure 3.66.   **EXAMPLE 39**

Sketch the graph of $\frac{1}{3} \csc(2x - \pi/2)$. Although we have not dealt specifically with cosecant in this section, we have pointed to lines of attack. The graph of $\frac{1}{3} \csc(2x - \pi/2)$ will have the general appearance of that cosecant. The earlier discussions of period and shift still apply. Since $\csc[2(x + 2\pi/2) - \pi/2] = \csc(2x - \pi/2 + 2\pi) = \csc(2x - \pi/2)$, the period is $2\pi/2 = \pi$, and since cosecant is undefined at 0, $\csc(2x + \pi/2)$ will be undefined when $2x - \pi/2 = 0$, or $x = \pi/4$. This together with the factor $\frac{1}{3}$ yields the graph of Figure 3.67.   **EXAMPLE 40**

Sketch the graph of $\frac{3}{2} \sin(-\frac{1}{2}x + \pi)$.   **EXAMPLE 41**

$$\frac{3}{2} \sin\left(-\frac{1}{2}x + \pi\right) = \frac{3}{2} \sin\left(-1\left(\frac{1}{2}x - \pi\right)\right) = -\frac{3}{2} \sin\left(\frac{1}{2}x - \pi\right).$$

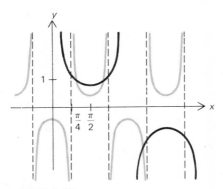

FIGURE 3-67

3.8.   *OTHER FUNCTIONS BASED ON THE CIRCULAR FUNCTIONS*   207

COMBINED EFFECTS
OF *a*, *b*, AND *c*,
IN $a \cdot \text{cir}(bx + c)$

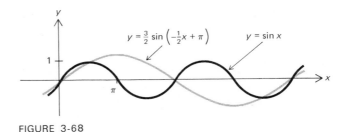

FIGURE 3-68

Here, then, the period is 4π; a standard period starts at 2π; and the amplitude is $\frac{3}{2}$. That $-\frac{3}{2} < 0$ means that the standard period will switch sides of the x-axis. Figure 3.68 shows the result.

3.9. SOME PROJECTS

In order to better understand both the circular functions themselves and the strategies used in developing their properties, the reader is asked in this section to conduct a similar investigation of some analogous functions. We admit that these functions are not as useful in the real world as the circular functions, but as mathematicians we need not let that be of very great concern—our concern is rather with developing structure. Besides, many mathematical discoveries and investigations have become useful only after the fact. The investigation closely parallels our study of the circular functions, so reference to earlier parts of this chapter will probably be helpful.

The functions we shall study are the *square functions*, which we now define. Let the square whose edges are parallel to the x- and y-axis and which intersect the axes at ± 1 be called the unit square. See Figure 3.69. Define the function Q with domain $\mathbb{R}$ as follows.

$$Q(w) = \text{the point on the unit square } w \text{ units from } (1, 0),$$

the distance being measured along the unit square with the positive

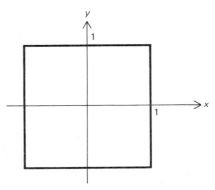

FIGURE 3-69

direction being counterclockwise. Next we define the coordinate functions α and β with domain the set of points in the plane by: if P is a point in the plane with coordinates (a, b), then

$$\alpha(P) = a \qquad \beta(P) = b.$$

Finally, the basic square functions friend and cofriend are defined by

$$\left.\begin{array}{l} \text{fri } x = (\beta \circ Q)(x) \\ \text{cof } x = (\alpha \circ Q)(x) \end{array}\right\} \text{for each } x \in \mathbb{R}.$$

We now break the investigation of friend, cofriend, and their relatives into phases.

Phase 1. In order to gain some initial insight into these functions, you are to graph each of them carefully. Discuss the periodicity of each, and if either is periodic, give its period.

Phase 2. Related to phase 1 is the problem of evaluating the functions for various arguments. Find methods similar to those of Section 3.3c that are valid for friend and cofriend. You should carefully explain both the methods and their derivation or justification. What values ought to be singled out as "special" and why?

Phase 3. Write a flowchart to compute values for fri t, at least for some reasonable subset of the domain.

Phase 4. Define four more square functions analogous to tangent, secant, cosecant, and cotangent; graph them; and comment on their behavior.

Phase 5. Find identities analogous to (5), (6), and (7) of Section 3.5 and Problem 52.

Phase 6. Find the inverses for all six square functions. Discuss the issue of function as it applies to these inverses.

Phase 7. Graph several specific functions of the form a sqr$(bx + c)$ where sqr represents any one of the six square functions.

Phase 8. Discuss the problem of formulas for fri$(x + y)$, fri$(x - y)$, cof$(x + y)$, and cof$(x - y)$.

Such an investigation might also be carried out with other functions than the square functions. Several are suggested below. For each set, one should be guided by the phases above, modifying each as seems appropriate for the specific functions.

1. The diamond functions. Consider the *unit diamond* shown in Figure 3.70. Let $M(t)$ be the point of the unit diamond t units form $(1, 0)$, where the distance is measured along the diamond with the positive direction being counterclockwise. Compose M with the coordinate functions α and β used to help define the square functions to define the diamond functions facet and sparkle as follows:

$$\text{facet: fac } t = (\alpha \circ M)(t)$$
$$\text{sparkle: spk } t = (\beta \circ M)(t).$$

(*Note:* One can write a rather simple equation for the unit diamond itself by using absolute values. Try it as Phase 0.)

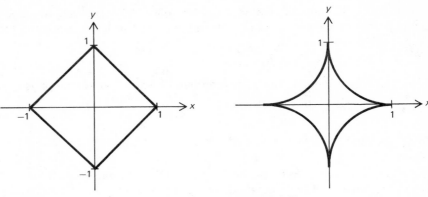

FIGURE 3-70   FIGURE 3-71

2. The "collapsed circular functions" or sorcular functions. Consider the *collapsed unit circle* of Figure 3.71, in which each of the four arcs is a quarter of a circle with radius 1 (for want of a better name, we dub such a figure a sorcle, and this one in particular the unit sorcle.) Define $N(t)$ as the point of the unit sorcle t units from $(1, 0)$, the distance being measured along the sorcle with the positive direction beginning counter-clockwise. Compose N with the coordinate functions α and β used to help define the square functions to define the sorcular functions wiggle and waggle as follows:

$$\text{wiggle}: \text{wig } t = (\alpha \circ N)(t)$$
$$\text{waggle}: \text{wag } t = (\beta \circ N)(t).$$

Discuss any especially close relationships these functions have with the circular functions (since a sorcle is composed of pieces of circles).

3. The "Skware" functions. Another square than that we have called the unit square and used for defining the square functions, and which also seems to be special is the square with sides formed by the axes and the lines $x = 1$ and $y = 1$. See Figure 3.72. We shall call this the funda-

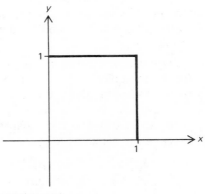

FIGURE 3-72

mental square. Let $K(t)$ be the point of the fundamental square t units from $(1, 0)$, where distance is measured along the square with the positive direction being counterclockwise. Compose K with the coordinate functions α and β used to define the square functions to define what we shall call the skware functions, zig and zag.

PROJECTS

$$\text{zig } t = (\alpha \circ K)(t)$$
$$\text{zag } t = (\beta \circ K)(t).$$

You may wish to answer the question of how these functions are related to the square functions.

4. *The "expanded circular functions" or round functions.* Instead of using the unit circle, use the circle of radius 2 centered at the origin. Develop the definition and names for the functions that result from this circle as the circular functions did from the unit circle. Calling these the round functions, find as much information as you can about them, not being bound by the phases suggested for the study of the square functions. Clearly explain all that you do.

3.10. AN OVERVIEW

So far, we have investigated polynomial functions and circular functions. The questions we have asked about these two sets of functions have not been the same. This is at least in part a result of the differences between the two sets. Zeros were not easy to find for polynomials and so demanded the bulk of our time and effort. Zeros were easy for the circular functions and so occupied little of our time. We looked more closely at inverses of the circular functions, since there are only six circular functions, and the task was manageable. Polynomial functions differ from each other to such an extent that a systematic analysis of their inverses was much too extensive a task. (See also Section 2.10.)

OVERVIEW

With the circular functions, we investigated various relationships among the functions. Although, at first glance, we may not appear to have done the same thing for polynomials, in fact we did in that any two polynomials p and q (where the degree of p is $\leq$ the degree of q) are related by the existence of polynomials d and r so that $p(x) = q(x)d(x) + r(x)$ for all x. We did not develop formulas for $p(x + y)$ because they are not at all simple enough to be useful, and because polynomials do not present the computational problems that in part motivated our search for such formulas among the circular functions.

In short, though the reasons for different emphases were not spelled out in detail all along the line, the difference in our study of polynomial and circular functions has been dictated not by whim but by the nature of the functions.

We might also note one instance of interplay between these two sets of elementary functions, which we have encountered but not examined as such. Notice that in using the formulas of Section 3.3a for sine and cosine, we actually computed approximations, using only finitely many terms. The use of only finitely many terms results in the use of a polynomial function. Thus if we use

$$t - \frac{t^3}{3!} + \frac{t^5}{5!} - \frac{t^7}{7!}$$

to get approximate values for sine, we are claiming that the seventh degree polynomial is fairly close to sine for at least some values. Such a use of polynomials to approximate other functions was mentioned in Chapter 2, but a full investigation requires the powerful tools of calculus.

Finally, the two sets of functions can be brought together to form a kind of hybrid that involves both. Such functions are, of course, more difficult and, one might say, less elementary. Section 3.8 examined the composition of linear functions with the circular functions, and we take a further brief look at other hybrids in Problems 83, 93 and 94 before going on to a new set of elementary functions.

SUMMARY OF IMPORTANT CONCEPTS AND NOTATION

UNIT CIRCLE: (Definition 3.1) The unit circle is the circle of radius 1, which has its center at the origin.

DEFINITION OF $P(t)$: (Definition 3.2) The point $P(t)$ is the point on the unit circle that is the terminal point of an arc, t units long, whose initial point is (1, 0). The positive direction for measuring arcs is counter-clockwise.

COSINE (Definition 3.3) The cosine of a real number t is the x-coordinate of the point on the unit circle that is t units from (1, 0), where the distance is measured along the circle and the positive direction is counterclockwise.

SINE: (Definition 3.3) The sine of a real number t is the y-coordinate of the point on the unit circle that is t units from (1, 0), where distance is measured along the circle and the positive direction is counterclockwise.

SPECIAL VALUES: (Section 3.3a)

FORMULAS FOR SIN t AND COS t: (Section 3.3a and Figure 3.10)

PERIODIC FUNCTION: (Definition 3.4) A periodic function is a function f for which there is some number p with the property $f(x+p) = f(x)$ for all $x \in D_f$.

PERIOD: (Definition 3.5) The smallest positive number p which satisifies $f(x+p) = f(x)$ for all $x \in D_f$ is called the period of the function f.

REDUCTION TO THE FIRST QUADRANT: (Section 3.3c and Figure 3.13)

GRAPHS OF SINE AND COSINE: (Section 3.4a and Figures 3.21 and 3.22)

OTHER CIRCULAR FUNCTIONS: (Definitions 3.6 and 3.7)

$$\tan t = \frac{\sin t}{\cos t}$$

$$\sec t = \frac{1}{\cos t}$$

$$\csc t = \frac{1}{\sin t}$$

$$\cot t = \frac{1}{\tan t}.$$

GRAPHS OF OTHER CIRCULAR FUNCTIONS: (Section 3.4c)
   Cosecant: (Figure 3.33)
   Secant: (Figure 3.34)
   Tangent: (Figure 3.36)
   Cotangent: (Figure 3.38)
IDENTITIES: (Section 3.5, including a partial summary of the identities near the end of Section 3.5a)
INVERSES OF CIRCULAR FUNCTIONS: (Section 3.6 and Figures 3.40 and 3.41) Sin^{-1}, Cos^{-1}, Tan^{-1}, Sec^{-1}, Csc^{-1}, Cot^{-1}, $\sin^{-1}$, $\cos^{-1}$, $\tan^{-1}$, $\sec^{-1}$, $\csc^{-1}$, $\cot^{-1}$.
RADIAN: (Definition 3.8) An angle has measure 1 radian if, as a central angle of a circle it subtends an arc whose length is equal to the length of the radius of the circle.
SOLUTION OF RIGHT TRIANGLES: (Section 3.7b)
TRIANGLE DEFINITIONS OF CIRCULAR FUNCTIONS: (Definitions 3.9a and 3.9b) In triangle ABC with $\angle ACB$ being a right angle,

$$\sin(\angle BAC) = \frac{\text{length of } BC}{\text{length of } AB},$$

$$\cos(\angle BAC) = \frac{\text{length of } AC}{\text{length of } AB},$$

$$\tan(\angle BAC) = \frac{\text{length of } BC}{\text{length of } AC}.$$

SOLUTION OF OBLIQUE TRIANGLES: (Definition 3.10 and Section 3.7c)
LAW OF COSINES: (Formula 3.1, Section 3.7c)

$$b^2 = a^2 + c^2 - 2ac \cos B$$
$$c^2 = a^2 + b^2 - 2ab \cos C$$
$$a^2 = b^2 + c^2 - 2bc \cos A.$$

LAW OF SINES: (Formula 3.2, Section 3.7c) $a/\sin A = b/\sin B = c/\sin C$.
FUNCTIONS OF THE FORM $a \cdot \text{cir}(bx + c)$: (Section 3.8)

PRACTICE EXERCISES

Section 3.2a

1. Draw a large graph of the unit circle; locate the following points as carefully as you can on the circle; and label them.

(a) $P(\pi)$ (b) $P(2\pi)$ (c) $P(3\pi)$
(d) $P(4\pi)$ (e) $P(8\pi)$ (f) $P(9\pi)$
(g) $P(\pi/2)$ (h) $P(3\pi/2)$ (i) $P(5\pi/2)$
(j) $P(7\pi/2)$ (k) $P(9\pi/2)$ (l) $P(3\pi/4)$
(m) $P(5\pi/4)$ (n) $P(-5\pi/4)$ (o) $P(\pi/6)$
(p) $P(2\pi/3)$ (q) $P(-\pi/3)$ (r) $P(-\pi/2)$
(s) $P(7\pi/6)$ (t) $P(1)$ (u) $P(2)$
(v) $P(3)$ (w) $P(4)$ (x) $P(5)$

Section 3.2b

2. Making use of your knowledge of the P function (e.g., Figure 3.3 or Problem 1) find exact values for the following.
(a) $\sin \pi$ (b) $\cos \pi$ (c) $\sin \pi/2$
(d) $\cos \pi/2$ (e) $\sin 0$ (f) $\cos 0$
(g) $\sin 3\pi/2$ (h) $\cos 3\pi/2$ (i) $\sin -\pi$
(j) $\cos -\pi$ (k) $\sin -\pi/2$ (l) $\cos -\pi/2$
(m) $\sin 8\pi$ (n) $\cos 9\pi$ (o) $\sin 11\pi/2$
(p) $\cos 17\pi/2$ (q) $\sin -9\pi/2$ (r) $\cos -13\pi/2$

Section 3.3a

3. Using the formulas on p. 159 for $\sin t$, compute approximations for $\sin t$ by using the first (a) two terms; (b) three terms; (c) four terms; and (d) five terms of each sum. Then compare the results with each other and the value given in Table 1, Appendix D. Use t equals (i) 1; (ii) -1; (iii) 1.5; (iv) -1.5; (v) 0; (vi) 5; and (vii) -5.
4. Redo Problem 3 for $\cos t$.

Section 3.3c

5. Using Table 1, Appendix D, find approximations for the following values.
(a) $\sin 0.15$ (b) $\cos 1.27$ (c) $\sin 2.20$
(d) $\cos 2$ (e) $\sin 3$ (f) $\cos 3$
(g) $\sin 4$ (h) $\cos 5$ (i) $\sin 6$
(j) $\cos 6$ (k) $\sin 7$ (l) $\cos 7$
(m) $\sin -1.35$ (n) $\cos -0.77$ (o) $\sin -5.5$
(p) $\cos -6$ (q) $\sin -10$ (r) $\cos -20$
(s) $\sin 15$ (t) $\cos 15$ (u) $\sin 10$
(v) $\cos 10$ (w) $\sin 13$ (x) $\cos 13$

6. Find exact values for each of the following.
(a) $\cos 7\pi/6$ (b) $\sin 3\pi/4$ (c) $\cos 3\pi/4$
(d) $\sin 11\pi/6$ (e) $\cos 11\pi/6$ (f) $\sin 4\pi/3$
(g) $\cos 4\pi/3$ (h) $\sin 5\pi/4$ (i) $\cos 5\pi/4$
(j) $\sin 2\pi/3$ (k) $\cos 2\pi/3$ (l) $\sin 7\pi/4$
(m) $\cos 7\pi/4$ (n) $\sin 5\pi/6$ (o) $\cos 5\pi/6$
(p) $\sin 7\pi/6$ (q) $\cos 9\pi/6$ (r) $\sin 5\pi/3$

(s) $\cos 5\pi/3$ (t) $\sin 15\pi/4$ (u) $\cos 25\pi/3$
(v) $\sin -7\pi/6$ (w) $\cos -13\pi/4$ (x) $\sin -11\pi/6$
(y) $\cos 17\pi/3$ (z) $\sin 21\pi/4$

Section 3.4b

7. Find exact values for the following.
   (a) $\tan 4\pi/3$ (b) $\cot 7\pi/6$ (c) $\sec 3\pi/4$
   (d) $\cos 7\pi/6$ (e) $\tan -\pi/3$ (f) $\cot 3\pi/2$
   (g) $\sec -3\pi/4$ (h) $\sin 10\pi/3$ (i) $\tan 5\pi/4$
   (j) $\sec 5\pi/4$ (k) $\cot 17\pi/6$ (l) $\tan 7\pi/6$
   (m) $\tan 7\pi/4$ (n) $\sin 7\pi/4$ (o) $\cot 2\pi/3$
   (p) $\sec 2\pi/3$ (q) $\cos 8\pi$ (r) $\cot 8\pi$
   (s) $\cos 13\pi/2$ (t) $\csc 13\pi/2$ (u) $\sin 11\pi/6$
   (v) $\sec 15\pi/3$ (w) $\csc 7\pi/3$ (x) $\cot 4\pi/3$
8. Find approximate values for each of the following.
   (a) $\sin 1$ (b) $\cos 20$ (c) $\sec 2$
   (d) $\csc 6.5$ (e) $\tan 5.7$ (f) $\cot 5$
   (g) $\cot 1.7$ (h) $\sec 5.2$ (i) $\tan 1.3$
   (j) $\sin 10$ (k) $\cot 2$ (l) $\cos 1$
   (m) $\sin 4.8$ (n) $\sec 6$ (o) $\csc 2$
   (p) $\cot 7.8$ (q) $\tan 1$ (r) $\cos 1.9$
9. Find values for the following, being exact wherever possible.
   (a) $\sin 8\pi/3$ (b) $\cot 16$ (c) $\tan -3.9$
   (d) $\csc -19\pi/4$ (e) $\sec 8$ (f) $\cos -6.7$
   (g) $\sin 21$ (h) $\cos -8$ (i) $\cot -3$
   (j) $\sin 2.73$ (k) $\sin 1.76$ (l) $\tan -9\pi/4$
   (m) $\sec -2.73$ (n) $\cot 36$ (o) $\cos -2.1$
   (p) $\tan 8$ (q) $\cot 30\pi/6$ (r) $\csc -3.6$
   (s) $\cos -1.5$ (t) $\sec -10\pi$ (u) $\sin -0.7$

Section 3.4c

10. Use the graphical techniques of Section 3.4c to observe the relation-
    ships and make careful graphs of
    (a) Sine and cosecant on the same axes.
    (b) Cosine and secant on the same axes.
    (c) Cotangent and tangent on the same axes.

Section 3.5a

11. Simplify each of the following as far as you can.
    (a) $1 + \tan(2t) \tan t$

    (b) $\dfrac{\sec x + \csc x}{1 + \tan x}$

    (c) $\dfrac{2 \tan a}{1 + \tan^2 a}$

(d) $\sec(t + 3\pi/2)$

(e) $\dfrac{3 \sin 2t}{\tan t} + 6 \sin^2 t$

(f) $(\cos B - \sin B)^2$

(g) $\csc t - \cot t \cos t - \sin t$

(h) $\dfrac{\tan u}{1 - \cot u} + \dfrac{\cot u}{1 - \tan u} - 1 - \tan u$

(i) $\sec^4 x - \tan^4 x - 1$

(j) $\sin t/(1 + \cos t) + \cot t$

(k) $\cot A(\cot A + \tan A)$

(l) $\cos \theta + \sin \theta \tan \theta$

(m) $\sin a \sec a \cot a$

(n) $\sec t + \csc t \cot t$

(o) $\dfrac{\sin x + \tan x}{1 + \sec x}$

(p) $\dfrac{\sin a \sec a}{\tan a + \cot a}$

(q) $\dfrac{1 - \cos 2x}{\sin 2x}$

12. Prove that the following are identically true.

(a) $\dfrac{1 + \tan^2 t}{\tan^2 t} = \csc^2 t$

(b) $\dfrac{\sin 2t}{\sin 4t} = \dfrac{1}{2} \sec 2t$

(c) $\tan x + \cot x = \sec x \csc x$

(d) $\dfrac{1 + \cot a}{\csc a} = \dfrac{1 + \tan a}{\sec a}$

(e) $\dfrac{1 - \cos x}{\sin x} = \dfrac{\sin x}{1 + \cos x}$

(f) $\dfrac{\sin x + \sin y}{\sin x - \sin y} = \dfrac{\csc y + \csc x}{\csc y - \csc x}$

(g) $\tan \dfrac{t}{2} = \dfrac{1 + \sin t - \cos t}{1 + \sin t + \cos t}$

(h) $\sec \theta - \tan \theta = \cos \theta/(1 + \sin \theta)$

(i) $\tan^2 x - \sin^2 x = \tan^2 x \sin^2 x$

(j) $\csc^4 w - \cot^4 w = \csc^2 w + \cot^2 w$

(k) $\dfrac{\sin^3 \theta - \cos^3 \theta}{\sin \theta - \cos \theta} = 1 + \sin \theta \cos \theta$

(l) $\dfrac{\sin 3x}{\sin x} - \dfrac{\cos 3x}{\cos x} = 2$

(m) $\sin 2x/(1 + \cos 2x) = \tan x$

(n) $\dfrac{\tan t + \sec t - \cos t}{\sec t + \tan t} = \sin t$

(o) $\sin^4 y - \cos^4 y + \cos 2y = 0$

(p) $\tan y + \cot y = 2 \csc 2y$

(q) $\dfrac{\sin 3t}{\sin t} = 2 \cos 2t + 1$

(r) $\tan 2\theta = \dfrac{2}{\cot \theta - \tan \theta}$

(s) $\dfrac{\sec t}{1 + \cos t} = \dfrac{\sec t - 1}{\sin^2 t}$

(t) $\dfrac{\tan \theta - \tan \phi}{\cot \theta - \cot \phi} + \tan \theta \tan \phi = 0$

(u) $\sec 2x = \dfrac{\sec^2 x}{2 - \sec^2 x}$

Section 3.5b

13. (i) Write as sums:
    (a) $\sin 4x \cos 3x$
    (b) $\sin 6x \cos 7x$
    (c) $\sin 8x \sin 8x$
    (d) $\cos 2x \cos 3x$
    (e) $\cos 3x \sin 2x$
    (f) $\sin -5x \cos 3x$
    (g) $\cos (-2x) \cos 2x$
    (h) $\cos 12x \sin (-3x)$
    (i) $\cos 3x \sin 3x$
    (j) $\sin (-2x) \sin (-3x)$

    (ii) Write as products:
    (a) $\sin 4x + \sin 3x$
    (b) $\cos 6x + \cos 7x$
    (c) $\sin 8x - \sin 8x$
    (d) $\cos 2x + \cos 3x$
    (e) $\cos 3x - \cos 2x$
    (f) $\sin (-3x) - \sin (-2x)$
    (g) $\cos 3x - \cos (-2x)$
    (h) $\sin (-2x) + \sin 5x$
    (i) $\sin 8x - \sin (-3x)$
    (j) $\cos (-2x) + \cos 2x$

Section 3.6

14. Evaluate each of the following.
    (a) $\text{Sin}^{-1} 0$
    (b) $\text{Sin}^{-1} \sqrt{3}/2$
    (c) $\text{Cos}^{-1} -\tfrac{1}{2}$
    (d) $\text{Cos}^{-1} \sqrt{3}/2$
    (e) $\text{Tan}^{-1} 1$
    (f) $\text{Tan}^{-1} -\sqrt{3}$
    (g) $\text{Csc}^{-1} 2$
    (h) $\text{Cot}^{-1} -1$
    (i) $\text{Sec}^{-1} -2$
    (j) $\text{Cos}^{-1} 1$
    (k) $\text{Sin}^{-1} -1$
    (l) $\text{Tan}^{-1} \sqrt{3}/3$
    (m) $\sin^{-1} -\tfrac{1}{2}$
    (n) $\cos^{-1} \sqrt{3}/2$
    (o) $\tan^{-1} 1$
    (p) $\cot^{-1} -\sqrt{3}$
    (q) $\sec^{-1} -1$
    (r) $\csc^{-1} 2/\sqrt{3}$

15. Evaluate each of the following, using Table 1.
    (a) $\text{Sin}^{-1} 0.3051$
    (b) $\text{Cos}^{-1} 0.7900$
    (c) $\text{Cot}^{-1} 1.140$
    (d) $\text{Csc}^{-1} 1.011$
    (e) $\text{Sec}^{-1} 2.833$
    (f) $\text{Tan}^{-1} 0.4586$
    (g) $\sin^{-1} 0.9915$
    (h) $\cot^{-1} 0.8267$
    (i) $\tan^{-1} -0.7602$
    (j) $\sec^{-1} 1.011$
    (k) $\csc^{-1} 2.048$
    (l) $\cos^{-1} 0.2675$

(m) $\cot^{-1} 0.6674$ (n) $\sec^{-1} -1.025$ (o) $\csc^{-1} 3.612$
(p) $\tan^{-1} 1.225$ (q) $\sin^{-1} -0.7833$ (r) $\cos^{-1} -0.3058$

Section 3.7a

16. Find the radian equivalents of the following angles:
    (a) 5°   (b) 10°   (c) $22\frac{1}{2}°$ (d) 130°   (e) 50°
    (f) 400°   (g) 1°   (h) 600°   (i) 36°
17. Find the degree equivalents of the following angles in radians.
    (a) $\pi/12$ (b) 1   (c) 23   (d) 6   (e) $\pi/7$
    (f) $7\pi/4$ (g) 30   (h) $\pi/8$ (i) $\pi/5$

Section 3.7b

18. Using Table 2, find approximate values for the following.
    (a) $\sin 15°$ (b) $\cos 37°$ (c) $\tan 42°$
    (d) $\sin 81°$ (e) $\cos 77°$ (f) $\tan 66°$
    (g) $\sec 2°$ (h) $\csc 70°$ (i) $\cot 12°$
    (j) $\csc 18°$ (k) $\cot 51°$ (l) $\sec 22°$
    (m) $\sin 11°20'$ (n) $\cos 23°50'$ (o) $\sec 62°30'$
    (p) $\csc 31°12'$ (q) $\tan 85°50'$ (r) $\cot 51°5'$
19. Using Table 2, Appendix D, find approximate values for the following.
    (a) $\mathrm{Sin}^{-1} 0.2924$ (b) $\mathrm{Cos}^{-1} 0.5585$ (c) $\mathrm{Tan}^{-1} 1.4281$
    (d) $\mathrm{Sec}^{-1} 5.241$ (e) $\mathrm{Csc}^{-1} 2.790$ (f) $\mathrm{Cot}^{-1} 2.7475$
    (g) $\mathrm{Sin}^{-1} 0.1650$ (h) $\mathrm{Cos}^{-1} 0.6756$ (i) $\mathrm{Tan}^{-1} 0.8000$
    (j) $\mathrm{Cot}^{-1} 10$ (k) $\mathrm{Sec}^{-1} 5$ (l) $\mathrm{Csc}^{-1} 1.5$
20. Using the standard triangle below, solve the triangle if:
    (a) $\alpha = 20°, b = 10$ ft   (b) $\alpha = 35°, a = 13$ ft
    (c) $\alpha = 10°, c = 5$ in.   (d) $\beta = 3°, a = 100$ mm
    (e) $\beta = 40°, b = 50$ ft   (f) $\beta = 15°, c = 25$ ft
    (g) $a = 10$ ft, $b = 10$ ft   (h) $a = 5$ ft, $b = 12$ ft
    (i) $a = 6$ ft, $c = 15$ ft   (j) $b = 19$ in, $c = 21$ in.
    (k) $c = 6.3$ in, $b = 3.1$ in.   (l) $a = 10$ ft, $c = 13$ ft
    (m) $\alpha = 22°, c = 21$ ft   (n) $\beta = 50°, a = 15$ ft
    (o) $\alpha = 61°, b = 2$ ft

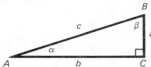

Section 3.7c

21. Labeling a triangle as shown, solve the triangle at the left if:
    (a) $a = 6, c = 10, B = 10°$ (b) $a = 6, c = 10, C = 10°$
    (c) $c = 6, a = 4, A = 30°$ (d) $a = 4, b = 5, c = 6$
    (e) $a = 5, b = 12, c = 15$ (f) $b = 20, A = 100°, B = 20°$
    (g) $A = C = 40°, b = 15$ (h) $A = 5°, b = 7, c = 20$
    (i) $b = 12, a = 20, A = 30°$ (j) $a = 2, b = 3, c = 20$

Section 3.8

For Problems 22–27, sketch graphs of the given functions on one set of axes.

22. (a) $\frac{1}{3}\cos x$ (b) $\cos x$ (c) $2\cos x$ (d) $4\cos x$
23. (a) $-\frac{1}{2}\cos x$ (b) $-\cos x$ (c) $-2\frac{1}{2}\cos x$ (d) $-3\cos x$
24. (a) $\sec x$ (b) $2\sec x$ (c) $\frac{1}{2}\sec x$ (d) $-2\sec x$
25. (a) $\csc x$ (b) $\frac{1}{3}\csc x$ (c) $-1\csc x$ (d) $2\csc x$
26. (a) $\tan x$ (b) $2\tan x$ (c) $\frac{1}{2}\tan x$ (d) $-\frac{1}{2}\tan x$
27. (a) $\cot x$ (b) $\frac{3}{2}\cot x$ (c) $-\cot x$ (d) $-2\cot x$
28. Sketch graphs of the following functions.
    - (a) $\cos 2x$
    - (b) $\cos 3x$
    - (c) $\cos \frac{1}{2}x$
    - (d) $\tan 2x$
    - (e) $\tan 3x$
    - (f) $\tan x/2$
    - (g) $\sec 2x$
    - (h) $\sec 3x$
    - (i) $\sec x/2$
    - (j) $\cos(x + \pi/3)$
    - (k) $\cos(x - \pi/2)$
    - (l) $\cos(x + \pi/4)$
    - (m) $\cot(x - \pi/6)$
    - (n) $\cot(x + \pi/4)$
    - (o) $\cot(x - \pi)$
    - (p) $\csc(x + \pi/2)$
    - (q) $\csc(x - \pi/3)$
    - (r) $\cos(x + \pi/6)$
29. Sketch graphs of the following functions.
    - (a) $\sin(2x + \pi)$
    - (b) $\cos(3x - \pi)$
    - (c) $\sin(2x + \pi/2)$
    - (d) $\cos(4x + \pi)$
    - (e) $\cos(2x + \pi)$
    - (f) $\sin(3x - \pi)$
    - (g) $4\cos(x + \pi)$
    - (h) $-2\sin(x - \pi/3)$
    - (i) $2\sin(x - \pi/6)$
    - (j) $\frac{1}{2}\cos(x + \pi/4)$
    - (k) $-\frac{1}{2}\cos(x + \pi/2)$
    - (l) $3\sin(x - \pi)$
    - (m) $3\sin 2x$
    - (n) $2\cos x/2$
    - (o) $\frac{1}{2}\cos 2x$
    - (p) $3\sin x/3$
    - (q) $-2\sin 3x$
    - (r) $\frac{1}{3}\cos 2x$
30. Sketch graphs of the following functions.
    - (a) $2\sin(3x + \pi)$
    - (b) $\frac{1}{3}\cos(2x - \pi/2)$
    - (c) $3\cos(x/2 + \pi)$
    - (d) $-3\sin(x/2 + \pi/3)$
    - (e) $(\frac{3}{2})\sin(5x - \pi)$
    - (f) $4\cos(4x - \pi)$
    - (g) $2\sin(-x + \pi/2)$
    - (h) $\frac{1}{2}\cos(4x + \pi)$
    - (i) $-2\sin(2x - \pi/3)$
    - (j) $2\cos(-x + \pi/2)$
    - (k) $\frac{1}{3}\sin(\pi/4 - x)$
    - (l) $2\cos(2x - \pi/3)$
    - (m) $(\frac{3}{2})\tan(x/2 + \pi)$
    - (n) $\frac{1}{3}\cot(2x - \pi/2)$
    - (o) $\pi\cos(\frac{3}{2}x + \pi)$
    - (p) $4\sin(3x/2 + \pi/4)$
    - (q) $\frac{1}{4}\sec(2x - \pi)$
    - (r) $2\cos(\pi x - 2\pi/3)$

PROBLEMS

31. State the quadrant(s) in which $P(t)$ may lie under each of the following conditions.
    - (a) $\sin t > 0$ and $\cos t < 0$
    - (b) $\sin t < 0$ and $\cos t < 0$
    - (c) $\sin t > 0$ and $\cos t > 0$
    - (d) $\sin t < 0$ and $\cos t > 0$
    - (e) $(\sin t)(\cos t) > 0$
    - (f) $(\sin t)(\cos t) < 0$
32. Prove that if f is a periodic function and g is any function for which $g \circ f$ can be formed, $g \circ f$ is also periodic.
33. Prove that no smaller positive number than 2π satisfies the conditions

for being the period for sin t or cos t and hence for $P(t)$. (Hint: For sin t, if there were a smaller number θ, it would have to work when $t=0$, i.e., $\sin(0+\theta)=\sin 0$. This leaves only one possibility for θ, and this can be eliminated by considering $t=\pi/2$. Once established for sine and cosine, show that it is true for P by contradiction (i.e., if there were a θ, it would have to work for sine and cosine, which is impossible).)

34. Adapt the argument that makes use of Figure 3.12 to deal specifically with the case that
    (a) $P(t)$ is in the third quadrant.
    (b) $P(t)$ is in the fourth quadrant.

35. On one set of axes, carefully draw the graph of each of the functions specified below, for $x \in [-1.6, 1.6]$. Use a full sheet of paper.
    (a) sec x (b) $f(x)=x^2+1$
    (c) $g(x)=x^2/10+1$ (d) $h(x)=3x^2+1$

36. Carefully explain why $\{\pi/2 + k\pi \mid k \in \mathbb{Z}\}$ is the set of zeros for cotangent.

37. The following sketches are each the graph of one period for some periodic function. Sketch a more complete graph of each function.

(a)
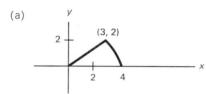

(b)
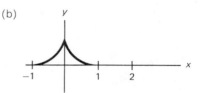

(c)
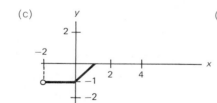

(d)
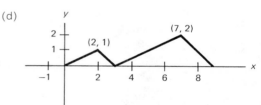

(e)
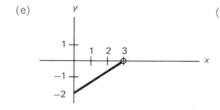

(f)
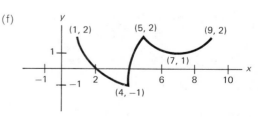

38. Some other circular functions one could define are given below:
    (a) widget: wid $t=(\sin t)(\cos t)$; (b) midget: mid $t=\sin t + \cos t$;
    (c) zidget: zid $t=\sin t - \cos t$; (d) alpha: alf $t=\sin^2 t$; (e) beta:
    bet $t=\cos^2 t$; and (f) gamma: gam $t=\tan^2 t$.

(i) Carefully draw a graph of each using the procedures followed in the text.

(ii) Determine the period of each and verify your claim as rigorously as you can.

(iii) What is the set of zeros for each?

(iv) Discuss other special points (maximums, minimums, "holes" in the domain, etc.)

39. Do the functions tangent and cotangent have maximum or minimum points? Explain carefully.

40. Repeat Problem 39 for secant and cosecant.

41. Classify each of the six circular functions as being odd, even, or neither. Verify your claims. (See Problems 89–91, Chapter 1.)

42. Derive the identity $\cot^2 t + 1 = \csc^2 t$.

43. Use the identity $\sin^2 t + \cos^2 t = 1$ to find values of the following expressions under the given conditions.

(a) $\sin t$ if $\cos t = \frac{4}{5}$ and $P(t)$ is in quadrant I.

(b) $\sin t$ if $\cos t = \frac{4}{5}$ and $P(t)$ is in quadrant IV.

(c) $\cos t$ if $\sin t = -\frac{5}{13}$ and $P(t)$ is in quadrant III.

(d) $\cos t$ if $\sin t = -\frac{5}{13}$ and $P(t)$ is in quadrant IV.

(e) What can you say about $\tan t$ if $\sin t = -\frac{3}{5}$.

44. Derive the identities $\cos 2t = 1 - 2\sin^2 t$ and $\cos 2t = 2\cos^2 t - 1$.

45. Find intervals for t for which

(a) $\sin \dfrac{t}{2} = \sqrt{\dfrac{1 - \cos t}{2}}$

(b) $\sin \dfrac{t}{2} = -\sqrt{\dfrac{1 - \cos t}{2}}$

(c) $\cos \dfrac{t}{2} = \sqrt{\dfrac{\cos t + 1}{2}}$

(d) $\cos \dfrac{t}{2} = -\sqrt{\dfrac{\cos t + 1}{2}}$

46. Derive a formula for each of the following:

(a) $\sin(t + u + x)$ (b) $\cos(t + u + x)$

(c) $\sin 3t$ (d) $\cos 3t$

(Use any of the formulas developed in the text.)

47. Derive a formula in terms of $\tan \alpha$ and $\tan \beta$, for

(a) $\tan(\alpha + \beta)$ (b) $\tan(\alpha - \beta)$

(Hint: Use the definitions and other known identities. A division of the numerator and denominator of a certain fraction by $\cos \alpha \cos \beta$ may prove of value, too.)

48. Is tangent odd, even, or neither?

(a) Answer by graphical observation.

(b) Answer by analytic proof, using facts about sine and cosine.

(c) Generalize part (b) by deciding about the oddness or eveness of f/g, where f is odd and g is even.

(See Problems 89–91, Chapter 1 for the definition of odd and even functions.)

49. Use Problems 47a and 48 to derive a formula for Problem 47b.

50. If f is periodic with period p, and g is also periodic with period p, what can be said about the periodicity of:

(a) $f + g$ (b) $f \cdot g$ (c) f/g (d) $f - g$

(Supply a proof or counterexample; answer whether periodic or not; and say as much as you can about the period in those cases which are periodic.)

51. Repeat Problem 50 if g has period m, where $m \neq p$.

52. Prove the following identities hold.
    (a) $\sin(\pi/2 - t) = \cos t$ (b) $\tan(\pi/2 - t) = \cot t$
    (c) $\cot(\pi/2 - t) = \tan t$ (d) $\sec(\pi/2 - t) = \csc t$
    (e) $\csc(\pi/2 - t) = \sec t$
    These, together with identity (6), in the text, exhibit relations between each function and its cofunction.

53. Use the identities to compute exact values for the following (Hint: $\pi/12 = \pi/3 - \pi/4$, etc.):
    (a) $\sin \pi/12$ (b) $\cos \pi/12$ (c) $\tan \pi/12$ (d) $\sin \pi/8$
    (e) $\cos \pi/8$ (f) $\tan \pi/8$ (g) $\sin 7\pi/12$ (h) $\cos 7\pi/12$
    (i) $\tan 7\pi/12$ (j) $\sec 5\pi/12$ (k) $\csc 5\pi/12$ (l) $\cot 5\pi/12$

54. Express each of the six circular functions just in terms of sine and constants.

55. Derive a formula for (a) $\tan t/2$ and (b) $\tan 2t$, in terms of tangents only.

56. A reasonable property to require of any system of measurement is that it be multiplicative, in the sense that increasing the item measured by a factor m should cause the measure to increase by the same factor. Assuming that this does hold for all systems of measurement you know, show that

    (a) The ratio of degrees/radians really is a constant.
    (b) An angle of t radians subtends an arc t units long.

    (Hint: From plane geometry, equal central angles subtend equal length arcs, so doubling a central angle doubles the length of arc subtended, and by extension, multiplying an angle by a factor σ causes a multiplication of the arc subtended by σ.)

57. Prove that if (x, y) is any point in the plane, θ is any of the angles formed by the positive x-axis and the ray from $(0, 0)$ through (x, y), and r is the distance from (x, y) to $(0, 0)$, then

    (a) $x = r \cos \theta$ (b) $y = r \sin \theta$
    (Hint: Use similar triangles.)

58. To find the height of a satellite above the earth, simultaneous sightings are made from two locations on earth. If the sightings are made at the instant that the satellite is directly above location A, location B is 4 miles from A on a line perpendicular to A's line of sight to the satellite, and the angle SBA is recorded to be 88°, how high is the satellite?

59. A house is 30 ft tall. What is the shortest a ladder could be to just reach the top of the house if the angle formed by the ladder and the ground is to be no greater than 80°?

60. At what angles should a piece of lumber be cut so as to be used as a brace for a vertical wall if the brace is to rest on the ground 5 ft from the wall and be fastened to the wall 7 ft above the ground? (See figure.)

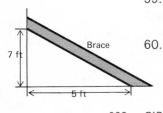

61. On a windless day, a little boy launches his glider from a window 22 ft from the ground. If he knows that the glider descends at an angle of 12° below the horizontal and that it will glide straight, how far from the house should he expect to find the glider?

62. An airliner at an altitude of 35,000 ft loses all its engines and glides to earth. If its glide path is 20° below the horizontal and the plane glides at a speed of 440 ft/sec, how long from the time the engines quit do the passengers and crew have to prepare for an unpremeditated landing?

63. A contractor has staked out a 36-ft-diameter circle in which to pile some gravel. If the angle of repose for the gravel (the largest angle between the horizontal and the slope of the pile for which no avalanches will occur) is 68°,
    (a) How high a pile of gravel can he get?
    (b) How many cubic yards of gravel will the pile contain?

64. Find the area of the lot shown.

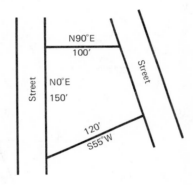

65. A radio transmitting tower is to be 150 ft high and is to have four guy wires to support it. If each wire is to make an angle of 70° with the ground, how much wire is needed?

66. If rancher Clipclop knows that his neighbor Znarq lives 6 miles due west of his house, and that the judge lives 13 miles due south of Znarq, at what angle away from due south should Clipclop ride to reach the judge's place when he discovers his top hand has been done-in?

67. A triangular corral with vertices X, Y, and Z has angle x = 30° and angle y = 50°. If side YZ is 100 yd long, what is the perimeter of the corral?

68. While riding a train across the great plains, a passenger saw a lone tree. When the tree was still ahead of the train (and since there was nothing else to see), he noted that the angle formed by his line of sight and the railroad tracks was 35°. Three miles later (still nothing else to see), the tree had been passed and the acute angle between the tracks and the line of sight was 30°. Obviously the tracks were in a straight line the whole time, so how close did the train come to the tree?

69. An air traffic controller sights airplane *A* and airplane *B* at the same altitude, headed toward each other. If plane *A* is 25 miles from the controller's radar station, *B* is 17 miles, and the angle between his two lines of sight is 28°, how far apart are the planes?

70. If a cartographer is doing aerial reconnaissance, and notes that the town of Galena and Hemitite are, respectively, $3\frac{1}{2}$ and $2\frac{3}{4}$ miles from his plane, and that the angle formed by the two towns with his plane as the vertex is 109°. How far apart are the towns (the straight line distance ignoring obstacles)?

71. Topographer Hengel Agehog wished to find out how high a certain mountain peak is above the plane. At one place in the plane, she measured the angle of her line of sight to the peak to be 22° above the horizontal. One half mile closer, she found the angle to be 27°. What is the altitude she seeks?

72. Find the area of lot (a).

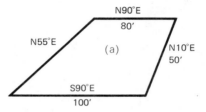

73. Find the area of parallelogram (b).

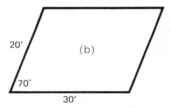

74. Star *A* is 25 light years from star *Q* and 40 light years from star *Z*. If the angle between the lines of sight from *Q* to the other two is 110°, how far apart are *Q* and *Z*?

75. If the angle in Problem 74 is from *A*, how far apart are *Q* and *Z*?

76. Leonia Milane is a sailing enthusiast. She needs to replace the sail on her boat. If the boom is perpendicular to the mast and is 4 m long, and the angle between the end of the boom and the top of the mast is 65°, how much will her new sail cost if sails cost $52.30 per sq m?

77. A civil engineer is to design a tunnel under a mountain. From the top of the mountain, there is a 52° angle between his lines of sight to the two places where the tunnel enters the mountain. A radar range finder tell him that the distance from his perch on top of the mountain to the entrance on one side is 200 m, and to the other entrance is 265 m.
    (a) How long will the tunnel be?
    (b) If, in addition, a plumb line divides the 52° angle into two angles,

one 30°, the other 22°, what can you say about the angle at which the tunnel departs from the horizontal?

78. If a plane heads due north at 150 miles per hour (mph), and a steady wind from the west blows at 10 mph,
    (a) How far has the plane actually traveled in 1 hour?
    (b) How far would the plane travel in 1 hour if the wind were coming from 20° south of west?

79. Prove that the area of a triangle with sides a, b, and c opposite angles α, β, and γ, respectively, is given by $A = \frac{1}{2}bc \sin \alpha$.

80. Complete the flowchart discussed in connection with Figure 3.54. It should input values for a, b, c, α, β, and γ, using zeros for quantities that are unknown, and output the actual values for all six quantities in in the same order.

81. Determine a valid rule for the function given by each of the following graphs.

(a)

(b)

(c)

(d)

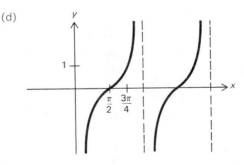

(e)

(f)

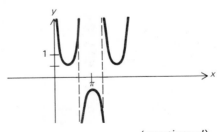

(*continued*)

(g)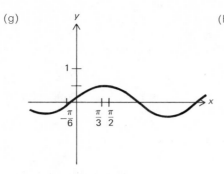

(h)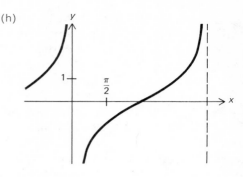

82. Sketch graphs of the following functions.
    - (a) $x \sin x$
    - (b) $\sin^2 x$
    - (c) $\sin^2 x + 2 \sin x + 1$
    - (d) $\sin x^2$
    - (e) $\sin(|x| - 2\pi/3)$
    - (f) $\sin^3 x$
    - (g) $\tan^2 x$
    - (h) $\tan^3 x$
    - (i) $\tan x^2$
    - (j) $\tan^2 x + 2 \tan x + 1$
    - (k) $\sec^2 x$
    - (l) $\sec^3 x$
    - (m) $\sec(|x| - 2\pi/3)$
    - (n) $\tan(|x| - 2\pi/3)$

    - (o) $\sec^2 x + 2 \sec x + 1$
    - (p) $\dfrac{\sin x}{1 + \cos x}$

    - (q) $\dfrac{1}{1 - \cos x}$
    - (r) $(\cos x - \sin x)^2$

    - (s) $\dfrac{\sin 2x}{\sin 4x}$
    - (t) $\dfrac{1}{2 - \cos x}$

83. Adapt the graphing program of Appendix C so as to graph (a) $a \sin(bx + c)$; (b) $a \sec(bx + c)$; and (c) $a \tan(bx + c)$ in a way that would help you see the effects of a, b, and c. (As it stands, these effects are hard to see; try a few examples.) The input should then be a, b, c, the endpoints of the interval, and the desired step size. Be sure to point out any limitations on a, b, c that you find it necessary to build into the program.

84. Discuss the domain and range of
    - (a) sine ∘ cosine
    - (b) cosine ∘ sine
    - (c) sine ∘ tangent
    - (d) tangent ∘ sine
    - (e) sine ∘ secant
    - (f) secant ∘ sine

85. Write a flowchart like that of Figure 3.10, but which computes approximate values of cosine using the first six terms of the cosine formula.

86. Write a flowchart to compute approximate values of sine, where both the argument and number of terms to be used are read in.

87. Write the flowchart of Problem 86 but make it suitable for computer implementation, i.e., with the computation of factorials detailed. Do you see any efficiencies; in other words, can you reduce the amount of multiplication and division entailed in the algorithm?

88. (a) Adapt the flowchart of Figure 3.13, so that it will input t and output both the value of t' and the proper sign for $\sin t$.

(b) Same as (a) but the sign for cos t.

(c) Same as (a) but the sign for tan t.

89. Prove (derive?) that

$$\tan \frac{t}{2} = \frac{1 - \cos t}{\sin t}$$

for all t for which both sides are defined.

90. Examine the effect of the constant d in $a \cdot \sin(bx + c) + d$, as was done for a, b, and c in Section 3.8. State both your conclusions and the documentation of your conclusions as clearly and precisely as you can.

91. Prove or disprove each of the following purported identities.

(a) $\dfrac{\cot x + \tan x}{\cot x - \tan x} = 1 - 2 \sin^2 x$

(b) $\dfrac{\cot \theta + \csc \theta}{\sin \theta + \tan \theta} = \dfrac{1}{2} \dfrac{\sin 2\theta \csc \theta}{\cos^2 \theta - \cos 2\theta}$

(c) $\dfrac{\tan x + \tan w}{\cot x + \cot w} = \dfrac{\tan x \tan w - 1}{1 - \cot x \cot w}$

(d) $\cos 2t = \cos^4 t - \sin^4 t$

(e) $\dfrac{\sec t - 1}{\sec t + 1} = \sec^2 \dfrac{t}{2} - 1$

(f) $\cot y - \tan y = \sec y \cot y$

(g) $(1 - \sin \beta)(\sec \beta + \tan \beta) = \cos \beta$

(h) $\dfrac{\tan t + \tan(w - t)}{1 + \tan t + \tan(t - w)} = \tan w$

(i) $\csc^2 \theta + \sec^2 \theta = \sec^2 \theta \csc^2 \theta$

(j) $\dfrac{\sec^2 t - \tan^2 t + \tan t}{\sec t} = \sec t + \csc t$

(k) $\dfrac{\sin(a + b)\sin(a - b)}{\cos^2 a \cos^2 b} = \sec^2 a - \sec^2 b$

(l) $\dfrac{\sin(a + b)}{\sin(a - b)} = \dfrac{\tan a + \tan b}{\tan a - \tan b}$

(m) $\dfrac{\cos(a + b)}{\cos(a - b)} = \dfrac{\cot a + \cot b}{\cot a - \cot b}$

(n) $\left(\dfrac{\sin 2w}{\sin w}\right)^2 + \left(\dfrac{\cos 2w}{\cos w}\right)^2 = 1$

(o) $\dfrac{\sin \theta + \sin 2\theta}{1 + \cos \theta + \cos 2\theta} = \tan \theta$

(p) $(\csc t - \cot t)(1 + \cos t) = \sin t$

(q) $\dfrac{\sin 2x - \sin x}{\cos 2x + \cos x} = \tan \dfrac{x}{2}$

(r) $\dfrac{\cos 3x - \cos x}{\cos 3x + \cos x} = -2 \tan 2x \tan x$

(s) $\sin 5\theta + \sin 3\theta = 8 \sin \theta \cos^2 \theta \cos 2\theta$

(t) $\dfrac{\cos 4x}{1 + \sin 4x} = \sec 4x - \tan 4x$

(u) $\dfrac{\cot \theta}{\csc \theta - 1} = \dfrac{\csc \theta + 1}{\cot \theta}$

(v) $1 + \tan 3x \tan 4x = \dfrac{2 \cos x}{\cos 7x + \cos x}$

92. Using the techniques of Section 3.8, and given the graphs of f in each problem below, determine the graph of $a \cdot f(bx + c)$ (where a, b, and c are specified in each problem) as efficiently as you can.

(a)

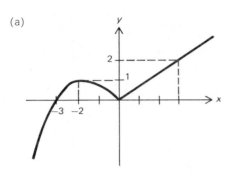

(b)

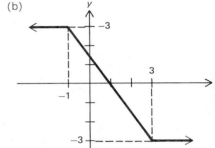

(c)

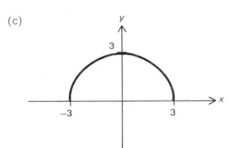

(d)

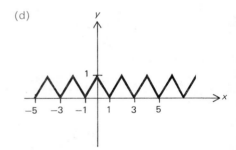

In Problems 93 and 94 consider finding zeros for a variety of functions built around the circular functions. Some include polynomial functions in various roles as well. It may help you to remember that finding the zeros of $f(x)$ is equivalent to solving the equation $f(x) = 0$ for x. In addition, remember that when we solve an equation, we choose among all the operations at our disposal to isolate the variable on one side of the equation and the constants on the other side. For example, in solving $5x + 3 = 13$, we first isolate $5x$ using subtraction and then complete the solution by division. If we performed an operation, say zap, on x and had zap $x + 3 = 13$, we would isolate zap x (zap $x = 10$ here) then try to "undo" the zap operation to "get back to x."

93. Find all the zeros for each of the following:

    (a) $m(x) = \sin x - \frac{1}{2}$ (b) $a(t) = \cos t + \sqrt{3}$

(c) $n(w) = \sec w - \sqrt{2}$

(d) $i(y) = \csc y + 2$

(e) $f(a) = \sin(a + 2)$

(f) $o(s) = \sin(s + 2) - \frac{1}{2}$

(g) $m(z) = \cos 4z$

(h) $d(u) = \tan(u + \pi/2) - 2$

(i) $n(t) = \tan t - 2$

(j) $u(x) = (3 \cot x) + \sqrt{3}$

(k) $m(w) = (\sec 2w) - \sqrt{2}$

(l) $m(z) = 2 \csc z + 2$

94. Find all the zeros for each of the following.

(a) $d(x) = \sin x \cos x$

(b) $i(z) = \cos z - \sec z$

(c) $m(t) = \sin^2 t + 2 \sin t + 1$

(d) $e(m) = 4 \cos^2 m - 1$

(e) $n(w) = \tan w(2 \sin w + 1)$

(f) $s(y) = 2 \sin^2 y - 3 \sin y + 1$

(g) $i(a) = \sin(a^2)$

(h) $o(k) = \cos(k^2 + 3k + 2) - 1$

(i) $n(j) = \sin(3j + 5) - \sin(4j - 2)$

(j) $a(p) = \cos(2p + 1) - \cos(3 - p)$

(k) $b(x) = 4 \sin^4 x - 12 \sin x \cos^2 x - 7 \cos^2 x + 9 \sin x + 5$

(l) $a(z) = (\sin \circ \sin)(z)$

(m) $c(t) = (\sin \circ \cos)(t)$

(n) $u(w) = \sin^2 w - 3 \cos^2 w$

(o) $s(x) = \sin x + \sin \dfrac{x}{2}$

(p) $p(t) = \csc^2 t - \cot^2 t - 1$

(q) $i(x) = 3 \tan x - 2 \cos x$

(r) $a(w) = (\cos w - \sin w)^2$

(s) $m(t) = \sin 2t / \sin 4t$

(t) $g(m) = \sin(m^2 - 2m - 4)$

(u) $o(a) = \sin a - \cot\left(a - \dfrac{3\pi}{2}\right)$

(v) $r(p) = \sin p + \cos p$

(w) $i(t) = \sin 3t + \sin^2 t + 1$

(x) $t(x) = 2 \sin x - \sin 2x$

(y) $h(x) = \sin 2x + \cos 2x - 1\frac{1}{2}$

(z) $m(w) = \cos 2w - 3 \sin w + 1$

95. Simplify, re-express, or reduce each of the following.

(a) $1 - \cos 3x$ (to a single term)

(b) $\sin 4x \cos 4x$

(c) $\cos(A + B)\cos A + \sin(A + B)\sin A$

(d) $\dfrac{1}{2} \sin \dfrac{\pi}{8} \cos \dfrac{\pi}{8}$

(e) $\sin^4 \dfrac{\theta}{2} \cos^4 \dfrac{\theta}{2}$ (to a form without exponents)

(f) $4 \cos 5x - 4$ (to a single term)

(g) $8 \sin^2 \dfrac{\theta}{2} \cos^2 \dfrac{\theta}{2}$

(h) $1 + \cos 4x$ (to a form using only constants, $\sin x$, and/or $\cos x$)

(i) $\sin^2 x^2 + \cos^2 x^2$

(j) $\sin^2 2x \cos^2 2x$ (to a form without exponents)

(k) $2 \sin \dfrac{\pi}{12} \cos \dfrac{\pi}{12}$

(l) $3 \sin^2 3x + 3 \cos^2 3x$

(m) $\tan^2(-2x) - \sec^2(-2x)$

(n) $\cos^4 t - \sin^4 t$

(o) $1 - \cos 4\theta$

(p) $\sin^4 w$ (using only functions to the first power)

(q) $\dfrac{\cos t - 1}{\sin t}$

CHAPTER 4

EXPONENTIAL AND LOGARITHMIC FUNCTIONS

4.1. INTRODUCTION

Exponential functions are elementary functions in the sense that they give good descriptions of various everyday phenomena such as population growth and radioactive decay. *Elementary*, as used here, does not mean that these functions are easily defined or dealt with. In fact, two of the most satisfying and rigorous approaches to exponential functions are through the solution of a certain differential equation and in conjunction with a certain definite integral, both of which approaches require a knowledge of the calculus.† Even the path we shall follow depends in some sense on the calculus because an understanding of exponential functions grew along with the development of calculus. As a result, we shall be forced to forgo some proof and rely on the word of the mathematical community.

Logarithmic functions will develop naturally in the course of investigating the exponential functions.

---

† There is another approach available that constructs the exponential functions as descriptions of a certain kind of growth pattern. Although this approach also needs some calculus to be completely rigorous, the dependence is less obvious than the route we have chosen. The interested reader is referred to M. P. Fobes, *Elementary Functions: Backdrop for the Calculus*, Chapter 8 (New York: Macmillan Publishing Co., 1973) for an excellent treatment of this approach.

This chapter will be divided into two parts. Sections 4.2 and 4.3 deal primarily with the theory surrounding exponentials and logarithms; Section 4.4 deals primarily with applications. Those who plan to go on in mathematics should make every attempt to master Sections 4.2 and 4.3, while others may prefer to put their emphasis on the rest of the chapter.

4.2. EXPONENTIAL FUNCTIONS

One of the mechanisms learned in algebra is that of exponents. Typically, one begins with natural numbers for exponents to indicate how many times a quantity is to be used as a factor in an expression, and certain laws of exponents are then discovered. The notion of exponent is then extended to permit any integer to be an exponent by defining x^0 to be equal to 1 if $x \neq 0$, and x^{-n} to be equal to $1/x^n$. The laws of exponents are then extended to cover the expanded set of permissible exponents. Finally, one moves further away from the idea of an exponent as an indication of the number of identical factors present by defining $x^{p/q}$ to mean $(\sqrt[q]{x})^p$ for any rational number p/q. Once again, the laws of exponents are extended to cover this even larger set of permissible exponents.

At this stage, the laws are†

$$\left.\begin{array}{l} 1. \ x^r \cdot x^s = x^{r+s} \\ 2. \ (xy)^r = x^r y^r \\ 3. \ (x/y)^r = x^r/y^r \\ 4. \ (x^r)^s = x^{rs} \\ 5. \ x^r/x^s = x^{r-s} \\ 6. \ x^{-7} = 1/x^r \end{array}\right\} r, s \in \mathbb{Q}‡$$

Polynomial functions, you will remember, make use of nonnegative integers for exponents, e.g., $3x^2 + 2x - 1x^0$. These exponents are fixed numbers (2, 1, and 0 in our example). What would happen if we let the exponent be the argument of a function or relation? In our discussion so far, we have seen that an exponent may be any rational number so that we can define an exponential function, $\exp_a$, by

DEFINITION 4.1a

$\exp_a$ on $\mathbb{Q}$

$$\exp_a(r) = a^r, \qquad r \in \mathbb{Q}$$

The number a is called the *base* of the exponential function $\exp_a$. If $\exp_a$ is to have domain $\mathbb{Q}$, not just $\mathbb{N}$, we must restrict the values of the base a to be in $(0, \infty)$. If we let $a = 0$, we would have $\exp_a(r) = 0^r$, which is not defined for $r = 0$; if we let a be negative, we would have $\exp_{-1}(\frac{1}{2}) = (-1)^{1/2} = \sqrt{-1}$, which is undefined. Only a few isolated values of $\exp_a$ would be defined for $a \in (-\infty, 0)$ such as $\exp_{-2} 3 = (-2)^3 = -8$; however, these few values do not give us enough reason to allow the base to be negative.

---

† If you are weak on the subject of exponents, see Appendix A, Section A.16, to refresh or strengthen your expertise before going on.
‡ Actually, there are only four really different laws since (3) is a consequence of (2 and 6), and (6) can be viewed as a special case of (5).

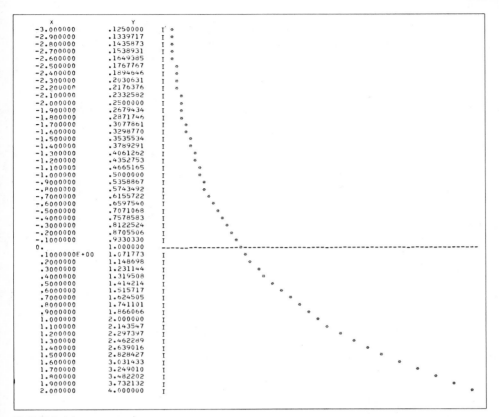

```
      x                 Y
 -3.000000         .1250000      I'  ⊛
 -2.900000         .1339717      I   ⊛
 -2.800000         .1435873      I   ⊛
 -2.700000         .1538931      I   ⊛
 -2.600000         .1649385      I   ⊛
 -2.500000         .1767767      I    ⊛
 -2.400000         .1894646      I    ⊛
 -2.300000         .2030631      I    ⊛
 -2.200000         .2176376      I    ⊛
 -2.100000         .2332582      I     ⊛
 -2.000000         .2500000      I     ⊛
 -1.900000         .2679434      I     ⊛
 -1.800000         .2871746      I      ⊛
 -1.700000         .3077861      I      ⊛
 -1.600000         .3298770      I       ⊛
 -1.500000         .3535534      I        ⊛
 -1.400000         .3789291      I        ⊛
 -1.300000         .4061262      I         ⊛
 -1.200000         .4352753      I         ⊛
 -1.100000         .4665165      I          ⊛
 -1.000000         .5000000      I           ⊛
  -.9000000        .5358867      I            ⊛
  -.8000000        .5743492      I            ⊛
  -.7000000        .6155722      I             ⊛
  -.6000000        .6597540      I             ⊛
  -.5000000        .7071068      I              ⊛
  -.4000000        .7578583      I               ⊛
  -.3000000        .8122524      I                ⊛
  -.2000000        .8705506      I                 ⊛
  -.1000000        .9330330      I                  ⊛
 0.                1.000000      -------------------⊛-----------------------------------
  .1000000E+00     1.071773      I                    ⊛
  .2000000         1.148698      I                     ⊛
  .3000000         1.231144      I                      ⊛
  .4000000         1.319508      I                       ⊛
  .5000000         1.414214      I                        ⊛
  .6000000         1.515717      I                         ⊛
  .7000000         1.624505      I                          ⊛
  .8000000         1.741101      I                            ⊛
  .9000000         1.866066      I                             ⊛
 1.000000          2.000000      I                              ⊛
 1.100000          2.143547      I                               ⊛
 1.200000          2.297397      I                                 ⊛
 1.300000          2.462289      I                                  ⊛
 1.400000          2.639016      I                                    ⊛
 1.500000          2.828427      I                                      ⊛
 1.600000          3.031433      I                                       ⊛
 1.700000          3.249010      I                                         ⊛
 1.800000          3.482202      I                                           ⊛
 1.900000          3.732132      I                                             ⊛
 2.000000          4.000000      I                                               ⊛
```

FIGURE 4.1a. Computer generated graph of $\exp_2$.

An attempt to graph $\exp_a$ is an exercise in frustration. Since the domain is ℚ not ℝ, one does not wish to let the graph include points whose abscissas are real but not rational, nor does one wish to omit any rational abscissas. The rational numbers are so densely scattered among the real numbers, however, that no such separation can be made except in the mind. We shall not be stopped so easily, however. Let us try to graph $\exp_2 x = 2^x$ as well as we can in spite of the difficulty just described. Using the graphing program of Appendix C, we could get the results shown in Figure 4.1(a) and 4.1(b) where we should note that we have only rational arguments.

These graphs should be an inspiration in the sense that the apparently good (unerratic) behavior for rational arguments suggests that a sensible definition for a^x ought to be obtainable for all real x, which would be compatible with the definitions to date.

In fact the domain of $\exp_a$ can be extended to ℝ, but the details and **NOTE**

X	Y
0.	1.000000
.2000000E-01	1.013959
.4000000E-01	1.028114
.6000000E-01	1.042466
.8000000E-01	1.057018
.1000000	1.071773
.1200000	1.086735
.1400000	1.101905
.1600000	1.117287
.1800000	1.132884
.2000000	1.148698
.2200000	1.164734
.2400000	1.180993
.2600000	1.197479
.2800000	1.214195
.3000000	1.231144
.3200000	1.248331
.3400000	1.265757
.3600000	1.283426
.3800000	1.301342
.4000000	1.319508
.4200000	1.337928
.4400000	1.356604
.4600000	1.375542
.4800000	1.394744
.5000000	1.414214
.5200000	1.433955
.5400000	1.453973
.5600000	1.474269
.5800000	1.494849
.6000000	1.515717
.6200000	1.536875
.6400000	1.558329
.6600000	1.580083
.6800000	1.602140
.7000000	1.624505
.7200000	1.647182
.7400000	1.670176
.7600000	1.693491
.7800000	1.717131
.8000000	1.741101
.8200000	1.765406
.8400000	1.790050
.8600000	1.815038
.8800000	1.840375
.9000000	1.866066
.9200000	1.892115
.9400000	1.918528
.9600000	1.945310
.9800000	1.972465
1.000000	2.000000

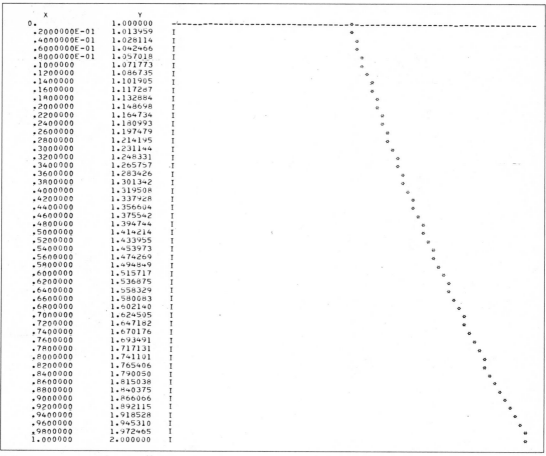

FIGURE 4.1b. Computer generated graph of $\exp_2$.

formalities require the calculus. Nevertheless, let us sketch the thinking that goes into this extension of domain. First, we show that if x is any real number, there is a sequence†

$$r_1, r_2, r_3, \cdots$$

of rational numbers with the property that r_i gets closer and closer to x as i gets larger and larger. For example, the sequence 1, 1.4, 1.41, 1.414, 1.4142, $\cdots$ gets closer and closer to $\sqrt{2}$. Next we consider the sequence

$$t_1, t_2, t_3, \cdots$$

where $t_i = a^{r_i}$ for each i and shows that this sequence also gets closer and closer to a single number as i gets larger and larger. The number that the t_i's approach, called the limit of the sequence

$$t_1, t_2, t_3 \cdots,$$

† See Chapter 5 for more discussion of sequences in general.

234 *EXPONENTIAL AND LOGARITHMIC FUNCTIONS*

X	Y
1.0000000000	2.0000000000
1.5000000000	2.8284271247
1.2500000000	2.3784142300
1.3750000000	2.5936791093
1.4375000000	2.7085110939
1.4062500000	2.6504732863
1.4218750000	2.6793350481
1.4140625000	2.6648650942
1.4179687500	2.6720902764
1.4160156250	2.6684752399
1.4150390625	2.6666695561
1.4145507813	2.6657671725
1.4143066406	2.6653160951
1.4141845703	2.6650905851
1.4142456055	2.6652033377
1.4142150879	2.6651469608
1.4141998291	2.6651187728
1.4142074585	2.6651328668
1.4142112732	2.6651399138
1.4142131805	2.6651434373
1.4142141342	2.6651451991
1.4142136574	2.6651443182
1.4142134190	2.6651438778
1.4142135382	2.6651440980
1.4142135978	2.6651442081
1.4142135680	2.6651441530
1.4142135531	2.6651441255
1.4142135605	2.6651441393
1.4142135642	2.6651441462
1.4142135624	2.6651441427
1.4142135615	2.6651441410
1.4142135619	2.6651441419
1.4142135622	2.6651441423
1.4142135623	2.6651441425
1.4142135623	2.6651441426
1.4142135624	2.6651441427
1.4142135624	2.6651441427
1.4142135624	2.6651441427
1.4142135624	2.6651441427
1.4142135624	2.6651441427

FIGURE 4.2

is then taken to be the value of a^x. Figure 4.2 shows the output from a program that used successive halving, generated a (finite) sequence of closer approximations for $\sqrt{2}$, and then computed $\exp_2(x_i)$ for each x_i in that sequence. An examination of that output should help you to see the reasonableness of the claim that a^x is definable for all real values of x. Finally, we show that with this definition, the traditional laws of exponents still hold.

We now define the exponential function $\exp_a$ by $\exp_a x = a^x$ for all $x \in \mathbb{R}$. Thus we have the whole set of functions $\{\exp_a \mid a \in (0, \infty)\}$ each with domain $\mathbb{R}$, satisfying the following properties for all $a, b \in (0, \infty)$ and all $x, y \in \mathbb{R}$:†

DEFINITION 4.1b

$\exp_a$ **on** $\mathbb{R}$

1. $\exp_a(x + y) = \exp_a(x) \cdot \exp_a(y)$, i.e., $a^{x+y} = a^x a^y$.
2. $\exp_a \cdot \exp_b = \exp_{a \cdot b}$, i.e., $a^x b^x = (ab)^x$.
3. $\exp_a / \exp_b = \exp_{a/b}$, i.e., $a^x / b^x = (a/b)^x$.
4. $(\exp_a x)^y = \exp_a(x \cdot y)$, i.e., $(a^x)^y = a^{xy}$.

† Do you see that these properties resemble the identities among the circular functions, in that they tell us how various functions relate to one another?

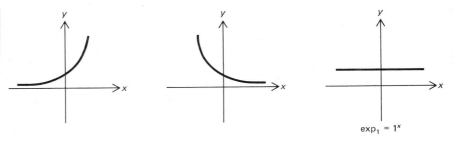

$\exp_1 = 1^x$

FIGURE 4-3

5. $\exp_a(x - y) = \exp_a x / \exp_a y$, i.e., $a^{x-y} = a^x/a^y$.
6. $\exp_a(-x) = 1/\exp_a(x)$, i.e., $a^{-x} = 1/a^x$.

We have glibly referred to $\exp_a$ as a function. The "single-valuedness" holds for rational arguments as a consequence of the definition of a^x for $x \in \mathbb{Q}$: $a^{p/q}$ is by definition, $(\sqrt[q]{a})^p$, which is a single specific number for each choice of p, q, and a. That the definition produces an unambiguous value for irrational arguments we shall have to take on faith.

Analogously, when we look for zeros, we can rigorously deal with rational arguments but not irrational arguments. If various exponential functions are graphed, three basic shapes of graphs occur, as sketched in Figure 4.3.

Notice that the third of these is a special case: $\exp_1 x = 1^x$. Since $1^x = 1$ for all x, $\exp_1$ has no zeros. It seems that the other two varieties do not have zeros either. Indeed the values of $\exp_a x$ are positive for all $x \in \mathbb{R}$, regardless of which value of a in $(0, \infty)$ is chosen.

In proving $\exp_a$ has no zeros, we shall try to prove that $\exp_a$ is always positive (and therefore never crosses the x-axis). We shall restrict our investigation to $x \in \mathbb{Q}$ for the first part of our discussion. In addition, we must remember that $a \in (0, \infty)$.

This is a parallel of the proof to the left. We have used exactly the same words, but have used specific numbers so that what we are saying in the proof might be clearer. This side is *not* a proof —only one example of the proof procedure. We start with the number $\frac{3}{2}$, but since the real proof looks at seven different cases in dealing with all rational numbers, we shall use the numbers $7/-4$, $-3/5$, $-6/-5$, $1/0$, $0/0$, $0/9$ to illustrate the other cases.

Since x is a rational number, it can be written as some p/q, where p and q are integers. In attempting to show that $a^{p/q}$ is always positive, we must deal with seven cases.

Since 1.5 is a rational number, it can be written as $\frac{3}{2}$, where 3 and 2 are integers. In attempting to show that $a^{3/2}$ is always positive, we must deal with seven cases.

Case 1. $p > 0$, $q > 0$. By the definition of rational exponents, $a^{p/q} = \sqrt[q]{a^p}$ or $(\sqrt[q]{a})^p$. Since a is greater than zero, $\sqrt[q]{a}$ is greater than zero. It is easy to show further that $(\sqrt[q]{a})^p > 0$. (See Problem 39, Chapter 5.) Therefore $a^{p/q} > 0$.

Case 2. $p > 0$, $q < 0$. Thus p/q is negative. At this point, we need to digress a moment. If $m < 0$, then $-m > 0$ because multiplying through the inequality by -1 changes its direction. Thus $-(p/q)$ is positive if p/q is negative, as we are assuming here.

Now $a^{p/q} = 1/a^{-p/q}$ and for this case, $a^{-p/q}$ is positive by case 1; therefore $a^{p/q} > 0$, since both the numerator and denominator of $1/a^{-p/q}$ are positive.

Case 3. $p < 0$, $q > 0$. Here p/q is negative, and this argument is just like case 2.

Case 4. $p < 0$, $q < 0$. Here $p/q > 0$ and the argument reverts to case 1. (We could replace p and q by $\hat{p}$ and $\hat{q}$, where $\hat{p} = -p$, $\hat{q} = -q$, so $\hat{p}$, $\hat{q} > 0$ and $\hat{p}/\hat{q} = p/q$, and case 1 applies directly.)

Case 5. $p \neq 0$, $q = 0$. This is impossible because p/q with $q = 0$ is undefined.

Case 6. $p = 0$, $q = 0$. Here $0/0$ is also undefined.

Case 7. $p = 0$, $q \neq 0$. ($q > 0$ or $q < 0$). Here $p/q = 0$. This is easy as $a^{p/q} = a^0 = 1 > 0$. Thus $a^{p/q} > 0$.

Case 1. $3 > 0$, $2 > 0$. By the definition of rational exponents, $a^{3/2} = \sqrt[2]{a^3}$ or $(\sqrt[2]{a})^3$. Since a is greater than zero, $\sqrt[2]{a}$ is greater than zero. It is easy to show further that $(\sqrt[2]{a})^3 > 0$. (See Problem 39, Chapter 5.) Therefore $a^{3/2} > 0$.

Case 2. $7 > 0$, $-4 < 0$. Thus $7/-4$ is negative. At this point we need to digress a moment. If $-5 < 0$, then $-(-5) > 0$ because multiplying through the inequality by -1 changes its direction. Thus $-(7/-4)$ is positive if $7/-4$ is negative, as we are assuming here.

Now $a^{7/-4} = 1/a^{-(7/-4)}$ and for this case, $a^{-(7/-4)} [= a^{7/4}]$ is positive by case 1; therefore $a^{7/-4} > 0$, since both the numerator and denominator of $1/a^{-(7/-4)}$ are positive.

Case 3. $-3 < 0$, $5 > 0$. Here $-3/5$ is negative, and this argument is just like case 2.

Case 4. $-6 < 0$, $-5 < 0$. Here $-6/-5 = 6/5 > 0$ and the argument reverts to case 1. (We could replace -6 and -5 by 6 and 5, where $6 = -(-6)$, $5 = -(-5)$, so $6, 5 > 0$ and $-6/-5 = 6/5$, and case 1 applies directly.)

Case 5. $-1 < 0$, $0 = 0$. This is impossible because $-1/0$ is undefined.

Case 6. $0 = 0$, $0 = 0$. Here $0/0$ is also undefined.

Case 7. $p = 0$, $q = 9 \neq 0$ ($9 > 0$ for our case). Here $0/9 = 0$. This is easy as $a^{0/9} = a^0 = 1 > 0$. Thus $a^{0/9} > 0$.

Thus, by cases 1–7, a^x for $x \in \mathbb{Q}$ is always positive. We can next argue that since $a^r > 0$ for all $r \in \mathbb{Q}$, if a^x were ≥ 0 for any irrational values of x, the smoothness we assumed for the graph of $\exp_a$ would be violated. Although this argument is not a proof, it describes in general terms what an actual proof says. Hence a search for zeros for exponential functions would be pointless, since the fact that $\exp_a(x) = a^x$ is always positive means that the graph is always above the x-axis, and as we said above can thus have no zeros.

The question of finding values for these functions for various arguments is one we shall defer until the next section. It should be fairly clear already that such functional evaluations are not easy, except for integer arguments and "nice" bases (such as $(\frac{1}{3})^2$ or 10^{10}).†

As for relationships among the functions, we have already observed that $\exp_a \exp_b = \exp_{a \cdot b}$ and $\exp_a/\exp_b = \exp_{a/b}$ for all a, $b \in (0, \infty)$. We can also show that $\exp_{1/a} = 1/\exp_a$ (see Problem 15). While exponentials do not have the wealth of interrelationships that exist among the circular functions, there is one that will allow us to relate any two exponential functions of our choice. We shall develop this relationship later in the next section.

Finally, we are not able to find any more definitive word on graphing without more powerful tools than we have at our disposal now.

4.3. LOGARITHMIC FUNCTIONS

LOGARITHMIC
FUNCTIONS

We can certainly form the inverse of $\exp_a$, and with the help of Section 1.5c we can sketch its graph. Figure 4.4 shows the graphs of both $\exp_2$ and its inverse $\exp_2^{-1}$. A look at the graph of the inverse gives one a glimmer of hope—perhaps the inverse of each $\exp_a$ is already a function and we will not need to do anything to make a function out of it. Certainly, $\exp_1 x = 1^x$ is an unwanted function for this whole matter of inverses (why?) so let us require $a \neq 1$. For the remaining $\exp_a$'s, we could be sure that $\exp_a^{-1}$ were a function if we could show that $\exp_a(x) = \exp_a(y)$ only if $x = y$. However, we can do this here only for rational arguments of $\exp_a$, and leave the extension to $\mathbb{R}$ for one's faith and/or further study in mathematics. The result, though, is true.

To prove $\exp_a^{-1}$ is a function for these values, means that we must show that if $\exp_a(p/q) = \exp_a(r/s)$, then $p/q = r/s$. Let us suppose (erroneously) that $p/q \neq r/s$. Then, using the assumption that $\exp_a(p/q) = \exp_a(r/s)$, we have (since $\exp_a(p/q) = a^{p/q}$ and $\exp_a(r/s) = a^{r/s}$,

$$a^{p/q} = a^{r/s} \qquad \text{or} \qquad \frac{a^{p/q}}{a^{r/s}} = 1.$$

Hence $a^{p/q-r/s} = 1$, which would mean that $a^{p/q-r/s} = a^0$, but we assumed that $p/q \neq r/s$ and, therefore, $p/q - r/s \neq 0$. Thus $p/q - r/s$ must be some element of $\mathbb{Q}$, say m/n, for nonzero integers m and n. Hence $a^{m/n} = 1$. Since 1 to any power is still 1, we can say that $a^{m/n} = 1^m$. Then $(\sqrt[n]{a})^m = 1^m$, by the definition of $a^{m/n}$. Since $\sqrt[n]{a} > 0$, we may take mth roots on both sides without ambiguity to get $\sqrt[n]{a} = 1$. Then we raise both sides to the nth power to get $a = 1$, which is a contradiction, since we had ruled out this value of a in our definition of $\exp_a^{-1}$. In other words, we are forced to conclude that if $a^{p/q} = a^{r/s}$, either $a = 1$ or $p/q = r/s$. Since we

† Computers make the job easy for other values, but how do they do it?

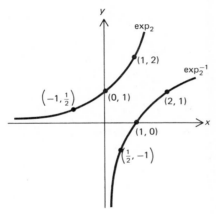

FIGURE 4-4

do not want $a = 1$, we must conclude $p/q = r/s$; hence $\exp_a^{-1}$ is a function. Now that you see where this proof was headed, it might help to go back and read it again.

The inverse $\exp_a^{-1}$ is then a function, but it is ordinarily not denoted by $\exp_a^{-1}$, but rather by $\log_a$. This notation is historical and grows out of the study of equations of the form $a^x = b$, where a and b are constants. In this setting, x was called the logarithm of b to the base a, and one wrote $x = \log_a b$. Thus the equations $y = a^x$ and $x = \log_a y$ are equivalent, much as $y = 3x + 4$ and $x = (y - 4)/3$ are equivalent, and $\log_a$ really is the inverse of $\exp_a$. The function $\log_a$ is called a logarithmic function; a is again called the base of the function.

DEFINITION 4.2
$\log_a$

Before going into the properties of $\log_a$, it is time to ask about the domain and range. Since $D_{\exp_a} = \mathbb{R}$, we know that $R_{\log_a} = \mathbb{R}$, but we never came to grips with $R_{\exp_a} = (0, \infty)$. To do this, consider $\exp_2$ again. See Figure 4.4. Let $n \in \mathbb{N}$. Then $\exp_2 n = 2^n$; as n gets larger, so does 2^n, without bound. Then $\exp_2(-n) = (\frac{1}{2})^n$, and as n gets larger, $(\frac{1}{2})^n$ gets analogously closer to 0. If we fill in the gaps between the integers, it appears that $\exp_a x$ can be made as large as one likes, merely by picking x large enough, and as close to zero as one likes by picking x sufficiently small (i.e., large in the negative direction). If $R_{\exp_a} = (0, \infty)$, then $D_{\log_a} = (0, \infty)$.

The logarithmic functions inherit certain structures from the exponential functions. For example, the fact that $\exp_a(x + y) = \exp_a x \cdot \exp_a y$ generates the corresponding fact that

**PROPERTIES OF
LOGARITHMIC
FUNCTIONS**

$$\log_a(r \cdot s) = \log_a r + \log_a s. \tag{1}$$

This may be shown by letting $x = \log_a r$, $y = \log_a s$, and $z = \log_a(r \cdot s)$, and by establishing $z = x + y$. Using the translations to exponential form, we may write $a^x = r$, $a^y = s$, $a^z = r \cdot s$. Therefore,

$$r \cdot s = a^z \qquad \text{and} \qquad r \cdot s = (a^x)(a^y) = a^{x+y},$$

whence $z = x + y$, as we needed.

In a similar way, one can prove that (see Problems 25–27) :

$$\log_a \frac{r}{s} = \log_a r - \log_a s. \tag{2}$$

$$\log_a r^p = p \cdot \log_a r. \tag{3}$$

$$\log_a = -\log_{1/a}. \tag{4}$$

Because $\log_a$ is the inverse of $\exp_a$, we have $\exp_a \circ \log_a$ and $\log_a \circ \exp_a$, each being the identity function although with different domains. Another way to say this is

$$a^{\log_a x} = x \; [x \in (0, \infty)] \tag{5}$$

and

$$\log_a(a^x) = x \; (x \in \mathbb{R}). \tag{6}$$

(Remember that inverse functions give you back what the original function started with; therefore, $(f \circ f^{-1})(x) = x$ and $((f^{-1}) \circ f)(x) = x$.)

From these facts one can also derive that

$$\log_b = (\log_b a) \cdot \log_a. \tag{7}$$

A search for zeros for $\log_a$ is less frustrating than the search for zeros of $\exp_a$ but remains quite uninspiring, since $\log_a 1 = 0$ for any $a \in (0, \infty)$. (See Problem 29.) Also, $\log_a x \neq 0$ if $x \neq 1$, 1 is the only zero for every one of the functions $\log_a$.

On the topic of graphing, we can do no better than we did for $\exp_a$. The two basic shapes for graphs of logarithmic functions are given in Figure 4.5.

We may now use the logarithmic functions to help us derive the relationship between any two exponential functions which we promised at the end of the previous section. We want to find a general relationship between $\exp_a$ and $\exp_b$ for any pair of positive real numbers a and b. What we need to do is to relate a^x and b^x. Hopefully, this can be done by relating a and b. Since a and b are both in $(0, \infty)$ and $R_{\exp_a} = R_{\exp_b} = (0, \infty)$, there is some real y so that $b^y = a$. In other words, we may express a as some power of b. In fact, $y = \log_b a$, using the definition of $\log_a b$. Hence $a^x = (b^y)^x = b^{yx} = (b^x)^y = (b^x)^{\log_b a}$. This translates into

$$\exp_a = (\exp_b)^{\log_b a} \tag{8}$$

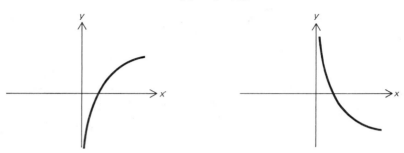

FIGURE 4-5

which does, in fact, provide a general relationship. For example, we may write $3^x = (4^x)^{\log_4 3}$, or $2^x = (10^x)^{\log_{10} 2}$ or $(\sqrt{2})^5 = (3^5)^{\log_3 \sqrt{2}}$, using the fact that $a^x = (b^x)^{\log_b a}$. As you can see, this allows us to change the base we are using.

This relationship is similar to the one among polynomial functions that if P and Q are *any* two polynomial functions, and say they are labeled so that the degree of $P \geq$ the degree of Q, then there are two other polynomials D and R so that $P = Q \cdot D + R$. Each of these enables one to give an explicit relationship between any two functions in the appropriate set of functions.

NOTE

As it turns out, some bases are more convenient than others for various purposes. A base of 10 proves useful in aiding certain computations done by hand, as we shall see in the section on applications. A base of 2 has already proved useful as an example to be graphed. But the best base for most analytic work is an irrational number denoted by e,† much as another special irrational number which is of interest in the study of the circular functions is denoted by π. The value of e is $2.718281828459045\cdots$. The function $\exp_e$ is often denoted exp and is called the natural exponential function.

e

DEFINITION 4.3
exp

That e is in any legitimate sense a "natural" choice for a base is most clearly seen in the calculus, but we can point to one evidence of its "naturalness" and consequent usefulness. As is the case for $\sin x$ and $\cos x$ (see Chapter 3) the value $\exp_e x$ can be expressed as a series (see Section 5.3). Specifically,

$$\exp_e x = 1 + x + \frac{x^2}{2!} + \frac{x^3}{3!} + \cdots,$$

then for any other base a, let us note that $a = e^{\log_e a}$ using property (5) of logarithmic functions, so

$$\exp_a x = a^x = (e^{\log_e a})^x = e^{x \log_e a}$$
$$= \exp_e(x \log_e a).$$

Substituting $x \log_e a$ into the series above, we have

$$= 1 + x \log_e a + \frac{(x \log_e a)^2}{2!} + \frac{(x \log_e a)^3}{3!} + \cdots.$$

Approximate values for the exponential functions can then be computed by using finitely many of these terms. Tables of $\log_e$ can be found in most books of mathematical tables, as can tables of $\log_{10}$. (The use of $\log_{10}$ tables is explained in Section 4.4b.)

That the formula above for $\exp_e$ should be so similar to those for sine and cosine is remarkable, considering the vastly different origins of the functions. But despite their varied backgrounds, the relationship between $\exp_e$ and the circular functions is even closer than an apparently circumstantial similarity in their series. This greater intimacy only becomes evident

---

† The choice of the letter e is not for exponential but for Euler, (pronounced oiler), the name of a great mathematician of the 18th century who did prolific work in many branches of mathematics, notably in analysis.

in the realm of complex numbers, for which the reader is directed to Section 7.5.

4.4. SOME APPLICATIONS

We divide this section into subsections according to the kind of application being considered.

4.4a. Applications of Exponential Functions

EXPONENTIAL GROWTH AND DECAY

Generalized exponential functions of the form $k \cdot \exp_a$ with k any real number are used to describe numerous phenomena. Among these are population growth and decline, radioactive decay, the progress of certain chemical reactions, and generalized compound interest.† Exponentials also express the relationship that occurs when the rate of change of some quantity is directly proportional to the present amount of that quantity. Let us look at just a few examples.

EXAMPLE 1

Psychologist Hengel Agehog believes that the rate of increase in a subject's depression is proportional to the amount of depression being experienced if depression is calibrated in d.u.'s (depression units) and no outside stimulus is brought to bear. Using her theory, if a patient is feeling 10 d.u.'s at the start of observation and 20 d.u.'s 5 hr later, how depressed should the patient be 12 hr after the start of observation?

By hypothesis, if $d(t)$ denotes depression d in d.u.'s experienced at t hours after the start of observation, $d(t)$ is directly proportional to $\exp_a(t)$, i.e., $d(t) = k \cdot \exp_a(t)$. We know that $10 = k \cdot \exp_a(0) = ka^0 = k$, since $d(t)$ at the start of the experiment ($t = 0$) is 10. Then, substituting the value of k in $d(t) = k \cdot \exp_a(t)$, we have $d(t) = 10 \exp_a t$. Next, for $t = 5$, we know that $d(t) = 20$, and we have

$$d(t) = 10 \exp_a (t)$$
$$20 = 10 \exp_a (5) = 10a^5$$
$$2 = a^5 \quad \text{and} \quad a = \sqrt[5]{2}.$$

Now, to find $d(12)$, we substitute into

$$d(t) = 10 \exp_{\sqrt[5]{2}}(t) = 10(\sqrt[5]{2})^t$$

and

$$d(12) = 10(\sqrt[5]{2})^{12}$$
$$= 10(\sqrt[5]{2^{12}}) = 10(\sqrt[5]{2^2 \cdot 2^{10}}) = 10(\sqrt[5]{2^2} \cdot 2^2)$$
$$= 10 \cdot \sqrt[5]{4} \cdot 4 = 40\sqrt[5]{4}.$$

This is a good mathematical answer, but an approximation would be of more use to Dr. Agehog. This could be approximated by computer, or by hand using a method described in Section 4.4b. In any case, $\sqrt[5]{4} \approx 1.32$ so that the subject should be feeling 52.8 d.u.'s if the theory is correct.

† That is, continuous compounding of interest.

EXAMPLE 2

EXPONENTIAL
GROWTH AND
DECAY

The rate of radioactive decay is known to be directly proportional to the amount of substance present. If there were 50 grams of a certain radioactive isotope present in 1960 and only 40 grams in 1980, how many grams were present in the sample when it was created in 1950? The function is $A(t) = ka^t$, and we need to find k and a before being able to answer the question. Here it seems sensible to let $t = 0$ either at 1950 or 1960. Let us try both approaches. First, let us set $t = 0$ at 1960. If $A(t)$ is the amount of isotope at time t, then at time $t = 0$, $50 = A(0) = ka^0$ so that $k = 50$. Our equation now reads $A(t) = 50a^t$ [or $50 \exp_a(t)$]. In 1980, $t = 20$ and $A(t) = 40$; therefore, we have $40 = 50a^{20}$ or $a^{20} = \frac{4}{5}$, whence $a = \sqrt[20]{4/5}$.

In an effort to simplify this answer, let us try again and measure time in decades. Then 1980 corresponds to $t = 2$; k still equals 50 but $40 = ka^2$ so that $a = \sqrt{4/5} \approx 0.894$. Then the amount in 1950 is $A(-1) = 50a^{-1} \approx 50(0.894)^{-1} = 50/0.894 \approx 56$ grams.

If we continue to measure t in decades but let $t = 0$ in 1950, then the amount $A(t)$ in 1960 is still 50 grams, but now $t = 1$, so $A(1) = 50$ and $50 = ka^1$. Similarly, in 1980, $k = 3$ and $A(t) = 40$, so $40 = ka^3$. Solving these simultaneously, we have

$$50 = ka^1 \quad \text{or} \quad a = \frac{50}{k} \tag{1}$$

$$40 = ka^3. \tag{2}$$

Substituting in Equation (2), we get

$$40 = k\left(\frac{50}{k}\right)^3$$

$$40 = \frac{50^3}{k^2}$$

$$k^2 = \frac{50^3}{40} = 50^2 \cdot \frac{50}{40} = 50^2 \cdot \frac{5}{2^2}$$

$$k = \frac{50}{2}\sqrt{5} = 25\sqrt{5} \approx 55.9 \text{ grams.}$$

EXAMPLE 3

The Last National Bank announces plans to offer continuous compounding of interest on its certificates of deposit. If the base interest rate is 5%, and Marvin Springspeed deposits $1000 in such an account at the start of January 1,

a. What would he receive if he closed the account on January 1 five years later?
b. How long must he leave his money in this account to have it double?

To solve this, we need to note first that because this is an instance of continuous compounding, a function $k \cdot \exp_a$ is appropriate to describe the compounding. Also, if we rewrite $k \exp_a = k(\exp_e)^{\log_e a}$, it is helpful to know that $\log_e a$ turns out to be the stated (or base) rate of interest per unit of time. In this example, this number is 0.05, but make careful note that this rate is *not* the net rate over the unit of time. Indeed, the net effect of a rate of 5% per year compounded continuously is an increase of approximately 5.127% over 1 year (see Problem 38.) Continuing with the example, since the argument is time in years, $\log_e a = 0.05$. Thus part(a) is answered by noting first that if $A(t) = ke^{0.05t}$, then at $t = 0$, $1000 = ke^0$, whence $k = 1000$, and then computing, $A(5) = 1000e^{0.05 \cdot 5} = 1000e^{0.25}$. Finding $e^{0.25} \approx 1.284$ from Table 4, Appendix D, we find that Marvin has approximately $1284.

To answer part (b), since to double his money $A(t)$ must equal 2000, we need to solve $2000 = 1000e^{0.05t}$ or $e^{0.05t} = 2$. If we apply $\log_e$ to both sides of the equation, we have $0.05t = \log_e 2$ or $t = \log_e 2/0.05$. To get an idea of the size of t, we use a table to find $\log_e 2 \approx 0.6932$ so that $t \approx 13.86$ years.

4.4b. Computational Application of Logarithmic Functions

To an insightful eye, the properties (1), (2), and (3) of logarithmic functions might suggest some computational aids. Property (1) states that

$$\log_a xy = \log_a x + \log_a y.$$

Thus if one had a complicated multiplication problem, say (4896)(37.81), this person might note that

$$\log_a(4896 \cdot 37.81) = \log_a(4896) + \log_a(37.81)$$

and that if he could easily find the two values on the right side of the above equation, he would only have to add to get $\log_a$, the desired product. If he could translate from values of $\log_a x$ back to x easily enough, he would have accomplished the multiplication by addition and some translations which we hope will be fairly easy. In a similar way, property (2) of logarithmic functions suggests doing divisions by subtractions, and property (3) suggests taking powers by multiplication.

Although he has no easy way to compute $\log_a x$, it might be a worthwhile investment of time and energy to make a table of values in order to simplify many tedious computations. Also, it is important to note that the translation from $\log_a x$ back to x is by applying $\log_a^{-1} = \exp_a$; were it not for the fact that both $\exp_a$ and $\log_a$ are functions, the route might be ambiguous (with several values to choose from) and this whole undertaking would be of little value. Since both are functions, there is no ambiguity so that we proceed.

There still seems to be a snag. If a table is to be useful, it will have to be huge to include all the numbers one might encounter in the multiplication, division, and power computations we hope to shorten. As yet, though, we have not chosen a specific base for the logarithmic function to be used, and the hope that a wise choice of base will help is fulfilled. The use of a place value notation for numbers allows us to write the number 39,876 as $3.9876 \cdot 10^4$ or the number 0.00372 as $3.72 \cdot 10^{-3}$. Any number can be written in this way as the product of a number between

1 and 10 and an integral power of 10. (This way of writing numbers is called scientific notation.) Thus, given any number x, there are numbers n and p, $n \in [1, 10)$, $p \in \mathbb{Z}$ so that $x = n \cdot 10^p$. Therefore, $\log_a x = \log_a n + \log_a 10^p$. The choice of 10 as the base for the logarithmic function permits the easy evaluation of $\log_a 10^p$—it is just p. If we use $\log_{10}$, we need only tabulate the values of $\log_{10} x$ for $x \in [1, 10)$. This is done in Table 3, Appendix D. The values of $\log_{10} x$ are usually called *common logarithms*.

DEFINITION 4.4
Common
Logarithms

Let us look at three examples to better understand what we have gained.

Find $(458)(761)$.

EXAMPLE 4

$$\begin{aligned}
\log_{10}[(458)(761)] &= \log_{10}(4.58 \cdot 10^2) + \log_{10}(7.61 \cdot 10^2) \\
&= \log_{10}(4.58) + \log_{10} 10^2 + \log_{10}(7.61) \\
&\quad + \log_{10}(10^2) \\
&\approx 0.6609 + 2 + 0.8814 + 2 \qquad \text{(from Table 3)} \\
&= 5.5423 \\
&= 5 + 0.5423 \\
&\approx \log_{10} 10^5 + \log_{10} 3.486,
\end{aligned}$$

by using the table in reverse to find the value of x whose logarithm is $0.5423 = \log_{10}(10^5 \cdot 3.486)$. Therefore,

$$(458)(761) \approx 348{,}600.$$

Note that the answer arrived at is approximate, since tabulated values are approximations and we used interpolation in finding the number 3.486. Because the results are not exact, such computations are usually used in science and engineering, where the error of the approximation can be kept within the accuracy permitted by the data being used. Ordinarily, the only work done is the adding of the values of the logarithms—the explanatory steps are omitted in ordinary usage. Nevertheless, we continue with the explanatory steps here.

Find $82900/334$.

EXAMPLE 5

$$\begin{aligned}
\log_{10}\left(\frac{89200}{334}\right) &= \log_{10}(8.92 \cdot 10^4) - \log_{10}(3.34 \cdot 10^2) \\
&= \log_{10} 8.92 + 4 - \log_{10} 3.34 - 2 \\
&\approx 0.9504 - 0.5237 + 2 \\
&= 2.4267 \\
&\approx \log_{10} 10^2 + \log_{10} 2.671.
\end{aligned}$$

Therefore, the quotient ≈ 267.1.

Find $(66.7)^5$.

EXAMPLE 6

$$\begin{aligned}
\log_{10}(66.7)^5 &= 5 \log_{10} 66.7 \\
&\approx 5(1.8241) \\
&= 9.1205 \\
&\approx \log_{10}(10^9 \cdot 1.32).
\end{aligned}$$

Then, $66.7^5 \approx 1{,}320{,}000{,}000$.

We have carefully avoided one kind of problem which although it follows the same pattern as above, introduces a convention. We might, for example, wish to know the common logarithm of 0.000368. According to techniques already devised we can say

$$\log_{10} 0.000368 = \log_{10} 3.68 \cdot 10^{-4} = \log_{10} 3.68 + \log_{10} 10^{-4}$$
$$= 0.5658 - 4.$$

Here we pause. Certainly we could perform the subtraction and get the value -3.4342. A quick look at the graph of $\log_{10} x$ would assure us that the negative value is perfectly legitimate. Consider the case, however, where having completed several calculations using common logs, we have $\log_{10}$ (final answer) ≈ -3.4342.

By now you have probably come to associate the digits to the right of the decimal point with the sequence of digits in the answer, and the number to the left of the decimal point with the location of the decimal point in the answer. If one has $\log_{10} x = -3.4342$, there is a great tendency to forget that the .4342 portion is negative as well as the 3; hence, one may make the error of finding x such that $\log_{10} x = 0.4342$ in the table, thinking that this is the correct sequence of digits for the answer to the problem.

To avoid this error, we consider the "fractional" and "integral" portions of the number separately, always writing the "fractional" portion as a positive number. This quantity is usually called the mantissa of the logarithm. Since to write -4.5658 in the above example would be misleading at best (as written, the mantissa is negative), we rewrite the corresponding integral portion (called the characteristic) adding and subtracting 10 in the following manner. We write $\log_{10} 3.68 \cdot 10^{-4}$ as $6.5658 - 10$. (You should now verify that this does have the correct value.) Similarly then, $\log_{10} 0.111 = 9.0492 - 10$ and $\log_{10}(6.92 \cdot 10^{-7}) = 3.8401 - 10$. If adding and subtracting 10 is not enough to get the first part of the characteristic positive, we use 20 or 30 or a higher multiple of 10. Hence $\log_{10} 5.31 \cdot 10^{-13} = 7.7251 - 20$.

Although common logarithms are still of some practical value, the increasingly widespread availability of computers and sophisticated desk calculators has seriously reduced the frequency of manual use of common logarithms. A partial compensation is the fact that computers make use of logarithms to do some computations, notably the computation of exponential expressions with nonintegral exponents.

4.4c.   A Hidden Application of $\log_{10}$ and $\exp_{10}$—The Slide Rule

What follows is intended to explain why a slide rule works, not how to use one. If you have some familiarity with the slide rule, you will probably find it valuable ; however, if you know nothing of slide rules you will either need another reference source or have to plan to spend some time working through the following discussion in more detail.

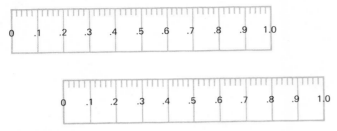

FIGURE 4-6

The slide rule is a simple calculation device based on the functions $\log_{10}$ and $\exp_{10}$. As noted in Section 4.4b, except for the location of the decimal point, all one uses are the mantissas of common logarithms. The slide rule adds and subtracts mantissas and produces the correct sequence of digits for the answer. The location of the decimal point is usually found by estimating the result.

We shall look at the construction of the slide rule in order to see how to use it. The slide rule uses lengths to correspond to numbers, similar to a number line. One could take two pieces of paper, say 10 in. long, and scale each to be a number line, starting at 0 and going to 1.0. (It will be instructive if you actually do this.) The various points on each could then correspond to the mantissas of common logarithms. See Figure 4.6. The multiplication of two numbers corresponds to addition of the logarithms, and division to subtraction of the logarithms; it is easy to add and subtract line segments. However, the logarithms (or their mantissas) are just a go-between, so if we labeled the scales by the sequence of digits corresponding to each mantissa, we could save ourselves the need of a table. Such coding does the work of applying $\log_{10}$ and then later $\exp_{10}$. Figure 4.7 shows the multiplication of $a = 19.3$ and $b = 3.28$. Adding lengths a and b to get length c corresponds to adding the mantissas of $\log_{10}(19.3)$ and $\log_{10}(3.28)$. The scaling indicated shows us that the length c corresponds to a number whose sequence of digits is approximately 633; since $(19.3)(3.28)$ is roughly $20 \cdot 3 = 60$, the actual product is approximately 63.3. In short, to multiply

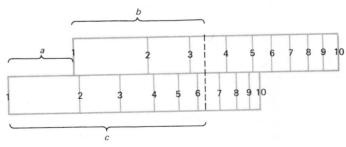

FIGURE 4-7

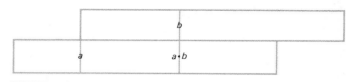

FIGURE 4-8

$a \cdot b$ on a slide rule, one finds the length labeled by b's sequence of digits on scale A and the length labeled by a's sequence of digits on scale B. Then one adds the lengths by placing the left end of scale B over the location just found on scale A, and reads the sequence of digits for the product on scale A under the location noted on scale B. See Figure 4.8.

In the case that the sum of the two mantissas would be greater than one (or the difference, for division, less than zero), one needs to modify the procedure a little. In the case of multiplication, one can envisage a second (or extended) lower scale, as in Figure 4.9. Although it could correspond to logarithms 1.0 through 2.0, there is no point in paying attention to the characteristic, since we have ignored it from the start. By the congruence of the scales, then, it appears that the right end point of the upper scale is directly above what would be labeled as a, and the answer is again read under b—the net effect is just to switch ends of the scale and appear to be subtracting. The two ends of the slide rule scale are called indices and the general rule for multiplication is to locate the first factor on one scale, place the appropriate index of the other scale at that location, and read the answer on the first scale below the location of the second factor. You are asked to work out a general procedure for division in the problems.

Commercial slide rules have many different scales for many purposes, and it is beyond the scope of this study to consider them, even though they too involve logarithms. The scales we have discussed above are the C and D scales on any commercial slide rule.

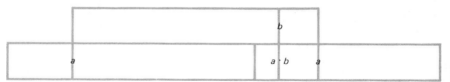

FIGURE 4-9

4.5. A PROJECT

In this section, you are asked to investigate briefly some new functions that are related to both the exponential and circular functions. We shall call the primary two functions tick and tock, and define them by

$$\text{tick}(x) = \frac{2^x + 2^{-x}}{2}, \qquad \text{tock}(x) = \frac{2^x - 2^{-x}}{2}.$$

248   *EXPONENTIAL AND LOGARITHMIC FUNCTIONS*

The relationship with exponential functions is quite clear, since the latter occur in the definitions. The relationship with the circular functions will appear shortly.

Phase 1. First sketch graphs of these two functions as carefully as you can. Note any interesting behavior you find (e.g., periodicity, maximum and minimum points, zeros, etc.).

Phase 2. Discuss the inverses of tick and tock as precisely as you can. Recall the investigations of other inverses so far in this book as a guide to what questions to try to answer. Where you cannot be precise or prove claims, at least state hypotheses and reasons for guessing them.

Phase 3. Prove each of the following equations is really an identity:

$$\text{tick}^2\, x - \text{tock}^2\, x = 1 \tag{1}$$

$$\text{tock}(x + y) = \text{tock}\, x\, \text{tick}\, y + \text{tick}\, x\, \text{tock}\, y. \tag{2}$$

Then conjecture a formula for tick $(x - y)$ and prove you are correct. (You may need to alter your original conjecture, as your attempt to prove it suggests modifications.)

Phase 4. Verify what your graphs from Phase 1 suggest concerning the oddness or evenness of tick and tock.

Phase 5. Define quotient and reciprocal functions and prove that they satisfy identities similar to the first identity of phase 3.

The functions tick and tock are closely related to another set of functions that are sometimes classed as elementary. These are the *hyperbolic functions*, which are defined by

hyperbolic sine: $\sinh x = \dfrac{e^x - e^{-x}}{2}$

hyperbolic cosine: $\cosh x = \dfrac{e^x + e^{-x}}{2}$

hyperbolic tangent: $\tanh x = \dfrac{\sinh x}{\cosh x}$

hyperbolic secant: $\text{sech}\, x = \dfrac{1}{\cosh x}$

hyperbolic cosecant: $\text{csch}\, x = \dfrac{1}{\text{sech}\, x}$

hyperbolic cotangent: $\coth x = \dfrac{1}{\tanh x}$.

These functions, while being defined by exponential functions, satisfy a great many identities similar to the trigonometric identities. The inverses of these functions turn out to be the results of some rather elementary manipulations in the calculus.†

---

† They are the results of some rather ordinary looking integrations.

4.6. AN OVERVIEW

Our approach to the exponential functions through an extension of the meaning of exponents might have suggested other paths to you. One such path would lead to functions whose rules do not have the variable in the exponent but would allow the variable to be raised to other powers than just the natural numbers. Among such rules would be

$$f(x) = x^{-2} = \frac{1}{x^2}$$

$$g(x) = x^{1/2} = \sqrt{x}$$

$$h(x) = x^{-7/5} = (1/\sqrt[5]{x})^7$$

plus combinations of such functions. These are investigated briefly in the project of Section 2.9. Actually, all of these functions can be viewed as combinations of polynomial functions and what we might term the *root functions*, f_n, where $f_n(x) = x^{1/n} = \sqrt[n]{x}$. These root functions in turn are just the functions taken from the inverses of the power functions. The graphs of these root functions are easily derived from the graphs of the power function, and such questions as domains and ranges are then quite straightforward. In short, there is very little of special interest in the root functions by themselves, and when one begins to build more complicated functions from them and the polynomials, the techniques of the calculus become necessary tools to have at one's disposal.

Although our investigations may have seemed to discriminate among the various questions one asks about functions, this has primarily been caused by the different natures of the functions and foreknowledge of later usefulness. We hope that the reader will review what we have discussed to date with an eye for similarities and differences with the goal of a better understanding of the nature of mathematical investigation and reasoning.

SUMMARY OF IMPORTANT CONCEPTS AND NOTATION

$\exp_a(\text{ON } \mathbb{Q})$: (Definition 4.1a) $\exp_a(r) = a^r, r \in \mathbb{Q}$.

BASE : (Definition 4.1a) The number a in $\exp_a(r)$ is called the base of the exponential function $\exp_a$.

$\exp_a(\text{ON } \mathbb{R})$: (Definition 4.1b) $\exp_a x = a^x$ for all $x \in \mathbb{R}$.

PROPERTIES OF EXPONENTIAL FUNCTIONS : (Section 4.2)

$\exp_a(x + y) = \exp_a(x) \cdot \exp_a(y)$ or $a^{x+y} = a^x a^y$.
$\exp_a \cdot \exp_b = \exp_{a \cdot b}$ or $a^x b^x = (ab)^x$.
$\exp_a/\exp_b = \exp_{a/b}$ or $a^x/b^x = (a/b)^x$.
$(\exp_a x)^y = \exp_a(x \cdot y)$ or $(a^x)^y = a^{xy}$.
$\exp_a(x - y) = \exp_a x/\exp_a y$ or $a^{x-y} = a^x/a^y$.
$\exp_a(-x) = 1/\exp_a(x)$ or $a^{-x} = 1/a^x$.

$\log_a$: (Definition 4.2) $\exp_a^{-1} = \log_a$, i.e., the $\log_a$ is the inverse function of the exponential function $\exp_a$.

PROPERTIES OF LOGARITHMIC FUNCTIONS: (Section 4.3)

$$\log_a(r \cdot s) = \log_a r + \log_a s.$$
$$\log_a(r/s) = \log_a r - \log_a s.$$
$$\log_a r^p = p \cdot \log_a r.$$
$$\log_a = -\log_{1/a}.$$
$$a^{\log_a x} = x \ (x \in (0, \infty)).$$
$$\log_a(a^x) = x \ (x \in \mathbb{R}).$$

e: (Section 4.3) $e = 2.718 \cdots$
exp: (Definition 4.3) $\exp = \exp_e$.
EXPONENTIAL GROWTH AND DECAY: (Section 4.4a)
COMMON LOGARITHMS: (Definition 4.4) The values of $\log_{10} x$ are called common logarithms.

PRACTICE EXERCISES

Section 4.2

1. Graph each of the following functions as carefully as you can:
   (a) $\exp_3$ (b) $\exp_2$ (c) $\exp_1$ (d) $\exp_{1/2}$ (e) $\exp_{1/3}$

Section 4.3

2. Sketch neat graphs of the following functions:
   (a) $\log_3$ (b) $\log_{1/2}$ (c) $\log_{1/3}$ (d) $\log_2$

Section 4.4a

3. A certain planet has a population of 10 million grogs at one time, and after one full revolution of its orbit about the sun, the population has increased to 12 million grogs. If the population of grogs grows at a rate proportional to the population present, what should the population be after one more revolution?

4. The rate at which the chemical gnath combines with the chemical smalluol to form gnathuol is proportional to the amount of gnath present. If a chemical engineer starts with 3 metric tons of gnath and 25 metric tons of smalluol, and 1 hr later finds that the solution contains only 1 metric ton of gnath, how much longer will he have to wait until 95% of the gnath has combined with smalluol?

5. A clever prospector marks the location of a find of gold by leaving a small quantity of a radioactive isotope whose half-life (the time required for any quantity of the substance to halve by radioactive disintegration) is 4 months. If the isotope is detectable as long as at least 10% of his sample remains, how long will his marker be of any use?

6. If the rate of production of neutrons in a nuclear reaction is proportional to the number of neutrons present, 1000 neutrons are introduced at

$t = 0$, and there are approximately 1 million at $t = 5$, how many neutrons will be there at $t = 60$?

7. If a population is growing at the rate of 2% per year and the generalized compound interest formula applies, how long will it take to double in size?

8. If an epidemic grows exponentially (i.e., the number of cases c at time t is given by $c = ka^t$), and there are 500 cases 1 week after the first case is contracted,
   (a) How many cases will there be 2 weeks later?
   (b) How long will it be until there are 1 million cases (provided the population involved is large enough)?

9. Newton's law of cooling states that a warm object loses heat to its cooler environment at a rate proportional to the difference in temperatures between the object and its environment. If a chunk of steel at 500° F is set to cool in an environment kept at 50° F, and it cools at a rate of 25% of the temperature difference per hour, how long will it be until the steel reaches a temperature of 100° F?

10. Maynard Basil Holywright has discovered that his enthusiasm for mathematics increases in proportion to how enthusiastic he already is. Suppose that enthusiasm is measured in eu's and that human beings cannot tolerate any more than 40 eu's without an (ecstatic) departure from this world. If Maynard experiences 10 eu's at the start of Mathematics 100 and 15 eu's at the end of Mathematics 100, what is the maximum number of mathematics courses beyond Mathematics 100 that he can safely take?

11. Suppose that the rate of increase of mass for any celestial body through accumulation of free floating matter in space is directly proportional to the present mass of the body. Further, suppose that the earth's mass increased 0.01% in the last 1 million years. How long will it take for the earth's mass to reach double its present value?

Section 4.4b

12. Find approximate values for the following, using Table 3, Appendix D.
    (a) $\log_{10} 1.21$
    (b) $\log_{10} 99.9$
    (c) $\log_{10} 39{,}000$
    (d) $\log_{10} 8{,}765$
    (e) $\log_{10} 1$
    (f) x if $\log_{10} x = 0.7284$
    (g) y if $\log_{10} y = 3.8976$
    (h) $\log_{10}(6.023 \cdot 10^{-23})$
    (i) w if $\log_{10} w = 2.9705$
    (j) z if $\log_{10} z = 0.2340$
    (k) $\log_{10} 21{,}740$
    (l) $\log_{10}(0.002174)$
    (m) $\log_{10}(0.0000333)$
    (n) $\log_{10} 0.912$
    (o) k if $\log_{10} k = 2.7135 - 3$
    (p) d if $\log_{10} d = 1.6284 - 5$

13. Find approximate values for the following, using Table 3, Appendix D.
    (a) $(345)(892)$
    (b) $(711)(543)$
    (c) $(23)(463)$
    (d) $(3480)/(29.6)$

(e) $(476,000)/(8.88)$
 (f) $(21)(106)$
(g) $10/361$
(h) $25/0.08$
 (i) $(39)(0.27)$
 (j) $261/0.32$
(k) $(398)(123)/(69.2)$
 (l) $\sqrt[5]{9730}$
(m) $(9230)(0.000159)$
(n) $(0.00671)(0.00318)$
(o) $(7.43 \cdot 10^{-4})/0.000\ 000\ 503$
(p) $(1/2)^{15}$
(q) $\sqrt{.000863}$
 (r) $(1,610,000)(98,900)/(6.02 \cdot 10^{23})$
(s) 9.51^7
 (t) $(21)(371)/16$
(u) $(29/0.0007)^2$
(v) $(0.3107)^3$
(w) $(0.29/0.73)^4$
(x) $(297)(0.006)(391)/1.7$
(y) $(36.7)(21,300,000)/0.00063$
(z) $(2.51)^{21}$

Section 4.3c

14. Find approximate values for the computations of Problem 13, using a slide rule.

PROBLEMS

15. Prove that $\exp_{1/a} = 1/\exp_a$ for all $a \in (0, \infty)$.
16. Show that the third law of exponents follows from the second and sixth laws.
17. Show that the sixth law of exponents is a special case of the fifth law.
18. Use the fact that $\exp_a(-x) = 1/\exp_a(x)$ to explain the behavior of $\exp_a(x)$ as x becomes large negative, where $a > 1$.
19. Write a flowchart that if implemented on a computer, would give you further understanding of $\exp_a x$ for x large positive and negative or would give evidence to support what you already know. Explain what your flowchart accomplishes along these lines.
20. Write a flowchart that will generate a sequence of rational numbers which close in on a given irrational number. In particular, let the given irrational number be $\sqrt{n}$, where n is not a perfect square, and let n be read in by the algorithm.
21. Expand the flowchart of Problem 20 to compute the sequence of values that is used to define $a^{\sqrt{n}}$.

22. Show that if $k \cdot \exp_a$ has domain restricted to $\mathbb{N}$, it forms a geometric progression.
23. Sketch precise graphs of the following functions.
   (a) $2 \cdot \exp_2$ (b) $\frac{1}{3} \exp_2$
   (c) g, where $g(x) = \exp_2 x + 2$ (d) h, where $h(x) = \exp_2 x - 3$
   (e) $\exp_2 + \exp_3$ (f) $\exp_2 x^2$
   (g) $(\exp_2 x)^2$
24. Why did we want to exclude $\exp_1$ from the discussion of $\exp_a^{-1}$?
25. Prove property (2) of logarithmic functions.
26. Prove property (3) of logarithmic functions.
27. Prove property (4) of logarithmic functions.
28. Prove property (7) of logarithmic functions.
29. Prove that
   (a) $\log_a a = 1$ for all $a \in (0, \infty)$ (b) $\log_a 1 = 0$ for all $a \in (0, \infty)$
30. Sketch precise graphs of the following functions:
   (a) $\log_2 + 1$ (b) $(\log_2)^2$ (c) $\log_2 x^2$
   (d) $\frac{1}{2} \log_2 x$ (e) $\log_2 x + \log_3 x$
31. Write a flowchart to compute approximate values of e^x by using only some of the terms of the series for e^x.
32. If $x > \frac{1}{2}$,

$$\log_e x = \frac{x-1}{x} + \frac{1}{2}\left(\frac{x-1}{x}\right)^2 + \frac{1}{3}\left(\frac{x-1}{x}\right)^3 + \cdots .$$

Write a flowchart to compute a table of values of $\log_e x$.
33. Expand the flowchart of Problem 32 to produce a table of $\log_{10} x$ for x between 1 and 10 if $\log_{10} e \approx 0.4342944819$. (See Problem 28.)
34. Compare the results of Example 3a with the amount received at 5% simple interest, and 5% compound interest compounded annually.
35. Marvin Springspeed is a genius of sorts, especially when it comes to multiplication. He can multiply scores of numbers in his head, but poor Marvin just cannot seem to grasp addition. On the basis of the material we have been studying, write an explanation for Marvin, showing him how he can do addition by using multiplication. You may use examples but remember that an example is not an explanation by itself.

   Note that Marvin is also taking a course in elementary functions and would greatly appreciate any explanation of functions involved. If he needs a table beyond Table 3, Appendix D, indicate why, what kind of table, and how he would use it. If Table 3 is sufficient, explain why and how he would use it. Marvin also appreciates clear, concise writing styles.
36. Referring to the preamble to Problems 93 and 94 of Chapter 3 find the zeros of the following functions.
   (a) $n(x) = \exp_3 x - 27$
   (b) $u(x) = \exp_4(-x) - 16$
   (c) $m(t) = \log_{10} t - 2 \log_{10} 4 - \log_{10} 32$
   (d) $e(z) = \log_{10}(3z + 2) - \log_{10}(z - 4) - 1$

(e) $r(a) = 2 \exp a - 3 \exp(-a) - 1$

(f) $i(w) = \exp_2(2w) - 2 \exp_2(w+1) - 12$

(g) $c(m) = \log_{2/3}(m) - 3$

(h) $a(m) = \exp_5(2m+3) - \exp_3 3$

(i) $l(p) = (\log_2 p)^2 + \log_2(8p) - 9$

(j) $a(x) = \log_{10}(x+1) - \log_{10}(x) - 1$

(k) $n(x) = \log_7(2x+1) + \log_7(4-x) - 2$

(l) $a(t) = \exp_6(2t) - 42 \exp_6 t + 216$

(m) $l(w) = \log_e(w^2 + 2) - \log_e(w+1) - \log_e 5$

(n) $y(z) = \exp_3(2z) - \exp_9(z+1) - 4$

(o) $s(x) = (\exp_2 x)^3 - 4(\exp_2 x)^2 + 5(\exp_2 x) - 2$

(p) $i(x) = \log_5(3-x) + \log_5(2x+1)$

(q) $s(g) = \log_5(3-g) + \log_{25}(2g+1)$

37. Use the greatest integer function to produce precise, succint definitions of characteristic and mantissa of a logarithm.

38. Verify the claim in Example 3 that a 5% rate of interest per year, continuously compounded, produces a net increase of 5.127% per year.

39. As suggested in Section 4.4c, devise and explain a general procedure for dividing with a slide rule.

CHAPTER 5

FUNCTIONS WITH DOMAIN ℕ

5.1. INTRODUCTION

Functions whose domains are ℕ or a subset of ℕ occur frequently. The outcome of a race can be expressed by a function f with $f(1)$ being the winner, $f(2)$ being the second place entry, and so on. The execution of many algorithms can be outlined by such functions. For example, the succession of midpoints in the successive halving algorithm of Section 2.7a [where $f(n) =$ the nth midpoint] provides a record of the approach toward a zero for the polynomial: $f(1)$ is the first midpoint, $f(2)$ the second, $f(20)$ the 20th, etc. Quantities that are indexed yearly, such as the gross national product, or daily, such as the maximum daily temperature, suggest functions whose domains are the successive years or days, respectively, and functional values that are the recorded numbers for such years or days. Thus if we number the days of the year consecutively and let $T(n)$ be the maximum temperature for a given day, we might have

$$T(1) = 15°$$
$$T(2) = 21°$$
$$T(3) = 5°$$
$$\vdots$$
$$T(180) = 65°$$

$$T(181) = 71°$$
$$\vdots$$
$$T(364) = 20°$$
$$T(365) = 31°.$$

We shall find that most of the questions we have asked about functions to date will not be useful questions for these functions, much as the same questions were not always appropriate for both polynomial and circular functions. In this case, the reason is largely because the only common aspect among these new functions is their domain. There are no requirements of any structural similarities that would help in establishing various relationships or properties. Zeros are not subject to systematic discovery at least in part because graphs consist of isolated points (since the domains are just $\{1, 2, 3, \cdots\}$), not smooth curves. Inverses do not often prove to be of much interest, since the domain is frequently used only to assign an order to the range elements, not because it has any intrinsic value.

To compensate at least in part for these "non-issues" is an important matter called convergence, which is of great importance in dealing with sequences. Although a rigorous, complete investigation of this topic is appropriate to courses in advanced calculus, we shall look into some aspects informally. All in all, our foray will be brief and will be concerned primarily with those aspects of most immediate usefulness.

5.2. SEQUENCES

5.2a. Notation

DEFINITION 5.1
Sequence

A function that has domain $\mathbb{N}$ is called a *sequence*. At times, it is convenient to discuss functions whose domains are subsets of $\mathbb{N}$ of the form $\{1, 2, 3, \cdots, n\}$. Such functions are called finite sequences, since they consist of only finitely many ordered pairs.

The notation for sequences often varies from ordinary functional notation, due to historical usage and reasons of convenience. For instance, to denote the range element corresponding to a given domain element, instead of enclosing the argument in parentheses after the function name, the argument is used as a subscript on the function name. Thus $f(3)$

DEFINITION 5.2
nth Term

would be written f_3 and $g(n)$ as g_n, and one calls a_n the *nth term* of the sequence a.

Also, it is common practice to refer to the set of range elements written in increasing order of domain element as the sequence, instead of the corresponding set, of ordered pairs. That is, we would refer to

$$\{f(1), f(2), f(3), \cdots\}$$

as the sequence, rather than

$$\{(1, f(1)), (2, f(2)), (3, f(3)), \cdots\}$$

or, if $g(n) = n^2$, we would call

$$\{1, 4, 9, \cdots\}$$

the sequence, instead of the more complete

$$\{(1, 1), (2, 4), (3, 9), \cdots\}.$$

Though not universally accepted, many mathematicians have come to use angled brackets $\langle \ \rangle$ (often called pointy brackets by mathematicians) in place of braces to enclose sequences. We shall adopt this notation and so write the above sequence of squares as $\langle 1, 4, 9, \cdots \rangle$. Notice that the domain elements are, in a sense, still there, in that they determine the order. Such notation is not used for ordinary functions because most domains do not permit an orderly listing of the individual elements of the domain.

Let s be the sequence defined by $s(n) = 1/n$. Then we may write the sequence as $\langle 1, \frac{1}{2}, \frac{1}{3}, \frac{1}{4}, \cdots \rangle$, and we ordinarily write $s_n = 1/n$. **EXAMPLE 1**

Also, $\langle 2, 4, 8, 16, \cdots \rangle$ is the sequence defined by $s_n = 2^n$, i.e., $\langle 2^1, 2^2, 2^3, 2^4, \cdots \rangle$. **EXAMPLE 2**

In order to safely use the direct listing of a sequence, the definition of the nth element should be clearly evident in those elements that are listed. Thus the listing $\langle 3, 3, 5, 4, 4, 3, 5, \cdots \rangle$ should not be used to describe a sequence s unless, from context, the definition is known. In this example, s_n equals the number of letters in the English word for the number n.

In cases in which the pattern is not clear, one should define s_n clearly. We may also write $\langle n^2 \rangle = \langle 1, 4, 9, 16, \cdots \rangle$ or $\langle (-1)^n n \rangle = \langle -1, 2, -3, 4, -5, 6, \cdots \rangle$; i.e., we may write the rule for s_n in the pointy brackets in place of s_n.

The specification of the nth term of a sequence is often given just in terms of n, as we have seen. Sometimes, however, it is advantageous to express the nth term in terms of the preceding term(s) of the sequence. For example, the sequence $\{2, 4, 16, 256, \cdots\}$ is efficiently specified by the requirement that $s_n = (s_{n-1})^2$. Thus we might write $s = \langle s_n \rangle = \langle (s_{n-1})^2 \rangle = \langle 2, 4, 16, 256, \cdots \rangle$. Rules or formulas that express s_n in terms of preceding terms are called *recursion formulas*.

There is some possibility of confusion between the sequence and its range. For example, let $d_n =$ the number of positive divisors of n that are $\leq n$:

$d_1 = 1$, since 1 can be divided only by 1.
$d_2 = 2$, since 2 can be divided by 2 and 1.
$d_3 = 2$, since 3 can be divided by 3 and 1.
$d_4 = 3$, since 4 can be divided by 4, 2, and 1.
$d_5 = 2$, since 5 can be divided by 5 and 1.
$d_6 = 4$, since 6 can be divided by 1, 2, 3, and 6.

Thus $\langle d_n \rangle = \langle 1, 2, 2, 3, 2, 4, 2, 4, 3, 4 \ 2, 6, \cdots \rangle$, and this could be confused with the set $\{1, 2, 3, \cdots\}$, since both collections appear to have the same elements. A little extra thought should suffice to clarify such confusions if they arise.

We shall next consider two types of sequences that are worthy of special note. Traditionally, they are called arithmetic and **geometric** progressions. Let us look first at arithmetic progressions.

5.2b.   Arithmetic Progressions

DEFINITION 5.3
Arithmetic Progression

A sequence $\langle a_n \rangle$ is called an *arithmetic progression* if there is a real number constant d such that $a_n = a_1 + (n-1)d$, or by recursion formula,† $s_n = s_{n-1} + d$. Examples include

$$\langle 2, 4, 6, 8, \cdots \rangle,$$

$$\langle 10, 10.3, 10.6, 10.9, 11.2, 11.5, \cdots \rangle,$$

$$\left\langle \frac{\pi}{2}, \frac{5\pi}{2}, \frac{9\pi}{2}, \cdots \right\rangle,$$

and

$$\left\langle 2, -1\frac{1}{2}, -7, -11\frac{1}{2}, \cdots \right\rangle.$$

Arithmetic progressions are frequent in practice and are relatively easy to work with, since the rule that specifies such a sequence is usually quite simple.

Indeed, arithmetic progressions are linear functions whose domain is $\mathbb{N}$ instead of $\mathbb{R}$. If the linear function f specified by $f(x) = mx + b$ has domain reduced to $\mathbb{N}$, what is left is $f(1) = m + b$, $f(2) = 2m + b$, $f(3) = 3m + b$, $\cdots$. This is the sequence $\langle n \cdot m + b \rangle$ or, in the form $\langle a_1 + (n-1)d \rangle$, we have $\langle (b+m) + (n-1)m \rangle$ with $a_1 = b + m$ and $d = m$. Linear functions are common in everyday life; in cases when nonintegral quantities are meaningless, these functions become sequences.

One kind of problem that frequently occurs in dealing with arithmetic progressions is that of reconstructing the whole sequence if one only knows a few parts. We shall deal with such problems by examples.

EXAMPLE 3   Suppose we have an arithmetic progression with $a_1 = 3$ and $d = 10$. From the definition, $\langle a_n \rangle = \langle a_1 + (n-1)d \rangle = \langle 3 + (n-1)10 \rangle$ or $\langle 3 + (1-1)10, 3 + (2-1)10, 3 + (3-1)10, \cdots \rangle = \langle 3, 13, 23, \cdots \rangle$.

EXAMPLE 4   If $a_3 = 5$ and $d = 5$, write out the sequence and find a_{10}. Again, from the definition, since $a_n = a_{n-1} + d$.‡

$$\begin{cases} a_3 = a_2 + 5 \\ a_2 = a_1 + 5, \end{cases}$$

$$\begin{aligned} \begin{cases} a_3 - 5 = a_2 \\ a_1 + 5 = a_2 \end{cases} \\ \hline a_3 - a_1 - 10 = 0 \end{aligned}$$

$$a_1 = a_3 - 10$$
$$a_1 = 5 - 10$$
$$a_1 = -5,$$

and

† See Problem 37.
‡ If the solution of simultaneous equations is not clear to you, see Appendix A, Section A-15.

$$\langle a_n \rangle = \langle a_1 + (n-1)d \rangle = \langle -5 + (n-1)5 \rangle$$
$$= \langle -5 + (1-1)5, -5 + (2-1)5, -5 + (3-1)5, \cdots \rangle$$
$$= \langle -5, 0, 5, \cdots \rangle$$

Then

$$a_{10} = a_1 + (10-1)5 = -5 + 9 \cdot 5 = 40.$$

(Although you may have seen a shorter way to solve this particular problem, this method paves the way to the solution of more difficult problems.)

If $\langle a_n \rangle$ is an arithmetic progression $a_7 = 35$ and $a_{10} = 44$, give a full specification for $\langle a_n \rangle$. Again, by the definition of arithmetic progression and use of algebra, we see that since $a_7 = a_1 + 6d$ and $a_{10} = a_1 + 9d$, we have the simultaneous equations

EXAMPLE 5

$$a_1 + 6d = 35$$
$$a_1 + 9d = 44.$$

These are easily solved to find that $d = 3$ and $a_1 = 17$. Hence $\langle a_n \rangle = \langle 17 + (n-1)3 \rangle = \langle 17, 20, 23, 26, \cdots \rangle$.

For further examples and applications, see the exercises at the end of this chapter.

By now you may have made one very helpful observation about arithmetic progressions, that to build any arithmetic progression, you simply add the same amount over and over to each successive term, e.g.,

1. In $\langle 2, 4, 6, 8, 10, \cdots \rangle$, we begin with 2 and repeatedly add 2.
2. In $\langle 2\pi, 5\pi/2, 3\pi, 7\pi/2, \cdots \rangle$, we begin with 2π and repeatedly add $\pi/2$.

Check out the earlier examples to be sure you see this. The quantity we add is, of course, the "d" in the definition. Notice also that this is what the recursion formula form of the definition of arithmetic progression emphasizes.

5.2c. Geometric Progressions

A sequence $\langle g_n \rangle$ is called a *geometric progression* if there is a real number r such that $g_n = g_1 r^{(n-1)}$, or alternatively, $g_n = g_{n-1} r$.† Examples include

GEOMETRIC PROGRESSIONS

DEFINITION 5.4
Geometric Progression

$$\langle 1, 2, 4, 8, 16, \cdots \rangle,$$
$$\langle 5, 5/4, 5/16, 5/64, \cdots \rangle,$$
$$\langle 1, -1, 1, -1, 1, -1, \cdots \rangle,$$

and

$$\langle 7, 7\sqrt{2}, 14, 14\sqrt{2}, 28, \cdots \rangle.$$

Whereas arithmetic progressions are closely related to linear functions,

† See Problem 38 for a proof that these definitions are equivalent.

geometric progressions are closely related to the exponential functions of Chapter 4. We shall discuss this further in the problems for this chapter, but for now suffice it to observe that, as with arithmetic progressions, the sequence can be reconstructed from a few pieces of information about the sequence.

EXAMPLE 6   If $g_1 = 3$ and $r = 2$, write out the sequence $\langle g_n \rangle$. Using the definition, $\langle g_n \rangle = \langle g_1 r^{(n-1)} \rangle$, we have

$$\langle g_n \rangle = \langle 3 \cdot 2^{(n-1)} \rangle = \langle 3 \cdot 2^{1-1}, 3 \cdot 2^{2-1}, 3 \cdot 2^{3-1}, 3 \cdot 2^{4-1}, \cdots \rangle \dagger$$
$$= \langle 3, 6, 12, 24, \cdots \rangle.$$

EXAMPLE 7   Given $r = \frac{1}{2}$ and $g_3 = 5$, what is $\langle g_n \rangle$? We use the definition

$$\begin{cases} g_3 = g_2(\frac{1}{2})^2 \\ g_2 = g_1(\frac{1}{2})^1. \end{cases}$$

Then we substitute,

$$g_3 = (g_1(\tfrac{1}{2}))(\tfrac{1}{4})$$

and

$$5 = g_1(\tfrac{1}{8}) \qquad \text{or} \qquad g_1 = 40.$$

and obtain

$$\langle g_n \rangle = \langle 40 \cdot (\tfrac{1}{2})^{n-1} \rangle$$
$$= \langle 40(\tfrac{1}{2})^{1-1}, 40(\tfrac{1}{2})^{2-1}, 40(\tfrac{1}{2})^{3-1}, 40(\tfrac{1}{2})^{4-1} \cdots \rangle$$
$$= \langle 40, 20, 10, 5, \cdots \rangle.$$

EXAMPLE 8   If $\langle g_n \rangle$ is a geometric progression with $g_4 = 270$ and $g_2 = 120$, give a full specification for $\langle g_n \rangle$. Again, the system of equations that results is not linear but can be solved by substitution, as in the preceding example or as follows:

$$g_1 \cdot r^3 = 270 \tag{1}$$
$$g_1 \cdot r = 120. \tag{2}$$

Dividing Equation (1) by Equation (2), we have

$$r^2 = \frac{270}{120} = \frac{9}{4}$$

so that

$$r = \pm \frac{3}{2}.$$

Then

$$g_1 = \frac{120}{r} = 80 \text{ or } -80.$$

Therefore,

$$\langle g_n \rangle = \langle 80(\tfrac{3}{2})^{n-1} \rangle = \langle 80, 120, 180, \cdots \rangle$$

† $2^{1-1} = 2^0 = 1$ by definition of $a^0 (=1)$.

or

$$= \langle -80(-\tfrac{3}{2})^{n-1}\rangle = \langle -80, +120, -180, \cdots \rangle.$$

The problem given us no reason to eliminate one of these two alternatives.

This time did you discover the "shortcut" yourself? If not, go back and try to find it before reading further. You should see that to get the next term of a geometric progression, simply multiply the term you are considering by r. In Example 7, with 40 as the first term, we find that $40(\tfrac{1}{2}) = 20$, the second term; $20(\tfrac{1}{2}) = 10$, the third term, etc. Be sure to include the the sign if r is negative, as in Example 8.

In real life, one sometimes encounters references to arithmetic and geometric progressions in the context of growth. Someone may say, for example, that the amount of garbage in this country is growing geometrically or that a certain company's profit is growing arithmetically. What is meant is that the amount of garbage is a geometric progression as a function of time or that the company's profit is an arithmetic progression as a function of time. By the nature of these two kinds of sequences you may already understand that the adjective *geometric* implies a rapid, accelerating growth, while the adjective *arithmetic* implies steady growth. These adjectives are rapidly being replaced, however, by *exponential* and *linear* in reference to exponential and linear functions, of which geometric and arithmetic progressions are just those whose domains are restricted to $\mathbb{N}$.

5.3. SERIES

5.3a. Series in General

It seems fairly natural if we are interested in, say, $\langle 1/n\rangle = \langle 1, \tfrac{1}{2}, \tfrac{1}{3}, \tfrac{1}{4}, \tfrac{1}{5}, \cdots \rangle$, that we should also be curious about $1 + \tfrac{1}{2} + \tfrac{1}{3} + \tfrac{1}{4} + \tfrac{1}{5} + \cdots$. In general, if we have $\langle a_n \rangle$, it seems natural to study $a_1 + a_2 + a_3 + a_4 + \cdots + a_n + \cdots$, or at least $a_1 + a_2 + a_3 + a_4 + \cdots + a_n$. These sums are called *infinite* and *finite* series, respectively. To facilitate dealing with series, we need some new notation.

For the finite series, the symbol $\sum_{i=k}^{m} t_i$, where $m, k \in \mathbb{Z}$, is read "the sum from i equals k to m of t_i" and is defined by

$$\sum_{i=k}^{m} t_i = t_k + t_{k+1} + t_{k+2} + \cdots + t_m.$$

The letter i is called the *index* of the sum and may be any letter. The non-negative integers k and m may be actual constants or variables and are called the *lower* and *upper limits* of the sum, and k must be $\leq m$. The capital Greek sigma is called the *summation sign*.

This symbolism provides a convenient shorthand for dealing with finite series. Analogously, for *infinite series*, we write

$$\sum_{i=a}^{\infty} t_i = t_a + t_{a+1} + t_{a+2} + t_{a+3} + \cdots,$$

DEFINITION 5.5
Series

where all the terminology above is the same except that here the upper limit is ∞, meaning that the sum goes on with no end.

Using these notations enables us to succinctly formalize the concept of series; we say that a *series* is a sum

$$\sum_{i=k}^{\beta} a_i,$$

where i and k represent nonnegative integers. If β is an integer, the series is a finite series. If β is the symbol ∞, the series is an infinite series.

The notations in the above definitions may seem overwhelming, but the following examples may help to clarify the use of the summation sign.

EXAMPLE 9

$$\sum_{i=4}^{7} i^2 = 4^2 + 5^2 + 6^2 + 7^2 = 16 + 25 + 36 + 49 = 126$$

EXAMPLE 10

$$\sum_{n=1}^{5} n! = 1 + 2 \cdot 1 + 3 \cdot 2 \cdot 1 + 4 \cdot 3 \cdot 2 \cdot 1 + 5 \cdot 4 \cdot 3 \cdot 2 \cdot 1$$
$$= 1 + 2 + 6 + 24 + 120 = 153$$

EXAMPLE 11

$$\sum_{j=5}^{5} 2^j = 2^5 = 32.$$

EXAMPLE 12

Consider $\sum_{k=1}^{20} 2$. This one may be initially confusing because the summand does not appear to depend on the index at all so that one's action does not seem clearly dictated. Actually, what this says is that the kth term is 2 for all values of k, i.e., we are dealing with a constant sequence. Thus

$$\sum_{k=1}^{20} 2 = \underbrace{2 + 2 + 2 + \cdots + 2}_{20 \text{ terms}} = 40.$$

One place we have already encountered series was in the evaluation of sine and cosine. You will remember that

$$\sin(x) = x - \frac{x^3}{3!} + \frac{x^5}{5!} - \frac{x^7}{7!} + \cdots + \frac{x^{2n-1}}{(2n-1)!} (-1)^{n+1} + \cdots$$

which would be written in our new notation as

$$\sum_{n=1}^{\infty} \frac{x^{2n-1}}{(2n-1)!} (-1)^{n+1}.$$

If we could actually find this sum for each x, we would have exact values for the sine. In practice, we usually deal with

$$\sum_{i=1}^{n} \frac{x^{2n-1}}{(2n-1)!} (-1)^{n+1}$$

to get approximate values such as those found in tables. Similarly, the series for cos x would be

$$\sum_{n=1}^{\infty} \frac{x^{2(n-1)}}{(2(n-1))!} (-1)^{n+1}.$$

In dealing with series, it is often valuable to talk about corresponding finite series such as just mentioned for sine. They are given a special name: the finite series $\sum_{i=1}^{n} a_i$ is called the *nth partial sum* of the infinite series $\sum_{i=1}^{\infty} a_i$.

DEFINITION 5.6
*n*th Partial Sum

5.3b. Convergence

When an infinite series has a specific sum, we say it *converges* to the sum. As we noted earlier, the techniques for dealing with convergence are necessarily left to the calculus. However, we can take a look at the ideas involved. If we consider

CONVERGENCE
DEFINITION 5.7
Convergence

$$\frac{1}{2} + \frac{1}{4} + \frac{1}{8} + \frac{1}{16} + \cdots = \sum_{i=1}^{\infty} \frac{1}{2^i}$$

it seems intuitively clear that this sum is 1. For instance, if someone served you $\frac{1}{2}$ of a pie, then $\frac{1}{4}$, then $\frac{1}{8}$, etc., and if he could do this infinitely many times, he would "eventually" serve you the entire pie.

Considering

$$\sum_{i=1}^{\infty} 2^i = 2 + 4 + 8 + 16 + 32 + \cdots,$$

it seems quite clear that this would never have a specific sum but would rather get larger and larger. This is, in fact, what happens in this case, and we say that the series diverges. More generally, one says that an infinite series *diverges* if it does not converge.

DEFINITION 5.8
Divergence

However, we must be careful because the following series may well appear to converge, but in reality it does not:

$$\sum_{i=1}^{\infty} \frac{1}{i} = 1 + \frac{1}{2} + \frac{1}{3} + \frac{1}{4} + \cdots.$$

(See Problems 48 and 49.)

Thus we must be cautious in using intuition and wait until we have more tools to discover more about convergence.

5.3c. Some Special Cases

Just as geometric and arithmetic progressions were special cases of sequences, we shall find that the related series are also special cases and that we can easily find values of their partial sums. First consider the partial sum of the elements of a geometric progression:

ARITHMETIC AND
GEOMETRIC SERIES

$$\sum_{i=1}^{n} g_i = \sum_{i=1}^{n} g_1 r^{i-1} = g_1 + rg_1 + r^2 g_1 + r^3 g_1 + \cdots + r^{n-1} g_1. \qquad (3)$$

Since $g_n = g_1 r^{n-1}$, by multiplying both sides of this expression by r, it is also true that

$$r \sum_{i=1}^{n} g_1 r^{i-1} = \sum_{i=1}^{n} g_1 r^i = rg_1 + r^2 g_1 + r^3 g_1 + \cdots + r^n g_1. \qquad (4)$$

Noting that Equations (3) and (4) have $n-1$ identical terms, we subtract Equation (4) from Equation (3), thus generating the equation

$$\sum_{i=1}^{n} g_1 r^{i-1} - \sum_{i=1}^{n} g_1 r^i = g_1 - r^n g_1$$

which is the same as

$$\left(\sum_{i=1}^{n} g_1 r^{i-1} \right) (1 - r) = g_1 (1 - r^n),$$

or

$$\sum_{i=1}^{n} g_1 r^{i-1} = g_1 \frac{(1 - r^n)}{1 - r}. \tag{5}$$

Equation (5) is just what we wanted: a neat formula for the sum of a finite number of terms of a geometric progression, indeed just in terms of g_1 and r. Although this formula is useful in itself, let us push on to a further observation. What happens to g_n as n gets larger and larger? Equation (5) gives us a way to give a useful answer. The only part of the formula that depends on n is the term r^n. If $|r| < 1$, then r^n gets closer and closer to zero. For example, if $r = |\frac{1}{2}|$, then $(\frac{1}{2})^3 = \frac{1}{8}$, which is smaller and therefore closer to zero than $\frac{1}{2}$. This is a good time to reexamine Figure 2.13 in Chapter 2 to see that the same is true for each $r \in (-1, 1)$. However, formal proof of this would take us too far afield, so we shall defer this to your study of the calculus.

What we are saying, then, is that if $|r| < 1$, then $\sum_{i=1}^{n} g_1 r^{i-1}$ gets closer and closer to

$$g_1 \frac{(1 - 0)}{(1 - r)} \quad \text{or} \quad \frac{g_1}{1 - r}.$$

This suggests both that an infinite sum such as

$$g_1 + g_1 r + g_1 r^2 + g_1 r^3 + \cdots$$

can, in fact, sensibly have a finite value, provided that $|r| < 1$, and that the sensible value is $g_1 / (1 - r)$. The overall reasoning we have used is actually quite sound, and we shall accept this formula for the sum of this infinite series:

$$\sum_{i=1}^{\infty} g_1 r^{i-1} = \frac{g_1}{1 - r}.$$

We may also find a formula for the value of the partial sum associated with an arithmetic progression. Since $a_n = a_1 (n - 1)d$,

$$\sum_{i=1}^{n} a_i = \sum_{i=1}^{n} (a_1 + (i - 1)d)$$
$$= a_1 + (a_1 + d) + \cdots + [a_1 + (n - 2)d] + [a_1 + (n - 1)d].$$

Or, added in the reverse order,

$$\sum_{i=1}^{n} (a_1 + (i - 1)d)$$
$$= [a_1 + (n - 1)d] + [a_1 + (n - 2)d] + \cdots + (a_1 + d) + a_1.$$

If we add vertically in the last two lines, we would have

$$2 \sum_{i=1}^{n} (a_1 + (i-1)d)$$

$$= [2a_1 + (n-1)d] + [2a_1 + (n-1)d] + \cdots + [2a_1 + (n-1)d]$$

$$\underbrace{\qquad\qquad\qquad\qquad\qquad\qquad\qquad\qquad\qquad}_{n \text{ such terms}}$$

$$= n \cdot 2a_1 + n(n-1)d.$$

Hence

$$\sum_{i=1}^{n} (a_1 + (i-1)d) = na_1 + \frac{n(n-1)d}{2}.$$

Another common form of this formula is generated by noting further that

$$\sum_{i=1}^{n} (a_1 + (i-1)d) = \frac{n}{2} [2a_1 + (n-1)d] = \frac{n}{2} (a_1 + [a_1 + (n-1)d])$$

$$= \frac{n}{2} [a_1 + a_n]$$

We note here that the sum

$$\sum_{i=1}^{\infty} a_n = \sum_{i=1}^{\infty} (a_1 + (i-1)d)$$

diverges, except when $d = 0$.

5.4.   MATHEMATICAL INDUCTION

Sequences are used a great deal in mathematics, although often in disguise. Often there are properties or formulas that involve a natural number as a variable, which one wishes to prove are true regardless of which natural number is involved in any specific instance. We have already encountered situations of this kind: one of the proofs omitted in our discussion of series could make use of the fact that if $|r| < 1$, then $|r^{n+1}| < |r^n|$ for all $n \in \mathbb{N}$ (e.g., if $r = \frac{1}{2} < 1$, then $(\frac{1}{2})^4 < (\frac{1}{2})^3$ or $\frac{1}{16} < \frac{1}{8}$, and similarly for all n). In discussing periodic functions, it seems sensible that if p is the period of a periodic function f, then $f(x + kp) = f(x)$ for each $k \in \mathbb{N}$ (indeed, for each $k \in \mathbb{Z}$). In discussing the distributive law of real number multiplication over real number addition, it would be nice to be able to conclude firmly that $a(b_1 + b_2 + \cdots + b_n) = ab_1 + ab_2 + \cdots + ab_n$ for all $n \in \mathbb{N}$, not just $n = 2$ or 3 or 4.

It might appear at first careful glance that to prove that a proposition $P(n)$ is true for all $n \in \mathbb{N}$ would be hopeless, in that one would have to prove that each of infinitely many individual cases is true. Fortunately, there is a technique of proof that is effective in many circumstances and which settles the infinity of cases in deceptively short order. It is called *mathematical induction*. This technique is appropriate in a chapter dealing with functions with domain $\mathbb{N}$, since if $P(n)$ represents a proposition that depends on the value of n, then P may be viewed as a sequence whose range is a subset of the set of all propositions. Our concern then is with the

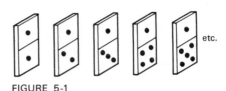

FIGURE 5-1

truth of each and every element in the sequence, and while such a question has not been relevant for the functions studied so far, it is for these sequences whose ranges are so unlike polynomials or circular functions, or exponential and logarithmic functions.

Imagine a very long line of dominoes set on end. See Figure 5.1. Children (of all ages) enjoy tipping over the first dominoe in the line and watching all the rest fall in succession. How can one be sure that all the dominoes will fall? An efficient way is to be sure that both the first one actually is knocked over; and that the dominoes have been arranged so that if any one domino tips over, then the next one will be forced to tip over, too.

Mathematical induction operates in much the same manner. One verifies that the first proposition, $P(1)$, is true (corresponding to knocking over the first domino) and then shows that if the kth proposition is true, then the $(k+1)$st must also be true [that the falling of the kth domino is guaranteed to knock over the $(k+1)$st]. Thus an induction proof is a two-part proof.

The process of mathematical induction can be likened to an infinite loop in a flowchart, and such an analogy should help you understand why induction works. If you *know* that the loop is entered and that *every* time the exit criterion is encountered, the decision is made to stay in the loop, it is clear that one will *never* leave the loop. In induction, phase 1, checking that $P(1)$ is true, corresponds to entry into the loop. Phase 2, checking that the truth of $P(k)$ guarantees the truth of $P(k+1)$, corresponds to being sure that every time the exit criterion is encountered, the decision is made to stay in the loop. Figure 5.2 exhibits an induction flowchart, showing the loop which, if shown to be infinite, would guarantee the truth of $P(n)$ for all n. In this loop, phase 1 would assign 1 to k, and check if $P(1)$ is true. If so, the loop is entered. Phase 2 checks that the loop instruction

> **THEN P(k + 1)**
> **IS ALSO TRUE**

always logically follows from the knowledge that $P(k)$ is true.

It is important that you realize that *both* phases of an induction proof are vital to actually prove any given proposition. To illustrate this, consider the following examples. Prove

EXAMPLE 13

$$\sum_{i=1}^{n} i^3 = \frac{n^2(n+1)^2 + 3}{4},$$

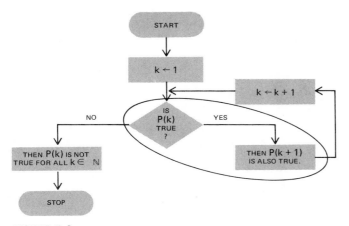

FIGURE 5-2

i.e.,

$$1^3 + 2^3 + 3^3 + \cdots + n^3 = \frac{n^2(n+1)^2 + 3}{4}$$

For $n = 1$,

$$1^3 = \frac{1^2(1+1)^2 + 3}{4}$$

$$1 = \frac{1 \cdot 2 + 3}{4}$$

$$1 = \frac{2 + 3}{4}$$

$$1 \neq 1\frac{1}{4}.$$

However, what if we had ignored the case of $n = 1$? Assume true for $n = k$ that

$$1^3 + 2^3 + 3^3 + \cdots + k^3 = \frac{k^2(k+1)^2 + 3}{4}. \qquad (6)$$

Now prove true for $n = k + 1$ if it is true for $n = k$, i.e., show that

$$1^3 + 2^3 + 3^3 + \cdots + k^3 + (k+1)^3 = \frac{(k+1)^2[(k+1)+1]^2 + 3}{4}.$$

Let us add $(k+1)^3$ to both sides of the equation (6):

$$1^3 + 2^3 + 3^3 + \cdots + k^3 + (k+1)^3 = \frac{k^2(k+1)^2 + 3}{4} + (k+1)^3$$

$$= \frac{k^2(k+1)^2 + 3 + 4(k+1)^3}{4}$$

$$= \frac{(k+1)^2(k^2+4(k+1))+3}{4}$$

$$= \frac{(k+1)^2(k^2+4k+4)+3}{4}$$

$$= \frac{(k+1)^2(k+2)^2+3}{4}$$

$$= \frac{(k+1)^2((k+1)+1)^2+3}{4},$$

which is what we wanted to show. Thus, if we had been careless about $P(1)$, we could easily have thought that we had proved a statement that was false.

EXAMPLE 14   Prove $n^2 - 3n + 2 = 0$ for all n. Careful observation would quickly show that this is false, but suppose we did not look carefully and proceded to apply mathematical induction. For $n = 1$,

$$1^2 - 3(1) + 2 = 0$$
$$1 - 3 + 2 = 0$$
$$0 = 0;$$

therefore, the proposition is true for $n = 1$.

Next, for phase 2, assume that the proposition is true for $n = k$: $k^2 - 3k + 2 = 0$. Prove that it is true for $n = k + 1$ if it is true for $n = k$, i.e., show that

$$(k+1)^2 - 3(k+1) + 2 = 0$$
$$k^2 + 2k + 1 - 3k - 3 + 2 = 0$$
$$(k^2 - 3k + 2) + 2k - 2 = 0.$$

The first term $(k^2 - 3k + 2)$ is 0 by our assumption that the proposition is true for $n = k$. However, for $k > 1$, the second term, $2k - 2$ is never 0, so that the proposition is not true.

Be careful here, too, though. In our example, the $(k+1)$st proposition is false if the kth one is true. But, in general, failure to complete phase 2 may only mean that you have not been insightful or clever enough to see your way through to the end. You need solid evidence to be able to assert disproof.

EXAMPLE 15   Last, let us consider $P(n) = n^2 - n + 41$, which is supposed to produce a prime number for each n. If you tried this for $P(1) = 1^2 - 1 + 41 = 41$, you would see that it produced a prime (41). However, if you next tried to carry out the second phase of induction, you would run into numerous technical difficulties. At that stage, you might say, "Oh well, I'll just check out a number of examples":

$$P(2) = 2^2 - 2 + 41 = 43$$
$$P(3) = 3^2 - 3 + 41 = 47$$
$$P(4) = 4^2 - 4 + 41 = 53$$
$$P(5) = 5^2 - 5 + 41 = 61.$$

At this stage, many people would be convinced that they just did not see a good strategy to use in the induction proof, and would assume the proposition to be true. The more diligent experimenter might try out 10 or 15 more examples. He would continue to get primes. In fact, this proposition is true for $P(1)$, $P(2)$, $\cdots$, $P(40)$. However, at $n = 41$, $P(41) = 41^2 - 41 + 41 = 41^2 \neq$ prime. This is not an isolated example. Experimental evidence, while often helpful in designing a full proof, is *never* sufficient by itself, no matter how abundant.

Now let us look at some samples of induction proofs that we can follow all the way through successfully.

Prove that if $0 < r < 1$, then $r^{n+1} < r^n$ for all $n \in \mathbb{N}$.

EXAMPLE 16

Phase 1: Let $n = 1$. We need to show that $r^2 < r^1$. We know that $r < 1$. Multiplying both sides of the inequality by r, we get an equivalent inequality:

$$r < 1$$
$$r \cdot r < r \cdot 1$$
$$r^2 < r.$$

Since $r > 0$, we know that the sense of the inequality is not changed so that $r^2 < r$ which is what we needed to show.

Phase 2: Assume that $r^{k+1} < r^k$, and prove that $r^{(k+1)+1} < r^{k+1}$. It is worth emphasizing that we may not assume that last statement; indeed, that is what we must show. Since we are assuming that $r^{k+1} < r^k$, we may multiply both sides by r and still have a valid inequality. It is

$$r^{k+1} \cdot r < r^k \cdot r$$

which is the same as

$$r^{k+2} < r^{k+1} \qquad \text{or} \qquad r^{(k+1)+1} < r^{k+1}$$

which is what we needed to show.

Prove

$$\frac{1}{2} + \frac{1}{4} + \frac{1}{8} + \cdots + \frac{1}{2^n} = 1 - \frac{1}{2^n}.$$

Phase 1: Prove for $n = 1$.

EXAMPLE 17

$$\frac{1}{2^1} = 1 - \frac{1}{2^1}$$

$$\frac{1}{2} = \frac{1}{2}$$

so that the formula is true for $n = 1$.

Phase 2: We assume that the formula is true for $n = k$, i.e.,

$$\frac{1}{2} + \frac{1}{4} + \frac{1}{8} + \cdots + \frac{1}{2^k} = 1 - \frac{1}{2^k}. \qquad (7)$$

Now we try to prove that it is true for $n = k + 1$:

$$\frac{1}{2}+\frac{1}{4}+\frac{1}{8}+\cdots+\frac{1}{2^k}+\frac{1}{2^{k+1}}=1-\frac{1}{2^{k+1}} \qquad (8)$$

Since Equation (7) is true, we can add the same thing to both sides of the equality, namely, $\frac{1}{2^{k+1}}$, getting

$$\left(\frac{1}{2}+\frac{1}{4}+\frac{1}{8}+\cdots+\frac{1}{2^k}\right)+\frac{1}{2^{k+1}}=\left(1-\frac{1}{2^k}\right)+\frac{1}{2^{k+1}}. \qquad (9)$$

Now the left side of Equation (9) is just like the left side of Equation (8). Let us see if we can manipulate the right side of Equation (9) so that it also looks like the right side of Equation (8). If we can, our proof is complete.

$$\left(1-\frac{1}{2^k}\right)+\frac{1}{2^{k+1}}=1-\frac{1}{2^k}+\frac{1}{2^{k+1}}$$

$$=1-\left(\frac{1}{2^k}-\frac{1}{2^{k+1}}\right)$$

$$=1-\left(\frac{2}{2^{k+1}}-\frac{1}{2^{k+1}}\right).$$

Putting the right-hand term over a lowest common denominator, we get

$$=1-\left(\frac{2-1}{2^{k+1}}\right)$$

$$=1-\frac{1}{2^{k+1}},$$

and our proof is complete.

EXAMPLE 18   Prove that

$$\sum_{i=1}^{n} i^2=\frac{n(n+1)(2n+1)}{6}.$$

(Note that this is proving a formula for another finite series whose parent sequence is neither an arithmetic nor geometric progression.)

Phase 1: We wish to show that

$$\sum_{i=1}^{1} i^2=\frac{1(1+1)(2\cdot 1+1)}{6}.$$

Evaluating each of the two expressions, we find that

$$\sum_{i=1}^{1} i^2=1^2=1 \quad \text{and} \quad \frac{1(1+1)(2+1)}{6}=\frac{1\cdot 2\cdot 3}{6}=1.$$

Both have the value 1; hence they are, in fact, equal.

Phase 2: We assume that

$$\sum_{i=1}^{k} i^2=\frac{k(k+1)(2k+1)}{6},$$

and need to show that we can deduce that

$$\sum_{i=1}^{k+1} i^2 = \frac{(k+1)[(k+1)+1][2(k+1)+1]}{6}.$$

The left-hand side can be rewritten as

$$\sum_{i=1}^{k+1} i^2 = \sum_{i=1}^{k} i^2 + (k+1)^2. \qquad (10)$$

Since we assume

$$\sum_{i=1}^{k} i^2 = \frac{k(k+1)(2k+1)}{6}$$

to be true for k (i.e., using the induction hypotheses), we can rewrite Equation (10) as

$$= \frac{k(k+1)(2k+1)}{6} + (k+1)^2. \qquad (11)$$

This can be rearranged using algebra as follows:

$$(k+1)\left[\frac{k(2k+1)}{6} + (k+1)\right] = (k+1)\left[\frac{2k^2+k+6(k+1)}{6}\right]$$

$$= \frac{k+1}{6}[2k^2+7k+6]$$

$$= \frac{k+1}{6}[(k+2)(2k+3)]$$

$$= \frac{(k+1)(k+2)(2k+3)}{6}$$

$$= \frac{(k+1)(k+1+1)(2k+2+1)}{6}$$

$$= \frac{(k+1)[(k+1)+1][2(k+1)+1]}{6}$$

$$= \sum_{i=1}^{k+1} i^2$$

Thus we have shown that $\sum_{i=1}^{k+1} i^2$ does equal the proposed formula, under the assumption that the formula held for $\sum_{i=1}^{k} i^2$; therefore, the proof is complete.

EXAMPLE 19

Prove that if $A, B_1, B_2, \cdots, B_n$ are all sets, then $A \cup (B_1 \cap B_2 \cap \cdots \cap B_n) = (A \cup B_1) \cap (A \cup B_2) \cap \cdots \cap (A \cup B_n)$ for all $n \in \mathbb{N}$. For $n = 1$, the statement is completely trivial: $A \cup (B_1) = A \cup B_1$ is clearly true. Indeed, this often leaves people feeling uneasy. In this case one may also check that $A \cup (B_1 \cap B_2)$ works all right, too (which it does by basic set algebra) but it is not necessary to do this. For $n = k$, we assume that $A \cup (B_1 \cap \cdots \cap B_k)$ does equal $(A \cup B_1) \cap \cdots \cap (A \cup B_k)$ and seek to verify that $A \cup (B_1 \cap \cdots \cap B_{k+1})$ behaves in a similar fashion. Then

$$A \cup (B_1 \cap \cdots \cap B_{k+1}) = A \cup [(B_1 \cap \cdots \cap B_k) \cap B_{k+1}]$$
$$= [A \cup (B_1 \cap \cdots \cap B_k] \cap [A \cup B_{k+1}].$$

(We treated $(B_1 \cap \cdots \cap B_k)$ as a single set and used the simple version of the distributive law.) We obtain

$$= [(A \cup B_1) \cap \cdots \cap (A \cup B_k)] \cap [A \cup B_{k+1}],$$

by our induction hypothesis for $n = k$,

$$= (A \cup B_1) \cap \cdots \cap (A \cup B_{k+1}).$$

which is what we needed to show.

Induction proofs often seem to be a form of "arm waving" to the beginner. It is best to complete a number of such proofs, thinking out the procedure at each step. Eventually, understanding will come. In general,

1. Prove the claim for $n = 1$.
2. Assume that the proposition is true for $n = k$ (and write out a statement of the proposition in terms of k).
3. Prove that the proposition is true for $n = k + 1$. Begin by stating the proposition for $n = k + 1$ and then use the statement for $n = k$ to prove it to be true.

At times, it proves useful to use the basic mechanism of induction to define sequences. Such definitions are called, naturally enough, inductive definitions. Typically, one defines the first object and then specifies how the $(n + 1)$st object is defined if the nth one has been defined. This suffices to define the full set of objects. Examples include the recursion formulas already given for arithmetic and geometric progressions, and the succession of midpoints in the successive halving algorithm of Section 2.7a.

SUMMARY OF IMPORTANT CONCEPTS AND NOTATION

SEQUENCE: (Definition 5.1) A function that has domain $\mathbb{N}$ is called a sequence.

nth TERM: (Definition 5.2) One calls a_n the nth term of the sequence a.

ARITHMETIC PROGRESSION: (Definition 5.3) A sequence $\langle a_n \rangle$ is called an arithmetic progression if there is a real number constant d such that $a_n = a_1 + (n - 1)d$.

GEOMETRIC PROGRESSION: (Definition 5.4) A sequence $\langle g_n \rangle$ is called a geometric progression if there is a real number r such that $g_n = g_1 r^{n-1}$.

SERIES NOTATION: (Section 5.3a)

Finite: $\sum\limits_{i=k}^{m} t_i = t_k + t_{k+1} + \cdots + t_{k+m}$.

Infinite: $\sum\limits_{i=k}^{\infty} t_i = t_k + t_{k+1} + \cdots$.

SERIES: (Definition 5.5) A series is a sum $\sum_{i=k}^{\beta} a_i$ where i and k represent nonnegative integers, and β is either a positive integer or the symbol ∞.

nth PARTIAL SUM: (Definition 5.6) The finite series $\sum_{i=1}^{n} a_i$ is called the nth partial sum of the series $\sum_{i=1}^{\infty} a_i$.

CONVERGENCE: (Definition 5.7) When an infinite series has a specific real number sum, we say it converges (to the sum).

DIVERGENCE: (Definition 5.8) An infinite series diverges if it does not converge.

GEOMETRIC SERIES, FORMULAS: (Section 5.3c)

$$\sum_{i=1}^{n} g_i = g_1 \frac{(1-r^n)}{1-r}, \qquad \sum_{i=1}^{\infty} g_i = \frac{g_1}{1-r} \qquad \text{where} \quad g_i = g_1 \cdot r^{i-1}$$

ARITHMETIC SERIES, FORMULA: (Section 5.3c)

$$\sum_{i=1}^{m} a_i = \frac{n}{2}(a_1 + a_n), \qquad \text{where} \quad a_i = a_1 + (i-1)d$$

MATHEMATICAL INDUCTION: (Section 5.4)

PRACTICE EXERCISES

Section 5.2a

1. Write out the first six terms of each of the following sequences:
   (a) $\langle 1/n^2 \rangle$　　　　(b) $\langle n+1 \rangle$　　　　(c) $\langle n/(n+1) \rangle$
   (d) $\langle 2n-1 \rangle$　　　　(e) $\langle n^2/(n+1) \rangle$　　(f) $\langle n^3 \rangle$
   (g) $\langle n - n/2 \rangle$

Section 5.2b

2. In each of the following problems, the sequence being considered is an arithmetic progression. Use the given information to find a complete specification of the sequence and any particular quantities requested.
   (a) $a_1 = 10, d = 3$　　　　　　(b) $a_3 = 15, a_{10} = 1$
   (c) $a_1 = \frac{1}{2}, d = \frac{1}{4}$　　　　　　(d) $a_1 = \pi, d = \pi/2$
   (e) $a_3 = 5, a_5 = 7$　　　　　　(f) $d = -\frac{3}{2}, a_6 = -1$
   (g) $a_3 = 5, d = -7$　　　　　　(h) $a_2 = 7, a_4 = 10$; find a_{25}
   (i) $a_2 = 1, a_3 = \frac{1}{2}$　　　　　　(j) $a_{17} = 50, d = 2$; what is a_{10}?
   (k) $a_{21} = -13, a_{20} = -7$　　(l) $a_1 = 0, a_{16} = 40$; find a_{50}.

Section 5.2c

3. In each of the following problems, the sequence being considered is a geometric progression. Use the given information to find a complete specification of the sequence and any particular quantities requested.
   (a) $g_1 = \frac{1}{2}, r = 1$　　　　　　(b) $g_1 = 3, r = 2$

(c) $g_1 = -6, r = -\frac{1}{2}$ (d) $g_5 = 5, g_6 = -5$

(e) $g_2 = 3, g_3 = 4$ (f) $g_4 = 1, g_2 = 4$

(g) $g_7 = 6, r = 2$ (h) $g_3 = 10, g_7 = 40$; what is g_8?

(i) $g_4 = \frac{1}{2}, r = -7$ (j) $g_4 = \frac{2}{3}, r = \frac{1}{3}$; what is g_2?

(k) $g_3 = -\frac{5}{8}, g_6 = \frac{5}{64}$ (l) $g_2 = 10, g_3 = 100$; find g_6.

Section 5.3

4. Evaluate each of the following.

(a) $\sum_{i=1}^{6} i + 1$ (b) $\sum_{j=7}^{10} j^2$

(c) $\sum_{k=1}^{5} 2^k$ (d) $\sum_{m=1}^{50} a_m$ if $a_m = 6 + (m-1)2$

(e) $\sum_{j=1}^{6} k_j$ if $k_1 = 4, k_n = \frac{3}{2} k_{n-1}$ (f) $\sum_{i=1}^{2000} (512 + (i-1)7)$

(g) $\sum_{j=1}^{500} (10{,}000 - (j-1)150)$ (h) $\sum_{n=1}^{6} \left(\frac{32}{2^n}\right)$

(i) $\sum_{k=1}^{6} 10^k$ (j) $\sum_{k=1}^{\infty} \frac{25}{10^k}$

(k) $\sum_{i=1}^{10} \frac{1}{i}$ (l) $\sum_{n=2}^{4} \frac{2}{n^2}$

(m) $\sum_{n=2}^{3} n^2 + 2n - 1$ (n) $\sum_{i=3}^{10} 10$

(o) $\sum_{n=3}^{6} \frac{10}{2n}$ (p) $\sum_{n=10}^{15} n^3$

(q) $\sum_{n=1}^{3} 2^n$ (r) $\sum_{i=5}^{7} 20$

(s) $\sum_{i=1}^{\infty} \frac{1}{3^i}$ (t) $\sum_{j=1}^{\infty} g_j$ if $\langle g_n \rangle$ is the geometric progression with $g_1 = 10, r = \frac{1}{2}$

(u) $\sum_{i=10}^{20} 2(i+7)$ (v) $\sum_{i=7}^{10} (3(i^2 - 1) + 6)$

Section 5.4

5. Prove that the following formulas are true for all $n \in \mathbb{N}$.

(a) $\sum_{i=1}^{n} i^3 = \frac{n^2(n+1)^2}{4}$

(b) $2 + 2^2 + \cdots + 2^n = 2(2^n - 1)$

(c) $2^2 + 4^2 + \cdots + (2n)^2 = \frac{2n(n+1)(2n+1)}{3}$

(d) $1 \cdot 2 + 2 \cdot 3 + 3 \cdot 4 + \cdots + n(n+1) = \dfrac{n(n+1)(n+2)}{3}$

(e) $1 + 3 + 5 + \cdots + (2n-1) = n^2$

(f) $1^3 + 3^3 + 5^3 + \cdots + (2n-1)^3 = n^2(2n^2-1)$

(g) $1/1\cdot2 + 1/2\cdot3 + \cdots + 1/n(n+1) = n/(n+1)$

(h) $1 + 2 + 3 + \cdots + n = \dfrac{n(n+1)}{2}$

(i) $(1 - 1/2^2)(1 - 1/3^2) \cdot \cdots \cdot (1 - 1/(n+1)^2)$
$$= (n+2)/2(n+1)$$

(j) $1 + 3 + 3^2 + \cdots + 3^n = \dfrac{3^{n+1} - 1}{2}$

(k) $(1 - \tfrac{1}{2})(1 - \tfrac{1}{3})(1 - \tfrac{1}{4}) \cdot \cdots \cdot (1 - 1/(n+1)) = 1/(n+1)$

PROBLEMS

6. Write out the first six terms of each of the following sequences:
   (a) $\langle a_n \rangle$ if $a_1 = 1$ and $a_n = 1/(a_{n-1} + 1)$
   (b) $\langle t_n \rangle$ if $t_1 = 5$ and $t_n = 1/t_{n-1}$
   (c) $\langle z_n \rangle$ if $z_1 = 10$ and $z_n = z_{n-1} + (-1)^n$
   (d) $\langle k_n \rangle$ if $k_1 = 10$ and $k_n = (k_{n-1} + 2)/2^n$
   (e) $\langle s_n \rangle$ if $s_1 = -3$ and $s_n = s_{n-1}/n$
   (f) $\langle m_n \rangle$ if $m_1 = 1$, $m_2 = 2$ and $m_n = m_{n-1} - m_{n-2}$
   (g) $\langle 1/n \rangle$
   (h) $\langle w_n \rangle$ if $w_1 = 7$, $w_2 = -2$, $w_n = 2w_{n-2}$

7. One sequence of historical interest is the *Fibonacci sequence*. (See also Problem 47 of Chapter 1.) It is defined by $a_1 = 1$, $a_2 = 1$ and $a_n = a_{n-1} + a_{n-2}$ if $n > 2$.
   (a) Write out the first 10 terms of $\langle a_n \rangle$.
   (b) Write a flowchart to give n terms of the Fibonacci sequence.

8. A sequence of value in the study of probability and statistics (which has various other uses as well as we have seen in Chapters 3 and 4), is the factorial sequence. The nth term of this sequence is usually denoted $n!$ and is defined by

$$n! = n(n-1)(n-2) \cdot \cdots \cdot 3 \cdot 2 \cdot 1.$$

   (a) Write out the first six terms of this sequence.
   (b) Simplify $4!/3!$
   (c) Simplify $n!/(n-1)!$
   (d) Define this sequence using a recursion formula.
   (e) Write a flowchart to compute m terms of $n!$

9. (a) Show that compound interest forms a geometric progression in the sense that if a_n is the amount present at the end of n years, then $\langle a_n \rangle$ is a geometric progression. (For convenience, assume yearly compounding. Use initial principal P and yearly rate r).
   (b) Write a flowchart that computes the amount at the end of n years as described in part (a).

10. (a) Show that simple interest forms an arithmetic progression in the sense that if a_n is the amount of money present at the end of n years, then $\langle a_n \rangle$ is an arithmetic progression.

    (b) Write a flowchart to compute simple interest.

11. Are the conclusions of Problem 9 altered if a_n is the amount at the start of the nth year? Explain.

12. Are the conclusions of Problem 10 altered if a_n is the amount at the start of the nth year? Explain.

13. Find a sequence that is both an arithmetic and a geometric progression.

14. Is the sequence you found in Problem 13 the only such? If so, prove so. If not find another example: Are there more?

15. One can have sequences whose ranges are sets of functions instead of numbers:

    (a) If f_1 is specified by $f_1(x) = x$ and $f_n = f_1 \cdot f_{n-1}$, give a non-recursive definition of f_n.

    (b) What interesting features do you see in each of the following sequences of functions: (i) $\langle (1/n) \sin x \rangle$; (ii) $\langle \sin(x/n) \rangle$; (iii) $\langle nx \rangle$; (iv) $\langle (1/n) \cos x \rangle$; and (v) $\langle \cos(x/n) \rangle$.

    (Hints: (i) Consider each sequence at $x = 0$. (ii) compare with elements when x is further from zero; and (iii) what happens as n gets larger and larger?)

16. If a chemist wishes to remove some impurities from a substance, and he knows that 20% of the impurity present in the sample will be removed by rinsing with 20 ml distilled water,

    (a) How many rinsings are needed to remove 90% of the impurities?

    (b) Exhibit a geometric progression that could be of interest to the chemist, and show both that it is a geometric progression and that the chemist would be interested.

17. A bouncing ball reaches only $\frac{2}{3}$ the height on the nth bounce that it did on the $(n-1)$st bounce. Exhibit a sequence of interest and comment on its nature.

18. The vitality of geometric progressions is suggested by the following problem. A man was offered a month's job, for which he had a choice of pay scales. He could get $100 per day or 1 cent for the first day, 2 cents for the second, 4 cents for the third, 8 cents for the fourth, and so on. Determine which pay scale is preferable and give specific justification for your conclusion. (Assume 22 working days per month.)

19. Write a flowchart that inputs any two terms of an arithmetic progression (and any other *necessary* information) and outputs the first 50 terms of the progression.

20. Repeat Problem 19 for geometric progressions.

21. Write a flowchart to output the first 50 terms of any sequence that can be defined recursively using no more than a_{n-1} and a_{n-2} to determine a_n. You should only need to change one instruction to change the sequence to be outputted. The input should be the first two terms of the sequence.

22. Write a flowchart to determine what the constant daily wage in Problem 18 would have to be in order to be as profitable as the sliding pay scale.

23. Problem 18 suggests a new kind of sequence, in that particular problem, the term s_n would be the total wages through day n. Write a flowchart that would input elements of a sequence $\langle a_n \rangle$ and output elements of the new sequence $\langle s_n \rangle$, where $s_n = a_1 + a_2 + \cdots + a_n$.

24. If $\langle a_n \rangle$ is any sequence of real numbers, and $\langle s_n \rangle$ is the associated sequence of partial sums:
    (a) Find a representation for $\sum_{i=n}^m a_i$ that makes use of $\langle s_n \rangle$.
    (b) If $\langle a_n \rangle$ is an arithmetic progression, use (a) to generate a formula for $\sum_{i=n}^m a_i$.
    (c) Find the same as (b) but for geometric progressions.

25. (a) Write a flowchart to output the first 50 terms of the Fibonacci sequence (see Problem 7) and of its sequence of partial sums.

26. If $\langle s_n \rangle$ is the sequence of partial sums associated with the sequence $\langle a_n \rangle$, find the quantities requested under the conditions cited:
    (a) s_{100} if $\langle a_n \rangle$ is an arithmetic progression, $a_{10} = 10$ and $a_{20} = 34$.
    (b) A full specification of $\langle a_n \rangle$ if $\langle a_n \rangle$ is an arithmetic progression, $a_1 = 25$, $s_{10} = 300$.
    (c) A full specification of $\langle a_n \rangle$ if $\langle a_n \rangle$ is an arithmetic progression, $s_{20} = 670$ and $s_{30} = 1455$.
    (d) s_{2050} if $\langle a_n \rangle$ is an arithmetic progression if $a_{100} = 138.5$, and $s_{16} = 20$.
    (e) The general term of $\langle a_n \rangle$ and of $\langle s_n \rangle$ if $\langle a_n \rangle$ is an arithmetic progression, and $s_5 = 50$ and $s_6 = 25$.

27. Using a technique of the same kind as those used to find the series formulas for arithmetic and geometric progressions, prove that the following decimal expressions represent rational numbers by expressing them as fractions of integers.
    (a) $0.777777 \cdots$ (b) $0.343434 \cdots$
    (c) $0.553553553 \cdots$ (d) $-0.222222 \cdots$
    (e) $0.123123123 \cdots$ (f) $3.282828 \cdots$
    (g) $0.999999 \cdots$ (h) $4.23121212 \cdots$
    (i) $2.1565656 \cdots$ (j) $0.00191919 \cdots$
    (k) $0.912912912 \cdots$ (l) $7.34734734 \cdots$

28. Use mathematical induction to prove that if $\langle a_n \rangle$ is an arithmetic progression and $\langle s_n \rangle$ is the corresponding series, then $s_n = n/2(a_1 + a_n)$. (The proof in the text does not directly use induction.)

29. Prove that if $a(b + c) = ab + ac$ for any real numbers a, b, c, then

$$a \left(\sum_{i=1}^n b_i \right) = \sum_{i=1}^n (ab_i)$$

    for all $n \in \mathbb{N}$, provided that $\{a, b_1, b_2, \cdots, b_n\} \in \mathbb{R}$.

30. Prove that if f is a periodic function with period p,
    (a) $f(x + np) = f(x)$ for all $n \in \mathbb{N}$ and all $x \in D_f$.
    (b) $f(x + np) = f(x)$ for all $n \in \mathbb{Z}$ and all $x \in D_f$.

31. Prove that $n < 2^n$
    (a) For all $n \in \mathbb{N}$
    (b) For all $n \in \mathbb{Z}$.
32. Prove that if intersection distributes over union, then
    $$A \cap (B_1 \cup B_2 \cup \cdots \cup B_n) = (A \cap B_1) \cup (A \cap B_2) \cup \cdots \cup (A \cap B_n)$$
    for all $n \in \mathbb{N}$, provided that A and each B_i are sets.
33. Prove that if $f(\alpha x + \beta y) = \alpha f(x) + \beta f(y)$ for all real numbers α and β, and all $x, y \in D_f$, then
    $$f\left(\sum_{i=1}^n a_i x_i\right) = \sum_{i=1}^n a_i f(x_i)$$
    for all $n \in \mathbb{N}$, provided that $\{a_1, \cdots, a_n\} \in \mathbb{R}$, and $\{x_1, \cdots, x_n\} \in D_f$.
34. Prove the generalized DeMorgan's Laws for sets:
    (a) $(A_1 \cap A_2 \cap \cdots \cap A_n)' = A'_1 \cup A'_2 \cup A'_3 \cup \cdots \cup A'_n$
    (b) $(A_1 \cup A_2 \cup \cdots \cup A_n)' = A'_1 \cap A'_2 \cap A'_3 \cap \cdots \cap A'_n$
    Assume as given the usual form of these laws: $(A \cap B)' = A' \cup B'$ and $(A \cup B)' = A' \cap B'$.
35. Prove that $(1 + a)^n \geq 1 + na$ for all $n \in \mathbb{N}$.
36. Prove that $x^n - y^n$ is divisible by $x - y$ if $x \neq y$.
    (Hint: $x^{n+1} - y^{n+1} = x^{n+1} - xy^n + xy^n - y^{n+1}$.)
37. Prove that the recursive definition for an arithmetic progression is equivalent to the other definition for an arithmetic progression.
38. Repeat Problem 37 for geometric progressions.
39. Prove that if $x > 0$, then $x^n > 0$ for every $n \in \mathbb{N}$.
40. Prove that if $a_1 < a_2 < \cdots < a_n$, then
    $$a_1 < \frac{a_1 + a_2 + \cdots + a_n}{n} < a_n \qquad \text{for all } n \in \mathbb{N}.$$
41. Prove that if $0 < a < b$, then $a^n < b^n$ for $n \in \mathbb{N}$.
42. (a) Prove that if $0 < x < 1$, then $x^{n-1} > x^n$ for all $n \in \mathbb{N}$.
    (b) Prove that if $0 < x < 1$ and $m, n \in \mathbb{N}$, $m > n$, then $x^m < x^n$.
43. Prove that if $p_0 \in [a, b]$ and $g(x) \in [a, b]$ for all $x \in [a, b]$ (we say that g is a contractive mapping on $[a, b]$, then if $\langle p_n \rangle$ is defined by $p_n = g(p_{n-1})$, $n = 1, 2, 3, \cdots$, then $\langle p_n \rangle \subseteq [a, b]$.
44. Prove that if $|x_n - s| \leq k|x_{n-1} - s|$ for each $n \in \mathbb{N}$, and where k does not depend on n, then $|x_n - s| \leq k^n |x_0 - s|$ for all $n \in \mathbb{N}$.
45. Prove that $(x - a_1)(x - a_2) \cdot \cdots \cdot (x - a_n)$ is of degree n for all $n \in \mathbb{N}$.
46. To prove that polynomials have the graphical behavior mentioned in Section 2.6, it suffices to show that for each polynomial $p(x) = a_n x^n + \cdots + a_0$ there is a number $R \in \mathbb{R}$ so that if $|x| > R$, then $|a_n x^n| > |a_{n-1} x^{n-1} + \cdots + a_0|$. That is, if one goes out far enough, the lead term dominates the rest of the polynomial and so forces its "behavior near infinity" upon the whole function.
    (a) Prove that this inequality is true for all $n \in \mathbb{N}$.
    (b) Go on to explain qualitatively how to combine this fact with the oddness or evenness of n to arrive at the claimed result in Section 2.6.

47. Prove that a set with n elements has 2^n subsets (including $\emptyset$ and the set itself).

48. To examine more closely the claimed divergence of $\sum_{i=1}^{\infty} 1/i$, first write out the first 18 terms of the series, then group together the third and fourth term, the fifth through eighth terms, the ninth through sixteenth terms, and in general the (2^n+1)th through (2^{n+1})st terms, and argue convincingly (prove) that each such grouping has value $\geq \frac{1}{2}$. Then use this to argue that the series diverges.

49. To nurture skepticism about computer arithmetic, think of an imaginary computer that carries only one decimal digit (it computes numbers to two digits and rounds off to one). Write on paper this machine's computation of $\sum_{i=1}^{18} 1/i$ and from that deduce the machine's opinion of the value of $\sum_{i=1}^{\infty} 1/i$. Compare this with Problem 48. (This behavior merely takes longer to appear in a real computer, but the same kind of conclusion is reached.)

50. To see that there is more than one kind of divergence for series, discuss the series $\sum_{i=1}^{\infty} (-1)^i$ and clearly justify why one may say this diverges.

51. Show that the restriction of $k \cdot \exp_a(x)$ to domain $\mathbb{N}$ forms a geometric progression.

52. Prove that x^n is an odd function if, n is odd, and an even function if n is even (see Problems 89–91, Chapter 1).

53. Prove that

$$\sum_{k=1}^{n} (a_k + b_k) = \sum_{k=1}^{n} a_k + \sum_{k=1}^{n} b_k,$$

where each a_k and $b_k \in \mathbb{R}$.

CHAPTER 6

INTRODUCTION TO ANALYTIC GEOMETRY

6.1. INTRODUCTION

Analytic geometry has been mentioned on several occasions in previous chapters. Two such instances were in our claims that (1) all linear functions have straight-line graphs and (2) all quadratic functions have parabolic graphs. In this chapter we plan to revisit those topics, develop the tools we need to study them, and extend the ideas beyond these two specific situations so that the reader can move on to a study of the calculus with confidence.

There are two other reasons we might cite for undertaking a study of analytic geometry. First of all, it is a useful subject. Some geometric proofs that are quite long and involved, using only the techniques of Euclidian (plane) geometry, turn out to be very simple when analytic geometry is applied. In addition, many of the results and techniques that analytic geometry provide find their way into such fields as engineering.

Second, you may well find analytic geometry fun. It can be fascinating to sketch a simple diagram, set up some rather messy algebraic equivalences, and then apply a few facts and a little algebraic manipulation to produce a neat result.

6.2. STRAIGHT LINES

Our goal in this section is to verify that the graphs of all linear functions are indeed straight lines. In the interest of efficient exposition we approach the issue indirectly by starting not with lines, but with points and line segments. With this preparation we shall then be able to study lines effectively and finally return to the issue which prompted these investigations. We begin by establishing some basic information about points, line segments, and then lines.

6.2a. Distance and Midpoint Formulas

DISTANCE   When we first discussed the Cartesian coordinate system, the first thing we did was to plot points. The next step might naturally be to look into what properties we might be able to discover about a line segment between any two points.† Suppose we have $P_1 : (x_1, y_1)$ and $P_2 : (x_2, y_2)$,

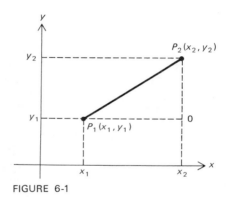

FIGURE 6-1

any two points in the plane (Figure 6.1). How might we go about finding the distance between P_1 and P_2, denoted $|P_1 P_2|$?

In order to find $|P_1 P_2|$, we can construct a right triangle, $P_1 P_2 O$, as shown in Figure 6.1, by dropping a perpendicular to the x-axis from P_2, and then constructing a line parallel to the x-axis through P_1. The intersection of these lines we label O. (Is this legitimate if P_1 is "above" P_2 instead of P_2 "above" P_1, as in Figure 6.1?)

We can easily find the length of two of the sides of triangle $P_1 P_2 O$. The length $|P_1 O|$ is the same as the distance between the points (x_1, O) and (x_2, O) on the x-axis. In Figure 6.1, this distance appears to be $x_2 - x_1$, since x_2 appears to be the distance from the origin to (x_2, O) and x_1 the distance from the origin to (x_1, O), and the desired length is just their difference. However, if P_2 were to the left of P_1, i.e., $x_2 > x_1$, and/or either of the points were not in quadrants I or IV, the above appear-

---

† Remember that a line extends infinitely in both directions (↔); a ray extends infinitely in one direction: (→); and a line segment is of finite (length —).

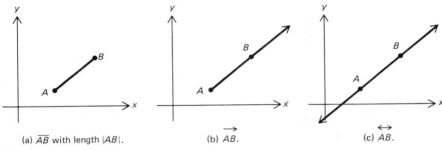

(a) $\overline{AB}$ with length $|AB|$.　(b) $\overrightarrow{AB}$.　(c) $\overleftrightarrow{AB}$.

FIGURE 6-2

ances would change somewhat. Nevertheless, there is a strong nugget of truth in our initial appraisal: the distance turns out to be $|x_2 - x_1|$ (or $|x_1 - x_2|$, which is the same—do you see why?), no matter where P_1 and O lie. This can be verified by checking out each of the possible relative positions and signs of x_1 and x_2. Indeed, for any two points on any one horizontal line, i.e., $R(x_1, a)$, $S(x_2, a)$, the distance between them is $|x_2 - x_1|$, and for any two points on any one vertical line, i.e., $T(b, y_1)$, $U(b, y_2)$, the distance between them is $|y_2 - y_1|$.

DISTANCE

Therefore, in our triangle, we observe that $|P_2 O| = |y_2 - y_1|$. Then, since $P_1 P_2 O$ is a right triangle, we may use the Pythagorean Theorem to get

$$|P_1 P_2|^2 = |P_1 O|^2 + |O P_2|^2$$
$$= |x_2 - x_1|^2 + |y_2 - y_1|^2$$

and

$$|P_1 P_2| = \sqrt{(x_2 - x_1)^2 + (y_2 - y_1)^2}.$$

FORMULA 6.1
Distance Formula

This is called the *distance formula*.

Note that we may drop the absolute value bars inside the square root since we are squaring each of $|x_2 - x_1|$ and $|y_2 - y_1|$ and squares of real numbers are always nonnegative, whether or not the original numbers are. Note also that the distance formula includes the special cases of P_1 and P_2, being on a horizontal or vertical line.†

NOTE

Let us summarize the notation we have developed associated with line segments and, in addition, introduce two new notations:

1. $\overline{AB}$ refers to the line segment with endpoints A and B; $|AB| = |BA|$ refers to the distance between A and B, i.e., the length of $\overline{AB}$.
2. $\overrightarrow{AB}$ is the ray beginning at A and passing through B; $\overleftrightarrow{AB}$ is the line passing through A and B.

See Figure 6.2.

---

† Many texts also introduce the concept of directed distance on lines that are parallel to either of the coordinate axes. Such a distance imparts a directionality—it is the distance from P to Q (denoted PQ), not the distance between them, and the directionality is communicated by sign. Thus if PQ is positive, QP is negative.

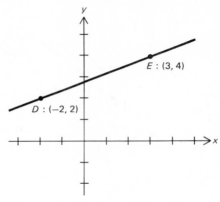

FIGURE 6-3

EXAMPLE 1

Now let us consider an example using our knowledge so far. Suppose that we want to know the length of $\overline{DE}$ in Figure 6.3. Using the distance formula, we would have

$$|DE| = \sqrt{(x_2 - x_1)^2 + (y_2 - y_1)^2}$$
$$= \sqrt{(3 - (-2))^2 + (4 - 2)^2}$$
$$= \sqrt{5^2 + 2^2}$$
$$= \sqrt{29}$$

or, if you want an approximate answer,

$$|DE| = \sqrt{29} \approx 5.385.$$

However, one should use approximate values only when called for or by context.

MIDPOINT

One more useful piece of information about line segments falls quite naturally into our discussion. We will find it is often quite useful to be able to find the midpoint of any line segment. Let us return to $P_1 : (x_1, y_1)$ and $P_2 : (x_2, y_2)$ again. In Figure 6.4, let $M : (x, y)$ be the midpoint of

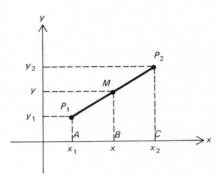

FIGURE 6-4

$\overline{P_1 P_2}$. Then, since all the perpendiculars to the x-axis are parallel to each other,

$$\frac{|AB|}{|P_1 M|} = \frac{|BC|}{|MP_2|} \qquad (1)$$

since parallel lines cut proportional segments on any two transversals. But we defined M such that

$$|P_1 M| = |MP_2|. \qquad (2)$$

Thus, clearing fractions in Equation (1), we get

$$|MP_2| \cdot |AB| = |BC| \cdot |P_1 M|$$

and substituting Equation (2) produces

$$|P_1 M| \cdot |AB| = |BC| \cdot |P_1 M|.$$

Next dividing by $|P_1 M|$,

$$|AB| = |BC|,$$

which is to say that

$$|x - x_1| = |x_2 - x|. \qquad (3)$$

If $x_1 \leq x_2$, then $x_1 \leq x \leq x_2$ and Equation (3) becomes

$$x - x_1 = x_2 - x \qquad \text{or} \qquad x = \frac{x_2 + x_1}{2}.$$

If $x_2 < x_1$, then $x_2 \leq x \leq x_1$ and Equation (3) becomes

$$x_1 - x = x - x_2 \qquad \text{or} \qquad x = \frac{x_2 + x_1}{2}.$$

Therefore, in every case, the x-coordinate of the midpoint is

$$x = \frac{x_2 + x_1}{2}.$$

By a very similar argument,

$$y = \frac{y_2 + y_1}{2}.$$

Thus the coordinates of the midpoint of a line segment $\overline{P_1 P_2}$ with $P_1 : (x_1, y_1)$ and $P_2 : (x_2, y_2)$ are given by

$$M : \left(\frac{x_2 + x_1}{2}, \frac{y_2 + y_1}{2} \right).$$

FORMULA 6.2
Midpoint Formula

This formula is aesthetically sound to most people (as well as being technically sound) because it asserts that the coordinates of the midpoint (a sort of "average" of the endpoints) are the averages of coordinates of the endpoints.

Thus if $A : (3, 1)$ and $B : (7, -3)$, the midpoint of $\overline{AB}$, is given by **EXAMPLE 2** coordinates

$$M : \left(\frac{7 + 3}{2}, \frac{-3 + 1}{2} \right) \qquad \text{or} \qquad M : (5, -1).$$

MIDPOINT It would be quite straightforward to develop formulas for the co-ordinates of points of trisection or even *n*-section for a line segment. We choose not to do this here, but you will have a chance to try it in the problems. Could you conjecture what the formula might be for trisection?

6.2b. Inclination and Slope

SLOPE A question we might ask about straight lines is "How can we measure the slant of a line?" Although many different schemes could be suggested, perhaps the most natural is to use the angle formed by the line and the *x*-axis. This angle is called the *inclination* of the line, and the definition is formalized as the smallest positive angle formed by the line and the *x*-axis if such an angle is formed, and zero if the line is parallel to the *x*-axis.

DEFINITION 6.1
Inclination

Unfortunately, inclination is an awkward quantity to use so that we turn to a less obvious index of slant, which will prove more useful. Let $P: (x_1, y_1)$ and $Q: (x_2, y_2)$ be any two points on the line. The *slope* of the line is defined to be

DEFINITION 6.2
Slope

$$m = \frac{y_2 - y_1}{x_2 - x_1}; \qquad x_2 \neq x_1.$$

If the line is vertical, the line is not assigned a slope.

Figure 6.5 shows this to be the ratio of two sides of a right triangle whose hypotenuse is $\overline{PQ}$. Facts from plane geometry regarding similar triangles assure us that the definition for slope is independent of the specific choice of *P* and *Q* on the line.

NOTE A brief word of warning is in order here—be sure you subtract co-ordinates in the same order in both numerator and denominator.

EXAMPLE 3 Let us consider a few examples to see if and how slope gives us information that corresponds to our intuitive notions of what it means for a line to be slanted. Consider Figure 6.6. For (a), the slope is $\frac{2-1}{3-1} = \frac{1}{2}$; for (b), the slope is $\frac{3-1}{3-1} = 1$; and for (c), the slope is $\frac{4-1}{3-1} = \frac{3}{2}$.

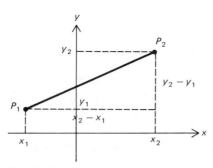

FIGURE 6-5

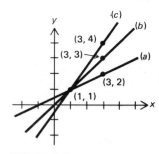

FIGURE 6-6

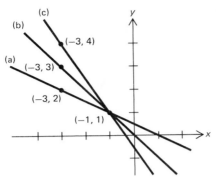

FIGURE 6-7

For parts (a)–(c) we see that the value of the slope increases as the lines get more slanted.

Now let us consider Figure 6.7, where for (a), $m = \dfrac{2-1}{-3-(-1)} = -\tfrac{1}{2}$; **EXAMPLE 4**

for (b), $m = \dfrac{3-1}{-3-(-1)} = -1$; and for (c), $m = \dfrac{4-1}{-3-(-1)} = -\tfrac{3}{2}$.

The absolute values of m for each of the lines in Figure 6.7 still indicate when a line is more slanted, but the minus sign seems to tell us the line is slanted in the other way. These examples should suggest a couple of clues to use when you do problems involving slope:

1. The larger the absolute value of m, the nearer the line comes to being vertical.
2. The sign gives a clue as to which way the line is sloped: positive slopes correspond to slants from lower left to upper right, while negative slopes correspond to slants from upper left to lower right.

Using Figure 6.8, we can find that there is a very close relationship between slope and inclination. You should see that

$$\tan \phi = m = \frac{y_2 - y_1}{x_2 - x_1}$$

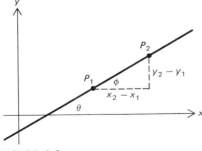

FIGURE 6-8

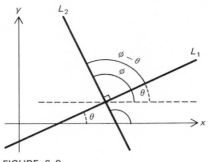

FIGURE 6-9

and, from plane geometry, $\theta = \phi$, so we see that tan $\theta = m$. From this, we can add to observations (1) and (2) above that:

3. When the inclination is between 0° and 45°, the slope is less than 1; when the inclination is between 45° and 90°, the slope is greater than 1.

We have already discussed sign, but a look at the behavior of tan θ for $90° \le \theta \le 180°$ confirms what our examples above suggested, i.e.,

4. When the inclination is between 90° and 135°, the slope is negative and less than -1 [i.e., in $(-\infty, -1)$]. When the inclination is between 135° and 180°, the slope is negative and greater than -1 [i.e., in $(-1, 0)$].

CONDITION FOR PERPENDICULARITY

The information we have so far gives us another very useful fact. Suppose that we have two lines L_1 and L_2, which are perpendicular to each other (Figure 6.9). Line L_1 has inclination θ and line L_2 has inclination ϕ. We can easily see that $\phi - \theta = 90°$ by our assumption that L_1 and L_2 are perpendicular.

For reasons that shall remain obscure for a moment, let us find $\cos(\phi - \theta)$.

$$\cos(\phi - \theta) = \cos \phi \cos \theta + \sin \phi \sin \theta$$

since $\phi - \theta = 90°$, and

$$\cos 90° = \cos \phi \cos \theta + \sin \phi \sin \theta$$

or

$$0 = \cos \phi \cos \theta + \sin \phi \sin \theta.$$

If we divide this equation through by cos ϕ cos θ, we get

$$0 = 1 + \tan \phi \tan \theta,$$

which conveniently tells us something of interest about the slopes of L_1 and L_2, namely, letting the slope of $L_1 = m_1$ and the slope of $L_2 = m_2$, we find, from

$$0 = 1 + \tan \phi \tan \theta,$$

FORMULA 6.3a
Perpendicular Lines that

$$0 = 1 + m_1 m_2 \qquad \text{or} \qquad m_1 m_2 = -1,$$

since $m_1 = \tan \theta_1$ and $m_2 = \tan \theta_2$. That is, when two lines are perpendicular, the product of their slopes is negative one.[†] The converse of this is also true: When the product of the slopes of two lines is -1, the lines are perpendicular. (Can you prove this? See Problem 21.)

CONDITION OF PERPENDICULARITY

Now that we have a definition of perpendicularity, it is natural to ask about parallel lines. The characterization turns out to be easier: two lines are parallel if the corresponding angles formed by a transversal are congruent. If the transversal is chosen to be the x-axis, the inclinations of the parallel lines must be the same, and also the slopes. The only time the x-axis cannot be used as the transversal is if the lines are parallel to the x-axis and hence both have slope 0; in other words, parallel lines always have equal slopes. Reversing the argument is easy: equal slopes quarantee parallel lines.

CONDITION FOR PARALLELISM

FORMULA 6.3b
Parallel Lines

6.2c.   Standard Equations of a Straight Line

We are nearly ready to show that graphs of linear functions are always straight lines. In fact, we shall discover more, that these are the only functions with straight line graphs. We approach this by building on the previous subsections and asking "What equations do straight lines satisfy?"

EQUATIONS OF STRAIGHT LINES

We start by assuming that we know two points on the line, $P_1(x_1, y_1)$ and $P_2(x_2, y_2)$. See Figure 6.10. How can we characterize an arbitrary third point $P(x, y)$ on this line? That is, what condition(s) can we demand that will assure that P be on the line? One would be that the slope of $\overleftrightarrow{PP_1}$ be the same as the slope of $\overleftrightarrow{P_1P_2}$. If we write this symbolically, we find:

$$\frac{y_2 - y_1}{x_2 - x_1} = \frac{y - y_1}{x - x_1}$$

or letting $m = (y_2 - y_1)/(x_2 - x_1)$ and clearing fractions,

$$y - y_1 = m(x - x_1)$$

FORMULA 6.4
Point-Slope Form

This form of the equation of a line is called the *point-slope* form of a line because it gives us the equation of a line if we know its slope and any one point on it.[‡] Further, if we know the y-intercept, we know one point on the line, namely, (O, b) and can therefore write

$$y - b = m(x - 0)$$

or

$$y = mx + b,$$

FORMULA 6.5
Slope-Intercept Form

the so called *slope-intercept* form of a straight line (b is used to denote the y-intercept, just as m denotes slope).

Two special cases arise at this point: horizontal and vertical lines. For line L in Figure 6.11, let us apply the point slope formula. Since $m = 0$,

---

[†] Actually, there is one exceptional case: when the lines are parallel to the axes. But, then, one of the lines has no slope at all, so that it is meaningless to talk of the product of their slopes.
[‡] This equation is sometimes left with $(y_2 - y_1)/(x_2 - x_1)$ in place of m and is then called the *two-point form* of the equation.

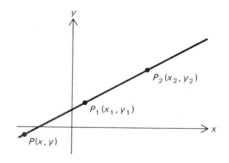

FIGURE 6-10

we have

$$y - y_1 = m(x - x_1)$$
$$y - d = 0(x - c),$$

using point (c, d), or

$$y = d.$$

Thus any horizontal line has the form $y = some\ constant$.

As vertical lines have no slope, we must take our clue from horizontal lines and see that all vertical lines have equations of the form $x = c$, some constant. Any other choice of equation would give us points with different x-coordinates and so allow us to find a slope, which would therefore not be a vertical line.

We have now discussed equations of all types of straight lines: slanted, horizontal, and vertical. We can write all these lines in one form, namely,

$$Ax + By + C = 0,$$

where A and B may not both be zero. Let us be sure. We can write all slanted lines as

$$(y - y_1) = m(x - x_1),$$

which can be rewritten as

$$(-m)x + y + (mx_1 - y_1) = 0,$$

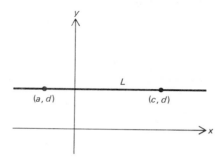

FIGURE 6-11

where $A = -m$, $B = 1$, and $C = mx_1 - y_1$. Vertical lines are in the form

$$x = k,$$

which we can rewrite as

$$x + 0y - k = 0,$$

where $A = 1$, $B = 0$, and $C = -k$. Horizontal lines are in the form

$$y = c \qquad \text{or} \qquad 0x + y - c = 0,$$

where $A = 0$, $B = 1$, and $C = -c$. Thus all straight lines have equations of the form

$$Ax + By + C = 0.$$

The next question is "Do all equations of the form $Ax + By + C = 0$ with A and B not both zero, represent straight lines?" First of all, we can rewrite $Ax + By + C = 0$ as follows:

$$Ax + By + C = 0$$

$$\frac{Ax}{B} + \frac{By}{B} + \frac{C}{B} = 0 \qquad \text{if} \quad B \neq 0$$

or

$$y = -\frac{A}{B}x - \frac{C}{B}$$

which is in the slope-y-intercept form with $m = -A/B$ and $b = -C/B$. Since this form gives us a straight line, so does $Ax + By + C = 0$.

If $B = 0$, we have

$$Ax + C = 0$$

or

$$x = \frac{-C}{A},$$

which is in the form of an equation for a vertical line.

Thus any equation of the form $Ax + By + C = 0$ with A and B not both zero represents some straight line. Now we can easily answer our original question. The linear function specified by $y = ax + b$ is also the linear function specified by $-ax + y - b = 0$. This equation is of the form $Ax + By + C = 0$ ($A = -a$, $B = 1$, $C = -b$) and so has a straight line graph.

6.3. PARABOLAS AND TRANSLATIONS IN THE PLANE

6.3a. Parabolas

Our second unverified claim was that the graphs of quadratic functions all have the same basic shape, called a parabola. As with the issue of straight lines and linear functions, we shall start not with the functions,

PARABOLAS

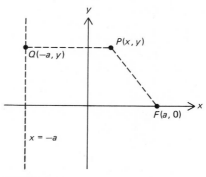

FIGURE 6-12

but with the curve. Specifically we start by defining parabola, then making some observations and eventually showing that the graphs of all quadratic functions have this shape.

DEFINITION 6.3
Parabola

A *parabola* is, by definition, the set of points in a plane that are equidistant from a line *L* (called the *directrix*) and a point *F* (called the *focus*) both residing in the plane and with *F* not on *L*.

Using this definition, let us see if we can combine some of the techniques of analytic geometry we discussed in the last section to get a general equation for the parabola.

We begin by setting up axes with the focus *F* on the *x*-axis at $(a, 0)$ with $a > 0$, and directrix $x = -a$. The point on the parabola that is the midpoint of the segment from the focus perpendicular to the directrix is called the *vertex* of the parabola, and with our choice of axes this point is, in fact, the origin.† If we let $P(x, y)$ denote an arbitrary point on the parabola, we can translate the definition of parabola into the condition that $|PF| = |PQ|$, where *Q* is the point $(-a, y)$. See Figure 6.12. Notice that the position of *Q* is dependent on how far above or below the *x*-axis *P* is. The condition

$$|PQ| = |PF|$$

can be rewritten using the distance formula as

$$\sqrt{(x - (-a))^2 + (y - y)^2} = \sqrt{(x - a)^2 + (y - 0)^2}$$

or

$$\sqrt{(x + a)^2} = \sqrt{(x - a)^2 + y^2}$$

or

$$(x + a)^2 = (x - a)^2 + y^2$$
$$x^2 + 2ax + a^2 = x^2 - 2ax + a^2 + y^2.$$

† Note that this is a new definition of vertex, not dependent on the usage in Chapter 2. You should see when we are done here that our use of vertex in Chapter 2 follows from our definition in this chapter.

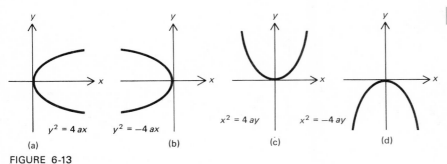

FIGURE 6-13

Simplifying, we get

$$y^2 = 4ax$$

as the equation of a parabola with focus at $(a, 0)$ on the positive x-axis and directrix $x = -a$ through the negative x-axis.

If the focus were on the negative x-axis, at $(-a, 0)$, and directrix $x = a$, we would have

$$y^2 = -4ax;$$

if F were on the positive y-axis, at $(0, a)$, and directrix $y = -a$,

$$x^2 = 4ay;$$

and if F were on the negative y-axis at $(0, -a)$, with directrix $y = a$,

$$x^2 = -4ay.$$

Thus we have, as four standard positions for parabolas, those listed in Figure 6.13.

Two more terms associated with the parabola are the *axis* of the parabola, which is the line through the focus perpendicular to the directrix and the *latus rectum*, which is the line segment perpendicular to the axis of the parabola through the focus with endpoints on the parabola. Figure 6.14 summarizes the terminology used in connection with parabolas.

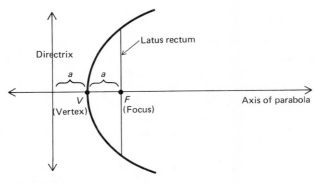

FIGURE 6-14

6.3. *PARABOLAS AND TRANSLATIONS IN THE PLANE*   295

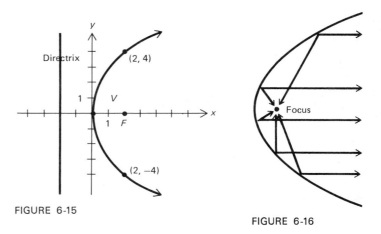

FIGURE 6-15

FIGURE 6-16

For parabolas in standard position, the number a is available in the equation and makes locating the focus and directrix an easy matter. Further, as Problem 27 asks you to verify, the length of the latus rectum is $4a$, which should prove quite useful in graphing a parabola.

EXAMPLE 5   We sketch $y^2 = 8x$ showing the focus, vertex, directrix, and latus rectum. This parabola is of the form $y^2 = 4ax$ and thus its axis is on the x-axis with vertex at the origin. Since $4a$ for this problem is 8, $a = 2$. Thus the directrix is $x = -2$. Likewise, the focus is at $(2, 0)$. The latus rectum has endpoints $(a, \pm 2a)$ or, in our example at $(2, 4)$ and $(2, -4)$. Thus without computing any points from the equation, we obtain Figure 6.15.

NOTE   We also mentioned in Chapter 2 that parabolas were used in mirrors of telescopes and similar instruments. The property of the parabola that is used in the telescope is the so-called reflector property: Any ray of light originating at the focus and reflected from a parabolic surface will be sent out parallel to the axis of the parabola (Figure 6.16). Also, incoming rays of light parallel to the axis of the parabolic reflector will be reflected through the focus. (The proof is beyond the scope of this book.)

6.3b.   Translations in the Plane

We know that the graphs of quadratic functions do not usually have their vertices at the origin; so, before we can resolve our question about their graphs, we need to develop enough machinery to move parabolas away from the origin. The techniques of such translation are valuable in a much wider context, however, so that we shall take some time here to develop the ideas more completely.

We shall first perform translations on a number of examples to illustrate the general procedure.

Let us begin by performing translations along the x-axis of a point

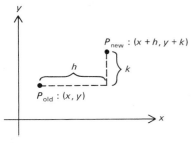

FIGURE 6-17

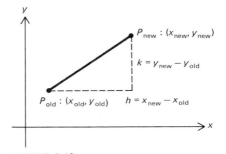

FIGURE 6-18

$O : (0, 0)$. We could move this along the x-axis h units so that its new coordinates would be $(0 + h, 0)$ or $(h, 0)$. Similarly, we could move it along the y-axis k units getting $(0, 0 + k)$ or $(0, k)$ as new coordinates. A natural next step is to move point O both ways at once: h units along the x-axis and k units along the y-axis. Thus O's new coordinates would be $(0 + h, 0 + k)$ or (h, k). Thus to move $O : (0, 0)$ 3 units along the x-axis and 4 units along the y-axis, O's new coordinates would be $(0 + 3, 0 + 4)$ or $(3, 4)$. To translate point $M : (3, -1)$ similarly, we would have the new coordinates of M as $(3 + 3, -1 + 4)$ or $(6, 3)$.

In general, then, to translate point $P(x, y)$, h units along the x-axis and k-units along the y-axis (we abbreviate this by saying we translate $P: (x, y)$ (h, k) units) we get P's new coordinates as $P : (x + h, y + k)$ (see Figure 6.17). Notice that although this appears as a two-step motion (move in x-direction, then y-direction), it is equivalent to a single movement along the line segment $\overline{P_{old} P_{new}}$.

Another way of looking at this is to see that

$$h = x_{new} - x_{old} \quad \text{and} \quad k = y_{new} - y_{old}$$

(see Figure 6.18) or

$$x_{old} = x_{new} - h$$
$$y_{old} = y_{new} - k.$$

We abbreviate this by

$$x = x_n - h$$
$$y = y_n - k$$

with the "sub-n" to denote new values of x and y.

Next we translate an entire line. Stop and think for a moment, and you will realize that if we translate every point of the line $(3, 4)$ units, then the entire line is translated $(3, 4)$ units. (Are you sure you see this?)

If we want to translate $y = 3x + 5$ $(2, 3)$ units, then we need first to observe that

EXAMPLE 6

$$x = x_n - 2$$
$$y = y_n - 3.$$

TRANSLATIONS IN THE PLANE

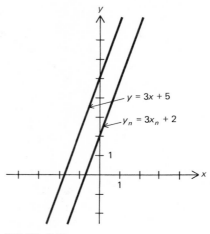

$y = 3x + 5$

$y_n = 3x_n + 2$

FIGURE 6-19

Thus the new equation becomes

$$(y_n - 3) = 3(x_n - 2) + 5$$
$$y_n = 3x_n - 6 + 5 + 3$$

or

$$y_n = 3x_n + 2.$$

EXAMPLE 7

See Figure 6.19. If we were going to work with this new equation, we would probably drop the "sub-n's," but we leave them for now to avoid confusion.

Let us next try translating a fairly simple polynomial such as

$$y = x^3 + 2x^2 - x - 2 = (x - 1)(x + 1)(x + 2).$$

We translate this one $(2, 1)$ units. Now

$$x = x_n - 2$$
$$y = y_n - 1$$

or

$$(y_n - 1) = (x_n - 2)^3 + 2(x_n - 2)^2 - (x_n - 2) - 2.$$

An easier way to do this would be to use the factored form:

$$(y_n - 1) = [(x_n - 2) - 1][(x_n - 2) + 1][(x_n - 2) + 2]$$
$$(y_n - 1) = (x_n - 3)(x_n - 1)(x_n + 0)$$
$$y_n - 1 = x_n^3 - 4x_n^2 + 3x_n$$

and thus

$$y_n = x_n^3 - 4x_n^2 + 3x_n + 1.$$

See Figure 6.20.

EXAMPLE 8

Suppose that we wanted to translate the exponential function $f(x) = 3^x$

298   *INTRODUCTION TO ANALYTIC GEOMETRY*

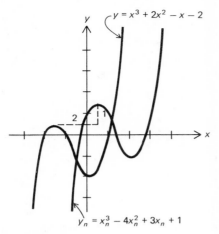

$$y = x^3 + 2x^2 - x - 2$$

$$y_n = x_n^3 - 4x_n^2 + 3x_n + 1$$

FIGURE 6-20

$(-2, -3)$ units. In this case

$$x = x_n + 2$$
$$y = y_n + 3.$$

Thus

$$y_n + 3 = 3^{x_n+2} \qquad \text{or} \qquad y_n = 3^{x_n+2} - 3 \qquad \text{or} \qquad y_n = 9 \cdot 3^{x_n} - 3.$$

See Figure 6.21.

In Section 3.8 in the discussion on circular functions, dealing with $a \cdot \text{cir}(bx + c)$, you encountered horizontal translation together with some stretchings and shrinkings.

Now let us try translating $y = \sin x$ through $(2, 4)$ units. **EXAMPLE 9**

Here $x = x_n - 2$, and $y = y_n - 4$, and we have $y_n - 4 = \sin(x_n - 2)$ or $y_n = \sin(x_n - 2) + 4$. Using Table 1, Appendix D, and the special values we know, we get Figure 6.22.

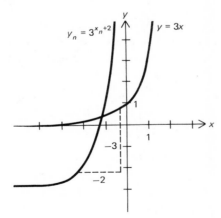

$$y_n = 3^{x_n+2}$$

$$y = 3x$$

FIGURE 6-21

6.3. *PARABOLAS AND TRANSLATIONS IN THE PLANE* 299

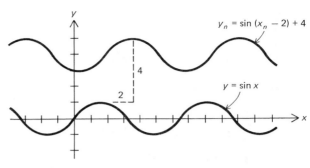

FIGURE 6-22

NOTE

Notice that we have inadvertently discovered a new graphing technique. If someone asked you to graph, say $y = 3^{x+2} - 1$, and you recognized that this was merely $y = 3^x$ translated $(-2, -1)$, you could sketch $y = 3^x$ and then merely translate it. Such observations often make graphing much easier than computing values for the original function.

6.3c. Translation and Parabolas

**TRANSLATION
AND PARABOLAS**

The discussion that follows would work for all four standard forms for parabolas, but we shall use just $x^2 = 4ay$. From our previous discussion, translation through (h, k) units would give us

$$x = x_n - h$$
$$y = y_n - k$$

and thus

$$(x_n - h)^2 = 4a(y_n - k).$$

From our knowledge of the parabola and translation we see that the new graph would be as shown in Figure 6.23.

Let us consider this in some detail. The old vertex $(0, 0)$ becomes $V_n : (h, k)$. The old focus $F : (0, a)$ becomes $F_n : (h, k + a)$. The old points of intersection with the latus rectum $(\pm 2a, a)$ become $(\pm 2a + h, a + k)$ or $(h + 2a, a + k)$ and $(h - 2a, a + k)$. The axis of the parabola, formerly $x = 0$, becomes $x = h$.

EXAMPLE 10

We may reverse our orientation to view $(x - 3)^2 = 4(y + 2)$ as the translations of $x^2 = 4y$ through $(3, -2)$ units.

The new axis would therefore be $x = +3$; the new vertex $(3, -2)$; the new focus $(3, -1)$, since $a = 1$ for $x^2 = 4y$; and the ends of the latus rectum at $(3 + 2, -2 + 1)$ and $(3 - 2, -2 + 1)$ or $(5, -1)$ and $(1, -1)$. See Figure 6.24.

If we expand the general translated form of the parabola, we get

$$4(y - k) = (x - h)^2$$
$$4y - 4k = x^2 - 2hx + h^2 \quad (continued)$$

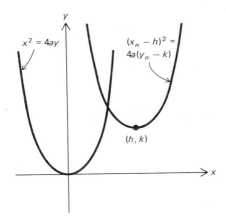

FIGURE 6-23

$$y = \frac{1}{4}x^2 - \frac{h}{2}x + \frac{h^2 + 4k}{4}.$$

This is in the general form of a quadratic: $y = ax^2 + bx + c$, with $a = \frac{1}{4}$, $b = -h/2$, and $c = \dfrac{h^2 + 4k}{4}$.

This should whet our appetites—the goal seems within reach. Can we now turn this around somehow to verify that $y = ax^2 + bx + c$ is always a translated parabola? Completing the square, we get

$$a\left(x^2 + \frac{b}{a}x + \frac{b^2}{4a^2}\right) = y - c + \frac{b^2}{4a}$$

$$a\left(x + \frac{b}{2a}\right)^2 = \left[y + \left(\frac{b^2 - 4ac}{4a}\right)\right]$$

$$\left(x + \frac{b}{2a}\right)^2 = \frac{1}{a}\left(y + \frac{b^2 - 4ac}{4a}\right).$$

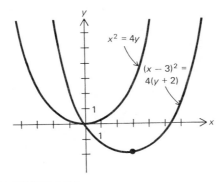

FIGURE 6-24

which really is the equation of a translated parabola (opening up if $a > 0$, down if $a < 0$). Note that the vertex is

$$\left(\frac{-b}{2a}, \ -\frac{b^2 - 4ac}{4a} \right)$$

which we derived by an almost identical argument in Section 2.2b, although we had not yet discussed translation.

In the problem section you are given an opportunity to investigate some other relations known as the conic sections, together with the application of translation to their graphs.

6.4. GEOMETRIC PROBLEMS TREATED A NEW WAY

As suggested in the introduction, we can often solve conventional geometric problems using the techniques of analytic geometry. We present two examples here to give you some idea of the techniques involved.

EXAMPLE 11

Prove that the line connecting the midpoints of two sides of a triangle is parallel to the third side and half its length.

We begin by placing an arbitrary triangle on the coordinate axes. There are many ways one can do this: you should choose the best position for constructing your proof, but one that does not inadvertently introduce a restriction (such as a right angle or two sides of equal length). Figure 6.25 shows one convenient location and labeling. Verify in your own thinking that the labeling allows vertex C to be in any quadrant and so allows ABC to represent any triangle.

Using the midpoint formula, we find the coordinates of the midpoints of the sides to be

$$\left(\frac{b+0}{2}, \ \frac{c+0}{2} \right), \quad \text{and} \quad \left(\frac{b+a}{2}, \ \frac{c+0}{2} \right)$$

or

$$D : \left(\frac{b}{2}, \ \frac{c}{2} \right), \quad \text{and} \quad E : \left(\frac{b+a}{2}, \ \frac{c}{2} \right).$$

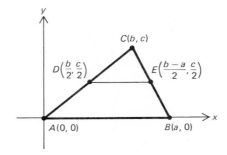

FIGURE 6-25

Now we need to prove that the line $\overleftrightarrow{DE}$ and the x-axis are parallel. The x-axis has slope 0 and the line segment $\overline{DE}$ has slope

$$m = \frac{c/2 - c/2}{(b+a)/2 - b/2} = 0$$

and thus is parallel to the third side of the triangle.

Next we need to show that $|DE| = \frac{1}{2}|AB|$.

$$|DE| = \sqrt{\left(\frac{c}{2} - \frac{c}{2}\right)^2 + \left(\frac{b+a}{2} - \frac{b}{2}\right)^2}$$

$$= \sqrt{\left(\frac{a}{2}\right)^2}$$

$$|DE| = \frac{a}{2} \qquad \text{(since } a > 0\text{)}.$$

We could use the distance formula again to find $|AB|$, but it should be obvious that $|AB| = a$. Thus $|DE| = |AB|/2$, and our proof is complete.

Prove that the diagonals of a parallelogram bisect each other. **EXAMPLE 12**

One valid assignment of coordinates is given in Figure 6.26. (Are you sure you understand why each point has the coordinates it does?)

Using the midpoint formula on $\overline{AD}$, we find the coordinates of O to be

$$\left(\frac{a+b+0}{2}, \frac{c+0}{2}\right) \qquad \text{or} \qquad O: \left(\frac{a+b}{2}, \frac{c}{2}\right).$$

Similarly, the midpoint of $\overline{CB}$ is

$$\left(\frac{a+b}{2}, \frac{c+0}{2}\right) = \left(\frac{a+b}{2}, \frac{c}{2}\right).$$

which is the same as the midpoint of $\overline{AD}$. Therefore, the point of inter- section (and there is only one) is the midpoint for each diagonal, as we were to prove.

As you will see, each problem will present its own difficulties, which

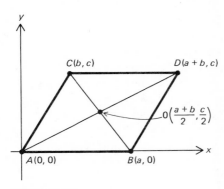

FIGURE 6-26

will have to be dealt with separately. If one setup of the figure does not work out well, try another—perhaps a solution will become apparent.

6.5. A PROJECT

PROJECT

In Chapter 2, we noted the need for methods of approximating zeros for polynomial functions. It should come as no surprise, then, that methods of approximation are needed to find zeros for a wide variety of other functions, particularly ones that involve functions of more than one type. For example, $f(x) = x^2 - \sin x$ is not subject to any standard strategies we have developed. Although a thorough investigation of such methods properly belongs in a numerical analysis course, we may profitably spend some of our time providing a foundation. Specifically, we shall use graphs to ascertain how many zeros to expect and their approximate values.

Phase 1: Rather than directly confront the equation $f(x) = 0$, it is often useful to decompose f into a sum of two functions, each of which is more manageable by itself, and consider a resulting equation $g(x) = h(x)$. (In the example suggested above, this would be $\sin x = x^2$.) Observe that the zeros we wished to find are now the x-coordinates of points of intersection of two graphs $y = g(x)$ and $y = h(x)$ (in the example, $y = x^2$ and $y = \sin x$). With such a strategy in mind, give as much information as you can concerning the number of zeros and approximate values of the zeros for each of the following functions.

(a) $f(x) = 2^x - x^2$.   (b) $g(x) = \sin x - x/2$.
(c) $h(x) = x^2 - \cos x$.   (d) $i(x) = \tan x - x$.
(e) $m(x) = \sin x - \log_2 x$.   (f) $t(x) = e^x + \sin x$.
(g) $k(x) = x - \sin x$.   (h) $b(x) = 2^x + x^2 - 2$.

Phase 2: In several of the examples in phase 1, you may have found that zeros were symmetrically located: each positive zero was matched by a zero that appeared to be its negative. Such observations can be verified if one develops some tests for symmetry. Such tests can also be used to assist the more general problem of graphing such functions.

a. There is an obvious kind of *balance* in any kind of symmetry, which is what the word implies. Using the language of analytic geometry, define the kinds of symmetry suggested in Figure 6.27 (the name for the type of symmetry is given in parentheses). You are likely to generate the best definitions by starting with points rather than entire graphs.

For example if we were concerned with symmetry in the x-axis, as illustrated by Figure 6.28, we would note that if the point (x, y) were on the graph, then so must be the point $(x, -y)$. Then, to test a relation defined by some equation in x and y for such symmetry (e.g., $xy^2 - |y| = x^3$), we would suppose that the equation were true for (a, b). (So, $ab^2 - |b| = a^3$ is supposed to be true.) Then we would determine whether $(a, -b)$ satisfied the equation $a(-b)^2 - |-b| = a^3$, which is equivalent to $ab^2 - |b| = a^3$, which we assumed to be true.

b. Using your definitions from (a), formulate and clearly state tests

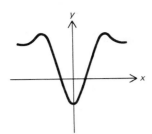

(a) Symmetry in the y-axis.

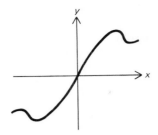

(b) Symmetry in the origin.

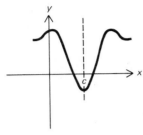

(c) Symmetry in the line $x = c$.

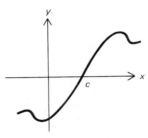

(d) Symmetry in the point $(c, 0)$.

FIGURE 6-27

that could be applied to functions to uncover such symmetries. Finally, try your tests on the following functions to see if they perform as expected.

1. $f(x) = x^3 - \sin x.$ (symmetry in origin)
2. $g(x) = x^2 + \cos x.$ (symmetry in y-axis)
3. $m(x) = e^x + x.$ (no symmetry)

4. $k(x) = \tan x + \dfrac{1}{x}.$ (symmetry in origin)

5. $m(x) = \sec x - \dfrac{1}{x^2}.$ (symmetry in y-axis)

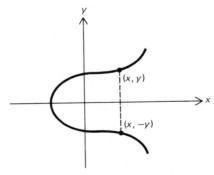

FIGURE 6-28

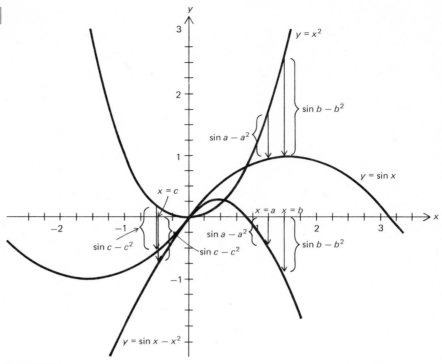

FIGURE 6-29

6. $a(x) = \cos(x - 2) + x^2 - 4x - 6.$ (symmetry in $x = 2$)

7. $g(x) = \left|\dfrac{1}{x}\right|.$ (symmetry in y-axis)

8. $t(x) = \tan x - x + \dfrac{\pi}{2}.$ $\left(\text{symmetry in } \left(\dfrac{\pi}{2}, 0\right)\right)$

9. $r(x) = x^2 - \sin x.$ (no symmetry)

10. $w(x) = \sin x + \cos x.$ $\left(\text{symmetry in } \left(-\dfrac{\pi}{4}, 0\right) \text{ and}\right.$

    symmetry in $x = \dfrac{\pi}{4}$ and

    many more

Phase 3: We began by asking for the zeros for $f(x)$ and immediately transformed f into $g - h$ so as to solve $g(x) = h(x)$. We now ask what connection, if any can be made between the graph of $y = f(x)$ and the graphs of $y = g(x)$ and $y = h(x)$, which we found more manageable. We refer back to the definition of subtraction (or addition) of functions: if we have values, even if only graphically, for g and h, we need only subtract them. The use of a straight edge held parallel to the y-axis would enable one to register such differences graphically without the need of

reading approximate values. Thus for $f(x) = \sin x - x^2$, we could sketch
a graph as in Figure 6.29. Analogously, one can use such devices to
approximate the solutions for certain inequalities. With this much assistance, sketch graphs of the functions in list A and then approximately solve
the inequalities in list B.

A	B
(a) $a(x) = x - \sin x$.	(a) $x \geq \sin x$.
(b) $b(x) = 2^x - x^2$.	(b) $2^x < x^2$.
(c) $c(x) = \cos x - x^2$.	(c) $\cos x > x^2$.
(d) $d(x) = \sin x - x$.	(d) $\sin x > x$.
(e) $e(x) = 2^x + x^2 - 2$.	(e) $2 - 2^x < x^2$.
(f) $f(x) = \sin x - \dfrac{x}{2}$.	(f) $\sin x \geq \dfrac{x}{2}$.
(g) $g(x) = \sin x + \log_2 x$.	(g) $\sin x < \log_2 x$

SUMMARY OF IMPORTANT CONCEPTS AND NOTATION

DISTANCE FORMULA: (Formula 6.1) The distance between two points
$P_1 : (x_1, y_1)$ and $P_2 : (x_2, y_2)$ is given by

$$|P_1 P_2| = (x_2 - x_1)^2 + (y_2 - y_1)^2.$$

MIDPOINT FORMULA: (Formula 6.2) The midpoint M of a line segment
joining $P_1 : (x_1, y_1)$ and $P_2 : (x_2, y_2)$ is given by

$$M : \left(\frac{x_2 + x_1}{2}, \frac{y_2 + y_1}{2} \right).$$

INCLINATION: (Definition 6.1) The inclination of a line is the smallest
positive angle from the x-axis to the line if such an angle is formed and
zero if the line is parallel to the x-axis.

SLOPE: (Definition 6.2) Let $P : (x_1, y_1)$ and $Q : (x_2, y_2)$ be any two
points on the line. The slope of the line is defined to be

$$\frac{y_2 - y_1}{x_2 - x_1} \qquad x_2 \neq x_1.$$

If $x_2 = x_1$ the line is not assigned a slope.

PERPENDICULAR LINES: (Formula 6.3) Two lines are perpendicular
if the product of their slopes is -1: $m_1 m_2 = -1$.

PARALLEL LINES: (Section 6.2b) Parallel lines always have equal
slopes and equal slopes guarantee parallel lines.

EQUATIONS OF LINES: (Section 6.2c)

POINT-SLOPE FORM: (Formula 6.4) $y - y_1 = m(x - x_1)$, where
(x_1, y_1) is the point and m is the slope.

SLOPE-INTERCEPT FORM: (Formula 6.5) $y = mx + b$, where m is
the slope and b is the y-intercept.

PARABOLA: (Definition 6.3) A parabola is the set of points in a plane

that are equidistant from a line *L* (the *directrix*) and a point *F* (the *focus*), both residing in the plane and with *F* not on *L*.
TRANSLATIONS IN THE PLANE: (Section 6.3b)
GEOMETRIC PROOFS: (Section 6.4)

PRACTICE EXERCISES

Section 6.2a

1. Find the distance between the following sets of points.
   (a) $A(3, 2)$, $A'(5, 6)$
   (b) $B(4, 8)$, $B'(1, 3)$
   (c) $C(-3, 2)$, $C'(-4, 5)$
   (d) $D(2, -5)$, $D'(6, -7)$
   (e) $E(1, -3)$, $E'(-2, 5)$
   (f) $F(-1, -3)$, $F'(-2, 1)$
   (g) $G(-1, -5)$, $G'(1, 7)$
   (h) $H(-2, -5)$, $H'(-6, -9)$
   (i) $I(2, -7)$, $I'(-3, -9)$
   (j) $J(-2, 3)$, $J'(-2, 9)$
   (k) $K(2, -7)$, $K'(-2, -7)$
   (l) $L(3, 2)$, $L'(2, 3)$
   (m) $M(\frac{3}{4}, 2)$, $M'(-\frac{3}{4}, -5)$
   (n) $N(\frac{1}{2}, 1)$, $N'(\frac{5}{2}, -1)$
   (o) $O(0, 7)$, $O'(0, 0)$
   (p) $P(0, \frac{1}{3})$, $P'(-3, \frac{4}{3})$
   (q) $Q(a, b)$, $Q'(a, -b)$
   (r) $R(a, b)$, $R'(-a, -b)$
   (s) $S(a, b)$, $S'(2a, 2b)$
   (t) $T(a, b)$, $T'(a/2, b/2)$
   (u) $U(a, b)$, $U'(b, a)$
   (v) $V(a, a)$, $V'(b, b)$

2. Find the midpoint of the line segment between the sets of points given in Problem 1.

Section 6.2b

3. Find the slope of the line connecting each pair of points in Problem 1.
4. Find the inclination of the line connecting each pair of points in Problem 1.

Section 6.2c

5. Find the equation of the line connecting each pair of points in Problem 1.
6. Find the equation of the lines with the following parameters:
   (a) $m = \frac{1}{2}$, $b = 0$
   (b) $m = 3$, $b = 4$
   (c) $m = -1$, $b = -1$
   (d) $m = 3$, $b = 5$
   (e) $m = 0$, $b = 4$
   (f) $m = -6$, $b = 2$
   (g) $m = -3$, $b = -\frac{3}{4}$
   (h) m undefined, $b = 3$
   (i) m undefined, $b = 0$
   (j) $m = -\frac{1}{2}$, $b = \frac{1}{8}$
   (k) $m = 10$, $b = -3$
   (l) $m = 90$, $b = 2$.

Section 6.3a

7. Sketch the following parabolas without setting up a table of values. Show the axis, the vertex, the directrix, the endpoints of the latus rectum, and the focus for each.

(a) $y^2 = 16x$ (b) $y^2 = x$ (c) $y^2 = -20x$
(d) $y^2 = -3x$ (e) $x^2 = 4y$ (f) $x^2 = 7y$
(g) $x^2 = -16y$ (h) $x^2 = -5y$

Section 6.3b

8. Translate each of the curves below through the amount indicated, and sketch both the old and new graphs on the same axes, clearly indicating which is which.

(a) $y = 3x + 4$ (2, 3)
(b) $2y + 2x = 3$ $(-1, -6)$
(c) $y = 4x - 7$ $(-1, 3)$
(d) $y = (x - 2)(x + 1)(x - 6)$ (2, 3)
(e) $y = (x + 3)(x - 5)(x + 1)$ $(-1, 2)$
(f) $y = (x - 2)^2(x + 1)^2$ $(-3, 2)$
(g) $y = 3^x$ (2, 4)
(h) $y = 2^x$ $(-1, -6)$
(i) $y = 2^x - 1$ $(2, -7)$
(j) $y = \cos x$ (2, 3)
(k) $y = \tan x$ $(-\pi, -6)$
(l) $y = \sin x$ $(\pi/2, 7)$

Section 6.3c

9. Sketch each of the following without plotting points.

(a) $(y - 3)^2 = 4(x - 2)$ (b) $(y + 6)^2 = 4(x + 3)$
(c) $(y - 1)^2 = x + 16$ (d) $(y + 1)^2 = 8(x - 2)$
(e) $2(x + 1) = (y - 2)^2$ (f) $(y + 4)^2 = 6(x + 1)$

10. Rewrite each of the following quadratic functions in the form of a translated parabola:

(a) $y = x^2 - 4x + 1$ (b) $y = x^2 + 2x - 5$
(c) $y = -x^2 + x + 1$ (d) $y = x^2 - 2x - 3$
(e) $y = x^2 + 4x + 1$ (f) $y = 2x^2 - 4x + 1$

Section 6.4

11. Using the techniques of analytic geometry, prove the following.

(a) The medians of an equilateral triangle intersect in a point.
(b) The medians of any triangle intersect in a point.
(c) The lengths of the diagonals of any rectangle are equal.
(d) The sum of the squares of the distances of any point from two opposite vertices of any rectangle is equal to the sum of the squares of the distance from the other two.
(e) The midpoint of the hypotenuse of any right triangle is equidistant from the three vertices.
(f) If the diagonals of a rectangle are perpendicular, then the rectangle is actually a square.

(g) The diagonals of a rhombus are perpendicular.

(h) If the diagonals of a quadrilateral bisect each other, then the quadrilateral is actually a parallelogram.

(i) The altitudes of a triangle intersect in a point.

(j) The midpoints of the four sides of any quadrilateral form a parellelo-gram.

PROBLEMS

12. Define notions of directed distance on lines parallel to the coordinate axes through the use of restrictions and formulas involving coordinates. (See footnote on p. 285.)

13. Which of the following pairs of lines are perpendicular, which parallel, and which neither? Explain why.

(a) $y = 3x + 7$
$y = -3x - 7$

(b) $y = x/2 + 5$
$y = -2x + 5$

(c) $y = -7 + 0.2x$
$y = 3 - 5x$

(d) $y = 3 + 3x$
$y = 4 + x/3$

(e) $y = 10x - 5$
$y = 10 - 5x$

(f) $3x = 2y + 5$
$-2x = 3y + 8$

(g) $3x/2 + y/3 = 12$
$2x - 9y = 0$

(h) $7 = 7x + 7y$
$1/7 = x/7 + y/7$

(i) $y = 12x + 7$
$x = 12y - 7$

(j) $x + 2y + 3 = 0$
$2y + 3 = x$

(k) $3y = 18x - 5$
$y = 4 - 6x$

(l) $3x + 2y = 88$
$2y + 3x = 10$

(m) $x + y + 1 = 0$
$-2x + y - 1 = 0$

(n) $3x + 7 = y$
$2x - 3 = -6y$

(o) $y = 3 - 7x$
$y = 4 - x/7$

14. Write equations for the lines satisfying the given conditions. If there is more than one such line for any problem or if there are none, clearly indicate such facts.

(a) Through (2, 2), parallel to $y = 3x + 7$

(b) Through (8, −7), parallel to $7x + 2y = 8$

(c) Through (0, 0), parallel to $y = x/2 + 2$

(d) Through (8, 1), perpendicular to $y = 3x + 7$

(e) Through (−2, −3), perpendicular to $y = x$.

(f) Through (−6, 3) perpendicular to $x + 2y = 3$.

(g) Parallel to $y = 2x + 1$, through y-intercept of $5x - 7y = 14$

(h) Perpendicular to $y = 2x + 1$ through y-intercept of $5x - 7y = 14$

(i) Parallel to $x + y + 5 = 0$, through x-intercept of $2x - 3y = 5$

(j) Parallel to $2x + 5y - 4 = 0$, through x-intercept of $10x + 4y + 3 = 0$

(k) Through the x-intercept of $2x + 4y = 10$ and the y-intercept of $x - y + 7 = 0$.

15. What value(s) of k (if any) will make the following pairs of line (i) parallel or (ii) perpendicular?

(a) $kx + 3y = 7$
$2x - y = 12$

(b) $2x + ky = 18$
$-x - 3y + 3 = 0$

(c) $7x - 2y + k = 0$
$2x + 3y - 5 = 0$

(d) $kx + ky - 7 = 0$
$5x + 2y = 10$

(e) $kx + 7y + 2 = 0$
$3x + ky - 8 = 0$

(f) $8x + ky = 4$
$kx - 8y = 5$

16. Write a flowchart that will input coefficients A, B, and C of equation $Ax + By + C = 0$, and the coordinates (R, S) of a point, and output the coefficients of the line through (R, S).

(a) Parallel to $Ax + By + C = 0$;
(b) Perpendicular to $Ax + By + C = 0$, or
(c) Does both parts (a) and (b).
(Does your flowchart have adequate protection against indeterminate cases?)

17. Determine the general form of equation for a parabola that opens sideways.

18. Write a flowchart that inputs the coordinates of two points P and Q, and outputs the length $|PQ|$; the midpoint of $\overline{PQ}$; the slope of $\overleftrightarrow{PQ}$; the coefficients A, B, and C of an equation for $\overleftrightarrow{PQ}$ in $Ax + By + C = 0$ form; and the y-intercept of $\overleftrightarrow{PQ}$.

19. Prove that the development of the distance formula will proceed unscathed if P_1 and P_2 are in relative positions different from those shown in Figure 6.1.

20. (a) Find formulas for the coordinates of the point one third of the distance from $P(x, y)$ to $Q(r, s)$.
(b) Find formulas for the coordinates of the point one mth of the distance from $P(x, y)$ to $Q(r, s)$.
(c) Find formulas for the coordinates of the point n/mths of the distance from $P(x, y)$ to $Q(r, s)$.
You should include your derivation and/or proof that your formulas are correct.

21. Prove that if the product of the slopes of two lines is -1, they are perpendicular.

22. What are the slope and y-intercept of $Ax + By + C = 0$?

23. What is the slope of any line perpendicular to $Ax + By + C = 0$?

24. Using the techniques of analytic geometry, prove that any angle inscribed in a semicircle is a right angle.

25. Write a flowchart to take a parabola in the form $y^2 = 4ax$ and print out the length of the latus rectum, the equation of the directrix, and the coordinates of the focus and vertex.

26. Write a flowchart that inputs the coefficients of a quadratic function and outputs the same information outputted in Problem 25.

27. Prove that
    (a) The coordinates of the endpoints of the latus rectum in a parabola with the focus $(a, 0)$, directrix $x = -a$, are $(a, \pm 2a)$.
    (b) The length of the latus rectum is $4a$.
28. Prove that any parabola resulting from translating $y^2 = 4ax$ has an equation of the form $Ax^2 + By^2 + Cx + Dy + E = 0$.
29. Another standard form of equation for a straight line is called the intercept form. Derive the equation $x/a + y/b = 1$ for the line with x-intercept $(a, 0)$ and y-intercept $(0, b)$.
30. Give a more complete argument, based on the triangle of Figure 6.8, that negative slopes correspond to slants from the lower left to the upper right, while positive slopes correspond to slants from the upper left to the lower right.
31. Find the equation of the graph consisting of all points equidistant from the points (a, b) and (r, s).
32. (a) Find a reasonably neat equation containing no radicals, for the graph consisting of all points $P(x, y)$ so that the sum of the distance from P to $(c, 0)$ and the distance from P to $(-c, 0)$ is always $2a$. [This graph is called an ellipse, and the points $(c, 0)$ and $(-c, 0)$ are its foci.]
    (b) Translate the graph of part (a) through (h, k) units (using its equation) to find one form of the equation for an ellipse.
    (c) Rewrite the equation of part (b) to show that it is of the form $Ax^2 + By^2 + Cx + Dy + E = 0$.
33. (a) Find an equation that does not involve radicals, for the graph consisting of all points $P(x, y)$ so that the distance from P to $(0, 0)$ is r. (This graph is called a circle of radius r centered at the origin.)
    (b) Translate the graph of part (a) through (h, k) units (using its equation) to find one form of the equation for a general circle.
    (c) Rewrite the equation of part (b) to show that it is of the form $Ax^2 + Ay^2 + Cx + Dy + E = 0$.
34. (a) Find a reasonably neat equation containing no radicals for the graph of all points $P(x, y)$ whose distance from $(c, 0)$ is $2a$ larger or smaller than its distance from $(-c, 0)$. [This is a hyperbola with foci $(c, 0)$, and $(-c, 0)$ and center $(0, 0)$.]
    (b) Translate the equation of part (a) so as to correspond to a hyperbola with center (h, k).
    (c) Rewrite the equation of part (b) to show that it is of the form $Ax^2 + By^2 + Cx + Dy + E = 0$.
35. Find a reasonably neat equation involving no radicals for the graph consisting of all points $P(x, y)$ that are twice as far from $(0, 0)$ as they are from $(a, 0)$.
36. Consider a three dimensional coordinate system with x, y, and z axes dividing space into 8 octants with 3 coordinate planes (xy, yz, xz). Using two right triangles, one after the other, derive a formula for the distance between the two points (x, y, z) and (a, b, c).

37. Formulate self-contained, precise definitions for:
    (a) focus   (b) directrix   (c) axis
    (d) vertex   (e) latus rectum
    of a parabola.
    In Section 2.7, we considered the matter of approximating the zeros of (polynomial) functions. The tools of analytic geometry provide a better opportunity for investigating such algorithms. Problems 38 and 39 guide you to two such algorithms.

38. As with successive halving, suppose that one starts with the points $(l, f(l))$, $(r, f(r))$ for which $f(l)$ and $f(r)$ are of opposite sign. Then there must be at least one zero somewhere in (l, r). Let c be the zero of the linear function determined by $(l, f(l))$ and $(r, f(r))$; i.e., we approximate the function by a linear function. We then replace either l or r with c so that the new interval $[l, r]$ still contains a sign change for f. This process is repeated until the interval containing the zero is small enough. Write a flowchart that details this algorithm. (This method, called the method of false position, is also the subject of Problem 56 of Chapter 2.)

39. Suppose that one has two approximations x_0 and x_1 for a zero for the function f (here we need not assume a sign change). Approximate f by the straight line through $(x_0, f(x_0))$ and $(x_1, f(x_1))$ and let x_2 be the zero of this linear function. Repeat this procedure with x_1 and x_2 to get x_3, etc., until $f(x_i)$ is close enough to zero to suggest that x_i is close to a zero, or $|x_i - x_{i-1}|$ is small enough to suggest closeness to a zero, or a certain number of cycles have been completed. Write a flowchart that details this algorithm (called the Secant Method).

40. Prove the converse of the Pythagorean Theorem: if the sum of the squares of two sides of a triangle equals the square of the third side, then the triangle is a right triangle.

CHAPTER 7

THE COMPLEX NUMBERS

7.1. INTRODUCTION

The first numbers one learns about are the natural numbers, $\mathbb{N}$. Then one usually is introduced to the positive rationals, and although this is often justified by pieces of pie and sharing candy bars, it can also be viewed as a desire to be able to solve all equations of the form $cx = d$, where c and d are natural numbers, i.e., to be able to solve equations such as $3x = 4$. In effect, one invents fractions, or rational numbers, to solve equations that could not be solved using only natural numbers.

Next one invents the negative rationals and zero so as to be able to solve equations as $3x + 4 = 4$ and $\frac{2}{3}x + \frac{3}{8} = \frac{6}{101}$. Again, this invention is motivated by physical phenomena that are easily described by such equations. As it turns out, $\mathbb{Q}$ is sufficient to solve $bx + c = d$ for b, c, and d any rational numbers instead of just the positive rationals.

The jump to $\mathbb{R}$ is less easily described, although it can be motivated by wanting to fill in all the holes in the number line left after $\mathbb{Q}$ has been placed on it, such as $\sqrt{2}$, π, and e. The real numbers allow solution of a much wider set of equations, but they still have limitations.† For example, as

---

† Although $\mathbb{R}$ permits the solution of more equations, it is not invented by the equation solution approach—that yields only the set of so-called algebraic numbers which includes numbers like $\sqrt[37]{-82.5}$ and $1/(1 + \sqrt{5})$ but does *not* include numbers like π and e.

noted in Chapter 2, not all quadratic functions have zeros in $\mathbb{R}$ (i.e., not all quadratic equations have solutions in $\mathbb{R}$). To remedy such defects, we invent a still larger number system. After inventing the system and looking briefly at it, we shall see how it gives us some new perspective on the elementary functions we have already studied, and how the elementary functions can aid in our working with this new number system—a nice symbiotic relationship!

7.2. THE COMPLEX NUMBERS

7.2a. Notation and the Complex Plane

COMPLEX PLANE

Let us begin by investigating a rather specific problem, that of solving the equation $x^2 = -r$, where $r \in \mathbb{R}$ and $r > 0$. Within $\mathbb{R}$, we know that there are no numbers whose squares are negative, merely by noting that the product of two real numbers with the same sign is positive, and the square of any real number is certainly the product of two numbers with the same sign.

In an effort to invent some new numbers, it might seem that we should have to invent a raft of new numbers separately. Whatever the new numbers are, we should like them to satisfy as many of the properties of $\mathbb{R}$ as possible. Thus, if we are going to make it possible to take square roots, we should like to retain the properties of radicals. Thus, in particular, $\sqrt{-r}$ might be expressible as $\sqrt{-1} \cdot \sqrt{+r}$. But $\sqrt{r}$ is a real number, so at this point it appears we really only need one new number, $\sqrt{-1}$, as any negative real number can be expressed as -1 times the absolute value of that real number.

NOTE

As is generally true in mathematics, notation is a matter of *convention*. The notation $\frac{2}{3}$ for the solution of $3x = 2$ is not dictated by the gods but only by convention. There is no reason why rational numbers could not have been written in an entirely different way, say for example, (2, 3) or $3 \infty 2$ or $\frac{3}{2}$ instead of $\frac{2}{3}$.†

For the journey into complex numbers, there are two common notational conventions. One of these chooses to denote the new number $\sqrt{-1}$ by (0, 1), and then reclothe the real numbers so that if $r \in \mathbb{R}$, then we denote r as $(r, 0)$ in the new system. This is useful in some respects, and those who pursue mathematics very far are likely to encounter the complex numbers

DEFINITION 7.1

i

in this costume. The notation we shall use denotes $\sqrt{-1}$ by the lower case *i*. That is, *i* stands for a new number defined by the requirement that $i^2 = -1$.

To represent *i* geometrically, we cannot put it on the number line anywhere, since that is entirely filled with real numbers. A natural choice is to move into a plane because we want to retain $\mathbb{R}$ in the the new system of numbers, and keep it as the number line we are familiar with. One esthetic

---

† Actually, we commonly use several different notations for rational numbers: $2 \div 3$, 2/3, $\frac{2}{3}$, and .66$\overline{6}$ all denote the same rational number.

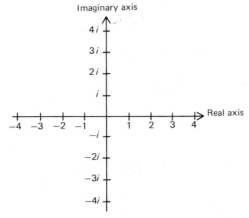

Imaginary axis

Real axis

FIGURE 7-1

placement of a point for i would be directly above the zero point on the real number line, and at the same distance from zero as 1 is from zero on the real number line, so that i retains some flavor of 1 as we might feel appropriate for $\sqrt{-1}$ (after all $\sqrt{+1} = 1$). This is a convenient choice for a variety of reasons.†

So far, we have dealt specifically with only one new number, i. We should find out what other new numbers this generates. From our initial concern with $x^2 = -r$, we should want $\sqrt{r} \cdot i$, which would call for all positive multiples of i. This can be represented by drawing the ray from 0 through i, and marking it off in integral multiples of i, although all points on the ray have meaning ($\sqrt{2}i$ is a solution of $x^2 = -2$; $\frac{3}{4}i$ is a solution of $x^2 = -\frac{9}{16}$, etc.). It is natural to think of extending the ray into a full line and obtaining negative multiples of i below the real line. These are perfectly sensible, for if arithmetic in this new set of numbers is to have a reasonable structure, we should be able to write $(-2i)^2 = (-2)^2 i^2 = 4 \cdot -1 = -4$. Hence $-2i$ and $2i$ would both be square roots of -4 [since $(2i)^2 = -4$, too], giving us $\pm 2i$ as square roots of -4, just as ± 2 are square roots of $+4$.

So at this stage we have two axes of numbers, as shown in Figure 7.1. The horizontal axis is called the *real axis*, and the vertical axis is called the *imaginary axis*. The latter terminology is an unfortunate holdover from former times when complex numbers were viewed with a kind of suspicion. Actually, they are no more imaginary than the so-called real numbers and have physical interpretations in such fields as electricity and magnetism.

From the geometric view developed so far, it seems sensible to identify all the points in the plane as complex numbers by noting their real and imaginary coordinates; but instead of using an ordered pair notation, we

† This geometric representation gives some cause to represent complex numbers as ordered pairs, as suggested earlier, since one is already accustomed to using ordered pairs to denote points in the plane.

7.2. *THE COMPLEX NUMBERS* 317

continue to let the i label the imaginary component and use the addition symbol to join the coordinates. Thus we have $3 + 4i$ for the point $(3, 4)$ in our complex plane, $-7 + \sqrt{2}i$ for $(-7, \sqrt{2})$, and so on as complex numbers. At this point, the addition sign is just a notation, not a genuine arithmetic operation, although we shall find that it is compatible with the extension of addition to the new system.

DEFINITION 7.2
Complex Numbers

We can now define the set of *complex numbers* as $\{a + bi \mid a, b \in \mathbb{R}$ and i denotes $\sqrt{-1}\}$, denote this set by $\mathbb{C}$, and proceed to investigate its arithmetic and algebra.

7.2b. Operations with Complex Numbers

OPERATIONS WITH
COMPLEX
NUMBERS

DEFINITION 7.3
Equality

Our first concern is to establish the meaning of equality among complex numbers. The fact that $\frac{6}{4} = \frac{3}{2}$ in $\mathbb{Q}$ should caution one from jumping too quickly to a definition that $a + bi = c + di$, provided that $a = c$ and $b = d$. However, a look at the geometrical representation of $\mathbb{C}$ suggests that this is a perfectly good definition, and in the end we do adopt this component type of definition of *equality*.

Of second concern should be the arithmetic operations. What, for example, is a sensible meaning for the sum of $(a + bi)$ and $(c + di)$? Not having added points in the plane before, our picture is no help, but thinking of polynomials in which $(a + bx) + (c + dx) = (a + c) + (b + d)x$, we might decide to add complex numbers component-wise. This, in fact, is what we do, and define

DEFINITION 7.4
Addition

$$(a + bi) + (c + di) = (a + c) + (b + d)i \qquad \text{for all } a, b, c, d \in \mathbb{R}.$$

Since $a + c \in \mathbb{R}$ and $b + d \in \mathbb{R}$, the right-hand side is again in $\mathbb{C}$, as we should wish. Thus

$$(3 + 4i) + (7 + 9i) = (3 + 7) + (4 + 9)i = 10 + 13i$$

and

$$\left(2 + \frac{i}{2}\right) + \left(\frac{3}{4} + (-4)i\right) = \left(2 + \frac{3}{4}\right) + \left(\frac{1}{2} + (-4)\right)i$$
$$= 2\frac{3}{4} + \left(-3\frac{1}{2}\right)i.$$

DEFINITION 7.5
Subtraction

Similarly, *subtraction* is defined by

$$(a + bi) - (c + di) = (a - c) + (b - d)i \qquad \text{for all } a, b, c, d \in \mathbb{R}.$$

Therefore,

$$(3 + 4i) - (7 + 9i) = (3 - 7) + (4 - 9)i = -4 - 5i.$$

Using our definition of addition, we could add the real number a to the imaginary number bi by writing them as two complex numbers. We should have $(a + 0i) + (0 + bi) = (a + 0) + (0 + bi) = a + bi$ so that the use of the " $+$ " symbol in denoting complex numbers really is compatible with addition as we have defined it. A somewhat longer argument (discussed later in this subsection) verifies that $a - bi$ and $a + (-b)i$ are also equivalent.

For multiplication, the above definitions might suggest component-wise multiplication, which we *could* do, but a definition that is more useful is given by again referring to polynomials. The polynomial product $(a + bx)(c + dx)$ is, by the grace of the distributive law, equal to $ac + (bc + ad)x + bdx^2$. Wishing to preserve the distributive law for complex arithmetic, we are motivated to define

$$(a + bi)(c + di) = ac + (bc + ad)i + adi^2,$$

or, since $i^2 = -1$,

$$(a + bi)(c + di) = (ac - bd) + (ad + bc)i \qquad \text{for all } a, b, c, d \in \mathbb{R}.$$

DEFINITION 7.6
Multiplication

Again the product is in $\mathbb{C}$, as we should wish. Thus

$$(3 + 4i)(7 + 9i) = (3 \cdot 7 - 4 \cdot 9) + (3 \cdot 9 + 4 \cdot 7)i$$
$$= (21 - 36) + (27 + 28)i$$
$$= -15 + 55i.$$

Division is more difficult. One approach is to see that if we could define $1/(c + di)$, i.e., the reciprocal of a complex number, then we could resort to multiplication to find the general formula for division. The reciprocal $1/(c + di)$ of $c + di$ should be a complex number $(x + yi)$ $[=1/(c + di)]$ so that $(c + di)(x + yi) = 1 = 1 + 0i$.† Using the definition for multiplication, we find that if

$$(c + di)(x + yi) = 1$$

then

$$(cx + dyi^2 + dxi + cyi) = 1$$

i.e., we need to have

$$(cx - dy) + (dx + cy)i = 1 + 0i,$$

which is equivalent to the system of equations below by virtue of the definition of equality of complex numbers.

$$cx - dy = 1$$
$$dx + cy = 0.$$

Solving these equations for x and y, we find that $x = c/(c^2 + d^2)$ and $y = -d/(c^2 + d^2)$. Then, since we are looking for a nice expression for $x + yi$ in terms of c, d, and i to serve as $1/(c + di)$, substitution of the values just computed gives us

$$\frac{1}{c + di} = x + yi = \frac{c}{c^2 + d^2} - \frac{d}{c^2 + d^2} i = \frac{c - di}{c^2 + d^2}.$$

Therefore

$$\frac{a + bi}{c + di} = (a + bi) \frac{1}{(c + di)} = (a + bi) \left(\frac{c - di}{c^2 + d^2} \right)$$

---

† This is because $1 + 0i$ is the identity for complex multiplication as well as real multiplication. See fact (6), p. 321.

$$= (a + bi)\left(\frac{c}{c^2 + d^2} - \frac{di}{c^2 + d^2}\right).$$

DEFINITION 7.7
Division

Multiplying this, we find that

$$\frac{(a + bi)}{(c + di)} = \frac{ac + bd}{c^2 + d^2} + \frac{bc - ad}{c^2 + d^2}\, i. \tag{1}$$

[It would be a good idea for you to try this multiplication and see if you get Equation (1).] We choose this formula as our definition of the division of complex numbers. As an example,

$$\frac{3 + 4i}{7 + 9i} = \frac{(3 \cdot 7) + (4 \cdot 9)}{7^2 + 9^2} + \frac{(4 \cdot 7) - (3 \cdot 9)}{7^2 + 9^2}\, i$$

$$= \frac{21 + 36}{49 + 81} + \frac{28 - 27}{49 + 81}\, i$$

$$= \frac{57}{130} + \frac{1}{130}\, i$$

NOTE   Look again at the system of equations used to find $x + yi$ in the development of a division rule. It would be valuable for you to direct your attention to the transformation of one complex number equation into two equivalent real number equations, using the definition of equality in $\mathbb{C}$. This is often a useful device for solving equations that involve complex numbers.

We should emphasize that we define the four arithmetic operations this way so as to maintain as many of the properties from the real number operations as we can.

NOTE   One noteworthy consequence of these definitions is that the use of complex arithmetic with complex numbers $r + 0i$, which are just real numbers in disguise, yields the same results as the use of real arithmetic on the undisguised real numbers. In other words if $\square$ represents any of the four arithmetic operations,

$$(r + 0i)\ \square\ (s + 0i) = (r\ \square\ s) + 0i.$$

This property is important enough so that no definitions of $+$, $-$, $\cdot$, and $\div$ would be acceptable if they failed to satisfy the property. Because this property is true, there is no ambiguity in the meaning of each arithmetic operation applied to real numbers, and so we allow ourselves to write $r + 0i$ simply as r.

With these definitions we can now verify that a variety of algebraic relationships that are valid in $\mathbb{R}$ are still valid in $\mathbb{C}$.

1. *Commutativity:*
   (a) Of addition:

$$(a + bi) + (c + di) = (a + c) + (b + d)i \qquad \text{by definition}$$
$$= (c + a) + (d + b)i \qquad \text{by commutativity in } \mathbb{R}$$
$$= (c + di) + (a + bi) \qquad \text{by definition.}$$

(b) Of multiplication: see Problem 10.

2. *Associativity:*
    (a) Of addition: see Problem 10.
    (b) Of multiplication:

$$[(a + bi)(c + di)](e + fi)$$
$$= [(ac - bd) + (bc + ad)i](e + fi)$$
$$= [(ac - bd)e - (bc + ad)f] + [(bc + ad)e + (ac - bd)f]i$$
$$= [ace - bde - bcf - adf] + [bce + ade + acf - bdf]i,$$

while

$$(a + bi)[(c + di)(e + fi)]$$
$$= (a + bi)[(ce - df) + (de + cf)i]$$
$$= [a(ce - df) - b(de + cf)] + [b(ce - df) + a(de + cf)]i$$
$$= [ace - adf - bde - bcf)] + [bce - bdf + ade + acf]i.$$

Since the two different groupings yield the same end result, they are equal.

3. *Distributivity:*

$$(a + bi)[(c + di) + (e + fi)] = (a + bi)(c + di) + (a + bi)(e + fi)$$

is true in general. Problem 10 asks you to prove this.

4. The real number 0 is still the *additive identity,* i.e., if $z \in \mathbb{C}$, $0 + z = z$.
    Proof: $z \in \mathbb{C}$ means $z = a + bi$ for some $a, b \in \mathbb{R}$, so that

$$0 + z = 0 + (a + bi) = (0 + 0i) + (a + bi)$$
$$= (0 + a) + (0 + b)i = a + bi = z.$$

5. $-(a + bi) = -a - bi$.†
    Proof: $-(a + bi) = 0 - (a + bi)$
$$= (0 - a) + (0 - b)i = -a - bi.$$

6. The real number 1 is still the *multiplicative identity,* i.e., if $z \in \mathbb{C}$, $1 \cdot z = z$.
    Proof: $1 \cdot z = 1 \cdot (a + bi) = (1 + 0i)(a + bi)$
$$= (1 \cdot a - 0 \cdot b) + (0 \cdot a + 1 \cdot b)i = a + bi = z.$$

DEFINITION 7.8
Conjugation

The complex number system contains a new unary operation called conjugation. If $a + bi \in \mathbb{C}$, then the *conjugate* of $a + bi$ is $a - bi$. The operation of conjugation is denoted by a bar over the number. That is, if $z = a + bi$, $\bar{z} = a - bi$. One of the special values of conjugation comes from the fact that $(a + b)(a - b) = a^2 - b^2$. So $z \cdot \bar{z} = (a + bi)(a - bi) = a^2 - (bi)^2 = a^2 + b^2$; i.e., the product of a complex number and its conjugate is always a real number (as a^2 and b^2 are both real). The algebra of conjugation is explored in Problems 12–15.

While an order relation ($\leq$) made sense on the real line, it cannot be extended to $\mathbb{C}$, at least not without losing some of its important properties.

---

† Without proof, it is not clear that the negative of a complex number really is formed by negating the two components.

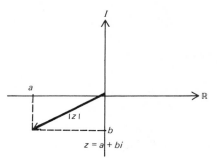

FIGURE 7-2

Problem 20 provides an opportunity to understand this better through an example, but a full axiomatic treatment of inequalities is beyond the scope of our investigation, and so we will not pursue this topic here. Instead, we retreat a step or two and note that another concept of size does extend,

DEFINITION 7.9
Absolute Value

namely, that of absolute value. The *absolute value* of a complex number z is the distance between the origin $0 + 0i$ and z without regard to sign. It can be given an algebraic formula by using the distance formula or the Pythagorean Theorem. See Figure 7.2. The formula that results is

$$|z| = \sqrt{a^2 + b^2},$$

where $z = a + bi$. Note that the range of this function is a subset of $\mathbb{R}$, therefore, one can compare absolute values of complex numbers with inequalities even though one cannot do so with the complex numbers themselves.

7.3. THE COMPLEX NUMBERS AND POLYNOMIAL FUNCTIONS

POLYNOMIALS

Since the four arithmetic operations are defined for complex numbers, there is no reason why one cannot extend the definition of polynomial functions to allow their domain to be $\mathbb{C}$ not just $\mathbb{R}$. Thus we may say that

DEFINITION 7.10
Polynomials

a function p with domain $\mathbb{C}$ (or a subset of $\mathbb{C}$) is called a *polynomial* if it can be specified by a rule

$$p(z) = a_n z^n + \cdots + a_1 z + a_0,$$

where each $a_i \in \mathbb{C}$.† Such an extension allows the possibility of complex zeros.

7.3a. Linear Functions

For the linear function f specified by $f(z) = c_1 z + c_2$, the only zero is the solution of the equation

$$c_1 z + c_2 = 0,$$

† The symbol z is usually used when the argument is complex, much as x is often used for real arguments.

which is $z = c_2/c_1$. If c_1 and c_2 are real, the zero is real so that complex numbers are not needed to find zeros in this case. If either coefficient is complex, then the zero may be complex.

COMPLEX POLYNOMIALS

Thus

$$3z + 4 = 0$$

has real zero

$$z = -\tfrac{4}{3},$$

and

$$(3 + 4i)z + 4 = 0$$

has complex zero

$$z = -\frac{4}{3} + 4i$$

$$= \frac{-4 + 0i}{3 + 4i} = \frac{-12 + 0 \cdot 4}{9 + 16} + \frac{0 \cdot 3 + 16}{9 + 16} i$$

or

$$z = -\frac{16}{25} + \frac{19}{25} i.$$

Similarly,

$$(3 + 4i)z - (2 + i) = 0$$

has complex zero

$$z = \frac{2 + i}{3 + 4i} = \frac{6 + 4}{9 + 16} + \frac{3 - 8}{9 + 16} i$$

$$z = \frac{2}{5} - \frac{1}{5} i.$$

7.3b. Quadratic Functions

For quadratic functions f, specified by $f(z) = az^2 + bz + c$, the process of completing the square (which we used to derive the quadratic formula) is valid whether a, b, and c are real or complex. This time, however, restricting a, b, and c to be real does not guarantee real zeros. If $b^2 - 4ac < 0$, let us denote $b^2 - 4ac$ by $-d$; we see from the development of Section 2.2b that the zeros of f are

$$\frac{-b \pm \sqrt{-d}}{2a} = \frac{-b \pm \sqrt{d}\,i}{2a}$$

which are two complex numbers. In fact they are conjugates of one another. So if f is a quadratic function with real coefficients, then it has either two distinct real zeros, one real zero of multiplicity 2, or two complex conjugate zeros. Thus, if we have

$$f(z) = z^2 + 2z + 4$$

7.3. *THE COMPLEX NUMBERS AND POLYNOMIAL FUNCTIONS*   323

$$z = \frac{-2 \pm \sqrt{2^2 - 4 \cdot 1 \cdot 4}}{2 \cdot 1}$$

using the quadratic formula; and

$$z = \frac{-2 \pm \sqrt{-16}}{2} = \frac{-2 \pm 4\sqrt{-1}}{2} = \frac{-2 \pm 4i}{2}$$

$$z = -1 \pm 2i.$$

If the coefficients are allowed to be complex, one may still complete the square and get the two zeros

$$\frac{-b + \sqrt{b^2 - 4ac}}{2a} \quad \text{and} \quad \frac{-b - \sqrt{b^2 - 4ac}}{2a},$$

but because a, b, and c may be complex, these expressions for the zeros cannot easily be written in the standard $\alpha + \beta i$ form. In addition, there is a possible question concerning the meaning of $\sqrt{b^2 - 4ac}$ if $b^2 - 4ac$ is not real. We shall return to the problem of roots of complex numbers in the next section, but we note here that square roots do exist for all complex numbers so that there are two zeros above.

It is tempting to ask for a graph of a quadratic function in order to better see and understand the zeros. Unfortunately, since the domain and range are both $\mathbb{C}$, and $\mathbb{C}$ needs two dimensions to be represented graphically, we should need four dimensions to graph $f(z)$. Since very few people can visualize a four-dimensional geometry, this idea is sadly abandoned.

7.3c. General Polynomials

As for general polynomials, the Factor Theorem and its converse still hold and synthetic division may also be used (although the complex multiplications involved are more tedious). The proofs and derivations of Chapter 2 for these theorems and algorithms are valid if one reads everything as complex instead of real, since the arguments were purely arithmetic and the arithmetic of $\mathbb{C}$ has no more restrictions than that of $\mathbb{R}$.

Also, as noted in Section 2.4b, if one extends the domains of the polynomials to $\mathbb{C}$, the Fundamental Theorem of Algebra guarantees that a polynomial of degree n has precisely n complex zeros when multiplicities are counted. (The proof of this result remains beyond the scope of this book.) Thus $\mathbb{C}$ is sufficient for finding all the zeros of all polynomials, not just the simple quadratic ones we started with in Section 7.2 of this chapter. That adding new numbers to solve one type of second degree equation should provide *all* the numbers needed for *all* polynomials is truly remarkable.

One final observation, which does not immediately suggest itself, but which is useful in finding zeros for polynomials, is that if a polynomial function has all real coefficients, then its nonreal complex zeros come in conjugate pairs. Stated a little differently, if p is a polynomial with all real coefficients and z is a complex zero of p, then so is $\bar{z}$. The proof is simple once one knows that $\overline{(\bar{z})} = z$ and $\overline{(a+b)} = \bar{a} + \bar{b}$ (see Problems 12 and 13 at the end of the chapter). If $p(z) = 0$, then $\overline{p(z)} = \bar{0} = 0$, but

$$\overline{p(z)} = \overline{a_n z^n + \cdots + a_0} = \overline{a_n z^n} + \cdots + \overline{a_0} = \overline{a_n} \overline{z^n} +$$
$$\cdots + \overline{a_0} = a_n (\bar{z})^n + \cdots + a_0 = p(\bar{z})$$

so that $p(\bar{z}) = 0$ as claimed. Geometrically, this says that the distribution of zeros in the complex plane for a polynomial with real coefficients is symmetric with respect to the real axis.

Find as many zeros as you can of $f(x) = x^4 + x^3 - 11x^2 + 9x + 20$, given that $2 + i$ is one zero. Since $2 + i$ is one zero, we may use synthetic division to reduce the problem; and we know that $2 - i$ is also a zero, since all the coefficients are real. This zero permits a second reduction. These steps are

EXAMPLE 1

$$
\begin{array}{r|rrrrr}
-(2+i)) & 1 & 1 & -11 & 9 & 20 \\
& & -2-i & -5-5i & 17-4i & 20+0i \\
\hline
-(2-i)) & 1 & 3+i & 6+5i & -8+4i \; | & 0 \\
& & -2+i & -10+5i & -8+4i \\
\hline
& 1 & 5 & 4 \; | & 0
\end{array}
$$

So the remaining zeros are zeros of $x^2 + 5x + 4$, which by factoring or the quadratic formula are -1 and -4. So we have found all four zeros, namely, $2 + i$, $2 - i$, -1, and -4.

7.4. THE CIRCULAR FUNCTIONS AND COMPLEX NUMBERS

The primary concern of this section is not to use complex numbers to better understand the circular functions but to use the circular functions to better understand the complex numbers. It is an application of the circular functions as functions of angles.

If $a + bi$ is a complex number, it is located graphically at the point (a, b) in the real-imaginary plane. See Figure 7.3. Let us draw the ray from

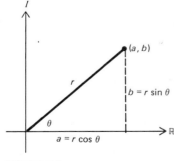

FIGURE 7-3

TRIGONOMETRIC FORM

the origin through (a, b), and let θ denote one of the angles formed by the positive x-axis and the new ray. We may write $a = r \cos \theta$ and $b = r \sin \theta$, using what we know of triangles and the circular functions. In addition, since r is the distance from the origin to $a + bi$, it is just $|a + bi| = \sqrt{a^2 + b^2}$.

Similarly, if $c + di$ is another complex number, $c = s \cos \phi$ and $d = a \sin \phi$, where ϕ is an angle formed by the positive x-axis and the ray from 0 through $c + di$ and $s = |c + di|$. If we multiply $a + bi$ and $c + di$ using the new representations, we have

$$(a + bi)(c + di) = (r \cos \theta + r \sin \theta \, i)(s \cos \phi + s \sin \phi \, i)$$
$$= (rs \cos \theta \cos \phi - rs \sin \theta \sin \phi)$$
$$+ (rs \cos \theta \sin \phi + rs \sin \theta \cos \phi)i,$$

by multiplying as we would polynomials of the form $(ax + by)(cx + dy)$. Also,

$$(a + bi)(c + di) = rs[\cos(\theta + \phi) + \sin(\theta + \phi)i]$$

by using identities for the sum of two angles in reverse. As we shall see, this points to easier multiplications of complex numbers.

First let us establish enough vocabulary to assist our deliberations. In writing $a + bi$ as $(r \cos \theta) + (r \sin \theta)i$, or as is more common, $r(\cos \theta + i \sin \theta)$, (or even using a common shorthand, r cis θ for $r(\cos \theta + i \sin \theta)$), r is called the *modulus*† of the number $a + bi$ and θ the *amplitude* of $a + bi$. From the work just done, we can see that

DEFINITION 7.11

Modulus and Amplitude

$$r(\cos \theta + i \sin \theta) \cdot s(\cos \phi + i \sin \phi) = rs[\cos(\theta + \phi) + i \sin(\theta + \phi)]$$

or, in other words, the product of two complex numbers has as its modulus the product of the original moduli and has as amplitude the sum of the original amplitudes.

COMPUTATION OF *n*th ROOTS

This representation is especially useful in finding nth roots of complex numbers. First we consider nth powers. A seemingly logical extension of the product rule just derived would be that $[r(\cos \theta + i \sin \theta)]^n$ would be $r^n(\cos n\theta + i \sin n\theta)$, since there are n multiplications instead of two, and the moduli and amplitudes of each factor are the same. We can easily prove that this is so using mathematical induction.‡

1. For $n = 1$, it is certainly true that

$$[r(\cos \theta + i \sin \theta)]^1 = r^1(\cos 1 \cdot \theta + i \sin 1 \cdot \theta).$$

2. Assume that

$$[r(\cos \theta + i \sin \theta)]^k = r^k(\cos k\theta + i \sin k\theta).$$

We need to show that

$$[r(\cos \theta + i \sin \theta)]^{k+1} = r^{k+1}(\cos(k+1)\theta + i \sin(k+1)\theta).$$

---

† As we have noted, r is also the absolute value of $a + bi$. Two words for one concept may seem confusing, but their usage overlaps and it would be wise to be familiar with both.
‡ For those who have not studied Section 5.4, actual trial of $n = 2$, 3, and 4 is suggested as a means of convincing yourself of the validity of this result for all n, even though such trials do *not* constitute a proof.

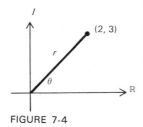

FIGURE 7-4

We do so by straightforward multiplication and use of the induction hypotheses.

$$[r(\cos \theta + i \sin \theta)]^{k+1} = [r(\cos \theta + i \sin \theta)]^k [r(\cos \theta + i \sin \theta)]$$
$$= [r^k(\cos k\theta + i \sin k\theta)][r(\cos \theta + i \sin \theta)]$$
$$= r^{k+1}[(\cos k\theta \cos \theta - \sin k\theta \sin \theta)$$
$$+ (\sin k\theta \cos \theta + \cos k\theta \sin \theta)i]$$
$$= r^{k+1}[\cos(k+1)\theta + i \sin(k+1)\theta],$$

as needed. This result is usually called *de Moivre's Theorem*:

$$[r(\cos \theta + i \sin \theta)]^n = r^n(\cos n\theta + i \sin n\theta).$$

THEOREM 7.1

De Moivre's Theorem

Suppose we want to find $(2 + 3i)^4$. Rather than multiply this out the long way, $(2 + 3i)(2 + 3i)(2 + 3i)(2 + 3i)$, we can use de Moivre's Theorem.

For $z = 2 + 3i$,

$$r = |2 + 3i| = \sqrt{4+9} = \sqrt{13}.$$

The angle θ is the angle between the x-axis and the ray through $2 + 3i$. See Figure 7.4. Then

$$z = 2 + 3i = \sqrt{13}(\cos \theta + i \sin \theta)$$

and thus

$$z^4 = (2 + 3i)^4 = (\sqrt{13})^4(\cos 4\theta + i \sin 4\theta),$$

which is equivalent to $169(\cos 4\theta + i \sin 4\theta)$. We shall discuss actually finding θ and the procedure for translating this back to the $a + bi$ form later.

Now let us consider the equation $z^n = c$, where c is any complex number. The solutions to this equation are the zeros of the nth degree polynomial $f(z) = z^n - c$. Also, by assertion of the Fundamental Theorem of Algebra, we know that there are precisely n zeros for this function, and hence n roots for the equation, counting multiplicities. To make use of trigonometric form in solving this equation, we need to reformulate the definition of *equality* of complex numbers in terms of moduli and amplitudes. A quick observation confirms that the moduli of the two equal numbers must be the same and that the amplitudes must differ by an integral number of complete revolutions.† The measurement of angles by degrees

† Problems 35–38 outline a formal proof of this result.

proves convenient in this application so that we may write

$$r(\cos\theta + i\sin\theta) = s(\cos\phi + i\sin\phi)$$

if and only if $r = s$ and $\theta = \phi + k \cdot 360°$ for some $k \in \mathbb{Z}$. So, if $s(\cos\phi + i\sin\phi)$ is a solution of the equation $z^n = c$, and say $c = R(\cos\theta + i\sin\theta)$, then

$$s^n(\cos n\phi + i\sin n\phi) = R(\cos\theta + i\sin\theta),$$

whence $s^n = R$ or $s = \sqrt[n]{R}$ and $n \cdot \phi = \theta + k \cdot 360°$ or $\phi = \theta/n + k(360°/n)$. Furthermore, the amplitudes

$$\frac{\theta}{n} + 0\left(\frac{360°}{n}\right), \quad \frac{\theta}{n} + 1\left(\frac{360°}{n}\right), \quad \frac{\theta}{n} + 2\left(\frac{360°}{n}\right), \quad \cdots, \quad \frac{\theta}{n} + (n-1)\left(\frac{360°}{n}\right)$$

are all distinct, so that the corresponding n complex numbers with moduli $\sqrt[n]{R}$ form the complete solution set of $z^n = c$. These n numbers are located on the circle of radius $\sqrt[n]{R}$ in the complex plane and are equally spaced around the circle, the angle formed by any consecutive two numbers and the origin being $360°/n$. These n different roots will be clearer if we consider a specific example, after which you should reread the above development.

EXAMPLE 2   Find all five fifth roots of 2. Since $2 = 2 + 0i$, $r = 2$, and $\theta = 0$ for this example, then $2 = 2(\cos 0° + i\sin 0°)$. Thus the modulus of a fifth root is the real number $\sqrt[5]{2}$, and the amplitude must be one of the numbers $0°/5 + k(360°/5)$, which is to say that one of the numbers $0 \cdot 72° = 0°$, $1 \cdot 72° = 72°$, $2 \cdot 72° = 144°$, $3 \cdot 72° = 216°$, or $4 \cdot 72° = 288°$. Note that with $k \geq 5$ or $k < 0$, we have amplitudes equivalent to the five just found: $5 \cdot 72° = 360°$, which is equivalent to $0°$; $-2 \cdot 72° = -144°$, which is equivalent to $216°$; and so on. The fifth roots of 2 are then the five numbers shown in Figure 7.5.

We have not yet attempted translations between $a + bi$ and trigonometric form for genuinely complex numbers. Let us briefly consider three examples of such translations to see how they can be accomplished.

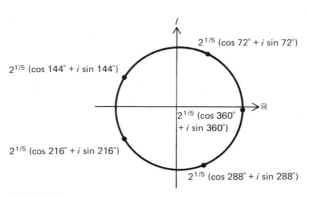

FIGURE 7-5

Express $4 + 3i$ in trigonometric form. To do this, we need to find values   **EXAMPLE 3**
for r and θ. Since $r = |a + bi|$, we have

$$r = \sqrt{4^2 + 3^2} = \sqrt{16 + 9} = 5.$$

As for θ, note than $\sin \theta = 3/r = \frac{3}{5}$ and $p(\theta)$ is in the first quadrant. Using
Table 2, Appendix D, we find $\theta = 37°$ to the nearest degree. Therefore,
$4 + 3i \approx 5(\cos 37° + i \sin 37°)$.

Express $2(\cos(-45°) + i \sin(-45°))$ in $a + bi$ form. Since   **EXAMPLE 4**

$$\cos(-45°) = \sqrt{2}/2 \quad \text{and} \quad \sin(-45°) = -\sqrt{2}/2,$$

$$2(\cos(-45°) + i \sin(-45°)) = 2(\sqrt{2}/2 - (\sqrt{2}/2)i) = \sqrt{2} - \sqrt{2}i,$$

which is in the desired form.

Find $(4 + 3i)^5$. From Example 3, $z = 4 + 3i \approx 5(\cos 37° + i \sin 37°)$ **EXAMPLE 5**
so that

$$
\begin{aligned}
z^5 &\approx 5^5(\cos 185° + i \sin 185°)\\
&= 5^5(-\cos(185° - 180°) + i(-\sin(185° - 180°))\\
&= 5^5(-\cos 5° - i \sin 5°) \approx 3125(-0.9962 - i(0.0872))\\
&= (3125)(-0.9962) - i(0.0872)(3125).
\end{aligned}
$$

It is worth observing before we leave this application of the circular
functions to complex numbers that we have also made use of mathematical
induction, a topic from the study of functions with domain $\mathbb{N}$, and have
used results from, and have added results to, the study of polynomial
functions. Such interrelationships among seemingly distinct topics in
mathematics give added dimension to the structures discovered and
created. They also show that the different branches of mathematics are
not really separate from one another. In the next section, we shall bring
together several concepts from circular functions, series, and exponential
functions.

7.5. THE PSYCHODELIC REALM OF COMPLEX FUNCTION THEORY

This section is included to demonstrate some of the close relationships   **COMPEX SERIES**
among the elementary functions that only appear when one considers
complex numbers. Although we shall reason about the results, we shall
not, in a strict sense, be proving them, nor will all the motivations be clear.

We noted in Section 4.3 that e^x could be expressed by a series, namely,

$$e^x = 1 + x + \frac{x^2}{2!} + \frac{x^3}{3!} + \cdots.$$

Since we can now add, subtract, multiply, and divide complex numbers, we
might be tempted to define

$$e^z = 1 + z + \frac{z^2}{2!} + \frac{z^3}{3!} + \cdots.$$

That the infinite sum actually has a value for all complex numbers z is perhaps startling but nevertheless true. Using properties of exponential functions, we can then extend the domain of $\exp_a$ to $\mathbb{C}$.

$$\exp_a(z) = a^z = e^{z \log_e a}$$
$$= 1 + (z \log_e a) + \frac{(z \log_e a)^2}{2!} + \frac{(z \log_e a)^3}{3!} + \cdots.$$

Actually, though, our real interest is with e^z. First, let us consider e^{ix}, where $x \in \mathbb{R}$. This restricts e^z to the imaginary axis instead of the real axis. Using the series definition, we see that

$$e^{ix} = 1 + ix + \frac{(ix)^2}{2!} + \frac{(ix)^3}{3!} + \frac{(ix)^4}{4!} + \frac{(ix)^5}{5!} + \frac{(ix)^6}{6!} + \frac{(ix)^7}{7!} + \cdots$$

$$= 1 + ix - \frac{x^2}{2!} - \frac{ix^3}{3!} + \frac{x^4}{4!} + \frac{ix^5}{5!} - \frac{x^6}{6!} - \frac{ix^7}{7!} + \cdots$$

If it is legal to regroup (and it is here), we could rewrite this as

$$\left(1 - \frac{x^2}{2!} + \frac{x^4}{4!} - \frac{x^6}{6!} + \cdots\right) + i\left(x - \frac{x^3}{3!} + \frac{x^5}{5!} - \frac{x^7}{7!} + \cdots\right).$$

This in turn sparks our memory of the series formulas for sine and cosine, we find that

$$e^{ix} = \cos x + i \sin x.$$

This almost unbelievable connection between exponential and circular functions is, indeed, true. It also leads to the remarkable fact that

$$e^{i\pi} = -1,$$

since $e^{i\pi} = \cos \pi + i \sin \pi = -1$. That the three strange numbers e, i, and π can be combined to generate such a simple number as -1 is incredible.

Taking another direction, we find a way to extend the domain of the circular functions to $\mathbb{C}$. Noting that the cycle of signs and instances of i in the series for e^{ix}, we consider

$$e^{-ix} = 1 - ix - \frac{x^2}{2!} + \frac{ix^3}{3!} + \frac{x^4}{4!} - \frac{ix^5}{5!} - \cdots$$

so that

$$e^{ix} + e^{-ix} = 2 - \frac{2x^2}{2!} + \frac{2x^4}{4!} - \frac{2x^6}{6!} + \cdots = 2 \cos x$$

and

$$e^{ix} - e^{-ix} = 2i \sin x$$

in the same vein. Rearranging slightly, we have

$$\sin x = \frac{e^{ix} - e^{-ix}}{2i} \qquad \text{and} \qquad \cos x = \frac{e^{ix} + e^{-ix}}{2}.$$

We could then define

$$\sin z = \frac{e^{iz} - e^{-iz}}{2i} \qquad \text{and} \qquad \cos z = \frac{e^{iz} + e^{-iz}}{2}$$

for all $z \in \mathbb{C}$. To be sure, we could have replaced x with z in the series for sine and cosine, but this route reminds us of the functions tick and tock of Section 4.5 and leads to the definitions of the hyperbolic functions as follows:

$$\left. \begin{aligned} \sinh z &= \frac{e^z - e^{-z}}{2} \\ \cosh z &= \frac{e^z + e^{-z}}{2} \end{aligned} \right\} \text{ for } z \in \mathbb{C}$$

or

$$\left. \begin{aligned} \sinh x &= \frac{e^x - e^{-x}}{2} \\ \cosh x &= \frac{e^x + e^{-x}}{2} \end{aligned} \right\} \text{ for } x \in \mathbb{R}.$$

Therefore, the hyperbolic functions are closely related to both the exponential and circular functions, and it comes as no surprise that they share properties with each set of functions. Although we shall not pursue this further, we note also that

$$\sin(z) = \sin(a + bi) = \sin a \cos bi + \cos a \sin(bi)$$
$$= \sin a \cosh b + i \cos a \sinh b,$$

which suggests a whole new set of relationships.

Finally, we return to the complex exponential function to discover that

$$e^{z + 2\pi i} = e^z e^{2\pi i} = e^z (e^{\pi i})^2 = e^z (-1)^2 = e^z$$

so that e^z is a periodic function with period $2\pi i$, or 2π in the imaginary direction. With this final example of unexpected behavior we conclude our discussion of complex function theory and the ways in which it can link together exponential, circular, and hyperbolic functions.

SUMMARY OF IMPORTANT CONCEPTS AND NOTATION

i: (Definition 7.1) The number i is a number satisfying $i^2 = -1$.
COMPLEX NUMBERS: (Definition 7.2) The set of complex numbers, denoted by $\mathbb{C}$, is $\{a + bi \,|\, a, b \in \mathbb{R} \text{ and } i \text{ denotes } \sqrt{-1}\}$.
OPERATIONS WITH COMPLEX NUMBERS: (Section 7.2b)
EQUALITY: (Definition 7.3) $a + bi = c + di$, provided that $a = c$ and $b = d$.

ADDITION: (Definition 7.4) $(a + bi) + (c + di) = (a + c) + (b + d)i$, $a, b, c, d \in \mathbb{R}$.

SUBTRACTION: (Definition 7.5) $(a + bi) - (c + di) = (a - c) + (b - d)i$, $a, b, c, d, \in \mathbb{R}$.

MULTIPLICATION: (Definition 7.6) $(a + bi)(c + di) = (ac - bd) + (ad + bc)i$, $a, b, c, d \in \mathbb{R}$.

DIVISION: (Definition 7.7)

$$\frac{(a + bi)}{(c + di)} = \frac{ac + bd}{c^2 + d^2} + \frac{bc - ad}{c^2 + d^2} i.$$

CONJUGATION: (Definition 7.8) The conjugate of $a + bi$ is $a - bi$.

ABSOLUTE VALUE: (Definition 7.9) For $z = a + bi$, $|z| = \sqrt{a^2 + b^2}$.

POLYNOMIALS: (Definition 7.10 and Section 7.3)

TRIGONOMETRIC FORM: (Section 7.4) r cis $\theta = r(\cos \theta + i \sin \theta)$.

MODULUS: (Definition 7.11) The modulus of a complex number z is $|z|$.

AMPLITUDE: (Definition 7.11) The amplitude of a complex number z is the smallest positive angle from the $\mathbb{R}$ axis to the line determined by z and the origin.

DE MOIVRE'S THEOREM: (Theorem 7.1)

$$[r(\cos \theta + i \sin \theta)]^n = r^n[\cos n\theta + i \sin n\theta].$$

COMPUTATION OF nTH ROOTS: (Section 7.4).

PRACTICE EXERCISES

Section 7.2a

1. Locate the following complex numbers in the complex plane.
   - (a) $1 + i$
   - (b) $2 + 3i$
   - (c) $2 - 3i$
   - (d) $-2 + 4i$
   - (e) $-5 - 4i$
   - (f) $0 + 3i$
   - (g) $-4 + 0i$
   - (h) $5 + 2i$
   - (i) $-4 + 3i$
   - (j) $-2 - 5i$

Section 7.2b

2. Find the value of each of the following as a single complex number. Write each answer in as simple a form as you can.
   - (a) $(3 + 4i) + (7 - 5i)$
   - (b) $(3 + 4i)(7 - 5i)$
   - (c) $(3 + 4i) - (7 - 5i)$
   - (d) $(3 + 4i)/(7 - 5i)$
   - (e) $(1 - 2i)[(-10 - 12i) + (-2 + 2i)]$
   - (f) $(\frac{3}{2} + 2i)(\frac{5}{4} - (\frac{1}{3})i)$
   - (g) $(\frac{3}{2} + 2i) + (\frac{5}{4} - (\frac{1}{3})i)$
   - (h) $(1 + 2i)(3 + 4i)$

(i) $(-1-2i)(-3-4i)$

(j) $(1+2i)/(3+4i)$

(k) $[(3-2i)(-4+i)]/(1+i)$

(l) i^2

(m) i^3

(n) i^4

(o) i^5

(p) i^6

(q) i^7

(r) i^8

(s) $1/i$

(t) $1/i^2$

(u) $1/i^3$

(v) $1/i^4$

(w) $|3-4i|$

(x) $|(2-3i)+(8-2i)|$

(y) $|(2-3i)(8-2i)|$

(z) $|(2-3i)||(8-2i)|$

(aa) $\overline{5+6i}$

(bb) $\overline{2-4i}$

(cc) $\overline{(-3+2i)}\,\overline{(8+5i)}$

(dd) $\bar{i}$

(ee) $|\overline{4+3i}|$

Section 7.3a

3. Find all the zeros of the following functions.

   (a) $a(z) = (3+4i)z + (4+3i)$

   (b) $b(z) = (2+3i)z + 5$

   (c) $c(z) = 4z + (3-4i)$

   (d) $d(z) = 5z + 2i$

   (e) $e(z) = 3iz + 6i$

   (f) $f(z) = (2+2i)z - 3 - 4i$

   (g) $g(z) = 3z + (2+3i)$

   (h) $h(z) = 2z - (2+5i)$

   (i) $i(z) = (2+3i)z + (2+3i)$

   (j) $j(z) = (3+3i)z + 1$

Section 7.3b

4. (a) $e(x) = x^2 - 3x + 3$

   (b) $p(x) = 2x^2 + 5x + 4$

   (c) $s(y) = 2y - 2y^2 - 2$ $\Big\}$ Find all the zeros.

   (d) $a(x) = 3x^2 + x + 4$

   (e) $b(y) = y^2 + 2y - 11$

(f) $c(z) = 5z^2 - z + 4$
(g) $d(x) = 4x^2 + 2x + 1$
(h) $f(x) = 2x^2 + x + 7$ } Find all the zeros.
(i) $g(x) = x^2 + x + 11$

Section 7.3c

5. (a) $p(q) = q^3 - 2q^2 + 3$
   (b) $m(z) = z^3 - 1$
   (c) $d(t) = 3t^4 - 6t^3 - 4t^2 - 10t - 15$ } Find all the zeros.
   (d) $n(a) = a^4 + a^3 - 2a^2 + 4a - 24$
   (e) $s(b) = b^4 + 2b^2 + 1$

Section 7.4

6. Express in trigonometric form:

   (a) $3 + 4i$ (b) $-5 + 12i$ (c) $-8 - 6i$
   (d) $1 + i$ (e) $2 - 3i$ (f) $5i$
   (g) 6 (h) $\frac{3}{2} - 2i$ (i) $2 + 3i$
   (j) $-6 + 3i$ (k) $-2 - 3i$ (l) $i/2$
   (m) $-4 + 9i$ (n) 3 (o) $2 - 9i$

7. Express in $a + bi$ form:

   (a) $5(\cos 30° + i \sin 30°)$ (b) $2(\cos 225° + i \sin 225°)$
   (c) $\sqrt{2}(\cos 135° + i \sin 135°)$ (d) $5(\cos 60° + i \sin 60°)$
   (e) $6(\cos 270° + i \sin 270°)$ (f) $3(\cos 20° + i \sin 20°)$
   (g) $(\frac{1}{2})(\cos 100° + i \sin 100°)$ (h) $(\frac{3}{2})(\cos 215° + i \sin 215°)$
   (i) $4(\cos 200° + i \sin 200°)$ (j) $2(\cos 150° + i \sin 150°)$
   (k) $12(\cos 45° + i \sin 45°)$ (l) $3(\cos 10° + i \sin 10°)$
   (m) $(\frac{3}{4})(\cos 300° + i \sin 300°)$ (n) $21(\cos 390° + i \sin 390°)$
   (o) $10(\cos 250° + i \sin 250°)$ (p) $3(\cos 500° + i \sin 500°)$

8. Perform the following multiplications using trigonometric form. Check using the original definition of multiplication and translating as needed.

   (a) $(3 - 2i)(6 + i)$
   (b) $(2 + 2i)(3 - 3i)$
   (c) $(-3 - i)(-1 + 2i)$
   (d) $2(\cos 60° + i \sin 60°) \cdot 5(\cos 30° + i \sin 30°)$
   (e) $(1 + \sqrt{3}i)(2\sqrt{3} - 6i)$
   (f) $(\cos 45° + i \sin 45°) \cdot 4(\cos 180° + i \sin 180°)$
   (g) $\sqrt{2}(\cos 135° + i \sin 135°) \cdot 2(\cos(-45°) + i \sin(-45°))$
   (h) $8(\cos 120° + i \sin 120°) \cdot 3(\cos(-120°) + i \sin(-120°))$
   (i) $(3 + 5i)(2 + 6i)$
   (j) $(2 + 7i)(3 - 6i)$
   (k) $(3 + 8i)(-2 - i)$
   (l) $(4 + 5i)(-2 - 3i)$
   (m) $(2 - 3i)(2 + 3i)$

(n) $(4-7i)(12+11i)$

(o) 3 cis 60° · 2 cis 90°

(p) 4 cis 315° · 2 cis 300°

(q) 2 cis 10° · 10 cis 135°

9. Find the following roots and sketch a graph of each set of roots.

(a) Fourth roots of 4

(b) Third roots of 8

(c) Fifth roots of 32

(d) Sixth roots of 1

(e) Third roots of -8

(f) Fourth roots of i

(g) Fourth roots of $1/\sqrt{2} + (1/\sqrt{2})i$

(h) Eighth roots of 1

PROBLEMS

10. Prove that the following are true.

(a) The multiplication of complex numbers is commutative.

(b) The addition of complex numbers is associative.

(c) The multiplication of complex numbers distributes over addition of complex numbers.

(d) $0 \cdot z = 0$ for all $z \in \mathbb{C}$.

(e) Component multiplication of complex numbers would prohibit the distributive law from applying in complex arithmetic.

(f) $a - bi$ and $a + (-b)i$ really mean the same thing.

11. If $z = 2 + 3i$ and $w = -3 + i$, locate the following numbers in the complex plane.

(a) $z + w$	(b) $z - w$	(c) $w - z$	(d) $z \cdot w$
(e) z/w	(f) w/z	(g) z^2	(h) w^2
(i) $\bar{z}$	(j) $\bar{w}$	(k) $\overline{z + w}$	(l) $\bar{z} + \bar{w}$
(m) $\overline{z \cdot w}$	(n) $\bar{z} \cdot \bar{w}$	(o) $\overline{(\bar{z})}$	

12. Prove that $\overline{(\bar{z})} = z$ for all $z \in \mathbb{C}$.

13. Prove that conjugation distributes over addition, i.e., $\overline{(a+b)} = \bar{a} + \bar{b}$.

14. Formulate and prove a relationship between $\overline{(z \cdot w)}$ and the numbers $\bar{z}$ and $\bar{w}$.

15. Prove that $\bar{z} = z$ if and only if z is a real number.

16. Prove that $i^{j+4n} = i^j$ for all $j \in \mathbb{Z}$ and all $n \in \mathbb{Z}$.

17. Derive the quotient formula by multiplying and dividing the quotient by the conjugate of the denominator.

18. Derive the formula given for $|z|$.

19. Prove that an equivalent definition for $|z|$ is given by $|z| = \sqrt{z \cdot \bar{z}}$.

20. To convince yourself that a true inequality relation cannot hold in $\mathbb{C}$, suppose that (a) $i > 0 + 0i$; (b) $i < 0 + 0i$. In each case, see what the consequence of the relationship between i^2 (i.e., -1) and 0 would be, and i^4 (i.e., $+1$) and 0. Explain what you do: why and what it shows.

21. Solve the following system of equations, where w and z represent complex numbers.

$$3w - 4z = 5 + i$$
$$3iw + 2z = 2 + 2i.$$

22. Consider the complex equation $|z| = 3$.
    (a) Solve this equation as completely as you can.
    (b) Graph the solution set.
    (c) Solve $|x| = 3$ $(x \in \mathbb{R})$
    (d) Graph its solution set.
    (e) Write your observation concerning the preceding four parts.

23. Write a flowchart for complex addition. You will need to decide how best to input complex numbers. Assume that the computer does not know about complex numbers.

24. Solve Problem 23 for
    (a) Complex multiplication.
    (b) Complex division.

25. Using Section B.4e, Appendix B, and/or a reference manual in Fortran or Basic, learn enough about subprograms to write subprograms for complex arithmetic, and then do so. Make use of them in a program that computes a table of values for functions of a complex variable.

26. Using subprograms for complex arithmetic developed in Problem 25, write a program to find all the zeros of any quadratic equation.

27. If your computer installation has complex arithmetic capability built in,
    (a) Write a program to find the zeros of any quadratic function.
    (b) Write a program to evaluate functional values for any polynomial using synthetic division.

28. Prove that any third degree polynomial with real coefficients must have at least one real zero.

29. Where does the proof that $p(\bar{z}) = 0$ if $p(z) = 0$ break down if p does not have all real coefficients? Explain carefully.

30. Prove that a linear function with at least one genuinely complex coefficient may have a real zero.

31. Derive a formula for division of complex numbers using trigonometric form. (Hint: First conjecture a result, and then try to prove it).

32. Write a flowchart to translate from $a + bi$ form to trigonometric form.

33. Write a flowchart to translate from trigonometric form to $a + bi$ form.

34. Write a flowchart that will generate the nth roots of any positive real number. The input should include the value of n and the number whose roots are desired.

35. To formally prove that

$$r(\cos \theta + i \sin \theta) = s(\cos \phi + i \sin \phi)$$

if and only if $r = s$ and $\theta = \phi + k \cdot 360°$, start by showing that if $r = s$ and $\theta = \phi + k \cdot 360°$, then the corresponding complex numbers are equal in the original sense.

36. To continue the proof begun in Problem 35, assume that
$$r(\cos\,\theta + i\,\sin\,\theta) = s(\cos\,\phi + i\,\sin\,\phi).$$
Find two simultaneous equations in the variables r, s, θ, and ϕ, which must then be satisfied.

37. Solve the simultaneous equations of Problem 36 to find that, at least, $\phi = \theta + k \cdot 360°$ is a necessary condition.

38. Explain why r and s must both be greater than 0. Return to the equation of Problem 36 to verify that, indeed, $r = s$ and $\phi = \theta + 2k\pi$, thus concluding the proof started in Problem 35. (Explain why this is the conclusion.)

APPENDIX A

ALGEBRA REVIEW

This appendix is designed so that you can use it to remove any weaknesses you may have in basic algebraic techniques. It is *not* designed to teach basic algebra to those not already familiar with it; hence there will be few detailed explanations. The text will be broken into small parts. Each one will have several examples followed directly by practice problems, many of which will have answers. Try to work each problem without referring to the answers until you have finished all the work. (It might help to put a card over the answers until you are ready to look at them.) The answers are included so that you can check your work immediately.

A.1. ADDITION OF MONOMIAL ALGEBRAIC EXPRESSIONS

Example 1: $3x + 2x + 7x = (3 + 2 + 7)x = 12x$

Example 2: $2a + b + 3a + 6b = (2 + 3)a + (1 + 6)b = 5a + 7b$

Example 3: $2xy + y + 3xy = (3 + 2)xy + y = 5xy + y$

Add the following:

    (a) $2y + 3y + y$ $\hspace{4cm} = 6y$

    (b) $4r + 9r + r$ $\hspace{4cm} = 14r$

ADDITION OF
MONOMIALS

339

(c) $s + s + 2s$ $= 4s$
(d) $3x + 4y + x + 5y$ $= 4x + 9y$
(e) $2a + 3a + 4b + 5b$ $= 5a + 9b$
(f) $4ab + x + 3ab + y$ $= 7ab + x + y$
(g) $3xy + 4yz + xy + 2yz + 3xy$ $= 7xy + 6yz$
(h) $r + s + 3s + t + 10r + 2t$ $= 11r + 4s + 3t$
(i) $c + 2a + m + 3m + 2c + a + 2m$ $= 3a + 3c + 6m$
(j) $j + k + i + j + k + i + k + j$ $= 3j + 3k + 2i$

A.2. SUBTRACTION OF MONOMIAL ALGEBRAIC EXPRESSIONS

Example 1: $2x - x = (2 - 1)x = x$

Example 2: $4a - 2a - a = (4 - 2 - 1)a = a$

Example 3: $-10b - b - 5b = (-10 - 1 - 5)b = -16b$

Example 4: $4x - yz - 2x - 3yz = (4x - 2x) - yz - 3yz$
$$= (4 - 2)x + (-1 - 3)yz$$
$$= 2x + (-4)yz$$
$$= 2x - 4yz$$

Example 5: $-4xy - 2xy - z - 5z = (-4 - 2)xy + (-1 - 5)z$
$$= -6x - 6z$$

Simplify the following:

(a) $5z - 4z$ $= z$
(b) $10m - 2m - 3m$ $= 5m$
(c) $-20b - a - 15b - 11a$ $= -35b - 12a$
(d) $-x - 2y - z - 10x - 10y - 2z$ $= -11x - 12y - 3z$
(e) $2ab - 3ab - 10x - x$ $= -ab - 11x$
(f) $10y - 2z - 5z - z$ $= 10y - 8z$
(g) $mn - 10mn - n - 11n$ $= -9mn - 12n$
(h) $20z - 15z - x - 20x - ab - 10ab$ $= 5z - 21x - 11ab$
(i) $-4ab - 2ab - x - 10x$ $= -6ab - 11x$
(j) $-30x - z - 20x - 10z - y - 20y$ $= -50x - 21y - 11z$

A.3. MULTIPLICATION OF MONOMIAL ALGEBRAIC EXPRESSIONS

Example 1: $3(2x) = (3 \cdot 2)x = 6x$

Example 2: $(2x)(3x) = (2 \cdot 3)(x \cdot x) = 6x^2$

Example 3: $(2x)(3y) = (2 \cdot 3)(x \cdot y) = 6xy$

Example 4: $(-2x)(3y) = (-2 \cdot 3)(xy) = -6xy$

Example 5: $(2x)(3y)(-5z) = (2 \cdot 3 \cdot -5)(x \cdot y \cdot z) = -30xyz$

Example 6: $(-4x)(-5x)(2y) = (-4 \cdot -5 \cdot 2)(x \cdot x \cdot y)$
$$= (20 \cdot 2)(x^2 y) = 40x^2 y$$

Example 7: $(2x)(-3y)(z)(-2xz) = (2 \cdot -3 \cdot -2)(x \cdot y \cdot z \cdot xz)$
$$= 12x^2yz^2$$

Multiply the following:

(a) $-4(3z)$ $= -12z$
(b) $(-20x)(5x)$ $= -100x^2$
(c) $(x)(-2y)(-z)$ $= 2xyz$
(d) $(-3x)(-2x)(5z^3)$ $= 30x^2z^3$
(e) $-10(2x)(5x^2)y$ $= -100x^3y$
(f) $x \cdot y \cdot (-zx)$ $= -x^2yz$
(g) $(10x)(-2x)(3x)$ $= -60x^3$
(h) $-2x(3y)(z)(-2y)(-2z)$ $= -24xy^2z^2$
(i) $2x(2y)(2z)(-2xy)(xz)$ $= -8x^3y^2z^2$
(j) $-20x(10xy)(y^2)(-10x^2)$ $= 200x^4y^3$

A.4. DIVISION OF MONOMIAL ALGEBRAIC EXPRESSIONS

Example 1: $\dfrac{4x}{2} = \left(\dfrac{4}{2}\right)x = 2x$

Example 2: $\dfrac{-4x}{2} = \left(-\dfrac{4}{2}\right)x = -2x$

Example 3: $\dfrac{4x}{2x} = \left(\dfrac{4}{2}\right)\left(\dfrac{x}{x}\right) = 2 \cdot 1 = 2$

Example 4: $\dfrac{2xy}{x} = 2y\left(\dfrac{x}{x}\right) = 2y \cdot 1 = 2y$

Example 5: $\dfrac{-4x^2y^3}{2xy^2} = \left(-\dfrac{4}{2}\right)\left(\dfrac{x^2}{x}\right)\left(\dfrac{y^3}{y^2}\right) = -2xy$

Example 6: $\dfrac{-4x^2y^2}{y^2z^2} = \dfrac{-4x^2}{1} \cdot \dfrac{y^2}{y^2} \cdot \dfrac{1}{z^2} = \dfrac{-4x^2}{z^2}$

Divide the following:

(a) $10x/5$ $= 2x$
(b) $30x/6x$ $= 5$
(c) $-25xy/-5y$ $= 5x$
(d) $3x^2z/xy$ $= 3xz/y$
(e) $-4xz/-2z$ $= 2x$
(f) $20x^2y/5x^2y$ $= 4$
(g) $14abc^2/2a^2b^2$ $= 7c^2/ab$
(h) $16a^3x^2y/4axy$ $= 4a^2x$
(i) $-20x^2/5y^2$ $= -4x^2/y^2$
(j) $-2xyzabc/x^2yzb^2$ $= -2ac/xb$

A.5. COMBINATION OF OPERATIONS WITH MONOMIAL EXPRESSIONS, ORDER OF OPERATIONS

When doing problems in which a number of operations are involved, one needs to remember the rules governing the order of operations.

1. Remove all parentheses (from the inside out) using rules 1(a)–1(c) in order (see Section A.6.)

    (a) Perform all exponentiation left to right.

    (b) Perform all multiplications and divisions left to right.

    (c) Perform all additions and subtractions left to right

2. Perform all exponentiations left to right.

3. Perform all multiplications and divisions left to right.

4. Perform all additions and subtractions left to right.

Example 1: $4x + 2x - 3x - x = (4 + 2 - 3 - 1)x = 3x$ by rule 4

Example 2: $-4x + 2(3x) - x + 2x = -4x + 6x - x + 2x$ by rule 3

$$= (-4 + 6 - 1 + 2)x \qquad \text{by rule 4}$$

$$= 3x$$

Example 3: $(4x)^2 - 2x(3x) + \dfrac{6x^3}{2x} = 16x^2 - 2x(3x) + \dfrac{6x^3}{2x}$ by rule 2

$$= 16x^2 - 6x^2 + 3x^2 \qquad \text{by rule 3}$$

$$= (16 - 6 + 3)x^2 \qquad \text{by rule 4}$$

$$= 13x^2$$

Example 4: $3x^2(-y) + (-2x)^2y - 16x^2y + 2x(-4xy) - x(2y)$

$$= 3x^2(-y) + (4x^2)y - 16x^2y + 2x(-4xy) - x(2y) \qquad \text{by rule 2}$$

$$= -3x^2y + 4x^2y - 16x^2y - 8x^2y - 2xy \qquad \text{by rule 3}$$

$$= (-3 + 4 - 16 - 8)x^2y - 2xy \qquad \text{by rule 4}$$

$$= -23x^2y - 2xy$$

Simplify the following :

    (a) $2ab - 4ab + 16ab - ab$ $= 13ab$

    (b) $-x + 2x - 3xy + 3x + 4xy$ $= 4x + xy$

    (c) $x - z + 20x - 19z - 2z$ $= 19x - 22z$

    (d) $10s - 4s + 2s - 20s + st - 10st$ $= -12s - 9st$

    (e) $4x - 3xy + y - 3x - y + 12xy$ $= x + 9xy$

    (f) $2(3x^2) - 4x(2x) + 10x^3/2x$ $= 3x^2$

    (g) $10x(5y) - 20x^2y/5x - 13xy$ $= 33xy$

    (h) $2a(-b) + 10ab(2a) - 16b^2/2b$ $= -2ab + 20a^2b - 8b$

    (i) $12x^3y^2/xy - 2x^2(3y) - xy^2$ $= 61x^2y - xy^2$

    (j) $10xab(2yc^2) - 2abc(-5xyc) + 2ac^2(2xy) = 10abc^2xy + 4ac^2xy$

    (k) $(3x)^2 - 2(4x)^2 - 10x(-2x)^2/5x$ $= -31x^2$

    (l) $2a^2(-b)^2 - 20a^3(-2b)^2/(2a)^2$ $= 2a^2b^2 - 20ab^2$

    (m) $xy^2 - (-2y)^2 + 4x(2y)^2/2x$ $= 4y^2 + xy^2$

    (n) $(2xy)^2 - x^2(-2y)^2 - (2x)^2(-4y)^2$ $= -64x^2y^2$

(o) $(abc)^3 + (2a)^3(-b)^4(-c)^3/4b + (ab^2)^2/(-2ab)^2$
$$= -a^3b^3c^3 + b^2/4$$

A.6. REMOVING PARENTHESES FROM POLYNOMIAL EXPRESSIONS

Removing parentheses is the first step in simplifying algebraic expressions. (See Section A.5 for order of operations). Once parentheses are removed from any polynomial expressions in the problem, the problem becomes exactly like those discussed in Sections A.1–A.6.

REMOVING PARENTHESES

Example 1: $\quad 2(a - 2b) = 2a - 4b$

Example 2: $\quad -3x(-y + 3z) = 3xy - 9xz$

Example 3: $\quad 2a(x + 2)^2 = 2a(x^2 + 4x + 4) \qquad$ by rule 1(a)
$$= 2ax^2 + 8ax + 8a \qquad \text{by rule 1}$$

Example 4: $\quad -3x(x^2 - 2a(3b)) = -3x(x^2 - 6ab) \qquad$ by rule 1(b)
$$= -3x^3 + 18abx \qquad \text{by rule 1}$$

Example 5: $\quad -(x^2y + 2xy - 3x^2y) = -(2xy - 2x^2y) \qquad$ by rule 1(c)
$$= -1(2xy - 2x^2y)$$
$$= -2xy + 2x^2y \qquad \text{by rule 1}$$

Example 6: $\quad 2x(x^2y + (-3x)^2y^2) - 3x((y)^2(3x)^2 - 2x^2(2y)^2)$
$$= 2x(x^2y + (9x^2)y^2) - 3x(y^2(9x^2) - 2x^2(4y^2)) \qquad \text{by rule 1(a)}$$
$$= 2x(x^2y + 9x^2y^2) - 3x(9x^2y^2 - 8x^2y^2) \qquad \text{by rule 1(b)}$$
$$= 2x(x^2y + 9x^2y^2) - 3x(x^2y^2) \qquad \text{by rule 1(c)}$$
$$= 2x^3y + 18x^3y^2 - 3x^3y^2 \qquad \text{by rule 1}$$
$$= 2x^3y + 15x^3y^2 \qquad \text{by rule 4}$$

Simplify the following:

(a) $-(x - 2y)$ $\qquad\qquad = -x + 2y$
(b) $3(2x - 4z)$ $\qquad\qquad = 6x - 12z$
(c) $-2a(b - 3c)$ $\qquad\qquad = -2ab + 6ac$
(d) $2x(1 - 3z)$ $\qquad\qquad = 2x - 6xz$
(e) $-2(x + (2y)^2)$ $\qquad\qquad = -2x - 8y^2$
(f) $3a((2x)^3 - a^2)$ $\qquad\qquad = 24ax^3 - 3a^3$
(g) $-2b(ax^2 - (2ax)^2)$ $\qquad\qquad = -2abx^2 + 8a^2bx^2$
(h) $2x(-3y + (-y)^3)$ $\qquad\qquad = -6xy - 2xy^3$
(i) $-2(x + (2a)(-b)^3)$ $\qquad\qquad = -2x + 4ab^3$
(j) $3x((2a)^3(y) - 2y(-a)^3)$ $\qquad\qquad = 30a^3xy$
(k) $2ab((2b)^3(2y) - 3a(2ay))$ $\qquad\qquad = 32ab^4y - 12a^3by$
(l) $-3((2a)^2(3b)^2 - 2a^2)$ $\qquad\qquad = -108a^2b^2 + 6a^2$
(m) $2x(x^2(3y)^2 - (2x)^2(y)^2 - 2x)$ $\qquad\qquad = 10x^2y^2 - 4x^2$
(n) $-3x(x^4y^2 + (2x)^4(-y)^2 + (x^2y)^2)$ $\qquad\qquad = -54x^4y^2$
(o) $-3(2x(-y)^3 + (xy)(3y)^2 - 2x(-y)^2)$ $\qquad\qquad = -21xy^3 - 6xy^2$
(p) $-2a(b^3 + (2a)^2) - b)^3 - (2a)^2b^3)$ $\qquad\qquad = -2ab^3 - 12a^3b^3$

A.7. SIMPLIFYING ALGEBRAIC FRACTIONS

Example 1: $\dfrac{2ab}{b} = \dfrac{2a \cdot \cancel{b}}{\cancel{b}} = 2a$

Example 2: $\dfrac{3x^2y}{3xy} = \dfrac{\cancel{3xy} \cdot x}{\cancel{3xy}} = x$

Example 3: $\dfrac{x(2x+1)}{2x} = \dfrac{x \cdot (2x+1)}{x \cdot 2} = \dfrac{2x+1}{2}$

Example 4: $\dfrac{(a+b)(2x+3)}{2(a+b)} = \dfrac{\cancel{(a+b)}(2x+3)}{\cancel{(a+b)} \cdot 2(a+b)} = \dfrac{2x+3}{2(a+b)} = \dfrac{2x+3}{2a+2b}$

Example 5: $\dfrac{(2a+b)(3a^2+2b)(a+b^2)}{(2a+b)^2(a+b^2)} = \dfrac{(2a+\cancel{b})(a+\cancel{b^2})(3a^2+2b)}{(2\cancel{a}+b)(\cancel{a}+b^2)(2a+b)}$

$$= \dfrac{(3a^2+2b)}{(2a+b)}.$$

For more examples, see the last examples in Sections A.10–A.12.

(a) $2x^3y/2y$ $\qquad\qquad\qquad\qquad = x^2$

(b) $3xy/y$ $\qquad\qquad\qquad\qquad\quad\, = 3x$

(c) $2xz^2/xz$ $\qquad\qquad\qquad\qquad\; = 2z$

(d) $\dfrac{2x(3x-2)}{6x(2x-3)}$ $\qquad\qquad\qquad = \dfrac{3x-2}{6x-9}$

(e) $\dfrac{(a+b)(2x-3)}{2x(2x-3)^2}$ $\qquad\qquad = \dfrac{(a+b)}{2x(2x-3)}$

(f) $\dfrac{(2a+c)(a+3c)(a-b)^2}{(a+3c)^2(2a+c)^3}$ $\qquad = \dfrac{(a-b)^2}{(a+3c)(2a+c)^2}$

(g) $\dfrac{(x+y)(2x-y)^4(x+3y)}{2(x+y)^2(x+3y)^3}$ $\qquad = \dfrac{(2x-y)^4}{2(x+y)(x+3y)^2}$

(h) $\dfrac{(x-y)^3(x+y)^2(2x+1)}{(x+y)^3(2x+1)^2}$ $\qquad = \dfrac{(x-y)^3}{(x+y)^3}$

(i) $\dfrac{(2x+3)(x-3)^2(2x+y)^3}{(2x+y)^2(x-3)}$ $\qquad = (2x+3)(x-3)(2x+y)$

(j) $\dfrac{(3x+2)(2x-3)^3(3x+3y)^4}{(3x+2)^3(2x-3)^2}$ $\qquad = \dfrac{(2x-3)(3x+3y)^4}{(3x+2)^2}$

A.8. LOWEST COMMON DENOMINATOR OF ALGEBRAIC FRACTIONS

Example 1: $\dfrac{2}{x+5} + \dfrac{3}{x-2} \qquad$ L.C.D. $= (x-2)(x+5)$

$$= \dfrac{2}{x+5} \cdot \dfrac{(x-2)}{(x-2)} + \dfrac{3}{x-2} \cdot \dfrac{(x+5)}{(x+5)}$$

$$= \frac{2(x-2)}{(x+5)(x-2)} + \frac{3(x+5)}{(x-2)(x+5)}$$

$$= \frac{2(x-2)+3(x+5)}{(x+5)(x-2)} = \frac{2x-4+3x+15}{(x+5)(x-2)}$$

$$= \frac{5x+11}{(x+5)(x-2)}$$

Example 2: $\quad \dfrac{3}{(x+1)(x-2)} + \dfrac{2x}{(x-2)} \qquad$ L.C.D. $= (x+1)(x-2)$

$$= \frac{3}{(x+1)(x-2)} + \frac{2x}{(x-2)} \cdot \frac{(x+1)}{(x+1)}$$

$$= \frac{3}{(x+1)(x-2)} + \frac{2x(x+1)}{(x-2)(x+1)}$$

$$= \frac{3+2x(x+1)}{(x+1)(x-2)} = \frac{3+2x^2+2x}{(x+1)(x-2)} = \frac{2x^2+2x+3}{(x+1)(x-2)}$$

Example 3: $\quad \dfrac{3x}{2(x+1)} - \dfrac{2x-1}{(x-1)(x+2)} \quad$ L.C.D. $= 2(x+1)(x-1)(x+2)$

$$= \frac{3x}{2(x+1)} \cdot \frac{(x-1)(x+2)}{(x-1)(x+2)} - \frac{2x-1}{(x-1)(x+2)} \cdot \frac{2(x+1)}{2(x+1)}$$

$$= \frac{3x(x-1)(x+2)}{2(x+1)(x-1)(x+2)} - \frac{(2x-1)(2)(x+1)}{(x-1)(x+2)(2)(x+1)}$$

$$= \frac{3x(x-1)(x+2) - [(2x-1)(2)(x+1)]}{2(x+1)(x-1)(x+2)}$$

$$= \frac{3x(x^2+x-2) - [2(2x^2+x-1)]}{2(x+1)(x-1)(x+2)}$$

$$= \frac{3x^3+3x^2-6x - [4x^2+2x-2]}{2(x+1)(x-1)(x+2)}$$

$$= \frac{3x^3+3x^2-6x-4x^2-2x+2}{2(x+1)(x-1)(x+2)}$$

$$= \frac{3x^3-x^2-8x+2}{2(x+1)(x-1)(x+2)}$$

Put the following over a lowest common denominator and simplify.

(a) $\quad \dfrac{3}{(x-1)} - \dfrac{2}{(x+2)} \qquad\qquad = \dfrac{x+4}{(x-1)(x+2)}$

(b) $\quad \dfrac{2x}{(x-1)(2x+1)} + \dfrac{3x}{(x-1)} \qquad = \dfrac{6x^2+5x}{(x-1)(2x+1)}$

(c) $\dfrac{x-1}{(3x)(x+1)} - \dfrac{2x}{(x-2)}$ $= \dfrac{-6x^3 - 5x^2 - 3x + 2}{3x(x+1)(x-2)}$

(d) $\dfrac{3}{x^2+1} + \dfrac{2x}{x-1}$ $= \dfrac{2x^3 + 5x - 3}{(x^2+1)(x-1)}$

(e) $\dfrac{2x}{(x+1)(x-1)} - \dfrac{3}{2x+3}$ $= \dfrac{x^2 + 6x + 3}{(x+1)(x-1)(2x+3)}$

(f) $\dfrac{4}{2(x+4)} + \dfrac{3x}{x+4}$ $= \dfrac{4+6x}{2(x+4)}$

(g) $\dfrac{2x-3}{x-1} - \dfrac{3x+4}{2x+3}$ $= \dfrac{x^2 - x - 5}{(x-1)(2x+3)}$

(h) $\dfrac{4x-5}{x-6} + \dfrac{2x}{(x-6)(x+1)}$ $= \dfrac{4x^2 + x - 5}{(x-6)(x+1)}$

(i) $\dfrac{2x+1}{(x-1)^2} - \dfrac{3x}{x-1}$ $= \dfrac{-3x^2 + 5x + 1}{(x-1)^2}$

(j) $\dfrac{5x-7}{2} + \dfrac{3x}{(x-1)(x+1)}$ $= \dfrac{5x^3 - 7x^2 + x + 7}{2(x-1)(x+1)}$

A.9. MULTIPLICATION AND DIVISION OF POLYNOMIALS

The multiplication and division of polynomial expressions are similar to long multiplication and division of numbers, with monomials in the place of individual digits. The following two examples should suffice to remind you of the procedure.

$$
\begin{array}{l}
3x^2 - 5x + 2 \\
\underline{7x + 8} \\
21x^3 - 35x^2 + 14x \\
\underline{+ 24x^2 - 40x + 16} \\
21x^3 - 11x^2 - 26x + 16
\end{array}
\qquad \text{or} \qquad
\begin{array}{l}
3x^2 - 5x + 2 \\
\underline{7x + 8} \\
24x^2 - 40x + 16 \\
\underline{21x^3 - 35x^2 + 14x} \\
21x^3 - 11x^2 - 26x + 16
\end{array}
$$

$$
\begin{array}{r}
3x^2 + x + 5 \quad \text{Rem } 21 \\
2x-3\overline{)\, 6x^3 - 7x^2 + 5x + 9} \\
\underline{6x^3 - 9x^2} \\
2x^2 + 5x \\
\underline{2x^2 - 3x} \\
8x + 9 \\
\underline{8x - 12} \\
21
\end{array}
$$

A.10.   REMOVING MONOMIAL FACTORS

Example 1: $3ax + 2x = x(3a + 2)$

Example 2: $2abc + 4bc - 6c = 2c(ab + 2b - 3)$

Example 3: $6xyzw - 12xzw + 24wxzs = 6wxz(y - 2 + 4s)$

Example 4: $2xy + x - 10xyz = x(2y + 1 - 10yz)$

Example 5: $\dfrac{2x - 4x^2}{2(x^3 + 2x^2 - x)} = \dfrac{\cancel{2x}(1 - 4x^2)}{\cancel{2x}(x^2 + 2x - 1)} = \dfrac{1 - 4x^2}{x^2 + 2x - 1}$

(See Section 7.)

Remove the monomial factors from the following.

(a) $3xyz + 2xy - 5wxy$ $= xy(3z + 2 - 5w)$

(b) $10abc + 5adc + 15acd$ $= 5ac(2b + d + 3d)$

(c) $20wz - 10z + 30wxy$ $= 10z(2w - 1 + 3wy)$

(d) $5xw - 2x + 14xyz$ $= x(5w - 2 + 14yz)$

(e) $x^4 - 3x^3 + 2x^2 - x$ $= x(x^3 - 3x^2 + 2x - 1)$

(f) $15xy^2 + 5y - xy$ $= y(15xy - 5 - x)$

(g) $10abcd - 25abc + 30bcd$ $= 5bc(2ad - 5a + 6d)$

(h) $\dfrac{2x^3 - 3x^2}{2(x^4 - 2x^2)}$ $= \dfrac{2x - 3}{2(x^2 - 2)}$

(i) $\dfrac{3ab^2 + 6ab}{3(x^2 + 3)}$ $= \dfrac{ab(b + 2)}{x^2 + 3}$

(j) $\dfrac{2x^2y^2 - 16xy}{2x^3 - 8x^2}$ $= \dfrac{xy^2 - 8y}{x^2 - 4x} = \dfrac{y(xy - 8)}{x(x - 4)}$

A.11.   FACTORING GENERAL TRINOMIALS

Except for special cases (see Section A.12), this type of factoring is done primarily by inspection. The signs give some clue as to your general approach.

Example 1: $x^2 + 8x + 15 = (x + 3)(x + 5)$ $\Big\}$ Type 1 (see below)

Example 2: $2x^2 + 13x + 15 = (2x + 3)(x + 5)$

Example 3: $x^2 - 8x + 15 = (x - 3)(x - 5)$ $\Big\}$ Type 2 (see below)

Example 4: $2x^2 - 13x + 15 = (2x - 3)(2x - 5)$

Example 5: $x^2 + 2x - 15 = (x - 3)(x + 5)$ $\Big\}$ Type 3 (see below)

Example 6: $2x^2 + 7x - 15 = (2x - 3)(x + 5)$

Example 7: $x^2 - 2x - 15 = (x + 3)(x - 5)$ $\Big\}$ Type 4 (see below)

Example 8: $2x^2 - 7x - 15 = (2x + 3)(x - 5)$

Example 9: $\dfrac{x^2 + 10x + 25}{x^2 + 3x - 10} = \dfrac{(\cancel{x + 5})(x + 5)}{(\cancel{x + 5})(x - 2)} = \dfrac{x + 5}{x - 2}$

Example 10: $\dfrac{x^2 + 8x - 33}{x^2 + 5x - 24} = \dfrac{(\cancel{x - 3})(x + 11)}{(\cancel{x - 3})(x + 8)} = \dfrac{x + 11}{x + 8}$

$\left.\begin{array}{c} \\ \\ \\ \end{array}\right\}$ See Section A.7.

Type 1: $(\quad + \quad + \quad) = (\quad + \quad)(\quad + \quad)$
Type 2: $(\quad - \quad + \quad) = (\quad - \quad)(\quad - \quad)$
Type 3: $(\quad + \quad - \quad) = (\quad + \quad)(\quad - \quad)$ and the plus sign goes with the larger factor
Type 4: $(\quad - \quad - \quad) = (\quad + \quad)(\quad - \quad)$ and the minus sign goes with the larger factor

Factor the following problems.

(a) $x^2 + 9x + 20$ $= (x + 4)(x + 5)$
(b) $x^2 + 4x - 21$ $= (x + 7)(x - 3)$
(c) $x^2 - 15x + 56$ $= (x - 8)(x - 7)$
(d) $x^2 - 7x - 18$ $= (x - 9)(x + 2)$
(e) $x^2 - 9x - 10$ $= (x + 4)(x + 5)$
(f) $x^2 + 13x + 22$ $= (x + 11)(x + 2)$
(g) $x^2 + 7x - 8$ $= (x + 8)(x - 1)$
(h) $x^2 - 8x + 12$ $= (x - 6)(x - 2)$
(i) $x^2 + 2x + 1$ $= (x + 1)(x + 1)$
(j) $x^2 - 6x + 9$ $= (x - 3)(x - 3)$
(k) $x^2 + x - 2$ $= (x + 2)(x - 1)$
(l) $x^2 - 15x - 34$ $= (x - 17)(x + 2)$
(m) $2x^2 + 5x - 12$ $= (2x - 3)(x + 4)$
(n) $6x^2 - 5x - 4$ $= (3x - 4)(2x + 1)$
(o) $3x^2 - 2x + 25$ $= (x - 5)(3x - 5)$
(p) $6x^2 + 23x + 7$ $= (2x + 7)(3x + 1)$
(q) $5x^2 + 44x - 9$ $= (5x - 1)(x + 9)$
(r) $3x^2 - 7x - 6$ $= (3x + 2)(x - 3)$
(s) $6x^2 - 19x - 7$ $= (2x - 7)(3x + 1)$
(t) $2x^2 - 13x - 7$ $= (x - 7)(2x + 1)$

(u) $\dfrac{x^2 + 7x + 10}{x^2 + 10x + 25}$ $= \dfrac{(x + 2)}{(x + 5)}$

(v) $\dfrac{x^2 - 4x + 3}{2x^2 - x - 1}$ $= \dfrac{(x - 3)}{(2x + 1)}$

(w) $\dfrac{x^2 - 3x - 10}{x^2 + 6x + 8}$ $= \dfrac{(x - 5)}{(x + 4)}$

(x) $\dfrac{2x^2 + 7x + 3}{2x^2 - 3x - 2}$ $= \dfrac{(x + 3)}{(x - 2)}$

(y) $\dfrac{x^2 - 4x - 21}{2x^2 - 11x - 21}$ $= \dfrac{(x + 3)}{(2x + 2)}$

A.12. SPECIAL TYPES OF FACTORING

Perfect squares: $(x \pm y)^2 = x^2 \pm 2xy + y^2$
Example 1: $x^2 + 6x + 9 = (x + 3)(x + 3) = (x + 3)^2$
Example 2: $x^2 - 8x - 16 = (x - 4)(x - 4) = (x - 4)^2$

Example 3: $4x^2 + 4x + 1 = (2x+1)(2x+1) = (2x+1)^2$

Difference of two squares: $(x^2 - y^2) = (x-y)(x+y)$

Example 4: $x^2 - y^2 = (x-y)(x+y)$

Example 5: $a^2 - 4b^2 = (a-2b)(a+2b)$

Example 6: $9x^2 - 16y^2 = (3x-4y)(3x+4y)$

Sum or difference of two cubes: $(x^3 \pm y^3) = (x \pm y)(x^2 \mp xy + y^2)$

Example 7: $s^3 - t^3 = (s-t)(s^2 + st + t^2)$

Example 8: $s^3 + t^3 = (s+t)(s^2 - st + t^2)$

Example 9: $x^3 - 8 = (x-2)(x^2 + 2x + 4)$

Example 10: $8r^3 + 27t^3 = (2r+3t)(4r^2 - 6rt + 9t^2)$

See Section A.7 for further discussion of fractions:

Example 11: $\dfrac{x^3 - y^3}{x^2 - y^2} = \dfrac{(\cancel{x-y})(x^2 + xy + y^2)}{(\cancel{x-y})(x+y)} = \dfrac{x^2 + xy + y^2}{x+y}$

Example 12: $\dfrac{a^4 - b^4}{a^2 - 2ab + b^2} = \dfrac{(a^2 - b^2)(a^2 + b^2)}{(a-b)^2}$

$$= \dfrac{(\cancel{a-b})(a+b)(a^2 + b^2)}{(\cancel{a-b})(a-b)}$$

$$= \dfrac{(a+b)(a^2 + b^2)}{(a-b)}$$

Factor the following:

(a) $x^2 + 10x + 25 \quad = (x+5)^2$

(b) $x^2 - 20x + 100 \quad = (x-10)^2$

(c) $4x^2 - 12x + 9 \quad = (2x-3)^2$

(d) $x^2 + 22x + 121 \quad = (x+11)^2$

(e) $9x^2 - 6x + 1 \quad = (3x-1)^2$

(f) $p^2 - q^2 \quad = (p-q)(p+q)$

(g) $4r^2 - 1 \quad = (2r-1)(2r+1)$

(h) $x^2 - 16y^2 \quad = (x-4y)(x+4y)$

(i) $9m^2 - 25 \quad = (3m-5)(3m+5)$

(j) $4s^2 - 3r^2 \quad = (2s - \sqrt{3}r)(2s + \sqrt{3}r)$

(k) $m^3 - n^3 \quad = (m-n)(m^2 + mn + n^2)$

(l) $27x^3 - y^3 \quad = (3x-y)(9x^2 + 3xy + y^2)$

(m) $x^3 + 64y^3 \quad = (x+4y)(x^2 - 4xy + 16y^2)$

(n) $64x^3 - 8 \quad = (4x-2)(16x^2 + 8x + 4)$

(o) $10x^3 - 5y^3 \quad = (\sqrt[3]{10}x - \sqrt[3]{5}y)(\sqrt[3]{100}x + \sqrt[3]{15}xy + \sqrt[3]{25}y^2)$

(p) $\dfrac{x^3 - y^3}{x^4 - y^4} \quad = \dfrac{(x^2 + xy + y^2)}{(x+y)(x^2 + y^2)}$

(q) $\dfrac{x^2 - 10x + 25}{x^2 - x - 20} \quad = \dfrac{(x-5)}{(x+4)}$

(r) $\dfrac{x^2 + 12x + 36}{x^2 - 36} \quad = \dfrac{(x+6)}{(x-6)}$

$$\text{(s)} \quad \frac{27x^3 - y^3}{3x^2 - 2xy - y^2} = \frac{9x^2 + 3xy + y^2}{x + y}$$

$$\text{(t)} \quad \frac{125a^3 + 8}{5a^2 - 28a - 12} = \frac{25a^2 - 10a + 4}{a - 6}$$

A.13. SOLVING LINEAR EQUATIONS

SOLVING LINEAR EQUATIONS

Example 1: $\quad 2x = 10$

$$\frac{2x}{2} = \frac{10}{2}$$

$$x = 5$$

Example 2: $\quad \dfrac{x}{2} = 10$

$$2 \cdot \frac{x}{2} = 2 \cdot 10$$

$$x = 20$$

Example 3: $\quad x - 4 = 8$

$$x - 4 + 4 = 8 + 4$$

$$x + 0 = 12$$

$$x = 12$$

Example 4: $\quad 3x - 6 = 2x$

$$3x - 6 - 2x = 2x - 2x$$

$$3x - 2x - 6 = 0$$

$$x - 6 + 6 = 0 + 6$$

$$x + 0 = 6$$

$$x = 6$$

Example 5: $\quad 3x + 7 = 5x - 9$

$$3x + 7 - 5x = 5x - 9 - 5x$$

$$3x - 5x + 7 = 5x - 5x - 9$$

$$-2x + 7 = 0 - 9$$

$$-2x + 7 - 7 = -9 - 7$$

$$-2x + 0 = -16$$

$$-\tfrac{1}{2} \cdot -2x = -\tfrac{1}{2} \cdot -16$$

$$x = 8$$

Solve the following equations:

(a) $3x = 15$ $\qquad\qquad\qquad\qquad\qquad x = 5$

(b) $10x = 15$ $\qquad\qquad\qquad\qquad\quad x = \frac{3}{2}$

(c) $x/5 = 10$ $\qquad\qquad\qquad\qquad\quad x = 50$

(d) $2x/3 = 20$ $\qquad\qquad\qquad\qquad\; x = 30$

(e) $2x + 4 = 8$ $\qquad\qquad\qquad\qquad\; x = 2$

(f) $6x - 2 = 22$ $\qquad\qquad\qquad\qquad x = 4$

(g) $30 = 4x + 2$ $\qquad\qquad\qquad\qquad x = 7$

(h) $2x/5 - 10 = -20$ $x = -25$

(i) $3x + 4 = 2x$ $x = -4$

(j) $5x - 7 = -2x$ $x = 1$

(k) $8 - 2x = 14x$ $x = \frac{1}{2}$

(l) $3x/4 - 10 = x/2$ $x = 40$

(m) $2x - 8 = 10 + 6x$ $x = -\frac{9}{2}$

(n) $20 - 5x = 4x + 10$ $x = \frac{10}{9}$

(o) $2x/3 + 5 = 1 + x/6$ $x = -8$

(p) $\frac{2}{3} - x = 10x + \frac{5}{3}$ $x = -\frac{1}{11}$

(q) $2x + 10 - 3x = 20 + 10x$ $x = -\frac{10}{11}$

(r) $4x - 11 + x/4 = 5x - 10 + 3x/4$ $x = -\frac{2}{3}$

(s) $15 - 5x + 20 = 10x + 13$ $x = \frac{22}{15}$

(t) $2x/3 - 14 + 5x/3 = x/3 + 10 - 7x/3$ $x = \frac{72}{13}$

A.14. INEQUALITIES

Definitions:

$<$ means "less than," e.g., $3 < 4$, $-5 < -4$.

$\leq$ means "less than or equal to," e.g., $-3 \leq -2$, $2 \leq 2$.

$>$ means "greater than," e.g., $5 > 3$, $0 > -3$.

$\geq$ means "greater than or equal to," e.g., $-1 \geq -10$, $10 \geq 10$.

Adding or subtracting involving inequalities is the same as adding or subtracting involving equalities ($=$ signs).

Example 1: $\quad 3 - 4 < 10$

$\qquad -1 < 10$

Example 2: $\quad x + 5 > 20$

$\qquad x + 5 - 5 > 20 - 5$

$\qquad x + 0 > 15$

$\qquad x > 15$

Multiplication and division involving inequalities is the same as multiplication and division involving equalities when the signs are positive; however when multiplying or dividing an inequality by a negative number, the direction of the inequality is reversed.

Example 3: $\quad 2x < 10$

$\qquad 2x/2 < 10/2$

$\qquad x < 5$

Example 4: $\quad -2x < 10$

$\qquad -2x/-2 \quad 10/-2$

$\qquad x > -5$

Example 5: $\quad 3x - 10 \geq 5x$

$\qquad 3x - 10 - 5x \geq 5x - 5x$

$\qquad -2x - 10 \geq 0$

$\qquad -2x - 10 + 10 \geq 0 + 10$

$\qquad -2x + 0 \geq 10 \qquad$ (*continued*)

$$-2x \geq 10$$
$$-\tfrac{1}{2} \cdot -2x \leq 10 \cdot -\tfrac{1}{2}$$
$$x \leq -5$$

Solve the following inequalities.

(a)	$5x < 15$	$x < 3$
(b)	$-2x \leq 10$	$x \geq 5$
(c)	$x/5 > 20$	$x > 100$
(d)	$-x/6 \geq 3$	$x \leq 18$
(e)	$5x - 4 < 10$	$x < \frac{14}{5}$
(f)	$-4x + 3 \geq -1$	$x < 1$
(g)	$3x + 5 < 2x + 20$	$x > 15$
(h)	$4x - 5 \leq -7x + 6$	$x \leq 1$
(i)	$-2x/3 < 10$	$x > 15$
(j)	$-5x/4 \geq 1$	$x \leq -\frac{4}{5}$
(k)	$4x/5 > 10$	$x > \frac{25}{2}$
(l)	$-3x + 10 < -6x$	$x < -\frac{10}{3}$
(m)	$-10x + 15 \geq 4x + 10$	$x \leq \frac{5}{14}$
(n)	$-2x/3 + 5 < -5x/3 - 1$	$x < -6$

A.15. SYSTEMS OF LINEAR EQUATIONS

Example 1:

$$2x + 3y = 8 \qquad (1)$$
$$4x - 3y = -2 \qquad (2)$$

$$2x + 3y = 8 \qquad (1)$$
$$\underline{4x - 3y = -2} \qquad (2)$$
$$6x + 0 = 6 \qquad \text{by adding the equations}$$
$$6x = 6$$
$$x = 1$$

$$2x + 3y = 8 \qquad (1)$$
$$2(1) + 3y = 8$$
$$3y = 6$$
$$y = 2$$

Thus $x = 1$, $y = 2$.

Example 2:

$$2a - 3b = -8 \qquad (1)$$
$$4a + 5b = 28 \qquad (2)$$

$$-2(2a) - (-2)(3b) = (-2)(-8) \qquad (1)$$
$$4a \quad + \quad 5b \quad = 28 \qquad (2)$$

$$-4a + 6b = 16 \qquad (1)$$
$$\underline{4a + 5b = 28} \qquad (2)$$
$$11b = 44$$
$$b = 4$$

$$2a - 3(4) = -8$$
$$2a - 12 = -8$$
$$2a = 4$$
$$a = 2, \quad b = 4$$

Example 3: $t - 2w = 9$ (1)
$2t - 3w = 16$ (2)

$t = 2w + 9$ Substitute this version of (1) into (2)
$2(2w + 9) - 3w = 16$
$4w + 18 - 3w = 16$
$w + 18 = 16$
$w = -2$ Substitute this into (1).
$t - 2(-2) = 9$
$t + 4 = 9$
$t = 5, \quad w = -2$

Example 4: $x + y - z = 2$ (1)
$2x - y + 3z = 11$ (2)
$x + 2y + z = 7$ (3)

$\begin{array}{ll} x + y - z = 2 & (1) \\ 2x - y + 3z = 11 & (2) \\ \hline 3x + 2z = 13 & (a) \end{array}$ $\begin{array}{ll} x + y - z = 2 & (1) \\ x + 2y + z = 7 & (3) \end{array}$

$\begin{array}{ll} -2x - 2y + 2z = -4 & (1) \\ x + 2y + z = 7 & (3) \\ \hline -x + 3z = 3 & (b) \end{array}$

$\begin{array}{ll} 3x + 2z = 13 & (a) \\ -x + 3z = 3 & (b) \end{array}$

$\begin{array}{ll} 3x + 2z = 13 & (a) \\ -3x + 9z = 9 & (b) \\ \hline 11z = 22 \\ z = 2 \end{array}$

$3x + 2(2) = 13$ (a)
$3x = 9$
$x = 3$
$x + y - z = 2$ (1)
$3 + y - 2 = 2$

Substituting values for x and z, one obtains

$$y = 1$$

Thus $x = 3, y = 1, z = 2$.

Solve the following:

(a) $x + 2y = -5$
$3x - 2y = +1$ $x = -1, \quad y = -2$
(b) $2x - y = 3$
$x + 3y = -2$ $x = 1, \quad y = -1$
(c) $3x - 2y = 12$
$2x + 3y = -5$ $x = 2, \quad y = -3$
(d) $x + 4y = -3$
$2x + 3y = 4$ $x = 5, \quad y = -2$

(e) $2x + 3y = -1$
$4x + 5y = 1$ $\qquad$ $x = 4, \quad y = -3$

(f) $x - 2y = -9$
$5x + 3y = -6$ $\qquad$ $x = -3, \quad y = 3$

(g) $x - 2y + z = 5$
$x - 2z = 0$ $\qquad$ $x = 2, \quad y = -1, \quad z = 1$
$3x - 2y - z = 7$

(h) $2x - y + z = -8$
$x + 3y - 5z = 12$ $\qquad$ $x = -2, \quad y = 3, \quad z = -1$
$2x + 2y + 3z = -1$

A.16. ALGEBRA OF EXPONENTS

ALGEBRA OF
EXPONENTS

Fact 1: $x^a \cdot x^b = x^{a+b}$
$x^2 \cdot x^3 = x^5$
$x^{-2} \cdot x^{-3} = x^{-5}$
$x^{1/2} \cdot x^{3/2} = x^{4/2} = x^2$

Fact 2: $x^a/x^b = x^{a-b}$
$x^5/x^2 = x^3$
$b^{-3}/b^{-2} = b^{-3-(-2)} = b^{-1}$
$y^{1/4} \cdot y^{3/4} = y^{(1/4)-(3/4)} = y^{-1/2}$

Fact 3: $(xy)^a = x^a y^a$
$(x \cdot y)^7 = x^7 y^7$
$(2a)^4 = 2^4 a^4 = 16a^4$
$(3mn)^2 = (3m)^2 n^2 = 3^2 m^2 n^2 = 9m^2 n^2$

Fact 4: $(x^a)^b = x^{ab}$
$(x^2)^3 = x^6$
$(x^{-2})^{-3} = x^6$
$(s^{1/2})^{1/3} = s^{1/6}$

Fact 5: $x^{a/b}$ means $\sqrt[b]{x^a}$
$x^{2/3} = \sqrt[3]{x^2}$
$a^{1/2} = \sqrt{a}$

Fact 6: $x^{-a} = 1/x^a$
$x^{-5} = 1/x^5$
$y^{-2/3} = 1/y^{2/3} = \dfrac{1}{\sqrt[3]{y^2}}$

Summary Examples:

$$\frac{x^3 y^4 z^{-6}}{x^2 y^{-3} z^3} = x^{3-2} \cdot y^{4-(-3)} \cdot z^{(-6)-3} = xy^7 z^{-9} = xy^7/z^9$$

$$\frac{x^{2/3} y^4 z^{4/3}}{x^2 y^{4/3}} = x^{(2/3)-2} y^{4-(4/3)} z^{4/3} = x^{-4/3} y^{8/3} z^{4/3} = (x^{-4} y^8 z^4)^{1/3}$$

$$= (x^{-1} y^2 z^1)^{4(1/3)} = (y^2 z/x)^{4/3}$$

Simplify the following:

(a) $x^2 x^4$ $\qquad$ $= x^6$
(b) $y^3 y^{-5}$ $\qquad$ $= y^{-2}$

(c) $z^{-2}z^{-4}$ $\qquad = z^{-6}$

(d) $s^{1/2}s^3$ $\qquad = s^{3/2}$

(e) x^3/x^2 $\qquad = x$

(f) y^5/y^{-3} $\qquad = y^8$

(g) $3^{-2}/3^{-3}$ $\qquad = 3$

(h) $x^{1/2}/x^{-1/4}$ $\qquad = x^{3/4}$

(i) y^8/y^{-5} $\qquad = y^{13}$

(j) $(x^2)^3$ $\qquad = x^6$

(k) $(y^{-2})^{-5}$ $\qquad = y^{10}$

(l) $(z^{-3})^2$ $\qquad = z^{-6}$

(m) $(2^2)^{-2} = 2^{-4}$ $\qquad = \frac{1}{16}$

(n) $(x^{1/2})^3$ $\qquad = x^{3/2}$

(o) $(z^{1/2})^{1/4}$ $\qquad = z^{1/8}$

(p) $\dfrac{x^2x^4}{x^{-4}}$ $\qquad = x^{10}$

(q) $y^3x^2/y^{1/2}$ $\qquad = x^2y^{5/2}$

(r) $x^2y^3z^{-1}/x^{-3}y$ $\qquad = x^5y^2/z$

(s) $x^2y^3z^{-5}/y^{-3}z^{-1}$ $\qquad = x^2y^6/z^4$

(t) $(x^2y^{1/2}/y^3)^2$ $\qquad = x^4/y^5$

(u) $a^2b^{-3}c^{-3/2}/b^{-1}c^{-1/2}$ $\qquad = a^2/b^2c$

(v) $(a^2)^{-3}b^{1/2}/(b^{-1})^2$ $\qquad = b^{5/2}/a^6$

(w) $(z^5y^3x/a^{-5}b^{1/2})^3$ $\qquad = z^{15}y^9x^3a^{15}/b^{3/2}$

(x) $(x^3y^{-4}z^{-1/2})^4$ $\qquad = x^{12}/y^{16}z^2$

(y) $(a^3b^{-2}c^{-1/4}d^{-2/3})^6$ $\qquad = a^{18}/b^{12}c^{3/2}d^2$

A.17. BINOMIAL THEOREM

$$(a+b)^n = a^n + \frac{n}{1}a^{n-1}b + \frac{n(n-1)}{2\cdot 1}a^{n-2}b^2$$

$$+ \frac{n(n-1)(n-2)}{3\cdot 2\cdot 1}a^{n-3}b^3 + \cdots$$

$$+ \frac{n(n-1)(n-2)\cdots(n-(k+1))}{k\cdot k-1\cdot\cdots\cdot 3\cdot 2\cdot 1}a^{n-k}b^k + \cdots + b^n$$

$$= a^n + \binom{n}{1}a^{n-1}b + \binom{n}{2}a^{n-2}b^2 + \binom{n}{3}a^{n-3}b^3 + \cdots$$

$$+ \binom{n}{k}a^{n-k}b^k + \cdots + b^n$$

$$= \sum_{j=0}^{n}\binom{n}{j}a^{n-j}b^j$$

Some other common notations for the binomial coefficients $\binom{n}{j}$ are C^n_j, $_nC_j$ and $C(n,j)$. The coefficient $\binom{n}{j}$ can be computed by the formula

$$\binom{n}{j} = \frac{n^1}{j!\,(n-j)!}\quad \left[\binom{7}{3} = \frac{7!}{3!\,4!}\right].$$

One can also get their values from Pascal's triangle:

$$
\begin{array}{c}
1 \\
1\ 1 \\
1\ 2\ 1 \\
1\ 3\ 3\ 1 \\
1\ 4\ 6\ 4\ 1 \\
1\ 5\ 10\ 10\ 5\ 1 \\
1\ 6\ 15\ 20\ 15\ 6\ 1 \\
\vdots
\end{array}
$$

A.18.   COMPLETING THE SQUARE

COMPLETING THE SQUARE

The idea behind completing the square is to manipulate the constants in a quadratic expression to find, within it, a perfect trinomial square of the form $(ax + b)^2$.

Example 1: $x^2 + 4x + 1 = 0$
$$
\begin{aligned}
x^2 + 4x + 1 &= x^2 + 4x + 4 - 4 + 1 \\
&= x^2 + 4x + 4 - 3 \\
&= (x + 2)^2 - 3
\end{aligned}
$$

or

$$(x + 2)^2 = 3$$

and at this point we could easily solve this by taking the square root of each side.

Example 2: $x^2 - 5x + 2 = x^2 - 5x + \left(\dfrac{5}{2}\right)^2 - \left(\dfrac{5}{2}\right)^2 + 2$
$$
\begin{aligned}
&= x^2 - 5x + \tfrac{25}{2} + \tfrac{33}{4} \\
&= (x - \tfrac{5}{2})^2 + \tfrac{33}{4} \quad \text{or} \quad (x - \tfrac{5}{2})^2 = -\tfrac{33}{4}
\end{aligned}
$$

Example 3: $x^2 + y^2 + 2x - 3y + 10$
$$
\begin{aligned}
&= x^2 + 2x + \underline{\quad} + y^2 - 3y + \underline{\quad} + 10 - \underline{\quad} - \underline{\quad} \\
&= x^2 + 2x + 1 + y^2 - 3y + (\tfrac{3}{2})^2 + 10 - 1 - \tfrac{9}{4} \\
&= (x + 1)^2 + (y - \tfrac{3}{2})^2 + \tfrac{27}{4}
\end{aligned}
$$

(a) $x^2 + 8x + 1 = 0$	$(x + 4)^2 = 15$
(b) $x^2 - 8x + 1 = 0$	$(x - 4)^2 = 15$
(c) $x^2 + 8x + 20 = 0$	$(x + 4)^2 = -4$
(d) $y^2 - 5y - 5 = 0$	$(y - \tfrac{5}{2})^2 = \tfrac{45}{4}$
(e) $z^2 + 16z - 2 = 0$	$(z + 4)^2 = 18$
(f) $t^2 - t + 1 = 0$	$(t - \tfrac{1}{2})^2 = -\tfrac{3}{4}$
(g) $a^2 + a + 1 = 0$	$(a + \tfrac{1}{2})^2 = -\tfrac{3}{4}$
(h) $7 - 3x + x^2 = 0$	$(x - \tfrac{3}{2})^2 = -\tfrac{19}{4}$
(i) $16 + 2x + x^2 = 0$	$(x - 1)^2 = -15$
(j) $21 - 6x - x^2 = 0$	$(x + 3)^2 = 30$

A.19.   DIVISION BY ZERO

DIVISION BY ZERO

A division problem such as $\frac{21}{7} = 3$ is equivalent to the multiplication problem $21 = 3 \cdot 7$. In fact, division can be approached through equations using multiplication. Thus the quotient a/b is defined to be the solution to the equation $bx = a$.

DIVISION BY ZERO

Suppose now that one wishes to assign a value to $a/0$. He would then need to find a solution to the equation $0 \cdot x = a$. But $0 \cdot x = 0$ for all real x, so that there is no solution if $a \neq 0$. In other words, $a/0$ is not defined if $a \neq 0$.

If $a = 0$, then the relevant equation is $0 \cdot x = 0$, which is satisfied by *all* real numbers, and there is no way to decide how to pick one. So $0/0$ is indeterminate, or if you prefer, is also undefined, although for a rather different reason.

A.20. DISTANCE FORMULA

The distance between points $P(x_1, y_1)$ and $Q(x_2, y_2)$ in the coordinate plane is given by the formula

DISTANCE FORMULA

$$d = \sqrt{(x_2 - x_1)^2 + (y_2 - y_1)^2}$$

Example 1: $P(1, 5), Q(4, 8)$.

$$d = \sqrt{(4-1)^2 + (8-4)^2}$$
$$= \sqrt{9 + 16} = \sqrt{25} = 5$$

Example 2: $P(-2, 6), Q(1, 3)$.

$$d = \sqrt{(1-(-2))^2 + (3-6)^2}$$
$$= \sqrt{9 + 9} = \sqrt{18} = 3\sqrt{2}$$

Example 3: $P(0, 7), Q(-5, 3)$.

$$d = \sqrt{(-5-0)^2 + (3-7)^2}$$
$$= \sqrt{25 + 10} = \sqrt{41}$$

A derivation may be found in Section 6.2a.

A.21. INTERPOLATION

The basis of interpolation is the assumption that between any two tabulated values of a function, the behavior of the function is essentially that of a straight line. While not true, this gives reasonably accurate approximate values. The following examples for hypothetical functions demonstrate the process "in both directions" from nontabulated argument to corresponding value, and from nontabulated functional value to corresponding argument.

INTERPOLATION

Example 1: Find $h(5.23)$ if a table provides the values

$$h(5.2) = 6.135$$
$$h(5.3) = 7.072$$

(a) Set up a "mini-table"

		x	$h(x)$	
		5.2	6.135	
0.1	0.03	5.23	?	y
		5.3	7.072	0.937

(b) Form and solve a proportion:

$$0.03/0.1 = y/0.937$$
$$y = 0.281$$

A.21. *INTERPOLATION* 357

(c) Determine the desired value:

$$? \approx 6.135 + y \approx 6.135 + 0.281 \approx 6.416$$

Example 2: Find $k(28.6)$ if a table provides the values

$$k(28) = 891.2$$
$$k(29) = 803.7$$

(a) Set up a "mini-table"

x	$k(x)$
28	891.2
28.6	?
29	803.7

$1 \left\{ 0.6 \begin{cases} 28 \\ 28.6 \\ 29 \end{cases} \quad \begin{matrix} 891.2 \\ ? \\ 803.7 \end{matrix} \right\} z \right\} 87.5$

(b) Form and solve a proportion:

$$\frac{0.6}{1} = \frac{z}{875}$$

$$z = 52.50$$

(c) Determine the desired value:

$$? \approx 891.2 - z \approx 891.2 - 52.5 = 868.7$$

(Note: We subtracted here, since the tabulated values were decreasing.)

Example 3: Find x if zorp $x = 0.2190$ and a table informs us that

$$\text{zorp } 3.62 = 0.2148$$
$$\text{zorp } 3.63 = 0.2209$$

(a) Set up a "mini-table"

x	zorp x
3.62	0.2148
?	0.2190
3.63	0.2209

$0.01 \left\{ t \begin{cases} 3.62 \\ ? \\ 3.63 \end{cases} \quad \begin{matrix} 0.2148 \\ 0.2190 \\ 0.2209 \end{matrix} \right\} 0.0042 \right\} 0.0061$

(b) Set up and solve the proportion:

$$\frac{t}{0.01} = \frac{0.0042}{0.0061} \quad (= \tfrac{42}{61} \text{ if that helps you.})$$

$$t = 0.007$$

(c) Determine the desired value:

$$? \approx 3.62 + t \approx 3.62 + 0.007 = 3.627$$

A.22. SOME COMMON AREA-VOLUME FORMULAS

Unless otherwise noted, the letters mean:

a: altitude or height

$b:$ base or length
$r:$ radius of appropriate circle or sphere

(a) Plane area:

triangle: $\frac{1}{2}ba$

$\frac{1}{2}xy \sin z$:   sides x and y include angle z

$\sqrt{s(s-x)(s-y)(s-z)}$:   sides x, y, z, $s = \frac{1}{2}(x+y+z)$

rectangle: ba
parallelogram: ba
trapezoid: $\frac{1}{2}(b_1+b_2)a$:   bases b_1 and b_2
circle: πr^2
sector of a circle,

$$\frac{\pi r^2 \theta}{360} \quad \text{if } \theta \text{ is in degrees.}$$

$$\frac{r\theta}{2} \quad \text{if } \theta \text{ is in radians.}$$

(b) Volume:

sphere: $\frac{4}{3}\pi r^3$
cylinder: $\pi r^2 a$ if circular base
             Aa if some other base with area A.
cone: $\frac{1}{3}\pi r^2 a$
rectangular solid: $a \cdot w \cdot b$ ($w = $ width)

(c) Surface area:

sphere: $4\pi r^2$

A.23.   FIELD PROPERTIES OF THE REAL NUMBERS

$\mathbb{R}$ is *closed* under addition and multiplication (i.e., sums and products of real numbers are real numbers, too).

Addition and multiplication are *associative* (i.e., $a + (b+c) = (a+b) + c$; $a \cdot (b \cdot c) = (a \cdot b) \cdot c$ for all $a, b, c \in \mathbb{R}$.

Addition and multiplication are *commutative* (i.e., $a + b = b + a$; $b \cdot a = a \cdot b$ for all $a, b \in \mathbb{R}$).

The number 0 is the one and only *additive identity* ($0 + a = a$ for all $a \in \mathbb{R}$)

The number 1 is the one and only *multiplicative identity* ($1 \cdot a = a$ for all $a \in \mathbb{R}$).

For each real number a, there is a unique real number $(-a)$, which is its *additive inverse* ($a + (-a) = 0$ for all $a \in \mathbb{R}$).

For each real number a, except 0, there is a unique real number $(1/a)$, which is its *multiplicative inverse* ($a(1/a) = 1$ for all $a \in \mathbb{R}$).

Multiplication *distributes* over addition ($a(b+c) = ab + ac$ for all a, b, and $c \in \mathbb{R}$).

SUMMARY

APPENDIX B

BASIC AND FORTRAN

B.1.   INTRODUCTION

The following summaries of Basic and Fortran are intended to be a guide and reference for the learning of either or both of these common mathematically and scientifically oriented computer languages. They should be used in conjunction with Section 3 of the preliminary chapter of the text itself and will need to be supplemented by the details of procedure required by your computer center. We suggest particularly that an individual who wants to learn either of these languages start by skimming the description of the language, next study the examples of programs provided, and then try to code some of the flowcharts of Section 3,† referring to the language descriptions provided in this appendix as need arises. Once you have successfully run a few such programs, you are ready to head out on your own.

One word of warning—our descriptions of these languages are not complete—their intent is not to produce expert programmers but persons who can do enough programming to be able to write such programs as are suggested in the text. The student whose interest is sparked should refer to a more complete manual of the language.

---

† This includes both flowcharts written in the text and those assigned for you to write. Those problems in the preliminary chapter that are appropriate for such use are marked by an asterisk.

B.2. BASIC

Introductory Example

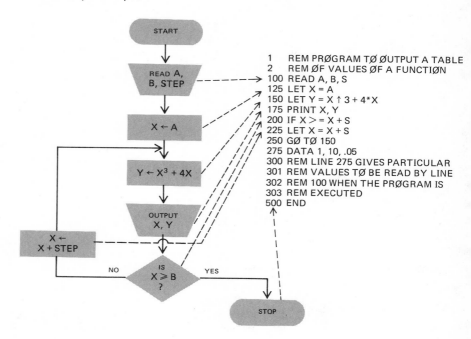

```
1    REM PRØGRAM TØ ØUTPUT A TABLE
2    REM ØF VALUES ØF A FUNCTIØN
100  READ A, B, S
125  LET X = A
150  LET Y = X ↑ 3 + 4*X
175  PRINT X, Y
200  IF X > = X + S
225  LET X = X + S
250  GØ TØ 150
275  DATA 1, 10, .05
300  REM LINE 275 GIVES PARTICULAR
301  REM VALUES TØ BE READ BY LINE
302  REM 100 WHEN THE PRØGRAM IS
303  REM EXECUTED
500  END
```

B.2a. Preliminaries

CHARACTERS

I. *Characters:* Basic uses the 26 letters of the alphabet (capitals only, no lowercase), the digits 0 through 9, and special symbols that include +, *, −, /, ↑, (,), ., ,, , ", =, >, and <.

Special care is needed not to confuse the letter **I** with the number one, or the letter **O** with the number zero. To minimize these confusions, **I** is usually written with bars at top and bottom while one is left plain **(1)**, and the letter **O** has a line through it **(Ø)** to distinguish it from zero.

VARIABLES

II. *Variables:* Basic variable names consist of either a letter or a letter followed by a single digit number. Thus **Q, M1, A3,** and **L** are legitimate variable names, while **IT, 2K,** and **W52** are *not* legitimate names.†

CONSTANTS

III. *Constants:* Constants must be in decimal form although a decimal point is needed only if some part of the number extends to the right of the decimal point. Commas must not be used (that is 100,000 must be written **100000**).

ARITHMETIC OPERATIONS AND EXPRESSIONS

IV. *Arithmetic Operations and Expressions:*

Addition is denoted by +.
Subtraction is denoted by −.

---

† For those familiar with Fortran, one does not have the mode problems that exist in Fortran—all numbers are real.

Multiplication is denoted by ∗ (*not* by a × or a dot or by writing variables next to each other as is done in algebra).

Division is denoted by /.

Negation is denoted by − (i.e., −t would be coded as −T and −3t would be coded as −3∗T).

Exponentiation is denoted by ↑ (i.e., r⁵ would be coded as R↑5).

ARITHMETIC OPERATIONS AND EXPRESSIONS

Arithmetic expressions may be constructed from variables and constants (using the above operation codes) just as one does in algebra. Parentheses should be used for grouping as in algebra, and need to be used more liberally, since expressions are written on one line. For example, $\frac{a+b}{c}$ must be coded (A + B)/C.† Parentheses should also be used around a negated expression if it occurs to the immediate right of an operation symbol. For example, $a \cdot (-b)$ should be coded as A∗(−B) not as A∗ − B.

B.2b. General Statement Form and Line Numbers

GENERAL STATEMENT FORM AND LINE NUMBERS

Basic programs consist of a sequence of Basic instructions. Each new instruction starts a new line, and each line begins with a line number consisting of from one to five digits and no decimal point. The line number serves two functions: (1) it provides a means of identifying the statement and (2) it indicates where in the program the statement belongs. When using an interactive terminal, lines may be typed in any order, but before being executed the computer reorders them according to line numbers. No two different statements may have the same line number. If more than one line is typed with the same line number, only the last one is used. This provides a convenient means for correcting a typing error—just retype the line correctly with the same line number.

Line numbers, while sequential, do not need to be consecutive. In fact, it is wise to leave several numbers between successive line numbers to permit the later addition of intermediate instructions should one wish to revise the program.

In giving the general form of Basic statements in the following paragraphs, we shall use *l.n.* to denote the need for a line number with the statement.

B.2c. I/Ø (Input/Output)

INPUT

I. *Input:* There are two Basic input instructions. One is

l.n. **READ** *list*

where *list* stands for a list of Basic variables (separated by commas) for

---

† When parentheses are absent, exponentiations and negations are done first, then multiplications and divisions (from left to right), and then additions and subtractions (from left to right). Hence A + B/C would be read as a + b/c and A/B·C would be read as (a/b) · c. (See order of operations, Appendix A, Section A.5)

which values are to be read. The values are then read from a **DATA** statement (See Section B.2f, II).

> *Examples:* 10 READ X,Y,Z
> 157 READ T1.P

The other instruction is

> *l.n.* INPUT *list*

where *list* again stands for a list of the variables for which values are to be read. This instruction causes values to be read from the keyboard of the computer terminal. In order to know when to type such data and which variables are to be read, the **INPUT** instruction should always be *preceded* by

> *l.n.* PRINT "WHAT VALUES DØ YØU WANT FØR *list*"

where *list* is identical to the list for the **INPUT** statement. If the list were A,X,Q, the statement causes output of

> WHAT VALUES DØ YØU WANT FØR A,X,Q?

on the terminal just prior to the computer's need for these numbers.

Examples:

> 13 PRINT "WHAT VALUES DØ YØU WANT FØR X,Y, AND Z?"
> 14 INPUT X,Y,Z
> 152 PRINT "PLEASE PRØVIDE, IN ØRDER, VALUES FØR L AND W".
> 153 INPUT L.W.

II. *Output:* To output information, one uses a **PRINT** statement that has the form

> *l.n.* PRINT *list*

where *list* now may contain any of the following items:

Variable names.
Computations whose results should be outputted.
Verbatim messages contained in quotation marks.

The individual items may be separated by commas, or, if one wishes more efficient use of paper (more numbers per line) by semicolons. For example, if **L**, **W**, and **H** have been assigned the values 2, 3, and 5 earlier in the program,

> 220 PRINT "L,W,H, AND VØLUME ARE"; L; W; H; L*W*H

will result in the following line of output:

> L,W,H AND VØLUME ARE 2 3 5 30

One can skip a line of output by using the statement

> PRINT

without any list.

B.2d. Non-I/Ø Executable Statements

I. *Assignment:* The Basic statement

> *l.n.* LET $v = expression$

corresponds to the flowchart assignment instruction V ← EXPRESSIØN in Basic, where *v* represents any variable names and *expression* stands for any valid arithmetic expression.†

ASSIGNMENT

Examples:

> 75 LET X = X + 1   means add one to X and put the result back in location X.
>
> 130 LET Y = 3*X↑2   means square X, then multiply the result by 3, and put the result in location Y.

II. GØ TØ: The Basic instruction

GØ TØ

$$l.n. \text{ GØ TØ } line\ number$$

where *line number* stands for any actual line number causes the computer to interrupt its normal movement to the statement with the next higher statement number and go instead to the statement with the specified line number. This kind of instruction is needed to accomplish loops (see also paragraph B.2d, III below).

Examples:

> 60 LET N = 0
> 65 READ X
> 70 LET X = X↑2
> 75 PRINT X
> 80 LET N = N + 1
> 85 IF N > = 10 THEN 95
> 90 GØ TØ 65
> 95 − − − −

The GØ TØ statement (line number 90) takes us back to read another number (X). (The IF-THEN in line 85 refers to the next paragraph)

III. IF-THEN: The fundamental decision instruction in Basic has the following form:

IF-THEN

$$l.n. \text{ IF } expression_1\ r\ expression_2 \text{ THEN } line\ number.$$

$Expression_1$ and $expression_2$ stand for valid arithmetic expressions; *line number* stands for an actual line number; and *r* stands for one of the following relations:

Basic Code	Meaning
=	equals
<	is less than
>	is greater than
< =	is less than or equal to
> =	is greater than or equal to

When this instruction is executed, $expression_1$ and $expression_2$ must have actual values. If it is true that $expression_1\ r\ expression_2$ for the relation

---

† "Arithmetic expression" here includes expressions that contain library functions.

chosen, then control is transferred to the line having the specified line number. If the relation is false, control passes naturally to the next statement in the program. This instruction is sometimes called a conditional GØ TØ, since it amounts to a GØ TØ if the given condition is true. Be careful in the use of equals ($=$) because computed values are not always exact.†

Examples:

$$300 \quad \text{IF} \quad B\uparrow 2 - 4*A*C < 0 \quad \text{THEN} \quad 900$$
$$12 \quad \text{IF} \quad X1 + X2 < = 10 \quad \text{THEN} \quad 22$$
$$29 \quad \text{IF} \quad Y\uparrow N > = 2*X \quad \text{THEN} \quad 5$$

END

IV. **END**: Corresponding to the flowchart **STOP** is the Basic instruction

$$l.n. \quad \text{END}$$

This instruction must be the last one in the program, i.e., must have the largest line number of any instruction in the program.

B.2e.   Subscripted Variables

SUBSCRIPTED VARIABLES

A variable may be subscripted by enclosing the subscript in parentheses after the variable name. (The variable name must be a single letter.) If a particular variable is to have more than 10 subscript values (i.e., $x_1, x_2, \cdots, x_{10}$) it must be included in a **DIM** statement (dimension statement) at the beginning of the program to reserve storage space. This has the form

$$l.n. \quad \text{DIM} \quad list$$

where *list* stands for a list of the names, separated by commas, of the variables that might possibly have more than 10 subscript values, each followed by a constant (no decimal point) in parentheses. The constants tell the largest numbers of subscript values expected for the variables.

A program using the subscripted variables X(I), Y(I), and M(I), each with the possibility of more than 10 subscript values, might have the **DIM** statement

$$1 \quad \text{DIM} \quad X(20), Y(300), M(55)$$

This would allow the program to use the variables X(1), X(2), $\cdots$, X(20), Y(1), Y(2), $\cdots$, Y(300), M(1), $\cdots$, M(55), but *not* X(21), Y(301), M(56), or any higher subscripts.

As another example, in a program to analyze test scores, if one wished

---

† You might see this more clearly by running the program in the introductory example (B2.2) several times with $=$ in place of $> =$ in the **IF-THEN** statement, using different data for **A, B,** and **S,** but keeping **S** between **0** and **1** (e.g., **0.3**).

to read in all the scores before any analysis were done, he might use

$$5 \text{ DIM } S(100)$$

if he would never be dealing with more than 100 scores at a time. One does not need to use all the subscript values permitted by a **DIM** statement.

In some problems, it is convenient to use more than one subscript on a particular variable. For example, $S_{i,j}$ could represent the score of the ith student in a class on the jth test of the term, thereby allowing one variable name to be used for all the test scores for a class for a whole term. Basic allows the use of such double subscripts. The only differences from the single subscripts discussed above is the appearance of two subscripts after the name, separated by commas, as in $S(I,J)$, and the need for two numbers in each **DIM** statement, such as

$$2 \text{ DIM } S(50,10).$$

The **DIM** statement is needed if either subscript ever exceeds **10**. One **DIM** statement may be used for all subscripted variables in a program. Thus, if the only subscripted variables to be used in a program were **S** (two subscripts), **T** (one subscript), **M** (one subscript), and **X** (two subscripts),

$$1 \text{ DIM } S(50,10),T(20),M(35),X(30,30)$$

would be legitimate.

B.2f. Other Features

I. *Blanks:* Blank spaces may be used to improve the readability of a program, since they are ignored except in verbatim messages in a **PRINT** statement. Thus one might use

$$100 \text{ LET } D = B{\uparrow}2 - 4*A*C$$

rather than

$$100 \text{ LETD} = B{\uparrow}2 - 4*A*C$$

II. *Data:* When inputting data using a **READ** statement, one uses one or more **DATA** statements as part of the program. These statements have the form

$$l.n. \text{ DATA } list$$

where *list* stands for a list of numbers separated by commas. These statements have line numbers and are read in order by line number so that it is important to label **DATA** statements in the correct order. The location of these within the program is unimportant, however. The data from all the **DATA** statements are put together and are read in sequence. Thus

$$100 \text{ DATA } 1,3,14, -40,71,1.414$$

is equivalent to

```
100  DATA  1
101  DATA  3,14, − 40
102  DATA  71,1.414
```

when the program is executed.

When using an **INPUT** statement, data should be typed according to the directions outputted by the corresponding **PRINT** statement.

ANY MORE DATA?

III. *"Any More Data?":* When using **READ** and **DATA** statements, *whenever* the program runs out of data during execution, it takes this as an indication that the program is done and so stops. If one wishes to do something other than stop, he can achieve the effect of "Any more data?" by reading only one number at a time and putting a number at the end of his data that is easily recognized as not being a piece of data. For example, in most programs the number −1234567 is unlikely to be a legitimate piece of data. Thus after reading in each number, one asks if it is this special number. If so, one has run out of data; if not, he proceeds with the number. For example, he could have

```
210  READ  A
211  IF  A = − 1234567  THEN  500
220  ————
```

if he wished to branch to line number **500** when he ran out of data, otherwise go to line **220**. One can easily modify this idea to fit particular algorithms.

REMARKS

IV. *Remarks:* One can put remarks in a Basic program whose functions are to provide explanations for a person reading the program but that are ignored by the computer. This is done with **REM** statements. These have the following form:

if **REM** *remark*

where *remark* stands for the message one wishes. The **REM** statements need line numbers. The **REM** statements are of particular value in programs that one intends to keep, to remind him of special features or requirements, as well as to identify the program.

LIST

V. *List:* One can get an up-to-date listing of his program at any time by typing the word **LIST** at the beginning of a line. The listing is in order by line numbers and has only the most recent version of any lines that have been corrected.

RUNNING A PROGRAM

VI. *Running a Program:* In general, it is wisest to write the program carefully on paper before going to the computer center terminal in order to minimize the time during which you are using it. The details of how to use the installation itself will vary from computer to computer; therefore, you should check with your computer center for these details.

B.2g. Examples

We present Basic versions of the flowcharts of Figures 12 and 19 of the preliminary chapter as examples of the use of Basic statements in actual programs. We use the device of paragraph B.2f III in place of **EXAMPLES** "Any more data?"

```
1       REM PRØGRAM TØ AVERAGE ANY NUMBER ØF TEST SCØRES
2       REM LIST ØF VARIABLE NAMES RELATED TØ FIGURE 12
3       REM     S FØR SCØRE
4       REM     T FØR SUM (TØTAL)
5       REM     C FØR CØUNT
6       REM     A FØR AVE
100     LET T = 0
110     LET C = 0
120     READ S
130     IF S = −123456 THEN 500
140     LET T = T + S
150     LET C = C + 1
160     GØ TØ 120
500     LET A = T/C
510     PRINT A
520     REM DATA STATEMENTS WØULD GØ IN HERE
521     REM AFTER THE LAST SCØRE ØNE MUST PUT THE NUMBER
522     REM −123456
10000 END
```

```
1       REM PRØGRAM TØ GRADE FIVE QUESTIØN MULTIPLE CHØICE TESTS
2       REM LIST ØF VARIABLE NAMES RELATED TØ FIGURE 19
3       REM     SUBSCRIPTED VARIABLE A FØR ANS
4       REM     SUBSCRIPTED VARIABLE K FØR KEY
5       REM     G FØR GRADE
6       REM     N FØR ID
7       REM NØTE NØ DIM STATEMENT IS NEEDED HERE
8       REM DATA ARRANGED: KEY FIRST, THEN STUDENT NUMBER AND
9       REM FIVE ANSWERS FØR EACH STUDENT
100     READ K(1), K(2), K(3), K(4), K(5)
110     READ N
120     IF N = −123456 THEN 500
130     READ A(1), A(2), A(3), A(4), A(5)
140     LET G = 0
150     LET I = 0
160     LET I = I + 1
170     IF A(I) = K(I) THEN 200
180     GØ TØ 250
200     LET G = G + 1
250     IF I < 5 THEN 160
260     PRINT N, G
270     GØ TØ 110
300     REM DATA STATEMENTS WØULD GØ HERE. −123456 MUST BE PUT
301     REM AFTER LAST STUDENT'S ANSWERS
500     END
```

B.3. FORTRAN

Introductory Example:

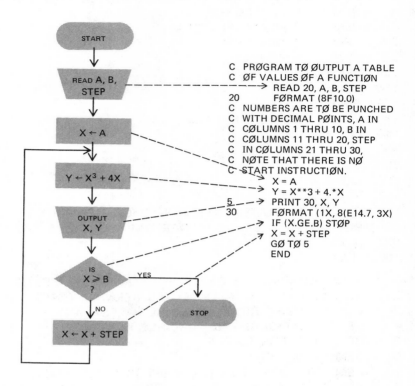

```
C   PRØGRAM TØ ØUTPUT A TABLE
C   ØF VALUES ØF A FUNCTIØN
        READ 20, A, B, STEP
20      FØRMAT (8F10.0)
C   NUMBERS ARE TØ BE PUNCHED
C   WITH DECIMAL PØINTS, A IN
C   CØLUMNS 1 THRU 10, B IN
C   CØLUMNS 11 THRU 20, STEP
C   IN CØLUMNS 21 THRU 30,
C   NØTE THAT THERE IS NØ
C   START INSTRUCTIØN.
        X = A
        Y = X**3 + 4.*X
5       PRINT 30, X, Y
30      FØRMAT (1X, 8(E14.7, 3X))
        IF (X.GE.B) STØP
        X = X + STEP
        GØ TØ 5
        END
```

Flowchart labels: START; READ A, B, STEP; X ← A; Y ← X³ + 4X; OUTPUT X, Y; IS X ≥ B ? (YES → STOP; NO → X ← X + STEP); STOP

B.3a. Preliminaries

CHARACTERS

I. *Characters:* Fortran uses the 26 letters of the alphabet (capitals not lower case), the digits 0 through 9, and some special symbols including $+$, $*$, $-$, $/$, $\$$, $(.)$, ., ., ", ?, and $=$. Special care is needed not to confuse the letter **I** with the number one, nor the letter **O** with the number zero. To help keep these confusions to a minimum, **I** is usually written with the bars at the top and bottom; one is left plain **(1)**; the letter " oh " has a bar through it **(Ø)** to distinguish it from zero **(0)**. We shall follow these conventions.

VARIABLES

II. *Variables:* Fortran variable names consist of no more than six characters, the first of which must be a letter, the rest being any combination of letters and numbers. Although one letter would usually suffice (**X, Y, A, B,** etc.), it is advisable to use names that suggest what the variable is used for, such as **GRADE, KEY2,** and **ID** in the grading problem in Section 3b of the preliminary chapter.

Fortran allows the use of two kinds of arithmetic, called *integer* and *real* (or fixed point and floating point.) Roughly speaking, integer arithmetic deals only with integers and not with fractions. Real arithmetic does not have this drawback (and also permits the use of much larger numbers than integer arithmetic). Most of our programming can be done using only real arithmetic—integer arithmetic should be used only with caution (but routinely for subscripts—see Section B.3e).

The type of arithmetic used in any computation is determined by the variable names and constants involved in the computation. Any variable that starts with one of the letters I, J, K, L, M, or N is considered to be an *integer variable*. All others are *real*. If we were to code the flowchart of Figure 18 or 19 in the preliminary chapter in Fortran, it would be wise to rename the variables **KEY1, KEY2, $\cdots$, KEY5,** and **ID** so as not to use any integer arithmetic. We might use **QEY1, QEY2, $\cdots$, QEY5,** and **STUNUM** (for **STU**dent **NUM**ber).

Care should be taken that no computation involves both integer and real variables. Although many computer installations can deal with such "mixed mode" expressions, not all can, and even with those that do you may not get what you intended.

Examples of Variable Names:

Valid	Invalid	Reason
CAT	EPSILØN	Too many characters
X	2MANY	Does not start with a letter
XSUB1		
GEØRGE	X.1	Uses a character other than a letter or number
ØNE		

III. *Constants:* Constants that are to be used in real arithmetic must have decimal points. Constants that are to be used in integer arithmetic must not have decimal points. Thus 10. is a real constant and 10 is an integer constant.

The only exception to this pair of rules is that any expression raised to an integer power should use an integer constant for that power, i.e., we write **X**$**$**2** *not* **X**$**$**2.** (see Section IV below). Commas may not be used (i.e., 100,000. must be written **100000.**).

IV. *Arithmetic Operations and Expressions:*

Addition is denoted by $+$.

Subtraction is denoted by $-$.

Multiplication is denoted by $*$ (*not* by an "$\times$" or a dot, or by writing variables next to each other as is done in algebra).

Division is denoted by /.

Negation is denoted by $-$; (i.e., $-t$ would be coded as $-$**T**)

Exponentiation is denoted by $**$ (i.e., r^5 would be coded **R**$**$**5** and $2x^3$ would be coded as **2.$*$X$**$3**).

Arithmetic expressions may be constructed as in algebra but using the

Fortran symbols. Parentheses should be used for grouping as in algebra, and should be used quite liberally. For example $\dfrac{a+b}{c}$ should be coded as (A + B)/C.† Parentheses should also be used if a negated expression occurs to the right of an operation symbol. For example, $a \cdot (-b)$ should be coded as A*(−B) not A* − B, and $3x^3 - 2x + 5$ would be 3.*X**3 − 2.*X + 5. in real arithmetic and 3*N**3 − 2*N + 5 in integer arithmetic.

B.3b. Placement of Statements on Punched Cards

Fortran programs consist of a deck of punched cards that contain instructions written in the Fortran language. Part of the grammar of the Fortran language is the placement of the statements on the cards. Each card has 80 columns, each of which can hold one character.

Columns 1–5 are reserved for statement numbers (see next heading).
Column 6 is reserved for "continuation." If a Fortran instruction takes more than one card, it may be continued onto successive cards, but each card after the first must have some nonblank character in column 6 (any character will do).
Columns 7–72 are used for the actual Fortran statement.
Columns 73–80 may be used for sequential numbering of cards or anything else one wishes—they are not read by the computer.

The following shows a card with statement number, **READ** statement and comment:

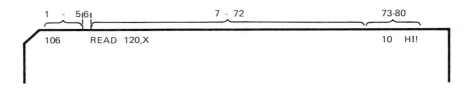

B.3c. Input and Output and FØRMAT

I. *Input:* There are two common forms of **READ** statement used to input information from punched cards. These have the form

$$\text{READ } (n_1, n_2) \; list \tag{a}$$

$$\text{READ } n_2, \; list \tag{b}$$

where n_1 stands for a number that identifies the input device (usually a card reader as opposed to a magnetic tape, paper tape, or other form of input) ; n_2 stands for the statement number of a **FØRMAT** statement, which tells the computer where on the card to find the information to be read and

---

† If parentheses are absent, exponentiations and negations are done first, then multiplications and divisions (left to right), and then additions and subtractions (left to right). So A + B/C would be interpreted as A + (B/C) and A/B*C would be interpreted as (A/B) · C. (See order of operations Appendix A, Section A.5.)

how to interpret it (as real, integer, alphabetic, etc.); and *list* stands for a
list of variable names (separated by commas) for which values are to be
read. The value of n_1 will vary from one computer installation to another.

Examples: Form (a)

> **READ (5, 1000) DEPTH, WIDTH, HEIGHT**
>> (Here "**5**" stands for card reader and **1000** refers to
>> statement **1000**—a format statement—see III
>> below. We read this " Read (from the card reader)
>> by format 1000, depth, width, height.")

> **READ (7, 20) X**
>> (If "**7**" stands for disk pack, then this reads
>> " Read **X** by format 20 from the disk.")

Form (b)

> **READ 1000, DEPTH, WIDTH, HEIGHT**
>>   ↗   (This form assumes that we are using
>> note   cards and can be used only with cards;
>> comma   one reads it " Read, by format **1000**,
>>   depth, width, height.")

> **READ 20, X**
>> " Read by format 20, **X**" or " Read **X** by format 20."

II. *Output:* There are two common forms of **WRITE** statement used to
output information on a typewriter or line printer. These have the form:

$$\textbf{WRITE} \ (n_1, n_2) \ list \qquad\qquad (a)$$

$$\textbf{PRINT} \ n_2, \ list \qquad\qquad (b)$$

where n_1 stands for a number that identifies the output device (printer,
card punch, etc.); n_2 stands for the statement number of a **FØRMAT** statement, which tells the computer where on the output page to print the
information and in what form (as an integer, a real number, alphabetic
information, etc.); and *list* stands for a list of variable names whose values
are to be printed out. The value of n_1 will vary from one computer installation
to another.

Examples: Form (a):

> **WRITE (6, 37) AREA, CØST** is read "Write using the
>> printer, by format 37, area and
>> cost," assuming that " 6 "
>> stands for the printer.

Form (b):

> **PRINT 37, AREA, CØST** This form assumes use of the
>> printer; so we read " Print by
>> format 37, area and cost."

III. **FØRMAT:** FØRMAT statements are simultaneously one of the
advantages and one of the nuisances of Fortran because they provide
flexibility in input and output but at the expense of having to learn a

FORMAT quantity of detail.† For the purposes of this course, one needs only enough FØRMAT to permit rudimentary input and output. We shall therefore side-step FØRMAT and direct you to your instructor or computer center for a recommendation of a few FØRMAT statements to use routinely in your programs. We avoid even giving a recommendation here because of the variability among computer centers and output devices as to what would be most efficient.

B.3d. Non-I/Ø Executable Statements

ASSIGNMENT I. *Assignment:* The Fortran statement that has the form

$$variable = expression$$

corresponds to the flowchart assignment instruction

VARIABLE ← EXPRESSION

where *variable* represents any valid variable name and *expression* stands for any grammatically correct arithmetic expression.‡ (See Arithmetic Operations and Expressions, B.3a, IV). In short, the leftward arrow is replaced by an equal sign, although the meaning remains that discussed for the arrow. The Fortran "=" is *not* the same, therefore, as an algebraic " =."

Examples: **X = X + 1.** means add one to **X** and put the result in location **X.**

Y = 3.*X**2 means square **X**, multiply the result by **3**, and put the result in location **Y.**

SUM = ØLD + NEW/10 means divide what is in location **NEW** by **10**, add the result to whatever is in location **ØLD**, and place the result in **SUM.**

STATEMENT NUMBERS II. **GØ TØ** *and Statement Numbers:* We have already seen the need to move to an instruction other than the next one in the list when we discussed flowcharts. For example, in Figure 12 of the preliminary chapter we needed to return to

from

COUNT ← COUNT + 1

We therefore need to be able to label instructions so that we can refer to them, as well as be able to move to an instruction from some other part of the program. Labeling is done using statement numbers. A statement

---

† The flexibility comes in the programmer's being permitted to specify the location on the page of the entire output, piece by piece, and the form in which the numbers are to be printed.
‡ "Arithmetic expression" here includes expressions that contain library subprograms, function subprograms, and similar Fortran expressions which are not arithmetic in the usual sense.

number is any number from 1 through 5 digits in length punched in columns 1 through 5. No decimal point may be used, and the location in columns 1 through 5 is unimportant. Thus the following two statements would be considered identical in Fortran:

```
Column number: 1 2 3 4 5 6 7 8 9 10 11 12 13 14 15 16 17 18
               3 5 7          G Ø   T   Ø        7   5
                 3 5 7   G Ø   T   Ø        7   5
```

Statement numbers need not be in any special order in a program, as they are merely labels.

No two statements in any one program may have the same statement number, and not all statements need to have statement numbers.

The Fortran statement:

$$GØ \ TØ \ n_2$$

where n_2 stands for a valid statement number causes the computer to go to the statement that has statement number n_2 for the next instruction to follow.

```
        N = 0
12      READ 10, X                  ⎫
10      FØRMAT (8F10.0)             ⎪  Notice that order of
        X = X**2                    ⎬    statement numbers is
        PRINT 5, X                  ⎪    not important.
5       FØRMAT (1X, 8(E14.7,3X))    ⎪
        N = N + 1                   ⎭
        IF (N.EQ.25)STØP    (See next paragraph)
        GØ TØ 12            Gets us back to read another card
```

III. *Logical* IF: The logical IF statement is used when a decision instruction is called for in a flowchart. It has the form:

$$IF \ (expression_1 \ r \ expression_2) \ statement$$

where $expression_1$ and $expression_2$ stand for valid arithmetic expressions of the same mode (both real or both integer); *statement* stands for an executable† Fortran statement; and r stands for one of the following relations:

Fortran Code	Meaning
.EQ.	is equal to
.NE.	is not equal to
.LT.	is less than
.GT.	is greater than
.LE.	is less than or equal to
.GE.	is greater than or equal to

† Roughly speaking, this means a statement that does something. Usually, it is an input, output, assignment, GØ TØ, or STØP statement. It must not be a FØRMAT or DIMENSIØN statement.

Some valid logical **IF** statements are:

IF (SUM .LE. 10.) READ 10, X
IF (AVE .GT. SCØRE) PRINT 15, SCØRE
IF (BIG .EQ. 4.) Y = BIG
IF (Y .GE. X) GØ TØ 20
IF (Z .LT. .002) STØP

When the logical **IF** is executed if the resulting statement inside the parentheses is true, the statement to the right of the parentheses is executed next. If the result is false, the statement to the right is ignored. Look again at the above examples and be sure you see what will happen in each case.

The sample segment in the flowchart below illustrates the logical **IF** in relation to flowchart form. One should avoid the use of **.EQ.** and **.NE.** when real arithmetic is being used, since one may not get exact equality when he expects to.†

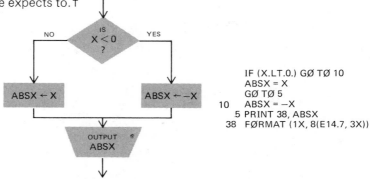

IF (X.LT.0.) GØ TØ 10
   ABSX = X
   GØ TØ 5
10   ABSX = −X
  5 PRINT 38, ABSX
 38   FØRMAT (1X, 8(E14.7, 3X))

IV. **STØP** *and* **END**: The Fortran statement **STØP** is just like the flowchart

$$\boxed{\text{STØP}}$$

It causes termination of the program and may appear any place in the program (and as many times as is convenient for the algorithm). But in addition to **STØP**, each Fortran program must have an **END** statement as the last statement in the program. Its function is to signal that there are no more Fortran statements in the program.

Be sure to note that like all Fortran statements, **STØP** and **END** begin in column 7.

B.3e.   Subscripts

It is often desirable to use subscripted variables. In Fortran, subscripts must be integer (not real) and are enclosed in parentheses after the variable name. Thus $SCORE_k$ in a flowchart would become **SCØRE(K)** in Fortran.

---

† You might see this better by running the program in the introductory example several times with **.EQ.** in place of **.GE.** in the IF Statement, using different values of **A, B,** and **STEP,** but with **STEP** being between 0 and 1 (e.g., 0.3).

The subscripts must be integers, may be constants or variables, and must be greater than zero. The same rules are followed in constructing subscripted variable names as for unsubscripted variables (i.e., those discussed to this point), but no variable names may be used for both subscripted and unsubscripted variables in the same program. Also, a **DIMENSIØN** statement must be placed at the beginning of the program to identify subscripted variables and to save enough memory locations for all the variables which the subscript allows. The **DIMENSIØN** statement has the form:

$$\text{DIMENSIØN } \textit{list}$$

where *list* stands for a list of the names of different subscripted variables to be used in the program, each followed by an integer constant in parentheses, with items in the list separated by commas. The integer constants tell the largest number of variables that would ever be needed using that name.

A program using subscripted variables $X(I)$, $Y(I)$, and $M(I)$ might have the **DIMENSIØN** statement

$$\text{DIMENSIØN } X(20), Y(300), M(4)$$

This allows the program to use the variables $X(1)$, $X(2)$, $\cdots$, $X(20)$, $Y(1)$, $Y(2)$, $\cdots$, $Y(300)$, $M(1)$, $M(2)$, $M(3)$, or $M(4)$, but not $X(21)$, $Y(301)$, $M(5)$, or any higher subscripts.

As another example, in a program to analyze test scores, if one wished to read in all the scores before any analysis was done, he might have

$$\text{DIMENSIØN SCØRE } (100)$$

if he would never have more than 100 different scores to handle at once. One does not need to use all the variables reserved by a **DIMENSIØN** statement.

In some problems, it is convenient to use more than one subscript on a particular variable. For example, $s_{i,j}$ could represent the score of the ith student in the class on the jth test of the term, thereby allowing one variable name to be used for all the test scores for a class for a whole term. Fortran permits the use of two, or even three, subscripts like this. The only differences from the above is the appearance of more than one subscript after the name such as $MATRIX(I,J)$, or $PØINT(L,M,N)$ and more than one dimension specification in the **DIMENSIØN** statement such as

$$\text{DIMENSIØN MATRIX } (10,25).$$

One **DIMENSIØN** statement may be used for all subscripted variables in a program. Thus, if the only subscripted variables to be used in a program were **ZAX** (two subscripts), **NEW** (one subscript), **EWEB** (three subscripts), and **TØWIN** (one subscript),

$$\text{DIMENSIØN } ZAX(20,30), NEW(75), EWEB(10,10,15), TØWIN(50)$$

would be legitimate.

B.3f. Other Features

BLANKS

I. *Blanks:* Blanks are usually† ignored in Fortran so that one may leave blanks where it makes for easier reading, such as

$$Y \;=\; 3.*X**5 \;+\; 4.*X**2 \;-\; 5.$$

instead of

$$Y=3.*X**5+4.*X**2-5.$$

DATA CARDS

II. *Data Cards:* Unlike Fortran statement cards, data cards may make full use of all 80 columns on a card. It is the **FØRMAT** statement governing the appropriate **READ** instruction that dictates how the numbers are interpreted. For example,

$$\overline{|\,12345}$$

could be interpreted as the number 12,345, or as the five numbers 1, 2, 3, 4, and 5, or lots of other ways, depending on the **FØRMAT** involved. If you are using prescribed **FØRMAT** statements, be sure you find out the corresponding requirements governing the location of data on the cards.

ANY MORE DATA?

III. *"Any More Data?"* Many versions of Fortran allow some special way of asking the question "Any more data?" but not all do, and among those that do the ways vary. Check this with your computer installation or instructor.

If such an option is not available, one can achieve the same effect with little more effort by reading in an extra variable at the end of the list—suppose we call it **TAG**. On data cards for your program, this space would normally be left blank and thus be read in as zero. Then one places a card after the last data card in which one actually punches "1." in the appropriate section of the card. For example, suppose you were reading in five numbers at a time from each card. Then you would actually read six numbers, the last being **TAG**. The sixth data section on each card would be left blank on all legitimate data cards, while the last card would have "1." punched in the sixth data section and nothing in the first five. A flowchart for the general case would read as follows:

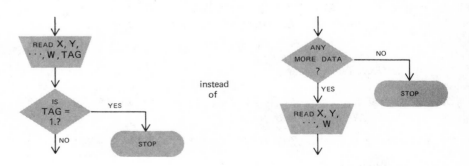

† The most notable exception is **FØRMAT** statements, but presumably you are to use a few prescribed **FØRMAT** statements and so need not worry about this.

One can easily invent useful variations on this idea for specific algorithms.

IV. *Comment Cards:* One can put comments in a Fortran program **COMMENT CARDS** whose sole function is to explain items to a person reading the program that are ignored by the computer. This is done by punching a **C** in column 1 and then the comment anywhere in columns 2 through 80. These are of particular value in programs that one intends to keep, to remind one of special features or requirements, as well as to identify the program.

V. *Job Structure:* In order to run a Fortran program on a computer, **JOB STRUCTURE** one needs more cards than just the program and data cards. These cards, called control cards, serve a variety of functions, depending on the computer center and the particular compiler being used. They may identify the user, direct billing for the running of the program, establish time limits for running the program, etc. They are written in an entirely different language and are read and handled by a supervisory program called a monitor, operating system, or driver. Thus a full job consists of one or more control cards, then the program (which you have written), then zero or more control cards, then your data cards (if any), then zero or more control cards. Check with your computer center for the exact form of control cards you will need. A deck looks like:

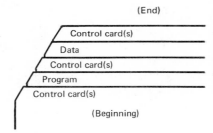

B.3g.   Examples

We present Fortran versions of the flowcharts in Figures 12 and 19 of **EXAMPLES** the preliminary chapter as examples of actual Fortran programs. We shall use the device of paragraph B.3f III above in place of "Any more data?"

```
C PRØGRAM TØ AVERAGE ANY NUMBER ØF TEST SCØRES
C SCØRES ARE TYPED ØNE PER CARD IN CØLUMNS 1 THRØUGH 10
C WITH DECIMAL PØINT. A CARD WITH 1. PUNCHED IN CØLUMNS
C 11 THRØUGH 20 IS TØ BE PLACED AFTER THE LAST DATA CARD.
      SUM = 0.
      CØUNT = 0.
10    READ 1000, SCØRE, TAG
1000  FØRMAT (8F10.0)
      IF (TAG .EQ. 1.) GØ TØ 25
      SUM = SUM + SCØRE
      CØUNT = CØUNT + 1.
      GØ TØ 10
25    AVE = SUM/CØUNT
```

```
      PRINT 1000,AVE
2000  FØRMAT (1X,8(E14.7,3X))
      STØP
      END

C PRØGRAM TØ GRADE FIVE QUESTIØN MULTIPLE CHØICE
C TESTS.
C NUMBERS ARE TØ BE INPUT WITH DECIMAL PØINT, ØNE
C NUMBER IN EACH SECTIØN ØF TEN CØLUMNS. INPUT KEY
C ØN FIRST CARD. THEN STUDENT NUMBER AND FIVE
C ANSWERS FØR EACH STUDENT ØN SUCCESSIVE CARDS.
      DIMENSIØN ANS(5), QEY(5)
      READ 111, QEY(1),QEY(2),QEY(3),QEY(4),QEY(5)
111   FØRMAT (8F10.0)
1     READ 111, STUNUM, ANS(1),ANS(2),ANS(3),ANS(4)
      ANS(5),TAG
C NØTE THAT BØTH READ STATEMENTS USE THE SAME
      FØRMAT STATEMENT.
      IF (TAG .EQ.1.) STØP
      GRADE = 0.
      I = 0
10    I = I + 1
      IF (ANS (I) .EQ. QEY (I)) GRADE = GRADE + 1.
      IF (I .LT. 5) GØ TØ 10
      PRINT 112, STUNUM, GRADE
112   FORMAT (1X,8(E14.7,3X))
      GØ TØ 1
      END
```

B.4. ADDITIONAL FEATURES OF BASIC AND FORTRAN

B.4a. Introduction

This section contains a few additional features of Basic and Fortran. The first (B.4b) is concerned with library functions. This is of more specific concern during or at the end of Chapter 1, with some of the specific functions listed coming from Chapters 3 and 4. Both Basic and Fortran allow one to augment the basic list of library functions. Subsection B.4c deals with such user-defined functions. There are of particular value in Chapter 2. The third (B.4d) briefly introduces the **FØR-NEXT** and **DØ** statements that facilitate loop constructions. The fourth (B.4e) is concerned with subprograms.

B.4b. Library Functions

Both Basic and Fortran have what are called library functions, which automatically perform the computation of very good approximations for various standard functions. One merely uses the Fortran or Basic name and encloses the desired argument in parentheses immediately following the name. Such expressions may be used in forming valid arithmetic expressions. To do so, both Fortran and Basic notation is similar to mathematical notation in that **F(ARG)** (**F** representing any valid function name and **ARG** representing any valid argument name for that function) represents the function value corresponding to the value of **ARG**, just as $f(x)$ denotes the function value y corresponding to x, not the whole function. We give below a list of the functions that are of interest to us in this book, and some examples of valid statements using some of these function.

Basic

Function Name	*Basic Name and Comments*
Absolute value	**ABS**
Square root	**SQR**—Argument must be ≥ 0
Greatest integer	**INT**—This is the function of Example 9, Chapter 1
Sine	**SIN** ⎫
Cosine	**CØS** ⎬ Arguments as real numbers (or radians but
Tangent	**TAN** ⎭ not degrees)
Tangent^{-1}	**ATN** Function values are real or as radians, not degrees.
Natural logarithm	**LOG**
Exponential (base e)	**EXP**

Examples:

LET Y = (SIN(X))↑2 + (COS(X))↑2 for $y \leftarrow \sin^2 x + \cos^2 x$
LET Z = SQR(3 + ABS(V)) for $z \leftarrow \sqrt{3 + |v|}$
If **P** has the value 36 and **Q** has the value 4, **R = SQR(P) − Q** would assign to **R** the value 2.

Fortran

Function Name	*Fortran Name and Comments*
Absolute value	**ABS**—For real arguments
	IABS—For integer arguments
Square root	**SQRT**—Needs real argument ≥ 0
Greatest integer	**IFIX**—Needs real argument, value is integer. (This is the function of Example 9, Chapter 1
Sine	**SIN** ⎫
Cosine	**CØS** ⎬ Arguments as real numbers (or in radians,
Tangent	**TAN** ⎭ *not* degrees)
Sine^{-1}	**ASIN** ⎫ Arguments real, function values are real or
Cosine^{-1}	**ACØS** ⎬ as radians, not degrees
Tangent^{-1}	**ATAN** ⎭
Natural logarithm	**ALØG** ⎫
Common logarithm	**ALØG10** ⎬ Real arguments
Exponential (base e)	**EXP** ⎭

Examples:

$Y = (SIN(X))**2 + (CØS(X))**2$ for $y \leftarrow \sin^2 x + \cos^2 x$

$Z = SQRT(3. + ABS(VALUE))$ for $z \leftarrow \sqrt{3 + |value|}$

If **P** has the value 36 and **Q** has the value 4,

$R = SQRT(P) - Q$ would assign to **R** the value 2.

B.4c. User Defined Functions—Basic

In Basic, one defines a function with a statement that may appear anywhere in the program. This statement has the form *l.n.* **DEF FN***a* $(x) =$ *expression* where *l.n.* is a line number; *a* may be any single letter, x is any single variable name; and *expression* is a valid arithmetic expression using the variable, constants, and the built-in functions. The variable is a dummy variable (i.e., need not be the name of arguments used in the program itself). The function is used in the program in the same way as the built-in functions.

EXAMPLE

10 DEF FNF (X) = 5*X↑5 − 14*X↑4 + 2*X↑3 − .5*X*X + X − 1

would permit one to compute values of $f(x) = 5x^5 - 14x^4 + 2x^3 - x^2/2 + x - 1$ merely by writing **FNF(X)** in an arithmetic expression. Thus one might have elsewhere in the program.

110 LET Y = FNF(5)

to compute f(5), and

150 LET Z = FNF(N1)

to compute f(**N1)** where **N1** has been read in or computed earlier in the program.

EXAMPLE

If a program contained

71 DEF FNA(T) = SIN(T) − COS(T)
28 DEF FNZ(Y) = 2↑Y − 5

then the program could also contain such statements as

50 LET Y = FNA(FNZ(X)) + 2

or

110 LET T = SQR(FNA(M1)↑3)

or

83 IF FNA(X + 1) > = FNA(12) THEN 200

provided that **X** and **M1** had been assigned values earlier in the program.

B.4d. User Defined Functions—Fortran

USER DEFINED
FUNCTIONS

The easiest way (though admittedly not the most flexible) is through the use of the arithmetic statement function. One defines such a function with a statement that precedes the first executable statement of the program. The definitional statement looks rather like an assignment statement in its form:

$$name \; (list) = expression$$

where *name* stands for the name of the function; *list* is a list of dummy arguments (they need not be the names of arguments actually used for the function in the program itself; and *expression* is a valid arithmetic expression using the dummy variables, constants, and library functions. The name of the function is constructed the same as a variable name, and the mode of the name dictates the mode of the functional values. The function is then used in the program in a manner similar to the library functions.

The definition:

EXAMPLE

PØLY(A3, A2, A1, A0,X) = A3*X**3 + A2*X**2 + A1*X + A0)

would permit one to compute the value of p(x) for p being any third degree (or less) polynomial by specifying at each use the coefficients and argument to be used. Thus one might have in the program, the assignment statements:

Y = PØLY(1.,−3.,4.,−2.5,7.)

which would compute p(7) for $p(x) = x^3 - 3x^2 + 4x - 2\frac{1}{2}$;

Z = PØLY(1.,−3.,4.,−2.5,−3.724)

which would compute p(−3.724) for $p(x) = x^3 - 3x^2 + 4x - 2\frac{1}{2}$;

W = PØLY(0.,5.,−4.,3.,3.14)

which would compute q(3.14) for $q(x) = 5x^2 - 4x + 3$.

If a program contained the definitions

EXAMPLE

F(C1,C2,C3,X) = C1*X**3 − C2*X**2 + C3
G(C1,C2,C3,X) = X**C1 − C2*SQRT(X) + C3

then the program could contain

Y = F(A,B,C,Z)

provided **A, B, C,** and **Z** all had been previously assigned values. It could also contain such statements as

Z = F(1.,2.,3.,4.)*G(1.,2.,3.,4.)

or

W = 3*X − 4*F(A,B,C,G(A1,A2,A3,X))

or

IF (F(A,B,C,X(1)) .GT. F(A,B,C,X(2))) GØ TØ 89

B.4. *ADDITIONAL FEATURES OF BASIC AND FORTRAN* 383

B.4e. Loop Instructions—Basic

For a flowchart segment such as the following,

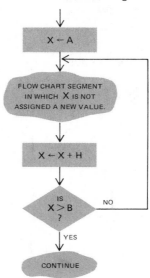

Basic permits a more efficient translation of the loop than direct uses of **IF-THEN** and **GØ TØ** statements. For the above example we would write

l.n. **FØR X = A TØ B STEP H**

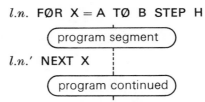

The **FØR, NEXT** combination initializes, increments, and tests the loop index (**X**, in the example) automatically. The **FØR** communicates the information as to where the loop starts, what variable is the index, what value it is to start with, what value it should end with, and the amount to add each time through the loop. The **NEXT** statement gives the location of the end of the loop and signals the test and increment of the index variable. One may omit "**STEP H**" in this form, in which case a step of 1 is assumed. (Occasionally, one may wish to "go backward," i.e., have $B < A$ and $H < 0$. In this case, the test is to see if $X < B$).

In summary, the general form of **FØR** and **NEXT** is:

l.n. **FØR** $var = val_1$ **TØ** val_2 **STEP** val_3

$$\vdots$$

l.n.' **NEXT** var

where var represents a variable name, and val_1, val_2, and val_3 may be constants or variables that will have been assigned values prior to execution of the loop. If **STEP** val_3 is omitted, val_3 is assumed to be 1.

Example:

"Ordinary" Program	*Same Program with* **FØR-NEXT**
10 READ L,R,S	10 READ L,R,S
60 LET X = L	60 FØR X = L TØ R STEP S
110 LET Y = X↑2 − 3	110 LET Y = X↑2 − 3
160 PRINT X,Y	160 PRINT X,Y
210 LET X = X + S	210 NEXT X
260 IF X > R THEN 360	260 PRINT "THE END"
310 GØ TØ 110	310 DATA ————
360 PRINT "THE END"	360 END
410 DATA ————	
460 END	

B.4f. Loop Instructions—Fortran

For a flowchart segment such as the following,

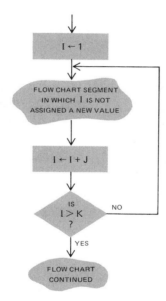

Fortran permits a more efficient translation than that provided by the use of **IF** and **GØ TØ** statements. For this example, we would write

DØ 1000 I = 1, K, J
(program segment)

1000 CONTINUE
(program segment)

The **DØ** statement automatically initializes, increments, and tests the loop counter, and tells the location of the end of the loop (statement number **1000**, in the example). All of this information is contained in the **DØ** statement. While the final statement in the loop may be almost any executable statement, it is customary (and less likely to produce error) to end **DØ** loops with **CONTINUE** statements. These do nothing more than signal the end of the loop. The value **J** (together with its preceding comma) may be omitted, in which case **J** is assumed to be 1.

The general form of the **DØ** statement, then, is

$$\text{D}\text{Ø} \ statement \ number \ ivar = ival_1, \ ival_2, \ ival_3$$

where *statement number* represents an integer constant giving the statement number of the final statement in the loop (usually, a **CONTINUE** statement); *ivar* represents an integer variable; while $ival_1$, $ival_2$, and $ival_3$ are integer constants or integer variables that will have been assigned values prior to execution of the loop. With the exception of a few compilers, all three numbers must be positive. If $ival_3$ is omitted (together with its comma), it is assumed to be 1.

Example:

"Ordinary Program"	*Same Program with* DØ
READ 5, PRINC, RATE	READ 5, PRINC, RATE
5 FØRMAT ———	5 FØRMAT ———
I = 1	DØ 200 I = 1,10,1
8 PRINC = PRINC(1 + RATE)	PRINC = (1 + RATE)∗PRINC
I = I + 1	200 CØNTINUE
IF (I.LE.10) GØ TØ 8	PRINT 10, PRINC
PRINT 10, PRINC	10 FØRMAT ———
10 FØRMAT ———	STØP
STØP	END
END	

B.4g.   Subprograms

Not infrequently in moderate and large-sized programs, it is helpful to have smaller programs linked to the main program, each of which accomplishes a procedure needed several times in the main program. For example, the evaluation of a polynomial by synthetic division, computation of n!, or ordering a set of numbers from high to low might be tasks done several times in a program, each requiring several steps to accomplish. Rather than rewrite each of these each time it is needed, one can write such "subalgorithms" as separate programs, and then have the main program refer to them when needed. Such auxiliary programs are called subprograms or subroutines, and both Fortran and Basic permit one to use subprograms.

The library functions (Section B.4b) provide examples of subprograms supplied by the compiler, and their use should help one understand the multiple-use-of-a single-subprogram concept just mentioned. Also the arithmetic statement function in Fortran and function in Basic (Section B.4c) provided examples of one-line subprograms incorporated into the main program.

B.4h. Subprograms—Basic

There is one type of subprogram separate from the main program. These are called subroutines. A statement of the form

$$l.n. \text{ GØSUB } l.n.'$$

directs transfer of control to line number $l.n.'$ (which must be a number, not a variable name) much as a **GØ TØ** statement does, except that record is kept as to where in the program the **GØSUB** statement was encountered. The subprogram starts on line number $l.n.'$ and proceeds as an ordinary program until the command **RETURN** is encountered. Control then transfers *automatically* back to the statement following the **GØSUB** statement (which is why record of its location was kept). The difference between this and the case of a **GØ TØ** is that one must include overt instructions for how to "return" from a **GØ TØ**. Therefore one may only "return" to one place, whereas a program may have many **GØSUB** statements directing control to the same subroutine, and have control returned to the "point of origin," whichever it was in any one execution of the subroutine.

The subroutine has full access to all the storage used in the main program and should be considered part of the main program as far as naming variables and the use of line numbers is concerned. Also, one subroutine may make use of another, but not of itself. The following example illustrates the use of a single subroutine to compute the binomial coefficient $\binom{n}{k} = n!/(k!(n-k)!)$.

```
10     READ N, K
20     PRINT
30     LET I = N
40     GØSUB 1000
50     LET T = F
60     LET I = K
70     GØSUB 1000
80     LET B1 = F
90     LET I = N - K
100    GØSUB 1000
110    LET B2 = F
120    LET A = T/(B1*B2)
130    PRINT A
140    GØ TØ 3000
900    REM    ********************
901    REM HERE IS THE SUBRØUTINE. IT CØMPUTES I FACTØRIAL
```

```
 902 REM   *******************
1000 LET  F = I
1010 LET  X = I - 1
1020 IF  X = 1, THEN  1050
1030 LET  F = F*X
1040 GØ TØ 1010
1050 RETURN
2222 REM   *******************
2223 REM  NØW THAT ALL IS NEEDED IS SØME DATA
3000 END
```

B.4i.   Subprograms—Fortran

Function subprograms, or functions, are designed to be used in the main program in essentially the same manner as library functions (see example below). Each starts with a statement of the form:

FUNCTIØN *name* (*list*)

where *name* stands for the name of the function—usual naming conventions apply and the mode of the functional value is dictated by the *name*—and *list* stands for a list of dummy argument names. The argument list is the only means of communication with the main program and so must include provisions for all such information needed by the function. Arrays may be referenced by name only, except that the function must contain its own DIMENSIØN statement. Although not required with all versions of Fortran, it is wisest to have the arrays in the subprogram be of the same size as those in the main program.

It is critically important that when used in the main program, the list of actual arguments supplied at that time be in precisely the same order as the corresponding dummy arguments. Thus if

FUNCTIØN PØLY (A,N,X)

computes the value of the Nth degree polynomial with coefficients contained in the array A, for the argument X, a main program use of the statement

Y = PØLY (CØEF,Z,IDEGRE)

would not be allowed. Instead,

Y = PØLY(CØEF,IDEGRE,Z)

would be needed and

Y = PØLY (B,5,3.)

would also be legitimate for finding p(3) for a fifth degree polynomial p with coefficients stored in the array B.

Aside from the FUNCTIØN and RETURN (see below) statements, functions are programs unto themselves. Variables used only by the sub-program are legitimate (and often necessary); input and output may be done; and furthermore there is no conflict if main program statement numbers or variable names are reused. Of course, the function does not act in isolation but in conjunction with a main program (or other subprogram) so that one does not wish to use **STØP** except in rare instances. Instead, when computation is complete, the statement

RETURN

causes transfer of control back to the main program. As with **STØP**, there may be several in any one subprogram.

To have the value computed transferred back appropriately, one must use an assignment statement.

$$name = value$$

where *name* is the function name and *value* is the desired functional value (perhaps in the form of a final computation).

Functions are to be placed after the **END** of the main program; if there is more than one, they are loaded one after the other (each having its own **END** statement), in any order.

The following function subprogram in integer mode computes n !.

```
   FUNCTIØN IFAC(N)
   IF (N .GT. 10) GØ TØ 20
C THIS IS A TEST TØ PREVENT ØVERFLØW, THE CHØICE ØF
C THE NUMBER 10 WAS ARBITRARY. IN REAL LIFE, THE
C VALUE WØULD DEPEND ØN THE LARGEST INTEGER
C ALLØWED BY YØUR CØMPUTER.
   M = N
   J = N
5  J = J − 1
   IF (J .LE. 1) GØ TØ 10
   M = M*J
   GØ TØ 5
10 IFAC = M
   RETURN
20 IFAC = 0
   PRINT 25, N
25 FØRMAT (46H ØVERFLØW AVERTED IN CØMPUTATIØN ØF IFAC,N =,
  *I10/26HIFAC ASSIGNED THE VALUE 0)
C THE CHØICE ØF 0 FØR IFAC HERE WAS AGAIN ARBITRARY
   RETURN
   END
```

This subprogram could then be used to compute binomial coefficients as shown in the following segment of a main program.

```
        READ 5, N, K
     5  FØRMAT (8I10)
        IANS = IFAC(N)/(IFAC(K)*IFAC(N − K))
        PRINT 10, IANS
    10  FØRMAT (1X, I10).
          ⋮
```

Fortran also permits one to use subroutine subprograms that are used in a different way and for purposes other than functions. The reader can use functions for these programs, since the list of dummy arguments allows one to transfer many values back for use by the main program, and functions may be used for input, output, etc.

B.5.   POSTSCRIPTS ON COMPUTING

POSTSCRIPTS

In this section, we shall look briefly at a few other aspects of computing and computer programs that are not part of any one programming language.

B.5a.   Compilers

A Fortran or Basic program must be translated to a more detailed "machine language" before being executed. Complicated programs called compilers do the translation. A punched card copy of the resulting machine language program is available from most compilers. This is called an *object deck*, as opposed to the original *source deck* written in Fortran or Basic. An object deck is valuable if one expects to run the program many times, as it eliminates the need for recompiling each time, and so saves time and money.

There is another kind of translating program called an *interpreter*. It does not produce an object deck. Instead it translates as it executes the program, not keeping a cumulative translation. Interpreters are most often used for student programs, since they are faster for short programs than a genuine compiler, and student programs rarely need an object deck.

NOTE   We shall use the word compiler to mean compiler or interpreter.

B.5b.   Debugging

Errors in computer programs are often called "bugs." The process of removing errors from programs is then called *debugging*. The compiler will detect violations of the grammar of the language and list these with a copy of the program; but only the programmer can check for logical errors of the program.

The process of debugging can be an art in itself. Although one sometimes has access to a program that will help to debug, the student will find most debugging a more personal task. We provide a few guidelines and suggestions to aid in such debugging.

First of all, careful planning of the program is essential. The more carefully thought-out a program is before it is coded (or even flow-charted), the less likely one is to need to overhaul the program later because of omission of special features or oversight of certain conditions the program could encounter. It would be hard to overestimate the value of the careful planning of a program.

Second, in order to eliminate bugs, it is often necessary to go looking for them. A major tool for this is the choosing of test data to insure the correctness of the algorithm. Data should be chosen that represents all the types of cases which could be encountered (sometimes including erroneous data which the program should detect as erroneous). For example, in a program to solve quadratic equations, one should test cases that yield two distinct real roots, a repeated real root, and complex conjugate roots. One might also include "erroneous" data such as a lead coefficient of zero, numbers punched in the wrong columns, or in the wrong mode, in order to see what happens, or to be sure features designed to deal with such cases do in fact handle them. The data chosen should be such that the correct answers are easily arrived at by hand.

Third, hand simulation or tracing of the program can reveal hidden bugs whose existence is known but whose location is not. Often this is combined with placing additional output statements in the program. The location of the error can be localized by checking the tracing with actually computed values and finding where incorrect results first occur.

Fourth, although the computer will detect certain kinds of errors and output its diagnosis of the error, do not always trust such error messages to give correct accounting of the bug. An error can often be propagated through several steps before causing a machine recognition of an error and, in so doing, appear at the time of machine recognition to be something entirely different from the actual cause. Here is one place that tracing becomes an indispensible tool.

Finally, a good programming habit to establish is to output whatever is input and to do so immediately after input. This serves as a quick check that the machine started with the numbers you intended and so guards against a keypunch error or incorrect **FØRMAT** specification, generating an error that could otherwise take a long time to track down. This procedure is sometimes called *echo checking*. If massive amounts of data are involved in a full-fledged run, the echo check output statements may be deleted after debugging, but they should be used with *very* few exceptions until debugging is complete.

B.5c. Computer Arithmetic

Computer arithmetic is usually not exact. This is because computers can store only finitely many digits for each number (the exact number of digits varies among computers.) As an example of how this finiteness can affect computational accuracy, consider computing $\frac{1}{3} \cdot 3$ on a machine that keeps six decimal digits per number. The actual result is 1, but in computing (1./3.)*3. our computer first computes 1./3. and gets .333333. It then multiplies to get .999999 instead of 1.000000. When only a few computations are involved, the error is minimal, but when many computations are involved the errors can propagate in such a way as to seriously contaminate the results obtained. This error caused by the finite representation of numbers is often called *roundoff error*; although it is beyond the scope of this book to go into detail about it, one should be aware that such a problem exists and realize that computer answers need not always be completely accurate.

The finiteness of machine arithmetic also leads to another problem, that of *overflow* or *underflow*, which are the conditions that occur when one attempts to handle a number that is too large or too small to be represented in the machine. For example, if a particular computer stores (integer) numbers in 16 digit binary form, then a number requiring more than 16 such digits (a decimal number greater than 65,536) would cause an overflow condition. The actual sizes of numbers that cause over- or underflow vary from computer to computer, but are usually more than adequate for any purposes we (or most other people) might have.

APPENDIX C

PROGRAM LISTINGS

This appendix contains listings of several programs referred to in the text and/or potentially useful in connection with the text. Each is written in Fortran for use with teletype terminals (more specifically, for use on a CDC 6400.) Only minor modifications should be needed to adapt these programs for use at your computer center, whether on time sharing terminals or in batch mode, (and you are welcome to reproduce them and/or transmit them in any form you wish and by any means you wish, whether mechanical or electrical, etc.)

The comment cards at the beginning of each program, together with the instructions given to the user by the program, serve to describe the action and needs of the program. Further, it is worth noting here that these programs are aimed at straightforward, uncomplicated algorithms rather than at efficient, elegant, but less transparent ones.

Finally, these programs are part of a modest file of programs being maintained by the mathematics department of the College of Wooster for use in connection with various courses offered in the department. At the time of this writing, copies of the entire file (in the form it exists at the time the copy is ordered) may be obtained from Advisory Services, Computer Center, Battelle Memorial Institute, 505 King, Columbus, Ohio, 43201, in a form of your choice (cards, tape, etc) for only the cost of the cards, tape, or whatever. You should ask for the MATHPF file from the College of Wooster.

PROGRAM 1. GRAPHING PROGRAM

```
        PROGRAM GRAPH (INPUT,OUTPUT,TAPE5=INPUT,TAPE6=OUTPUT,TAPE1)    000099
C     THIS PROGRAM GRAPHS ANY RATIONAL FUNCTION, FORTRAN LIBRARY        000100
C FUNCTION, OR COMBINATION OF RATIONAL OR LIBRARY FUNCTION WITH         000101
C THE EXCEPTION OF INTERVALS IN WHICH A DIVISION BY ZERO OCCUR--        000102
C IN WHICH CASE THE FUNCTION MAY BE GRAPHED IN TWO OR MORE PARTS,       000103
C EXCLUDING THE DISCONTINUITY(S).                                       000104
C     THE USER SUPPLIES THE FUNCTION TO BE GRAPHED IN LINE 1160         000105
C (160=:F(X)= USER'S FUNCTION IN FORTRAN NOTATION), THE LEFT AND        000106
C RIGHT ENDPOINTS, AND THE STEP SIZE BETWEEN POINTS.  ALL DATA          000107
C MUST BE INPUT WITH DECIMAL POINTS.  THE PROGRAM WILL GRAPH NO         000108
C NO MORE THAN 200 POINTS.  IT WILL REDEFINE THE STEP SIZE IF MORE      000109
C THAN 200 POINTS ARE ASKED FOR.                                        000110
C     WHEN THE FUNCTION TO BE GRAPHED HAS BEEN ENTERED, TYPE            000111
C     RUN,R     AND FOLLOW INSTRUCTIONS.                                000112
C                                                                       000113
C PROGRAM WAS WRITTEN BY TED BURROWES, DEPARTMENT OF MATHEMATICS        000114
C COLLEGE OF WOOSTER.                                                   000115
        DIMENSION X(200),Y(200),FCTN(30),PLOT(70)                       000135
        F(X)=X                                                          000160
        CALL CONNEC(5LTAPE5)                                            000162
        CALL CONNEC(5LTAPE6)                                            000163
        BLANK=1H                                                        000168
C INPUT:                                                                000169
        WRITE(6,200)                                                    000170
200     FORMAT(* DO YOU NEED INSTRUCTIONS ON THE USE OF THIS PROGRAM+   *) 000171
        READ(5,210) ANS                                                 000173
210     FORMAT(A1)                                                      000174
        IF(ANS.EQ.1HN)GO TO 1                                           000175
        WRITE(6,220)                                                    000176
220     FORMAT(* YOU SHOULD FIRST PUT THE FUNCTION YOU WISH TO STUDY*   000177
       +/1X,*IN LINE 160 AS FOLLOWS:*/1X,                              000178
       +*     160=:F(X)=   YOUR FN IN FORTRAN NOTATION      */1X,      000179
       +*HAVE YOU ENTERED YOUR FUNCTION AS DESCRIBED+   *)             000180
        READ(5,230) ANS                                                 000181
230     FORMAT(A1)                                                      000182
        IF(ANS.EQ.1HY) GO TO 250                                        000183
        WRITE(6,240)                                                    000184
240     FORMAT(* WHEN THE TELETYPE TYPES ≠..≠, YOU SHOULD ENTER YOUR*/1X, 000185
       +*FUNCTION, TYPE  RUN,FTN   ,AND ASK FOR INSTRUCTIONS AGAIN.*)   000186
        GO TO 9988                                                      000187
250     WRITE(6,260)                                                    000188
260     FORMAT(5X,*THIS PROGRAM WILL GRAPH ANY FUNCTION WHICH CAN*/1X,  000189
       +*BE WRITTEN IN FORTRAN NOTATION UNLESS THERE IS A DIVISION*/1X, 000190
       +*BY ZERO -- IN WHICH CASE THE FUNCTION CAN BE GRAPHED IN */1X,  000191
       +*SEVERAL PARTS.*/1X,                                            000192
       +5X,*YOU WILL BE ASKED A NUMBER OF QUESTIONS: FOR IDENTIFICATION*/ 000193
       +* OF THE FUNCTION YOU ARE GRAPHING, FOR THE ENDPOINTS YOU WISH*) 000194
        WRITE(6,261)                                                    000195
261     FORMAT(* TO USE, AND FOR THE STEP SIZE BETWEEN ARGUMENT VALUES.*/ 000196
       +* TYPE EACH ANSWER FOLLOWED BY ≠RETURN≠ AND EACH NUMBER*/       000197
       +* WITH A DECIMAL POINT, FOLLOWED BY ≠RETURN≠*/                  000198
       +5X,*IF YOU ASK FOR MORE THAN 200 POINTS, THE PROGRAM WILL*/1X,  000199
       +*REDEFINE YOUR STEP SIZE AND PRINT OUT A MESSAGE TELLING YOU*/1X, 000200
       +*THAT IT HAS DONE SO*/1X,                                       000201
       +5X,*YOU ALSO HAVE THE OPTION OF HAVING YOUR OUTPUT PRINTED*/1X, 000202
       +*OUT ON THE LINE PRINTER.  IF YOU DECIDE, TO DO THIS,*/1X,      000203
       +*FOLLOW THE INSTRUCTIONS IN THE PROGRAM AND THEN GO TO THE*/1X, 000204
       +*COMPUTER CENTER AND PICK UP YOUR OUTPUT USING THE*/1X,         000205
       +*IDENTIFICATION NUMBER YOUR TELETYPE PRINTS OUT.*/              000206
       +5X,*NOTE:  A LONG GRAPH TAKES SOME TIME TO PRINT ON THE*/1X,    000207
       +*TELETYPE, SO IF YOU'RE IN A HURRY, USE THE ABOVE OPTION.*      000208
       +/5X,*IN ADDITION, YOU HAVE THE OPTION OF CHOOSING TO GET*/      000210
       +* JUST A TABLE OF VALUES, JUST A GRAPH, OR BOTH.    ANSWER*/    000211
       +* YES OR NO WHEN ASKED ≠TABLE+≠ AND ≠GRAPH+≠.*)                000212
1       WRITE(6,8009)                                                   000253
8009    FORMAT(* IDENTIFY FUNCTION:  *)                                 000258
        READ(5,1001) FCTN                                               000280
1001    FORMAT(30A2)                                                    000290
335     WRITE(6,8003)                                                   000320
8003    FORMAT(* LEFT ENDPOINT:  *)                                     000330
        READ(5,8004) A                                                  000350
8004    FORMAT(F10.0)                                                   000360
        WRITE(6,8005)                                                   000370
8005    FORMAT(* RIGHT ENDPOINT:  *)                                    000380
        READ(5,8004) B                                                  000390
        WRITE(6,8007)                                                   000410
8007    FORMAT(* STEP SIZE:  *)                                         000420
        READ(5,8004) H                                                  000430
        WRITE(6,8008)                                                   000431
8008    FORMAT(* OUTPUT AT COMPUTER CENTER+  *)                         000433
        READ(5,8010) ANS                                                000434
8010    FORMAT(A1)                                                      000435
```

```
            IF(ANS.EQ.1HY) GO TO 22                                  000436
            NN=6$GO TO 25                                            000438
22          WRITE(6,8011)                                            000439
8011        FORMAT(* WHEN YOUR TELETYPE PRINTS ≠..≠.*/1X,            000440
           +*TYPE      DISPOSE, TAPE1,PR=IAD.*)                      000441
            NN=1                                                     000446
25          IF(B.GE.A) GO TO 15                                      000450
            HOLD=B                                                   000460
            B=A                                                      000470
            A=HOLD                                                   000480
            WRITE(6,1003)                                            000490
1003        FORMAT(1X,*ENDPOINTS REVERSED TO USUAL ORDER*)           000500
15          IF(H.GT.0.) GO TO 20                                     000510
            IF(H.LT.0.) GO TO 16                                     000520
            WRITE(6,1004)                                            000530
1004        FORMAT(/1X,*SINCE STEP SIZE = 0 NO REASONABLE GRAPH WILL RESULT*) 000540
            GO TO 9988                                               000550
16          H=-H                                                     000560
            WRITE(6,1005)                                            000570
1005        FORMAT(/1X,*STEP SIZE WAS NEGATIVE, SO IT HAS BEEN*/     000580
           +* REPLACED BY ABS(STEP SIZE)*)                           000590
20          IF((B-A)/H.GT.200.) GO TO 48                             000600
            IF(H.LT.B-A) GO TO 50                                    000610
            H=(B-A)/10.                                              000620
            WRITE(6,1006) H                                          000630
1006        FORMAT(/* SINCE STEP SIZE GREATER THAN LENGTH OF INTERVAL* 000640
           //1X,*ITS VALUE HAS BEEN CHANGED TO *,E14.7)             000650
            GO TO 50                                                 000660
48          H=(B-A)/200.                                             000670
            WRITE(6,1007) H                                          000680
1007        FORMAT(/1X,*SINCE YOUR VALUE FOR THE STEP SIZE WOULD HAVE*/ 000690
           +* RESULTED IN MORE THAN 200 POINTS BEING PLOTTED, ITS VALUE*/ 000691
           +* HAS BEEN INCREASED TO *1X,E14.7)                       000700
C CREATES A TABLE OF VALUES:                                         000710
50          I=1                                                      000720
            X(1)=A                                                   000730
            Y(1)=F(A)                                                000740
            DO 51 I=2,200                                            000745
            X(I)=A+H*(I-1)                                           000746
            Y(I)=F(X(I))                                             000747
            IF(X(I).GE.B) GO TO 56                                   000750
51          CONTINUE                                                 000760
C PREPARE FOR GRAPHING                                               000785
56          NUM=I                                                    000800
            XMAX=Y(1)                                                000810
            XMIN=XMAX                                                000820
            DO 60 J=2,NUM                                            000830
            IF(Y(J).GT.XMAX) XMAX=Y(J)                               000840
60          IF(Y(J).LT.XMIN) XMIN=Y(J)                               000850
            IF(XMIN.GT.0.)XMIN=0.                                    000860
            IF(XMAX.LT.0.)XMAX=0.                                    000870
            SCALE=69./(XMAX-XMIN)                                    000880
            DO 70 I=1,70                                             000890
70          PLOT(I)=1H                                               000900
            ISUB=SCALE*(-XMIN)+1                                     000910
            PLOT(ISUB)=1HI                                           000920
            WRITE(6,7000)                                            000921
7000        FORMAT(* TABLE+   *)                                     000922
            READ(5,705) ANS                                          000923
705         FORMAT(A1)                                               000924
            IF(ANS.EQ.1HN) GO TO 710                                 000925
            WRITE(NN,720) FCTN                                       000926
720         FORMAT(///* TABLE OF VALUES FOR *,30A2)                  000927
            WRITE(NN,1008)                                           000930
1008        FORMAT(9X,*X*,17X,*Y*)                                   000940
C PRINT TABLE                                                        000945
            WRITE(NN,1010) (X(I),Y(I),I=1,NUM)                       000950
710         WRITE(6,730)                                             000951
730         FORMAT(//* GRAPH+   *)                                   000952
            READ(5,705) ANS                                          000953
            IF(ANS.EQ.1HN) GO TO 332                                 000954
C IDENTIFY GRAPH                                                     000955
            WRITE(NN,1011) FCTN                                      000960
            TAG=0                                                    000965
            DO 90 I=1,NUM                                            000970
C CREATES AND PRINTS GRAPH                                           000975
            J=SCALE*(Y(I)-XMIN)+1.                                   000980
            TEMP=PLOT(J)                                             000983
            IF(ABS(X(I)).GE..5*H) GO TO 768                          000985
            TAG=3                                                    000986
            DO 770 K=1,70                                            000987
770         PLOT(K)=1H-                                              000988
768         PLOT(J)=1H*                                              001000
            WRITE(NN,1009) PLOT                                      001010
            IF(TAG.NE.3) GOTO 90                                     001011
            DO789 K=1,70                                             001012
```

```
789    PLOT(K)=1H                                                         001013
       PLOT(ISUB)=1HI                                                     001014
       TAG=0                                                              001015
90     PLOT(J)=TEMP                                                       001020
1009   FORMAT(1X,70A1)                                                    001030
1010   FORMAT(2(1X,G14.7))                                                001040
1011   FORMAT(//1X,*GRAPH OF *,30A2)                                      001050
332    WRITE(6,333)                                                       001052
333    FORMAT(* DO YOU WANT TO TRY ANOTHER INTERVAL+   *)                 001054
       READ(5,334) ANS                                                    001056
334    FORMAT(A1)                                                         001058
       IF(ANS.EQ.1HY) GO TO 335                                           001059
9988   STOP                                                               001060
       END                                                                001070
```

PROGRAM 2. PROGRAM TO APPROXIMATE ZEROS OF FUNCTIONS

```
       PROGRAM FUNZER(INPUT,OUTPUT,TAPE5=INPUT,TAPE6=OUTPUT)              000100
C THIS PROGRAM SEARCHES FOR REAL ZEROS OF FUNCTIONS BY                   000110
C SUCCESSIVE HALVING.  AN ARITHMETIC STATEMENT FUNCTION                  000120
C IN LINE 200 CONTAINS THE FUNCTION TO BE STUDIED -- IT MUST             000130
C BE CHANGED EACH TIME A NEW FUNCTION IS STUDIED.  A TOLERANCE           000140
C OF .000001 IS ASSUMED, BUT CAN BE CHANGED BY ALTERING LINE             000150
C 930.                                                                   000160
C THIS PROGRAM WAS WRITTEN BY TED BURROWES, COLLEGE OF WOOSTER,          000170
C WOOSTER, OHIO.                                                         000180
C                                                                        000190
       DIMENSION NAME(24)                                                000195
       F(X)=X                                                            000200
       CALL CONNEC(5LTAPE5)                                              000210
       CALL CONNEC(5LTAPE6)                                              000220
C INSTRUCTIONS FOR USING THE PROGRAM:                                    000230
C                                                                        000240
       WRITE(6,100)                                                      000250
100    FORMAT(* DO YOU NEED INSTRUCTIONS ON THE USE OF THIS PROGRAM+  *) 000260
       READ(5,110)ANS                                                    000280
110    FORMAT(A1)                                                        000290
       IF(ANS.EQ.1HN) GO TO 1                                            000300
       WRITE(6,120)                                                      000310
120    FORMAT(1X,*THIS PROGRAM IS DESIGNED TO FIND ZEROS OF ANY*/1X,     000320
      +*RATIONAL FUNCTION INCLUDING FORTRAN LIBRARY FUNCTIONS.*/1X,      000330
      +*IT IS BEST USED WHEN ONE KNOWS A ZERO EXISTS IN SOME*/1X,        000340
      +*INTERVAL, BUT WANTS A MORE PRECISE VALUE (AS MIGHT BE*/1X,       000350
      +*THE CASE WHEN USING PROGRAM #GRAPH#).  THE USER FIRST*/1X,       000360
      +*TYPES HIS FUNCTION IN LINE 200:  */1X,                          000370
      +*       200=:F(X)= #USER#S FUNCTION IN FORTRAN NOTATION##*/1X,    000380
      +*IF YOU HAVE NOT DONE THIS, TYPE #NO#. INSERT YOUR FN*/1X,        000390
      +*TYPE  RUN,FTN  ,AND ASK FOR INSTRUCTIONS AGAIN.  IF YOU*/        000400
      +* HAVE ALREADY ENTERED YOUR FUNCTION TYPE #YES#.  *)              000401
       READ(5,699) ANS                                                   000402
699    FORMAT(A1)                                                        000403
       IF(ANS.EQ.1HN) GO TO 999                                          000404
       WRITE(6,698)                                                      000405
698    FORMAT(*      WHEN ASKED, TYPE THE LEFT AND RIGHT ENDPOINTS*/     000408
      +* WITH DECIMAL POINTS.  WHEN ASKED #FUNCTION#+, TYPE THE*/        000410
      +* THE INFORMATION YOU WISH TO USE TO IDENTIFY YOUR FUNCTION.*)    000420
C INPUT AND ECHO CHECK:                                                   000450
C                                                                        000460
1      WRITE(6,130)                                                      000470
130    FORMAT(1X,*LEFT ENDPOINT+  *)                                     000480
       READ(5,1000) A                                                    000490
1000   FORMAT(F10.0)                                                     000500
       WRITE(6,140)                                                      000510
140    FORMAT(1X,*RIGHT ENDPOINT+  *)                                    000520
       READ(5,200) B                                                     000530
200    FORMAT(F10.0)                                                     000540
       WRITE(6,260)                                                      000550
260    FORMAT(1X,*FUNCTION+  *)                                          000560
       READ(5,170) NAME                                                  000570
170    FORMAT(24A2)                                                      000580
C                                                                        000630
C TESTS FOR IMPROPER AND TRIVIAL CONDITIONS:                             000640
C                                                                        000650
       W1=F(A)                                                           000660
       W2=F(B)                                                           000670
```

```
          IF(W1*W2.LE.0.) GO TO 10                                  000680
          WRITE(6,1002) A,W1,B,W2                                    000690
 1002     FORMAT(1X,*THERE IS NO DETECTABLE ZERO IN THIS CASE*/1X,   000700
         +*SINCE F(L) AND F(R) HAVE THE SAME SIGN: L = */1X,         000710
         +G14.7,*  F(L) = *,G14.7,*  R = *,G14.7,*  F(R) = */1X,     000720
         +G14.7,* (L=LEFT ENDPOINT; R=RIGHT ENDPOINT)*/)             000730
          GO TO 150                                                  000740
C                                                                    000750
C ALGORITHM PROPER:                                                  000760
C                                                                    000770
 10       IF(W1*W2.LT.0.) GO TO 20                                   000780
          IF(W1.EQ.0.) WRITE(6,1003) A                              000790
          IF(W2.EQ.0.) WRITE (6,1004) B                             000800
 1003     FORMAT(* VALUE OF LEFT ENDPOINT IS A ZERO FOR THE FN.*/)   000810
 1004     FORMAT(* VALUE OF RIGHT ENDPOINT IS A ZERO FOR THE FN.*/)  000830
          GO TO 150                                                  000850
 20       K=1                                                        000860
 25       C=(A+B)/2.                                                 000870
          W3=F(C)                                                    000880
          IF(W3.NE.0.) GO TO 30                                      000890
          WRITE(6,1007) C                                            000900
 1007     FORMAT(1X,G14.7,* IS A ZERO FOR THE FUNCTION.*)            000910
          GO TO 150                                                  000920
 30       IF(ABS(A-C).GT..000001) GO TO 40                           000930
          WRITE(6,1008) A,W1,C,W3,B,W2                               000940
 1008     FORMAT(1X,*CHOOSE YOUR ZERO FROM AMONG THE FOLLOWING*/1X,  000950
         +*ORDERED PAIRS:*/1X,3(*(*,G14.7,*, *,G14.7,*)*/1X,))       000960
          GO TO 150                                                  000970
 40       IF(K.LT.40) GO TO 50                                       000980
          WRITE(6,1008) A,W1,C,W3,B,W2                               000990
          GO TO 150                                                  001000
 50       K=K+1                                                      001010
          IF(W1*W3.LT.0.) GO TO 55                                   001020
          A = C                                                      001030
          W1=W3                                                      001040
          GO TO 25                                                   001050
 55       B = C                                                      001060
          W2=W3                                                      001070
          GO TO 25                                                   001080
 150      WRITE(6,160)                                               001090
 160      FORMAT(* DO YOU WANT TO TRY ANOTHER PAIR OF ENDPOINTS+  *) 001100
          READ(5,180) ANS                                            001120
 180      FORMAT(A1)                                                 001130
          IF(ANS.EQ.1HY) GO TO 1                                     001140
 999      STOP                                                       001150
          END                                                        001160
```

PROGRAM 3. PROGRAM TO COMPUTE FUNCTION VALUES

```
          PROGRAM FNVAL(INPUT,OUTPUT,TAPE5=INPUT,TAPE6=OUTPUT)       000100
C THIS PROGRAM IS DESIGNED FOR INTERACTIVE USE IN FINDING           000110
C VALUES OF ANY RATIONAL FUNCTION INCLUDING FORTRAN LIBRARY         000120
C FUNCTIONS.  THE FUNCTION TO BE STUDIED IS PUT IN LINE 200.        000130
C THE VALUES OF THE VARIABLE ARE INPUT WITH A DECIMAL POINT         000140
C AND THE VALUE OF THE FUNCTION IS COMPUTED IMMEDIATELY             000150
C ALLOWING ANALYSIS BEFORE ANOTHER VALUE OF THE VARIABLE IS         000160
C INPUT.                                                            000170
C THIS PROGRAM WAS WRITTEN BY SHARON BURROWES, C/O COLLEGE OF       000180
C WOOSTER, WOOSTER, OHIO.                                           000190
          F(X)=X                                                     000200
          CALL CONNEC(5LTAPE5)                                       000210
          CALL CONNEC(5LTAPE6)                                       000215
          WRITE(6,10)                                                000220
 10       FORMAT(1X,*DO YOU NEED INSTRUCTIONS FOR USING THIS PROGRAM+  *) 000230
          READ(5,20) ANS                                             000250
 20       FORMAT(A1)                                                 000260
          IF(ANS.EQ.1HN) GO TO 110                                   000270
          WRITE(6,30)                                                000280
 30       FORMAT(1X,*YOU MUST FIRST TYPE THE FUNCTION YOU WISH TO STUDY*/1X,000290
         +*IN LINE 200 AS FOLLOWS: *)                               000300
          WRITE(6,40)                                                000310
 40       FORMAT(1X,*        200=:F(X)= YOUR FN IN FORTRAN NOTATION        *) 000320
          WRITE(6,50)                                                000330
 50       FORMAT(1X,*HAVE YOU ENTERED YOUR FUNCTION AS DESCRIBED ABOVE+  *) 000340
          READ(5,60) ANSW                                            000360
 60       FORMAT(A1)                                                 000370
          IF(ANSW.EQ.1HY) GO TO 80                                   000380
          WRITE(6,70)                                                000390
```

```
  70      FORMAT(1X,*YOU WILL HAVE TO ENTER YOUR FUNCTION AND THEN*/1X,     000400
          +*TYPE   RUN,FTN   AND ASK FOR INSTRUCTIONS AGAIN.*)             000410
          GO TO 170                                                        000420
  80      WRITE(6,90)                                                      000430
  90      FORMAT(1X,*WHEN THE TELETYPE TYPES    X=    YOU SHOULD TYPE*/1X, 000440
          +*YOUR VALUE FOR THE ARGUMENT OF THE FUNCTION WITH A*)          000450
          WRITE(6,100)                                                     000460
 100      FORMAT(1X,*DECIMAL POINT.  THE PROGRAM WILL HANDLE ARGUMENT*/1X, 000470
          +*VALUES UP TO 14 CHARACTERS (INCLUDING SIGN AND DECIMAL).*)     000480
          WRITE(6,180)                                                     000490
 180      FORMAT(1X*WHEN THE TELETYPE TYPES ≠ANOTHER X≠≠ TYPE YES*/1X,     000500
          +*IF YOU WANT TO USE THE PROGRAM AGAIN; NO IF YOU WANT TO STOP.*/) 000510
 110      WRITE(6,120)                                                     000520
 120      FORMAT(1X,*X = *)                                                000530
          READ(5,130) X                                                    000540
 130      FORMAT(F14.7)                                                    000550
          Y=F(X)                                                           000560
          WRITE(6,140) Y                                                   000570
 140      FORMAT(10X,*F(X) = *,G21.14)                                     000580
          WRITE(6,150)                                                     000590
 150      FORMAT(1X,*ANOTHER X↓   *)                                       000600
          READ(5,160) ANSWE                                               000610
 160      FORMAT(A1)                                                       000620
          IF(ANSWE.EQ.1HY) GO TO 110                                       000630
 170      STOP                                                             000640
          END                                                             000650
```

PROGRAM 4. PROGRAM TO FIND RATIONAL ZEROS OF POLYNOMIAL FUNCTIONS WITH INTEGER COEFFICIENTS

```
          PROGRAM RATZER(INPUT,OUTPUT, TAPE5=INPUT,TAPE6=OUTPUT)          000100
C                                                                         000110
C THIS PROGRAM INPUTS THE INTEGER COEFFICIENTS OF A POLYNOMIAL OF         000120
C DEGREE NO GREATER THAN 10, AND CONSTRUCTS A LIST OF POTENTIAL          000130
C RATIONAL ZEROS IN DECIMAL FORM, DELETING REPETITIONS.  UPON            000140
C REQUEST, IT TESTS EACH AND OUTPUTS THE LIST OF ACTUAL ZEROS AND        000150
C ALSO VALUES WHICH APPEAR TO BE ZEROS BUT FOR WHICH P(X) IS NOT         000160
C STRICTLY EQUAL TO ZERO DUE TO ROUND-OFF ERROR.  (SYNTHETIC            000170
C DIVISION IS USED TO COMPUTE VALUES IN TESTING EACH POTENTIAL          000180
C ZERO.)                                                                 000190
C                                                                         000200
C THE ARRAYS TOP AND BOT STORE FACTORS OF A(N) AND A(0) AS              000210
C NUMERATORS AND DENOMINATORS OF POTENTIAL RATIONAL ZEROS.  THE         000220
C ARRAY TEMP STORES ALL RATIONAL ZEROS GENERATED BEFORE DUPLICATES      000230
C ARE DELETED AND BEFORE THE NEGATIVES HAVE BEEN ADDED, WHILE POT       000240
C CONTAINS THE COMPLETE, NON-DUPLICATED LIST OF POTENTIAL RATIONAL      000250
C ZEROS.  NO LIST IS KEPT OF ACTUAL ZEROS.  THE ARRAY X STORES          000260
C THE COEFFICIENTS READ IN FOR THE POLYNOMIAL BEING STUDIED.            000270
C                                                                         000280
C THIS PROGRAM WAS WRITTEN BY TED BURROWES, DEPARTMENT OF               000290
C MATHEMATICS, COLLEGE OF WOOSTER.                                      000300
C                                                                         000310
          DIMENSION TOP(20),BOT(20),TEMP(400),POT(800),M(11),X(11)       000320
          CALL CONNEC(5LTAPE5)                                            000330
          CALL CONNEC(5LTAPE6)                                            000340
C                                                                         000350
C INPUT AND ECHO CHECK:                                                  000360
C                                                                         000370
          WRITE(6,120)                                                    000380
 120      FORMAT(* DO YOU NEED INSTRUCTIONS ON THE USE OF THIS PROGRAM+  *)000390
          READ(5,121) ANSW                                               000400
 121      FORMAT(A1)                                                      000410
          IF(ANSW.EQ.1HN) GO TO 150                                      000420
          WRITE(6,122)                                                    000430
 122      FORMAT(* THIS PROGRAM IS DESIGNED TO STUDY POLYNOMIALS*/        000440
          +* OF DEGREE GREATER THAN 1 AND LESS THAN 10, AND WITH*/       000450
          +* INTEGER COEFFICIENTS.  WHEN ASKED, TYPE THE DEGREE*/        000460
          +* WITH A DECIMAL POINT.  THE COEFFICIENTS ARE ENTERED*/       000470
          +* FROM HIGHEST DEGREE TO LOWEST.  ENTER EACH COEFFICIENT WITH*/000480
          +* A DECIMAL POINT FOLLOWED BY PRESSING ≠RETURN≠.*/            000490
          +* YOU MAY ALSO ELECT WHETHER YOU WANT TO CHECK*/              000500
          +* THE LIST OF POTENTIAL ZEROS TO SEE IF ANY OF THEM ARE*/     000510
          +* ACTUALLY ZEROS BY TYPING YES OR NO WHEN ASKED.  NOW*/       000520
          +* FOLLOW THE INSTRUCTIONS BELOW:*/)                           000530
```

```
150      WRITE(6,100)                                                    000540
100      FORMAT(* DEGREE OF POLYNOMIAL+  *)                              000550
         READ(5,1000) Y                                                  000560
1000     FORMAT(F2.0)                                                    000570
         N=Y                                                             000580
2        IF(N.LE.10.AND.N.GT.1) GO TO 10                                 000590
         WRITE(6,1001) N                                                 000600
1001     FORMAT(* THE VALUES READ IN FOR THE DEGREE OF THE POLYNOMIAL*/  000610
        +* WAS UNACCEPTABLE:  YOU READ IN  *I3* FOR THE DEGREE*)         000620
         GO TO 150                                                       000630
10       NP = N + 1                                                      000640
         WRITE(6,101)                                                    000650
101      FORMAT(* COEFFICIENTS OF POLYNOMIAL+  *)                        000660
         DO 130 I=1,NP                                                   000670
         READ(5,1002) X(I)                                               000680
1002     FORMAT(F6.0)                                                    000690
130      M(I)=X(I)                                                       000700
         WRITE(6,1003) N,(M(I),I=1,NP)                                   000710
1003     FORMAT(* THE POLYNOMIAL BEING INVESTIGATED IS OF *,I2,*TH DEGREE*/000720
        +* WITH INTEGER COEFFICIENTS (FROM HIGHEST DEGREE TO */          000730
        +* LOWEST):  */5X,11(I6,2X)/)                                    000740
         IF(M(1).EQ.0) GO TO 501                                         000750
         GO TO 506                                                       000760
501      WRITE(6,502)                                                    000770
502      FORMAT(* POLYNOMIAL NOT OF DEGREE STATED ABOVE SINCE LEAD*/     000780
        +* COEFFICIENT IS ZERO.  TRY AGAIN.*)                            000790
         GO TO 1                                                         000800
506      IF(M(NP).EQ.0) GO TO 503                                        000810
         GO TO 505                                                       000820
503      WRITE(6,504)                                                    000830
504      FORMAT(* X=0 IS AN OBVIOUS ZERO; THE PROGRAM CANNOT FIND*/      000840
        +* OTHERS SINCE THE CONSTANT TERM EQUALS ZERO.  REDEFINE THE*/   000850
        +* FUNCTION BY REMOVING COMMON X FACTOR(S) AND TRY AGAIN.*)      000860
         GO TO 1                                                         000870
C                                                                        000880
C FINDING FACTORS OF A(N) AND A(0):                                      000890
C                                                                        000900
505      KB=0                                                            000910
         KT = 0                                                          000920
         JB = IABS(M(1))                                                 000930
         JT = IABS(M(NP))                                                000940
         DO 15 I=1,JT                                                    000950
         IF(JT/I*I.NE.JT) GO TO 15                                       000960
         KT = KT + 1                                                     000970
         IF(KT.GT.20) GO TO 23                                           000980
         TOP(KT) = I                                                     000990
15       CONTINUE                                                        001000
C                                                                        001010
C GENERATING POTENTIAL RATIONAL ZEROS:                                   001020
C                                                                        001030
         DO 20 I = 1,JB                                                  001040
         IF(JB/I*I.NE.JB) GO TO 20                                       001050
         KB = KB + 1                                                     001060
         IF(KB.GT.20) GO TO 23                                           001070
         BOT(KB) = I                                                     001080
20       CONTINUE                                                        001090
         GO TO 25                                                        001100
23       WRITE(6,1004)                                                   001110
1004     FORMAT(* THIS EXAMPLE HAS MORE FACTORS FOR THE LEAD COEFFICIENT*/001120
        +* AND/OR THE CONSTANT TERM THAN THE PROGRAM CAN HANDLE.*)       001130
         GO TO 1                                                         001140
25       K = 0                                                           001150
         DO 30 I = 1,KT                                                  001160
         DO 30 J = 1,KB                                                  001170
         K = K + 1                                                       001180
30       TEMP(K) = TOP(I)/BOT(J)                                         001190
         L = 1                                                           001200
         POT(1) = TEMP(1)                                                001210
         IF(KT*KB.EQ.1) GO TO 50                                         001220
         K = 2                                                           001230
31       J = 1                                                           001240
33       IF(TEMP(K).EQ.POT(J)) GO TO 40                                  001250
         IF(J.EQ.L) GO TO 35                                             001260
         J = J + 1                                                       001270
         GO TO 33                                                        001280
35       L = L + 1                                                       001290
         POT(L) = TEMP(K)                                                001300
40       IF(K.EQ.KT*KB) GO TO 50                                         001310
         K = K+1                                                         001320
         GO TO 31                                                        001330
50       DO 55 I = 1,L                                                   001340
55       POT(I+L) = -POT(I)                                              001350
         NUM = 2*L                                                       001360
         WRITE(6,1005) (POT(I), I=1,NUM)                                 001370
1005     FORMAT(* A LIST OF THE POTENTIAL RATIONAL ZEROS (IN DECIMAL*/   001380
        +* FORM) IS:  *,40(/,1X,6(F10.5,1X)))                            001390
         WRITE(6,103)                                                    001400
```

```
103     FORMAT(* DO YOU WANT TO CHECK THE LIST OF POTENTIAL ZEROS*/      001410
        +* TO SEE IF ANY OF THEM ARE ACTUALLY ZEROS+   *)                001420
        READ(5,104) ANS                                                  001430
104     FORMAT(A1)                                                       001440
        IF(ANS.EQ.1HN) GO TO 1                                           001450
C                                                                        001460
C TESTING FOR ACTUAL ZEROS:                                             001470
C                                                                        001480
        WRITE(6,1006)                                                    001490
1006    FORMAT(//* THE ACTUAL RATIONAL ZEROS ARE LISTED BELOW ALONG*/    001500
        +* WITH VALUES WHICH MAY BE ZEROS, BUT WHICH DO NOT (QUITE)*/    001510
        +* APPEAR SO DUE TO ROUND-OFF ERROR,*//3X,                       001520
        +*ACTUAL ZEROS*,12X,*POSSIBLE ZEROS-ASSOCIATED FUNCTIONAL VALUE*/ 001530
        +1X,//8X,*X*,24X,*X*,21X,*F(X)*/)                                001540
        DUMMY = 0.                                                       001550
        DO 60 I=1,NUM                                                    001560
        VAL = F(M,POT(I),N)                                              001570
        IF(VAL.NE.0.) GO TO 57                                           001580
        WRITE(6,1007) POT(I),VAL                                         001590
1007    FORMAT(1X,F15.7,39X,F4.1)                                        001600
        DUMMY = DUMMY + 1.                                               001610
        GO TO 60                                                         001620
57      IF(ABS(VAL).GE..0001.OR.VAL.EQ.0.) GO TO 60                      001630
        WRITE(6,1008)POT(I),VAL                                          001640
1008    FORMAT(26X,G14.7,9X,G14.7)                                       001650
        DUMMY = DUMMY + 1.                                               001660
60      CONTINUE                                                         001670
        IF(DUMMY.EQ.0.) WRITE(6,1009)                                    001680
1009    FORMAT(/* APPARENTLY THERE ARE NO RATIONAL ZEROS*)               001690
        WRITE(6,1010)                                                    001700
1010    FORMAT(///)                                                      001710
        GO TO 1                                                          001720
1       WRITE(6,106)                                                     001730
106     FORMAT(* DO YOU WANT TO STUDY ANOTHER POLYNOMIAL+   *)           001740
        READ(5,107) AN                                                   001750
107     FORMAT(A1)                                                       001760
        IF(AN.EQ.1HN) GO TO 999                                          001770
        GO TO 150                                                        001780
999     STOP                                                             001790
        END                                                              001800
C                                                                        001810
C POLYNOMIAL EVALUATION:                                                001820
C                                                                        001830
        FUNCTION F(M,X,N)                                                001840
        DIMENSION M(11),B(11)                                            001850
        J = N+1                                                          001860
        B(1) = M(1)                                                      001870
        DO 5 I = 2,J                                                     001880
5       B(I) = B(I-1)*X + M(I)                                           001890
        F = B(J)                                                         001900
        RETURN                                                           001910
        END                                                              001920
```

APPENDIX D

TABLES

TABLE 1

VALUES FOR THE CIRCULAR FUNCTIONS WITH REAL NUMBER ARGUMENTS

T	SIN(T)	COS(T)	TAN(T)	SEC(T)	CSC(T)	COT(T)
0.00	0.00000	1.0000	0.00000	1.000	UNDEFINED	UNDEFINED
.01	.01000	1.0000	.01000	1.000	100.00	100.00
.02	.02000	.9998	.02000	1.000	50.00	49.99
.03	.03000	.9996	.03001	1.000	33.34	33.32
.04	.03999	.9992	.04002	1.001	25.01	24.99
.05	.04998	.9988	.05004	1.001	20.01	19.98
.06	.05996	.9982	.06007	1.002	16.68	16.65
.07	.06994	.9976	.07011	1.002	14.30	14.26
.08	.07991	.9968	.08017	1.003	12.51	12.47
.09	.08988	.9960	.09024	1.004	11.13	11.08
.10	.09983	.9950	.1003	1.005	10.02	9.967
.11	.1098	.9940	.1104	1.006	9.109	9.054
.12	.1197	.9928	.1206	1.007	8.353	8.293
.13	.1296	.9916	.1307	1.009	7.714	7.649
.14	.1395	.9902	.1409	1.010	7.166	7.096
.15	.1494	.9888	.1511	1.011	6.692	6.617
.16	.1593	.9872	.1614	1.013	6.277	6.197
.17	.1692	.9856	.1717	1.015	5.911	5.826
.18	.1790	.9838	.1820	1.016	5.586	5.495
.19	.1889	.9820	.1923	1.018	5.295	5.200
.20	.1987	.9801	.2027	1.020	5.033	4.933
.21	.2085	.9780	.2131	1.022	4.797	4.692
.22	.2182	.9759	.2236	1.025	4.582	4.472
.23	.2280	.9737	.2341	1.027	4.386	4.271
.24	.2377	.9713	.2447	1.030	4.207	4.086
.25	.2474	.9689	.2553	1.032	4.042	3.916
.26	.2571	.9664	.2660	1.035	3.890	3.759
.27	.2667	.9638	.2768	1.038	3.749	3.613

T	SIN(T)	COS(T)	TAN(T)	SEC(T)	CSC(T)	COT(T)
.28	.2764	.9611	.2876	1.041	3.619	3.478
.29	.2860	.9582	.2984	1.044	3.497	3.351
.30	.2955	.9553	.3093	1.047	3.384	3.233
.31	.3051	.9523	.3203	1.050	3.278	3.122
.32	.3146	.9492	.3314	1.053	3.179	3.018
.33	.3240	.9460	.3425	1.057	3.086	2.919
.34	.3335	.9428	.3537	1.061	2.999	2.827
.35	.3429	.9394	.3650	1.065	2.916	2.740
.36	.3523	.9359	.3764	1.068	2.839	2.657
.37	.3616	.9323	.3879	1.073	2.765	2.578
.38	.3709	.9287	.3994	1.077	2.696	2.504
.39	.3802	.9249	.4111	1.081	2.630	2.433
.40	.3894	.9211	.4228	1.086	2.568	2.365
.41	.3986	.9171	.4346	1.090	2.509	2.301
.42	.4078	.9131	.4466	1.095	2.452	2.239
.43	.4169	.9090	.4586	1.100	2.399	2.180
.44	.4259	.9048	.4708	1.105	2.348	2.124
.45	.4350	.9004	.4831	1.111	2.299	2.070
.46	.4439	.8961	.4954	1.116	2.253	2.018
.47	.4529	.8916	.5080	1.122	2.208	1.969
.48	.4618	.8870	.5206	1.127	2.166	1.921
.49	.4706	.8823	.5334	1.133	2.125	1.875
.50	.4794	.8776	.5463	1.139	2.086	1.830
.51	.4882	.8727	.5594	1.146	2.048	1.788
.52	.4969	.8678	.5726	1.152	2.013	1.747
.53	.5055	.8628	.5859	1.159	1.978	1.707
.54	.5141	.8577	.5994	1.166	1.945	1.668
.55	.5227	.8525	.6131	1.173	1.913	1.631
.56	.5312	.8473	.6269	1.180	1.883	1.595
.57	.5396	.8419	.6410	1.188	1.853	1.560
.58	.5480	.8365	.6552	1.196	1.825	1.526
.59	.5564	.8309	.6696	1.203	1.797	1.494
.60	.5646	.8253	.6841	1.212	1.771	1.462
.61	.5729	.8196	.6989	1.220	1.746	1.431
.62	.5810	.8139	.7139	1.229	1.721	1.401
.63	.5891	.8080	.7291	1.238	1.697	1.372
.64	.5972	.8021	.7445	1.247	1.674	1.343
.65	.6052	.7961	.7602	1.256	1.652	1.315
.66	.6131	.7900	.7761	1.266	1.631	1.288
.67	.6210	.7838	.7923	1.276	1.610	1.262
.68	.6288	.7776	.8087	1.286	1.590	1.237
.69	.6365	.7712	.8253	1.297	1.571	1.212
.70	.6442	.7648	.8423	1.307	1.552	1.187
.71	.6518	.7584	.8595	1.319	1.534	1.163
.72	.6594	.7518	.8771	1.330	1.517	1.140
.73	.6669	.7452	.8949	1.342	1.500	1.117
.74	.6743	.7385	.9131	1.354	1.483	1.095
.75	.6816	.7317	.9316	1.367	1.467	1.073
.76	.6889	.7248	.9505	1.380	1.452	1.052
.77	.6961	.7179	.9697	1.393	1.437	1.031
.78	.7033	.7109	.9893	1.407	1.422	1.011
.79	.7104	.7038	1.009	1.421	1.408	.9908
.80	.7174	.6967	1.030	1.435	1.394	.9712
.81	.7243	.6895	1.050	1.450	1.381	.9520
.82	.7311	.6822	1.072	1.466	1.368	.9331
.83	.7379	.6749	1.093	1.482	1.355	.9146
.84	.7446	.6675	1.116	1.498	1.343	.8964
.85	.7513	.6600	1.138	1.515	1.331	.8785
.86	.7578	.6524	1.162	1.533	1.320	.8609
.87	.7643	.6448	1.185	1.551	1.308	.8437
.88	.7707	.6372	1.210	1.569	1.297	.8267
.89	.7771	.6294	1.235	1.589	1.287	.8100
.90	.7833	.6216	1.260	1.609	1.277	.7936
.91	.7895	.6137	1.286	1.629	1.267	.7774
.92	.7956	.6058	1.313	1.651	1.257	.7615
.93	.8016	.5978	1.341	1.673	1.247	.7458
.94	.8076	.5898	1.369	1.696	1.238	.7303
.95	.8134	.5817	1.398	1.719	1.229	.7151
.96	.8192	.5735	1.428	1.744	1.221	.7001
.97	.8249	.5653	1.459	1.769	1.212	.6853
.98	.8305	.5570	1.491	1.795	1.204	.6707
.99	.8360	.5487	1.524	1.823	1.196	.6563
1.00	.8415	.5403	1.557	1.851	1.188	.6421
1.01	.8468	.5319	1.592	1.880	1.181	.6281
1.02	.8521	.5234	1.628	1.911	1.174	.6142
1.03	.8573	.5148	1.665	1.942	1.166	.6005
1.04	.8624	.5062	1.704	1.975	1.160	.5870
1.05	.8674	.4976	1.743	2.010	1.153	.5736
1.06	.8724	.4889	1.784	2.046	1.146	.5604

T	SIN(T)	COS(T)	TAN(T)	SEC(T)	CSC(T)	COT(T)
1.07	.8772	.4801	1.827	2.083	1.140	.5473
1.08	.8820	.4713	1.871	2.122	1.134	.5344
1.09	.8866	.4625	1.917	2.162	1.128	.5216
1.10	.8912	.4536	1.965	2.205	1.122	.5090
1.11	.8957	.4447	2.014	2.249	1.116	.4964
1.12	.9001	.4357	2.066	2.295	1.111	.4840
1.13	.9044	.4267	2.120	2.344	1.106	.4718
1.14	.9086	.4176	2.176	2.395	1.101	.4596
1.15	.9128	.4085	2.234	2.448	1.096	.4475
1.16	.9168	.3993	2.296	2.504	1.091	.4356
1.17	.9208	.3902	2.360	2.563	1.086	.4237
1.18	.9246	.3809	2.427	2.625	1.082	.4120
1.19	.9284	.3717	2.498	2.691	1.077	.4003
1.20	.9320	.3624	2.572	2.760	1.073	.3888
1.21	.9356	.3530	2.650	2.833	1.069	.3773
1.22	.9391	.3436	2.733	2.910	1.065	.3659
1.23	.9425	.3342	2.820	2.992	1.061	.3546
1.24	.9458	.3248	2.912	3.079	1.057	.3434
1.25	.9490	.3153	3.010	3.171	1.054	.3323
1.26	.9521	.3058	3.113	3.270	1.050	.3212
1.27	.9551	.2963	3.224	3.375	1.047	.3102
1.28	.9580	.2867	3.341	3.488	1.044	.2993
1.29	.9608	.2771	3.467	3.609	1.041	.2884
1.30	.9636	.2675	3.602	3.738	1.038	.2776
1.31	.9662	.2579	3.747	3.878	1.035	.2669
1.32	.9687	.2482	3.903	4.029	1.032	.2562
1.33	.9711	.2385	4.072	4.193	1.030	.2456
1.34	.9735	.2288	4.256	4.372	1.027	.2350
1.35	.9757	.2190	4.455	4.566	1.025	.2245
1.36	.9779	.2092	4.673	4.779	1.023	.2140
1.37	.9799	.1994	4.913	5.014	1.021	.2035
1.38	.9819	.1896	5.177	5.273	1.018	.1931
1.39	.9837	.1798	5.471	5.561	1.017	.1828
1.40	.9854	.1700	5.798	5.883	1.015	.1725
1.41	.9871	.1601	6.165	6.246	1.013	.1622
1.42	.9887	.1502	6.581	6.657	1.011	.1519
1.43	.9901	.1403	7.055	7.126	1.010	.1417
1.44	.9915	.1304	7.602	7.667	1.009	.1315
1.45	.9927	.1205	8.238	8.299	1.007	.1214
1.46	.9939	.1106	8.989	9.044	1.006	.1113
1.47	.9949	.1006	9.887	9.938	1.005	.1011
1.48	.9959	.09067	10.98	11.03	1.004	.09105
1.49	.9967	.08071	12.35	12.39	1.003	.08097
1.50	.9975	.07074	14.10	14.14	1.003	.07091
1.51	.9982	.06076	16.43	16.46	1.002	.06087
1.52	.9987	.05077	19.67	19.69	1.001	.05084
1.53	.9992	.04079	24.50	24.52	1.001	.04082
1.54	.9995	.03079	32.46	32.48	1.000	.03081
1.55	.9998	.02079	48.08	48.09	1.000	.02080
1.56	.9999	.01080	92.62	92.63	1.000	.01080
1.57	1.0000	.000796	1255.8	1255.8	1.000	.000796

TABLE 2
VALUES FOR THE NATURAL TRIGONOMETRIC FUNCTIONS

θ IN DEGREES	SIN(θ)	COS(θ)	TAN(θ)	COT(θ)	SEC(θ)	CSC(θ)	
0	0.0000	1.000	0.0000	UNDEFINED	1.000	UNDEFINED	90
1	.01745	.9998	.01746	57.29	1.000	57.30	89
2	.03490	.9994	.03492	28.64	1.001	28.65	88
3	.05234	.9986	.05241	19.08	1.001	19.11	87
4	.06976	.9976	.06993	14.30	1.002	14.34	86
5	.08716	.9962	.08749	11.43	1.004	11.47	85
6	.1045	.9945	.1051	9.514	1.006	9.567	84
7	.1219	.9925	.1228	8.144	1.008	8.206	83
8	.1392	.9903	.1405	7.115	1.010	7.185	82
9	.1564	.9877	.1584	6.314	1.012	6.392	81
10	.1736	.9848	.1763	5.671	1.015	5.759	80
11	.1908	.9816	.1944	5.145	1.019	5.241	79
12	.2079	.9781	.2126	4.705	1.022	4.810	78
13	.2250	.9744	.2309	4.331	1.026	4.445	77
14	.2419	.9703	.2493	4.011	1.031	4.134	76
15	.2588	.9659	.2679	3.732	1.035	3.864	75
16	.2756	.9613	.2867	3.487	1.040	3.628	74
17	.2924	.9563	.3057	3.271	1.046	3.420	73
18	.3090	.9511	.3249	3.078	1.051	3.236	72
19	.3256	.9455	.3443	2.904	1.058	3.072	71
20	.3420	.9397	.3640	2.747	1.064	2.924	70
21	.3584	.9336	.3839	2.605	1.071	2.790	69
22	.3746	.9272	.4040	2.475	1.079	2.669	68
23	.3907	.9205	.4245	2.356	1.086	2.559	67
24	.4067	.9135	.4452	2.246	1.095	2.459	66
25	.4226	.9063	.4663	2.145	1.103	2.366	65
26	.4384	.8988	.4877	2.050	1.113	2.281	64
27	.4540	.8910	.5095	1.963	1.122	2.203	63
28	.4695	.8829	.5317	1.881	1.133	2.130	62
29	.4848	.8746	.5543	1.804	1.143	2.063	61
30	.5000	.8660	.5774	1.732	1.155	2.000	60
31	.5150	.8572	.6009	1.664	1.167	1.942	59
32	.5299	.8480	.6249	1.600	1.179	1.887	58
33	.5446	.8387	.6494	1.540	1.192	1.836	57
34	.5592	.8290	.6745	1.483	1.206	1.788	56
35	.5736	.8192	.7002	1.428	1.221	1.743	55
36	.5878	.8090	.7265	1.376	1.236	1.701	54
37	.6018	.7986	.7536	1.327	1.252	1.662	53
38	.6157	.7880	.7813	1.280	1.269	1.624	52
39	.6293	.7771	.8098	1.235	1.287	1.589	51
40	.6428	.7660	.8391	1.192	1.305	1.556	50
41	.6561	.7547	.8693	1.150	1.325	1.524	49
42	.6691	.7431	.9004	1.111	1.346	1.494	48
43	.6820	.7314	.9325	1.072	1.367	1.466	47
44	.6947	.7193	.9657	1.036	1.390	1.440	46
45	.7071	.7071	1.000	1.000	1.414	1.414	45
	COS(θ)	SIN(θ)	COT(θ)	TAN(θ)	CSC(θ)	SEC(θ)	θ IN DEGREES

TABLE 3
COMMON LOGARITHMS OF NUMBERS

x	0	1	2	3	4	5	6	7	8	9
1.0	0.0000	.0043	.0086	.0128	.0170	.0212	.0253	.0294	.0334	.0374
1.1	.0414	.0453	.0492	.0531	.0569	.0607	.0645	.0682	.0719	.0755
1.2	.0792	.0828	.0864	.0899	.0934	.0969	.1004	.1038	.1072	.1106
1.3	.1139	.1173	.1206	.1239	.1271	.1303	.1335	.1367	.1399	.1430
1.4	.1461	.1492	.1523	.1553	.1584	.1614	.1644	.1673	.1703	.1732
1.5	.1761	.1790	.1818	.1847	.1875	.1903	.1931	.1959	.1987	.2014
1.6	.2041	.2068	.2095	.2122	.2148	.2175	.2201	.2227	.2253	.2279
1.7	.2304	.2330	.2355	.2380	.2405	.2430	.2455	.2480	.2504	.2529
1.8	.2553	.2577	.2601	.2625	.2648	.2672	.2695	.2718	.2742	.2765
1.9	.2788	.2810	.2833	.2856	.2878	.2900	.2923	.2945	.2967	.2989
2.0	.3010	.3032	.3054	.3075	.3096	.3118	.3139	.3160	.3181	.3201
2.1	.3222	.3243	.3263	.3284	.3304	.3324	.3345	.3365	.3385	.3404
2.2	.3424	.3444	.3464	.3483	.3502	.3522	.3541	.3560	.3579	.3598
2.3	.3617	.3636	.3655	.3674	.3692	.3711	.3729	.3747	.3766	.3784
2.4	.3802	.3820	.3838	.3856	.3874	.3892	.3909	.3927	.3945	.3962
2.5	.3979	.3997	.4014	.4031	.4048	.4065	.4082	.4099	.4116	.4133
2.6	.4150	.4166	.4183	.4200	.4216	.4232	.4249	.4265	.4281	.4298
2.7	.4314	.4330	.4346	.4362	.4378	.4393	.4409	.4425	.4440	.4456
2.8	.4472	.4487	.4502	.4518	.4533	.4548	.4564	.4579	.4594	.4609
2.9	.4624	.4639	.4654	.4669	.4683	.4698	.4713	.4728	.4742	.4757
3.0	.4771	.4786	.4800	.4814	.4829	.4843	.4857	.4871	.4886	.4900
3.1	.4914	.4928	.4942	.4955	.4969	.4983	.4997	.5011	.5024	.5038
3.2	.5051	.5065	.5079	.5092	.5105	.5119	.5132	.5145	.5159	.5172
3.3	.5185	.5198	.5211	.5224	.5237	.5250	.5263	.5276	.5289	.5302
3.4	.5315	.5328	.5340	.5353	.5366	.5378	.5391	.5403	.5416	.5428
3.5	.5441	.5453	.5465	.5478	.5490	.5502	.5514	.5527	.5539	.5551
3.6	.5563	.5575	.5587	.5599	.5611	.5623	.5635	.5647	.5658	.5670
3.7	.5682	.5694	.5705	.5717	.5729	.5740	.5752	.5763	.5775	.5786
3.8	.5798	.5809	.5821	.5832	.5843	.5855	.5866	.5877	.5888	.5899
3.9	.5911	.5922	.5933	.5944	.5955	.5966	.5977	.5988	.5999	.6010
4.0	.6021	.6031	.6042	.6053	.6064	.6075	.6085	.6096	.6107	.6117
4.1	.6128	.6138	.6149	.6160	.6170	.6180	.6191	.6201	.6212	.6222
4.2	.6232	.6243	.6253	.6263	.6274	.6284	.6294	.6304	.6314	.6325
4.3	.6335	.6345	.6355	.6365	.6375	.6385	.6395	.6405	.6415	.6425
4.4	.6435	.6444	.6454	.6464	.6474	.6484	.6493	.6503	.6513	.6522
4.5	.6532	.6542	.6551	.6561	.6571	.6580	.6590	.6599	.6609	.6618
4.6	.6628	.6637	.6646	.6656	.6665	.6675	.6684	.6693	.6702	.6712
4.7	.6721	.6730	.6739	.6749	.6758	.6767	.6776	.6785	.6794	.6803
4.8	.6812	.6821	.6830	.6839	.6848	.6857	.6866	.6875	.6884	.6893
4.9	.6902	.6911	.6920	.6928	.6937	.6946	.6955	.6964	.6972	.6981
5.0	.6990	.6998	.7007	.7016	.7024	.7033	.7042	.7050	.7059	.7067
5.1	.7076	.7084	.7093	.7101	.7110	.7118	.7126	.7135	.7143	.7152
5.2	.7160	.7168	.7177	.7185	.7193	.7202	.7210	.7218	.7226	.7235
5.3	.7243	.7251	.7259	.7267	.7275	.7284	.7292	.7300	.7308	.7316
5.4	.7324	.7332	.7340	.7348	.7356	.7364	.7372	.7380	.7388	.7396
5.5	.7404	.7412	.7419	.7427	.7435	.7443	.7451	.7459	.7466	.7474
5.6	.7482	.7490	.7497	.7505	.7513	.7520	.7528	.7536	.7543	.7551
5.7	.7559	.7566	.7574	.7582	.7589	.7597	.7604	.7612	.7619	.7627
5.8	.7634	.7642	.7649	.7657	.7664	.7672	.7679	.7686	.7694	.7701
5.9	.7709	.7716	.7723	.7731	.7738	.7745	.7752	.7760	.7767	.7774
6.0	.7782	.7789	.7796	.7803	.7810	.7818	.7825	.7832	.7839	.7846
6.1	.7853	.7860	.7868	.7875	.7882	.7889	.7896	.7903	.7910	.7917
6.2	.7924	.7931	.7938	.7945	.7952	.7959	.7966	.7973	.7980	.7987
6.3	.7993	.8000	.8007	.8014	.8021	.8028	.8035	.8041	.8048	.8055
6.4	.8062	.8069	.8075	.8082	.8089	.8096	.8102	.8109	.8116	.8122
6.5	.8129	.8136	.8142	.8149	.8156	.8162	.8169	.8176	.8182	.8189
6.6	.8195	.8202	.8209	.8215	.8222	.8228	.8235	.8241	.8248	.8254
6.7	.8261	.8267	.8274	.8280	.8287	.8293	.8299	.8306	.8312	.8319
6.8	.8325	.8331	.8338	.8344	.8351	.8357	.8363	.8370	.8376	.8382
6.9	.8388	.8395	.8401	.8407	.8414	.8420	.8426	.8432	.8439	.8445
7.0	.8451	.8457	.8463	.8470	.8476	.8482	.8488	.8494	.8500	.8506
7.1	.8513	.8519	.8525	.8531	.8537	.8543	.8549	.8555	.8561	.8567
7.2	.8573	.8579	.8585	.8591	.8597	.8603	.8609	.8615	.8621	.8627
7.3	.8633	.8639	.8645	.8651	.8657	.8663	.8669	.8675	.8681	.8686
7.4	.8692	.8698	.8704	.8710	.8716	.8722	.8727	.8733	.8739	.8745

TABLE 3, *continued*

x	0	1	2	3	4	5	6	7	8	9
7.5	.8751	.8756	.8762	.8768	.8774	.8779	.8785	.8791	.8797	.8802
7.6	.8808	.8814	.8820	.8825	.8831	.8837	.8842	.8848	.8854	.8859
7.7	.8865	.8871	.8876	.8882	.8887	.8893	.8899	.8904	.8910	.8915
7.8	.8921	.8927	.8932	.8938	.8943	.8949	.8954	.8960	.8965	.8971
7.9	.8976	.8982	.8987	.8993	.8998	.9004	.9009	.9015	.9020	.9025
8.0	.9031	.9036	.9042	.9047	.9053	.9058	.9063	.9069	.9074	.9079
8.1	.9085	.9090	.9096	.9101	.9106	.9112	.9117	.9122	.9128	.9133
8.2	.9138	.9143	.9149	.9154	.9159	.9165	.9170	.9175	.9180	.9186
8.3	.9191	.9196	.9201	.9206	.9212	.9217	.9222	.9227	.9232	.9238
8.4	.9243	.9248	.9253	.9258	.9263	.9269	.9274	.9279	.9284	.9289
8.5	.9294	.9299	.9304	.9309	.9315	.9320	.9325	.9330	.9335	.9340
8.6	.9345	.9350	.9355	.9360	.9365	.9370	.9375	.9380	.9385	.9390
8.7	.9395	.9400	.9405	.9410	.9415	.9420	.9425	.9430	.9435	.9440
8.8	.9445	.9450	.9455	.9460	.9465	.9469	.9474	.9479	.9484	.9489
8.9	.9494	.9499	.9504	.9509	.9513	.9518	.9523	.9528	.9533	.9538
9.0	.9542	.9547	.9552	.9557	.9562	.9566	.9571	.9576	.9581	.9586
9.1	.9590	.9595	.9600	.9605	.9609	.9614	.9619	.9624	.9628	.9633
9.2	.9638	.9643	.9647	.9652	.9657	.9661	.9666	.9671	.9675	.9680
9.3	.9685	.9689	.9694	.9699	.9703	.9708	.9713	.9717	.9722	.9727
9.4	.9731	.9736	.9741	.9745	.9750	.9754	.9759	.9763	.9768	.9773
9.5	.9777	.9782	.9786	.9791	.9795	.9800	.9805	.9809	.9814	.9818
9.6	.9823	.9827	.9832	.9836	.9841	.9845	.9850	.9854	.9859	.9863
9.7	.9868	.9872	.9877	.9881	.9886	.9890	.9894	.9899	.9903	.9908
9.8	.9912	.9917	.9921	.9926	.9930	.9934	.9939	.9943	.9948	.9952
9.9	.9956	.9961	.9965	.9969	.9974	.9978	.9983	.9987	.9991	.9996
x	0	1	2	3	4	5	6	7	8	9

TABLE 4
VALUES OF EXP (X)

X	EXP(X)	X	EXP(X)	X	EXP(X)	X	EXP(X)
-2.00	.13534	-1.50	.22313	-1.00	.36788	-.50	.60653
-1.99	.13670	-1.49	.22537	-.99	.37158	-.49	.61263
-1.98	.13807	-1.48	.22764	-.98	.37531	-.48	.61878
-1.97	.13946	-1.47	.22993	-.97	.37908	-.47	.62500
-1.96	.14086	-1.46	.23224	-.96	.38289	-.46	.63128
-1.95	.14227	-1.45	.23457	-.95	.38674	-.45	.63763
-1.94	.14370	-1.44	.23693	-.94	.39063	-.44	.64404
-1.93	.14515	-1.43	.23931	-.93	.39455	-.43	.65051
-1.92	.14661	-1.42	.24171	-.92	.39852	-.42	.65705
-1.91	.14808	-1.41	.24414	-.91	.40252	-.41	.66365
-1.90	.14957	-1.40	.24660	-.90	.40657	-.40	.67032
-1.89	.15107	-1.39	.24908	-.89	.41066	-.39	.67706
-1.88	.15259	-1.38	.25158	-.88	.41478	-.38	.68386
-1.87	.15412	-1.37	.25411	-.87	.41895	-.37	.69073
-1.86	.15567	-1.36	.25666	-.86	.42316	-.36	.69768
-1.85	.15724	-1.35	.25924	-.85	.42741	-.35	.70469
-1.84	.15882	-1.34	.26185	-.84	.43171	-.34	.71177
-1.83	.16041	-1.33	.26448	-.83	.43605	-.33	.71892
-1.82	.16203	-1.32	.26714	-.82	.44043	-.32	.72615
-1.81	.16365	-1.31	.26982	-.81	.44486	-.31	.73345
-1.80	.16530	-1.30	.27253	-.80	.44933	-.30	.74082
-1.79	.16696	-1.29	.27527	-.79	.45384	-.29	.74826
-1.78	.16864	-1.28	.27804	-.78	.45841	-.28	.75578
-1.77	.17033	-1.27	.28083	-.77	.46301	-.27	.76338
-1.76	.17204	-1.26	.28365	-.76	.46767	-.26	.77105
-1.75	.17377	-1.25	.28650	-.75	.47237	-.25	.77880
-1.74	.17552	-1.24	.28938	-.74	.47711	-.24	.78663
-1.73	.17728	-1.23	.29229	-.73	.48191	-.23	.79453
-1.72	.17907	-1.22	.29523	-.72	.48675	-.22	.80252
-1.71	.18087	-1.21	.29820	-.71	.49164	-.21	.81058
-1.70	.18268	-1.20	.30119	-.70	.49659	-.20	.81873
-1.69	.18452	-1.19	.30422	-.69	.50158	-.19	.82696
-1.68	.18637	-1.18	.30728	-.68	.50662	-.18	.83527
-1.67	.18825	-1.17	.31037	-.67	.51171	-.17	.84366
-1.66	.19014	-1.16	.31349	-.66	.51685	-.16	.85214
-1.65	.19205	-1.15	.31664	-.65	.52205	-.15	.86071
-1.64	.19398	-1.14	.31982	-.64	.52729	-.14	.86936
-1.63	.19593	-1.13	.32303	-.63	.53259	-.13	.87810
-1.62	.19790	-1.12	.32628	-.62	.53794	-.12	.88692

X	EXP(X)	X	EXP(X)	X	EXP(X)	X	EXP(X)
-1.61	.19989	-1.11	.32956	-.61	.54335	-.11	.89583
-1.60	.20190	-1.10	.33287	-.60	.54881	-.10	.90484
-1.59	.20393	-1.09	.33622	-.59	.55433	-.09	.91393
-1.58	.20598	-1.08	.33960	-.58	.55990	-.08	.92312
-1.57	.20805	-1.07	.34301	-.57	.56553	-.07	.93239
-1.56	.21014	-1.06	.34646	-.56	.57121	-.06	.94176
-1.55	.21225	-1.05	.34994	-.55	.57695	-.05	.95123
-1.54	.21438	-1.04	.35345	-.54	.58275	-.04	.96079
-1.53	.21654	-1.03	.35701	-.53	.58860	-.03	.97045
-1.52	.21871	-1.02	.36059	-.52	.59452	-.02	.98020
-1.51	.22091	-1.01	.36422	-.51	.60050	-.01	.99005
0.00	1.00000	.50	1.64872	1.00	2.71828	1.50	4.48169
.01	1.01005	.51	1.66529	1.01	2.74560	1.51	4.52673
.02	1.02020	.52	1.68203	1.02	2.77319	1.52	4.57223
.03	1.03045	.53	1.69893	1.03	2.80107	1.53	4.61818
.04	1.04081	.54	1.71601	1.04	2.82922	1.54	4.66459
.05	1.05127	.55	1.73325	1.05	2.85765	1.55	4.71147
.06	1.06184	.56	1.75067	1.06	2.88637	1.56	4.75882
.07	1.07251	.57	1.76827	1.07	2.91538	1.57	4.80665
.08	1.08329	.58	1.78604	1.08	2.94468	1.58	4.85496
.09	1.09417	.59	1.80399	1.09	2.97427	1.59	4.90375
.10	1.10517	.60	1.82212	1.10	3.00417	1.60	4.95303
.11	1.11628	.61	1.84043	1.11	3.03436	1.61	5.00281
.12	1.12750	.62	1.85893	1.12	3.06485	1.62	5.05309
.13	1.13883	.63	1.87761	1.13	3.09566	1.63	5.10387
.14	1.15027	.64	1.89648	1.14	3.12677	1.64	5.15517
.15	1.16183	.65	1.91554	1.15	3.15819	1.65	5.20698
.16	1.17351	.66	1.93479	1.16	3.18993	1.66	5.25931
.17	1.18530	.67	1.95424	1.17	3.22199	1.67	5.31217
.18	1.19722	.68	1.97388	1.18	3.25437	1.68	5.36556
.19	1.20925	.69	1.99372	1.19	3.28708	1.69	5.41948
.20	1.22140	.70	2.01375	1.20	3.32012	1.70	5.47395
.21	1.23368	.71	2.03399	1.21	3.35348	1.71	5.52896
.22	1.24608	.72	2.05443	1.22	3.38719	1.72	5.58453
.23	1.25860	.73	2.07508	1.23	3.42123	1.73	5.64065
.24	1.27125	.74	2.09594	1.24	3.45561	1.74	5.69734
.25	1.28403	.75	2.11700	1.25	3.49034	1.75	5.75460
.26	1.29693	.76	2.13828	1.26	3.52542	1.76	5.81244
.27	1.30996	.77	2.15977	1.27	3.56085	1.77	5.87085
.28	1.32313	.78	2.18147	1.28	3.59664	1.78	5.92986
.29	1.33643	.79	2.20340	1.29	3.63279	1.79	5.98945
.30	1.34986	.80	2.22554	1.30	3.66930	1.80	6.04965
.31	1.36343	.81	2.24791	1.31	3.70617	1.81	6.11045
.32	1.37713	.82	2.27050	1.32	3.74342	1.82	6.17186
.33	1.39097	.83	2.29332	1.33	3.78104	1.83	6.23389
.34	1.40495	.84	2.31637	1.34	3.81904	1.84	6.29654
.35	1.41907	.85	2.33965	1.35	3.85743	1.85	6.35982
.36	1.43333	.86	2.36316	1.36	3.89619	1.86	6.42374
.37	1.44773	.87	2.38691	1.37	3.93535	1.87	6.48830
.38	1.46228	.88	2.41090	1.38	3.97490	1.88	6.55350
.39	1.47698	.89	2.43513	1.39	4.01485	1.89	6.61937
.40	1.49182	.90	2.45960	1.40	4.05520	1.90	6.68589
.41	1.50682	.91	2.48432	1.41	4.09596	1.91	6.75309
.42	1.52196	.92	2.50929	1.42	4.13712	1.92	6.82096
.43	1.53726	.93	2.53451	1.43	4.17870	1.93	6.88951
.44	1.55271	.94	2.55998	1.44	4.22070	1.94	6.95875
.45	1.56831	.95	2.58571	1.45	4.26311	1.95	7.02869
.46	1.58407	.96	2.61170	1.46	4.30596	1.96	7.09933
.47	1.59999	.97	2.63794	1.47	4.34924	1.97	7.17068
.48	1.61607	.98	2.66446	1.48	4.39295	1.98	7.24274
.49	1.63232	.99	2.69123	1.49	4.43710	1.99	7.31553
						2.000	7.38906

TABLE 5
DECIMAL EQUIVALENTS OF COMMON FRACTIONS

NUM / DENOM	1	2	3	4	5	6	7	8	9	10	11	12	13
1	1.0000000												
2	0.5000000	1.0000000											
3	0.3333333	0.6666667	1.0000000										
4	0.2500000	0.5000000	0.7500000	1.0000000									
5	0.2000000	0.4000000	0.6000000	0.8000000	1.0000000								
6	0.1666667	0.3333333	0.5000000	0.6666667	0.8333333	1.0000000							
7	0.1428571	0.2857143	0.4285714	0.5714286	0.7142857	0.8571429	1.0000000						
8	0.1250000	0.2500000	0.3750000	0.5000000	0.6250000	0.7500000	0.8750000	1.0000000					
9	0.1111111	0.2222222	0.3333333	0.4444444	0.5555556	0.6666667	0.7777778	0.8888889	1.0000000				
10	0.1000000	0.2000000	0.3000000	0.4000000	0.5000000	0.6000000	0.7000000	0.8000000	0.9000000	1.0000000			
11	0.0909091	0.1818182	0.2727273	0.3636364	0.4545455	0.5454545	0.6363636	0.7272727	0.8181818	0.9090909	1.0000000		
12	0.0833333	0.1666667	0.2500000	0.3333333	0.4166667	0.5000000	0.5833333	0.6666667	0.7500000	0.8333333	0.9166667	1.0000000	
13	0.0769231	0.1538462	0.2307692	0.3076923	0.3846154	0.4615385	0.5384615	0.6153846	0.6923077	0.7692308	0.8461538	0.9230769	1.0000000

NUM DENOM	14	15	16	17	18	19	20	21	22	23	24	25	26
14	1.0000000												
15	0.9333333	1.0000000											
16	0.8750000	0.9375000	1.0000000										
17	0.8235294	0.8823529	0.9411765	1.0000000									
18	0.7777778	0.8333333	0.8888889	0.9444444	1.0000000								
19	0.7368421	0.7894737	0.8421053	0.8947368	0.9473684	1.0000000							
20	0.7000000	0.7500000	0.8000000	0.8500000	0.9000000	0.9500000	1.0000000						
21	0.6666667	0.7142857	0.7619048	0.8095238	0.8571429	0.9047619	0.9523810	1.0000000					
22	0.6363636	0.6818182	0.7272727	0.7727273	0.8181818	0.8636364	0.9090909	0.9545455	1.0000000				
23	0.6086957	0.6521739	0.6956522	0.7391304	0.7826087	0.8260870	0.8695652	0.9130435	0.9565652	1.0000000			
24	0.5833333	0.6250000	0.6666667	0.7083333	0.7500000	0.7916667	0.8333333	0.8750000	0.9166667	0.9583333	1.0000000		
25	0.5600000	0.6000000	0.6400000	0.6800000	0.7200000	0.7600000	0.8000000	0.8400000	0.8800000	0.9200000	0.9600000	1.0000000	
26	0.5384615	0.5769231	0.6153846	0.6538462	0.6923077	0.7307692	0.7692308	0.8076923	0.8461538	0.8846154	0.9230769	0.9615385	1.0000000

NUM DENOM	1	2	3	4	5	6	7	8	9	10	11	12	13
14	0.0714286	0.1428571	0.2142857	0.2857143	0.3571429	0.4285714	0.5000000	0.5714286	0.6428571	0.7142857	0.7857143	0.8571429	0.9285714
15	0.0666667	0.1333333	0.2000000	0.2666667	0.3333333	0.4000000	0.4666667	0.5333333	0.6000000	0.6666667	0.7333333	0.8000000	0.8666667
16	0.0625000	0.1250000	0.1875000	0.2500000	0.3125000	0.3750000	0.4375000	0.5000000	0.5625000	0.6250000	0.6875000	0.7500000	0.8125000
17	0.0588235	0.1176471	0.1764706	0.2352941	0.2941176	0.3529412	0.4117647	0.4705882	0.5294118	0.5882353	0.6470588	0.7058824	0.7647059
18	0.0555556	0.1111111	0.1666667	0.2222222	0.2777778	0.3333333	0.3888889	0.4444444	0.5000000	0.5555556	0.6111111	0.6666667	0.7222222
19	0.0526316	0.1052632	0.1578947	0.2105263	0.2631579	0.3157895	0.3684211	0.4210526	0.4736842	0.5263158	0.5789474	0.6315789	0.6842105
20	0.0500000	0.1000000	0.1500000	0.2000000	0.2500000	0.3000000	0.3500000	0.4000000	0.4500000	0.5000000	0.5500000	0.6000000	0.6500000
21	0.0476190	0.0952381	0.1428571	0.1904762	0.2380952	0.2857143	0.3333333	0.3809524	0.4285714	0.4761905	0.5238095	0.5714286	0.6190476
22	0.0454545	0.0909091	0.1363636	0.1818182	0.2272727	0.2727273	0.3181818	0.3636364	0.4090909	0.4545455	0.5000000	0.5454545	0.5909091
23	0.0434783	0.0869565	0.1304348	0.1739130	0.2173913	0.2608696	0.3043478	0.3478261	0.3913043	0.4347826	0.4782609	0.5217391	0.5652174
24	0.0416667	0.0833333	0.1250000	0.1666667	0.2083333	0.2500000	0.2916667	0.3333333	0.3750000	0.4166667	0.4583333	0.5000000	0.5416667
25	0.0400000	0.0800000	0.1200000	0.1600000	0.2000000	0.2400000	0.2800000	0.3200000	0.3600000	0.4000000	0.4400000	0.4800000	0.5200000
26	0.0384615	0.0769231	0.1153846	0.1538462	0.1923077	0.2307692	0.2692308	0.3076923	0.3461538	0.3846154	0.4230769	0.4615385	0.5000000

APPENDIX E

ZEROS OF THE GENERAL, CUBIC, AND QUARTIC POLYNOMIALS

The three zeros x_1, x_2, and x_3 of the cubic polynomial

$$ax^3 + bx^2 + cx + d$$

(two of which may be complex) are given by the following formulas.

$$x_1 = \frac{1}{3a} \sqrt[3]{\frac{1}{2}(2b^3 - 9abc + 27a^2d) + \sqrt{\frac{1}{4}(2b^3 - 9abc + 27a^2d)^2 + (3a^2c - ab^2)^3}}$$

$$+ \frac{1}{3a} \sqrt[3]{\frac{1}{2}(2b^3 - 9abc + 27a^2d)^2 - \sqrt{\frac{1}{4}(2b^3 - 9abc + 27a^2d)^2 + (3a^2c - ab^2)^3}}$$

$$x_2 = -\frac{1}{6a} \sqrt[3]{\frac{1}{2}(2b^3 - 9abc + 27a^2d) + \sqrt{\frac{1}{4}(2b^3 - 9abc + 27a^2d)^2 + (3a^2c - ab^2)^3}}$$

$$+ \frac{-1}{6a} \sqrt[3]{\frac{1}{2}(2b^3 - 9abc + 27a^2d) - \sqrt{\frac{1}{4}(2b^3 - 9abc + 27a^2d)^2 + (3a^2c - ab^2)^3}}$$

$$+ \frac{-\sqrt{-3}}{6a} \sqrt[3]{\frac{1}{2}(2b^3 - 9abc + 27a^2d) + \sqrt{\frac{1}{4}(2b^3 - 9abc + 27a^2d)^2 + (3a^2c - ab^2)^3}}$$

$$+ \frac{-\sqrt{3}}{6a} \sqrt[3]{\frac{1}{2}(2b^3 - 9abc + 27a^2d) - \sqrt{\frac{1}{4}(2b^3 - 9abc + 27a^2d)^2 + (3a^2c - ab^2)^3}}.$$

$$x_3 = -\frac{1}{6a} \sqrt[3]{\frac{1}{2}(2b^3 - 9abc + 27a^2d) + \sqrt{\frac{1}{4}(2b^3 - 9abc + 27a^2d)^2 + (3a^2c - ab^2)^3} +}$$

$$+ \frac{-1}{6a} \sqrt[3]{\frac{1}{2}(2b^3 - 9abc + 27a^2d) - \sqrt{\frac{1}{4}(2b^3 - 9abc + 27a^2d)^2 + (3a^2c - ab^2)^3}}$$

$$+ \frac{-\sqrt{-3}}{6a} \sqrt[3]{\frac{1}{2}(2b^3 - 9abc + 27a^2d) + \sqrt{\frac{1}{4}(2b^3 - 9abc + 27a^2d)^2 + (3a^2c - ab^2)^3}}$$

$$+ \frac{\sqrt{-3}}{6a} \sqrt[3]{\frac{1}{2}(2b^3 - 9abc + 27a^2d) - \sqrt{\frac{1}{4}(2b^3 - 9abc + 27a^2d)^2 + (3a^2c - ab^2)^3}}.$$

The formulas for the zeros of the general fourth degree polynomial are entirely too complicated to write out as we did for the general cubic. Instead, we exhibit a flowchart to find the zeros z_1, z_2, z_3, and z_4 of the polynomial $ax^4 + bx^3 + cx^2 + dx + e$. Note the need to find a zero of a cubic polynomial in one step.

START

INPUT a, b, c, d, e

$$F \leftarrow -\frac{c}{a}$$
$$G \leftarrow \frac{bd - 4ea}{a^2}$$
$$H \leftarrow \frac{-b^2e + 4ace - ad^2}{a^3}$$

$$I \leftarrow \text{ANY ZERO OF } X^3 + FX^2 + GX + H$$

$$J \leftarrow \sqrt{\frac{b^2 - 4ac + Ia^2}{4a^2}}$$

IS J = 0 ?

YES

$$K \leftarrow \sqrt{\frac{3b^2 - 8ac}{4a^2} + 2\sqrt{I^2 - \frac{4e}{a}}}$$
$$L \leftarrow \sqrt{\frac{3b^2 - 8ac}{4a^2} - 2\sqrt{I^2 - \frac{4e}{a}}}$$

NO

$$K \leftarrow \sqrt{\frac{3b^2 - 4a^2J^2 - 8ac}{4a^2} + \frac{4abc - 8a^2d - b^3}{4a^3J}}$$
$$L \leftarrow \sqrt{\frac{3b^2 - 4a^2J^2 - 8ac}{4a^2} - \frac{4abc - 8a^2d - b^3}{4a^3J}}$$

$$z_1 \leftarrow -\frac{b}{4a} + \frac{J}{2} + \frac{K}{2}$$
$$z_2 \leftarrow -\frac{b}{4a} + \frac{J}{2} - \frac{K}{2}$$
$$z_3 \leftarrow -\frac{b}{4a} - \frac{J}{2} + \frac{L}{2}$$
$$z_4 \leftarrow -\frac{b}{4a} - \frac{J}{R} - \frac{L}{2}$$

OUTPUT z_1, z_2, z_3, z_4

STOP

APPENDIX F

ANSWERS TO ODD-NUMBERED EXERCISES

This appendix contains the answers to the odd-numbered exercises in the following sense : For each exercise in the Practice Exercise sections, answers are given for the " odd " parts of exercises (i.e., (a), (c), (e), (g), $\cdots$), while the traditional definition of odd and even is used in the Problem sections.

Answers are not ordinarily provided for exercises that ask for proofs or explanations.

Some answers depend upon the particular environment of the reader (such as " List the elements in the set of your siblings "), while others are in no way unique (such as " Give an example of a quadratic equation " or " Write a flowchart to computenet pay $\cdots$ "). Answers to such exercises are preceded by a bullet ($\bullet$). The former are then given a hypothetical answer, while the latter are given one of the many possible correct answers.

CHAPTER: PRELIMINARIES

Practice Exercises

1. $\bullet$(a) {Elementary functions, English composition, Art history, Physical Education}
   (c) {Mercury, Venus, Earth, Mars, Jupiter, Saturn, Uranus, Neptune, Pluto}

(e) {21, 23, 25, 27, 29, ···}

(g) {r, o, s, t, e}

(i) {1, 4, 9, 16, 25, 36, ···}

(k) {10, 11, 12, 13, 14, 15}

(m) {Alabama, Alaska, Arizona, ··· , Wisconsin, Wyoming}

2. (a) $\{x \mid x$ is a book in our school library$\}$.

(c) $\{x \mid x$ is a U.S. senator now in office$\}$.

(e) $\{x \mid x$ is a star in our galaxy$\}$.

(g) $\{x \mid x$ is a set containing precisely 3 elements$\}$.

(i) $\{x \mid x$ is a positive integer multiple of 10$\}$.

(k) $\{x \mid x$ is a negative integer$\}$.

(m) $\{x \mid x$ is a continent on our planet$\}$.

3. ● (a) The set containing Marvin Springspeed and no one else

  ● (c) The set containing Welfare Reform No. 36, Penal Code Revisions No. 3, Allocations to NASA, and no others

(e) The set containing all of the capital letters and all of the small letters

(g) The set of all positive numbers

(i) The set of all negative numbers

(j) The set containing all (familiar) English words containing the letters p, o, s, t, once and only once

(m) The set containing the names used to describe types of triangles

4. (a) No (c) Yes (e) No

(g) Yes (i) Yes (k) No

5. (a) F; $\subseteq$ (c) F; $\subseteq$ (e) F; $\in$

(g) T (i) T (k) T

(m) F; $\supseteq$ (o) F; $\in$

6. (a) $\{a\}, \{b\}, \{a, b\}$

(c) $\{*\}, \{?\}, \{"\}, \{@\}, \{*, ?\}, \{*, "\}, \{*, @\}, \{?, "\}, \{?, @\}, \{", @\}, \{*, ?, "\}, \{*, ?, @\},$ $\{*, ", @\}, \{?, ", @\}, \{*, ?, ", @\}$

7. (a) {yellow, green}

(c) {red, green, blue, lavender, brown, yellow}

(e) {green}

(g) {red, orange, yellow, green, blue, gold, black, lavender, brown}

(i) {blue, brown}

(k) {red, orange, yellow, green, blue, brown, black, gold}

8. (a) $\{4, 5, \cdots, 10\}$

(c) $\{d, e, f, \cdots, r, s\}$

(e) {2, 3, 4, 6, 7, 8, 9, 11, 12, 13, 14, 15} or $\{x \mid 1 \leq x \leq 15$ and $x \neq 1, 5,$ or $10\}$

(g) {blue, purple, brown, black}

(i) {red, orange, yellow, green, purple}

(k) {Saturday, Sunday}

(m) $\{y \mid y$ is an English word with more than three syllables$\}$

9. (a) {1, 3, 5, 7, 9}

(c) {2, 4}

(e) {3, 5}

(g) {2, 3, 4, 5}

(i) {7, 9}

(k) {7, 9}

(m) {1, 2, 3, 4, 5, 7, 9}

(o) $\{1, 2, 3, 4, 5, 6, 7, 8, 9, 10\} = U$

(q) {1, 3, 5}

(s) $\{7, 9\}$

(u) $\{1, 2, 3, 4, 5, 7, 9\}$

(w) $\{2, 4, 6, 8, 10\} = w$

(y) $\{6, 8, 10\}$

10. (a) $\{x \mid x$ is a female mathematics major$\}$.

(c) $\{x \mid x$ is a sophomore male mathematics major$\}$.

(e) $\{x \mid x$ is a sophmore (of either sex)$\}$.

(g) $\{x \mid x$ is a sophomore and not a mathematics major$\}$.

11. • (a) $A \cap A'$ for any set A; the set of snowflakes at 80° F; $\{x \mid x < 0$ and $x > 0\}$

12. (a) $(7, \infty)$ (c) $(-\infty, 0)$ (e) $[-16, \infty)$

(g) $[-1, \frac{4}{5})$ (i) $[-1, \frac{4}{5}]$

13. (a) $\{x \mid x > 2$ and $x < 3\}$ (c) $\{x \mid x \geq 3$ and $x \leq 21\}$

(e) $\{x \mid x < 3\}$ (g) $\{x \mid x > -2$ and $x < 30\}$

(i) $\{x \mid x > 10$ and $x < 15 \frac{1}{2}\}$

14. (a) Valid

(b) Not valid: $+\infty$ cannot be included, i.e., it should read $(0, +\infty)$.

(e) Valid

(g) Not valid: there are no numbers between 2 and 2 that do not include 2.

15. Output:

Bottom	Answer
1	1
2	$1/2 = .5$
3	$1/3 = .333 \cdots$
4	$1/4 = .25$
$\vdots$	$\vdots$
9	$1/9 = .111 \cdots$

Purpose: might be used to produce a table of decimal equivalents of reciprocals of integers through 9

17. Output:

X	Y	Z
1	1	1
2	4	8
3	9	27
4	16	64
5	25	125
$\vdots$	$\vdots$	$\vdots$
9	81	729

Purpose: to compute a table of squares and cubes of integers from 1 through 9

19. Output: $-2, 1, 3, 5, 6, 10$

Purpose: orders a list of six numbers from lowest to highest

21. Using data 70, 80, 90, output is:

SCORE$_1$	SCORE$_2$	SCORE$_3$	AVE
70	80	90	80

Purpose: to compute the average of three scores

Problems

23. $B = \{$chocolate, licorice, grape, raspberry, mint, cherry$\}$

25. Let $A = \{1, 2\}$, $B = \{\{1, 2\}, \{3, 4\}\}$, and $C = \{\{3\}, \{\{1, 2\}\}, \{3, 4\}, \{5, 6, 7\}\}$. Then $A \in B$, $B \in C$, but $A = \{1, 2\} \notin C$.

27. (a) False, they are in fact disjoint.   (b) No, intervals are not numbers.

(c) True   (d) True

(e) False, $2 \notin$ left side   (f) True

(g) False, 3 is not a set (h) True
(i) True

29. $\mathbb{N} \subseteq \mathbb{N}, \mathbb{Z} \subseteq \mathbb{Z}, \mathbb{Q} \subseteq \mathbb{Q}, \mathbb{R} \subseteq \mathbb{R}$
$\mathbb{N} \subseteq \mathbb{Z}, \mathbb{Z} \subseteq \mathbb{Q}, \mathbb{Q} \subseteq \mathbb{R}$
$\mathbb{N} \subseteq \mathbb{Q}, \mathbb{Z} \subseteq \mathbb{R}$
$\mathbb{N} \subseteq \mathbb{R}$

31. (a) $\{x \mid x > 0 \text{ and } x^2 = y\}$ (b) $\{x \mid x \in \mathbb{R} \text{ and } x^2 < 5\}$
(c) $\{x \mid x = n^2 \text{ for some } n \in \mathbb{N}\}$ (d) $\{x \mid x \text{ is an irrational number}\}$
(e) $\{x \mid x \text{ is a rational number} \in (0, 1)\}$

35. (a) $A \triangle B = (A \cap B') \cup (B \cap A')$ (b) Shaded area represents $A \triangle B$.

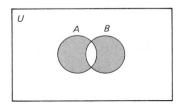

(c) Yes   (d) Yes
(e) Yes   (f) No

●37.

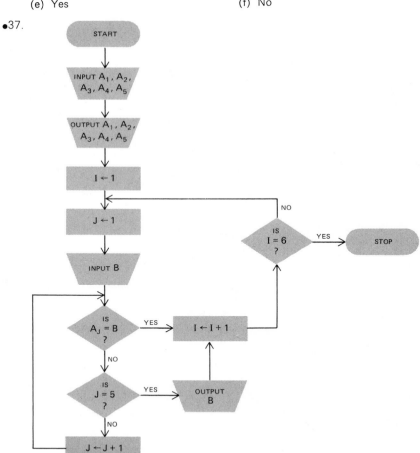

•39.

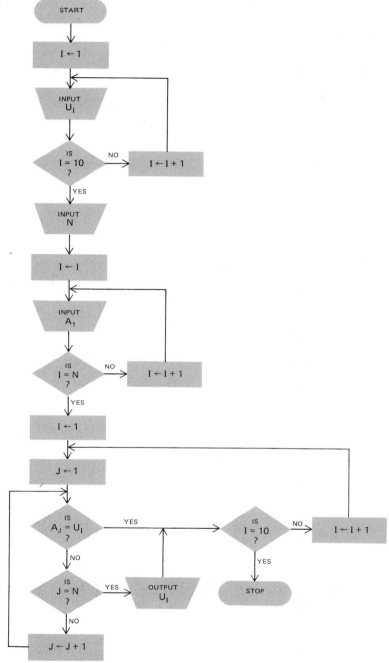

•41. Parts (a)–(d) are all quite similar; we exhibit only part (d).

(d)

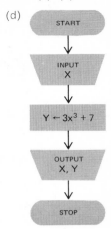

START

INPUT
X

$Y \leftarrow 3x^3 + 7$

OUTPUT
X, Y

STOP

•43. Parts (a)–(c) are intended to lead you to the complexity of part (d). We exhibit only part (d).

(d)

START

$I \leftarrow 1$

INPUT
X, N

$Y \leftarrow X$

$J \leftarrow 1$

$Y \leftarrow 2Y$

IS
J = N
?
→ NO → $J \leftarrow J + 1$

YES

OUTPUT
X, Y

IS
I = M
?
YES → STOP
NO → $I \leftarrow I + 1$

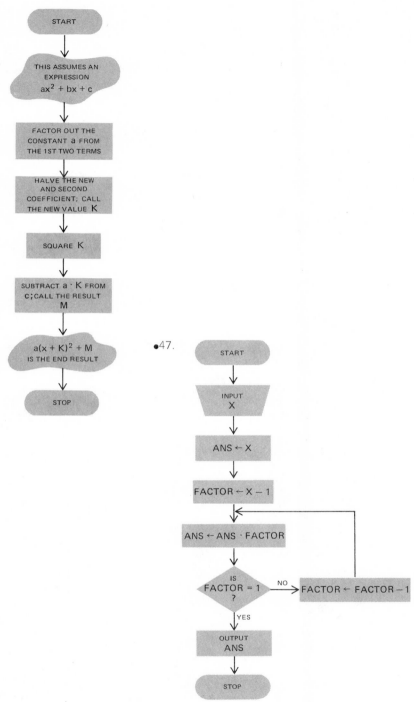

●45.

START

THIS ASSUMES AN
EXPRESSION
$ax^2 + bx + c$

FACTOR OUT THE
CONSTANT a FROM
THE 1ST TWO TERMS

HALVE THE NEW
AND SECOND
COEFFICIENT; CALL
THE NEW VALUE K

SQUARE K

SUBTRACT $a \cdot K$ FROM
c; CALL THE RESULT
M

$a(x + K)^2 + M$
IS THE END RESULT

STOP

●47.

START

INPUT
X

$ANS \leftarrow X$

$FACTOR \leftarrow X - 1$

$ANS \leftarrow ANS \cdot FACTOR$

IS
$FACTOR = 1$
?

NO → $FACTOR \leftarrow FACTOR - 1$

YES

OUTPUT
ANS

STOP

●49.

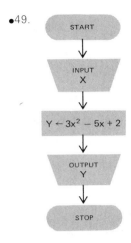

●51.

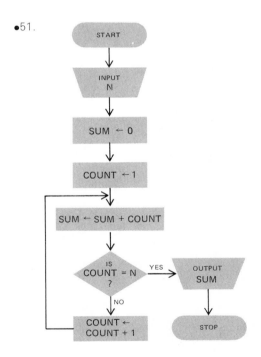

●53.

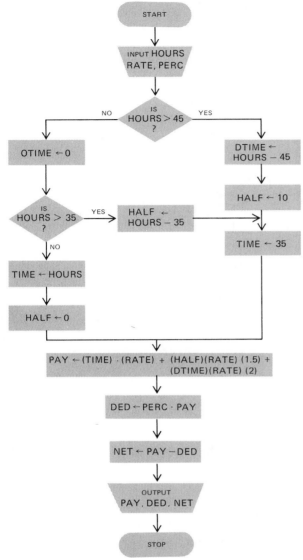

START

INPUT HOURS
RATE, PERC

IS
HOURS > 45
?

NO — OTIME ← 0

YES — DTIME ←
HOURS − 45

HALF ← 10

IS
HOURS > 35
?

YES — HALF ←
HOURS − 35

TIME ← 35

NO

TIME ← HOURS

HALF ← 0

PAY ← (TIME) ⋅ (RATE) + (HALF)(RATE) (1.5) +
(DTIME)(RATE) (2)

DED ← PERC ⋅ PAY

NET ← PAY − DED

OUTPUT
PAY, DED, NET

STOP

●55.

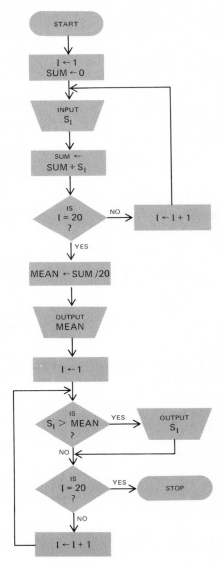

START

I ← 1
SUM ← 0

INPUT
S_I

SUM ←
SUM + S_I

IS
I = 20
?

NO → I ← I + 1

YES

MEAN ← SUM /20

OUTPUT
MEAN

I ← 1

IS
S_I > MEAN
?

YES → OUTPUT
S_I

NO

IS
I = 20
?

YES → STOP

NO

I ← I + 1

•57.

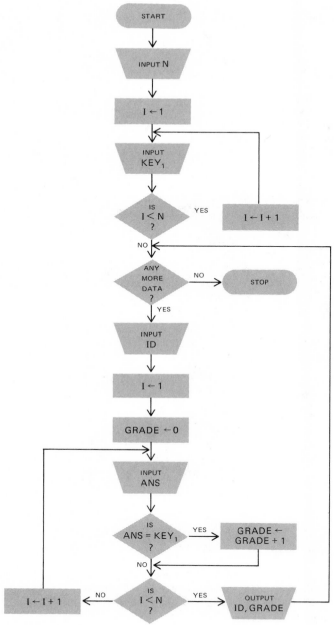

•59.

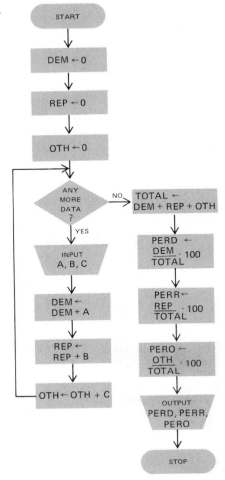

CHAPTER 1

Practice Exercises

1. The first set in each pair is the domain, the second the range.
   - (a) $\{w, c, 2, 3\}, \{x, z, 9, 4\}$
   - (c) $\{a, q, m, e\}, \{b, t, z\}$
   - (e) $\{Bach, Brahms, Beethoven\}, \{B\}$
   - (g) $\{1, 2, 5, 7\}, \{1, 2, 5, 7\}$
   - (i) $\{a, z, g\}, \{b, m, f, t, a\}$
   - (k) $\mathbb{R}, \mathbb{R}$
   - (m) $\mathbb{R}, \mathbb{R}$
   - (o) $\mathbb{R}, [0, +\infty)$
   - (q) $[-1, +\infty), [0, +\infty)$
   - (s) $\mathbb{R} - \{0\}, \mathbb{R} - \{0\}$
   - (u) $\mathbb{R} - \{0\}, (-\infty, -2] \cup [2, +\infty)$ though a verification of this is probably beyond your ability now.
   - (w) $\{x \mid x \subseteq \mathbb{R}\}, \{x \mid x \subseteq \mathbb{R}\}$
   - (y) $\mathbb{R}, \mathbb{R}$
2. (a) Yes
   - (c) Yes
   - (e) Yes
   - (g) Yes

(i) No; (a, b), (a, t)

(k) Yes

(m) Yes

(o) Yes

(q) Yes

(s) Yes

(u) Yes

(w) No; $(\{1\}, \mathbb{Z})$, $(\{1\}, \mathbb{N})$

(y) No; $(3, 5)$, $(3, -7)$

3. (a) 16

(c) 191

(e) 16

(g) $-\frac{1}{2}$

(i) 7

(k) 3

(m) 77

(o) -4872

(q) $\frac{17}{19}$

(s) 4.84

(u) $\frac{53}{23}$

(w) 22.09

(y) $-2\frac{2499}{2500}$

4. (a) $\dfrac{x^2}{x^2 - 3}$

(c) $x^2 + 4x + 4$

(e) $8x^3 - 18x^2 + 50x - 14$

(g) $4x^2$

(i) $3ab + 2$

(k) $a^2 + 2ab + b^2$

(m) $a - 3$

(o) 121

(q) $3x - 3y + 2$

5. (a) 3

(c) 64

(e) -7

(g) 0

(i) 1000

(k) 25

(m) -2

(o) $16x^4 - 96x^3 + 216x^2 - 216x + 81$

6. (a)

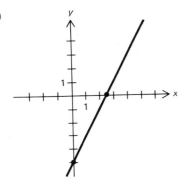

(c)

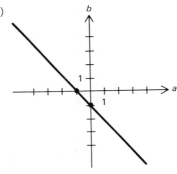

(e)

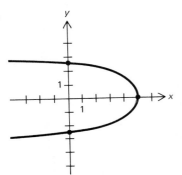

(f)

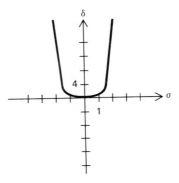

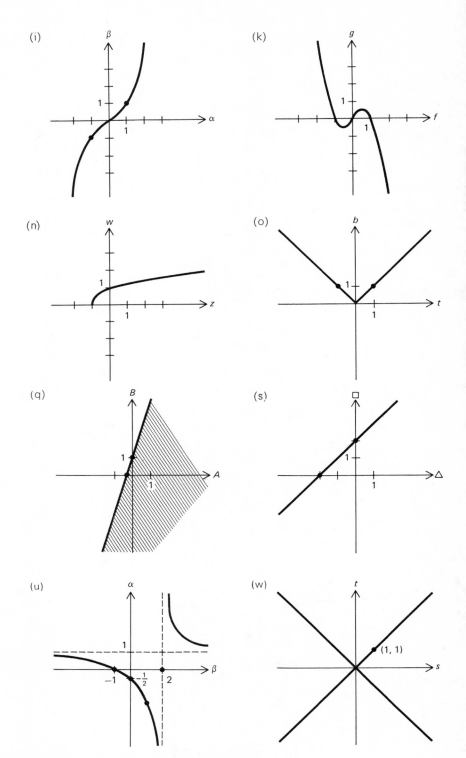

(y)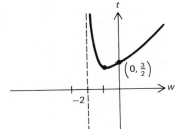

(aa)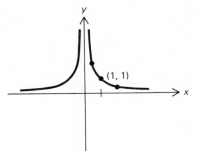

(cc)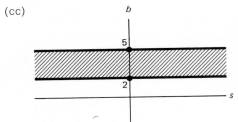

7. (a) $2 - 3\sqrt{2}$

(c) 2

(e) $\frac{11}{9}$

(g) $\sqrt{\dfrac{1}{x} - 3}$

(i) $\sqrt{\dfrac{2(2-x)}{x-1}}$

(k) $\dfrac{1}{3x^2(x-3)^2}$

(m) x (note $x \neq 1$)

(o) 4

(q) $3 + 2\sqrt{2}$

(s) 1

(u) 1 (note $x \neq 0$)

(w) $\dfrac{3-7x}{1-x}$ (note $x \neq 0$)

(y) $\dfrac{x - 1 + 2\sqrt{(2-x)(2x-3)}}{3x - 5}$ (note $x \neq 2$)

8. (a) $-\frac{3}{10}$

(c) -1

(e) $\frac{1}{3000}$

(g) 6

(i) $x - 3$ (note $x \neq 0$)

(k) $x^3 - 3x^2$ (note $x \neq 0$)

(m) $\dfrac{-2(x^2 - 4x + 2)}{x - 1}$

(o) $9 - 2x$

(q) $1/4x$

(s) 0

(u) $4\dfrac{1}{21609}$

(w) $\dfrac{1 - x}{2(x^2 - 4x + 2)}$

(y) $x^2 - 6x + 9$ (note $x \neq 0$)

9. (a) $\sqrt{x^2 - 5x + 2}$

(c) $\dfrac{4x - 3}{3x^3(x - 3)}$

(e) $\dfrac{3(11 - 4x)}{9 - 4x}$

(g) $\dfrac{x^2(1 + x)}{1(1 - x)(1 - 3x)}$ (note $x \neq 0, +3$)

(i) $\dfrac{4 - 15x}{x^2}$

(k) $\dfrac{x^3 + 3x^2 + 3x + 1}{x^3 - 3x^2 + 3x - 1}$

(m) $5 - 6x$

(o) $\frac{4}{3}$

(q) $x^2 + 2xw + w^2 - 3x - 3w$

(s) 4

10. (Answers here are not the only possible forms.)
    (a) $h(x) \cdot g(x)$, where $h(x) = x - 2$, $g(x) = x + 1$
    (c) $h(x) \cdot g(x)$, where $h(x) = 3$, $g(x) = x - 2$
    (e) $h(x) \cdot g(x)$, where $h(x) = x - 3$, $g(x) = x - 4$

11. (Answers here are not the only possible forms.)
    (a) $(h \circ g)(x)$, where $h(x) = x - 16$, $g(x) = x^2$
    (c) $(h \circ g)(x)$, where $h(x) = 1/x$, $g(x) = x + 2$
    (e) $(h \circ g)(x)$, where $h(x) = 1/\sqrt{x}$, $g(x) = 2 - x$

12. A few samples are given for each:
    (a) $f + h$, where $f(x) = x^2$, $h(x) = -16$
        $f \cdot h$, where $f(x) = x + 4$, $h(x) = x - 4$
        $f \circ h$, where $f(x) = x - 16$, $h(x) = x^2$
    (c) $f - g$, where $f(x) = x^3 + x$, $g(x) = -3x$
        $f \cdot g$, where $f(x) = x$, $g(x) = x^2 - 2$
        $h \cdot (f \circ g)$, where $h(x) = x$, $f(x) = x - 2$, $g(x) = x^2$
    (e) $g \circ h$, where $g(x) = \sqrt{x}$, $h(x) = x^2 - 7$
        $f \circ (g + h)$, where $f(x) = \sqrt{x}$, $g(x) = x^2$, $h(x) = -7$
        $f \cdot g$, where $f(x) = \sqrt{x + \sqrt{7}}$, $g(x) = \sqrt{x - \sqrt{7}}$

13. (a) $-\frac{21}{47}$, $-\frac{21}{47}$: functional composition is associative.
    (c) $234, 234$: multiplication of functions distributes over addition of functions.
    (e) $x^3 + 8x^2 + 19x + 12$, $x^3 + 8x^2 + 19x + 12$: multiplication of functions is associative.

    (g) $x^2 + 3x + \dfrac{3x}{2 - x^2}$, $x^2 + 3x + \dfrac{3x}{2 - x^2}$: addition of functions is commutative.

    (i) $\dfrac{3x(1 + x)}{x^2 - 2x - 1}$, $\dfrac{2 + 3x - x^2}{3x}$: functional composition is not commutative.

14. (a) i, iii, iv   (c) i, ii   (e) i, iii, iv
    (g) i, ii   (i) ii, iii, iv

15. (a) $\{(2, 1), (4, 3), (6, 5), (8, 7), \cdots\}$
    (c) $\{(2, 1), (4, 3), (3, 4), (1, 2)\}$
    (e) $\{(1, -1), (2, -2), \cdots, (99, -99)\}$
    (g) $\{(y, x) \mid x = (y - 2)/3\}$
    (i) $\{(w, k) \mid k$ is a senator from the state of $w\}$

16. (a) $\{\frac{7}{5}\}$ (c) $\{-1\}$ (e) $[4, 5)$

(g) $\{1, -1\}$ (i) $\{4, -6\}$ (k) $[0, 1)$

(m) $[8, 9)$ (o) $\varnothing$

17. (a) $(-\infty, -3], (-3, 2], (2, +\infty)$

(c) $(-\infty, -4), [-4, -2), [-2, 1), [1, 3], (3, +\infty)$

(e) $(-\infty, -40], (-40, -30), [-30, 10), (10, +\infty)$

(g) $\cdots (-7, -4], (-4, -3], (-3, 0], (0, 1], (1, 4], (4, 5], \cdots$

(a)

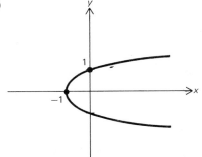

(c)

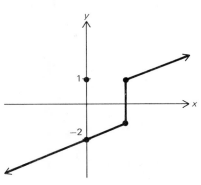

(e)

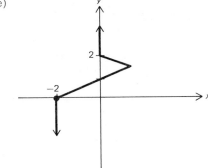

(g)

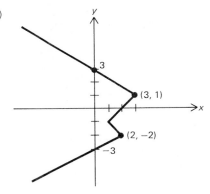

(i)

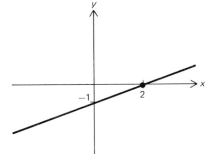

(k)

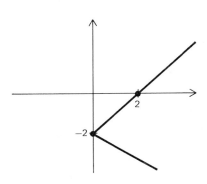

19. (a) f^{-1} (c) f (e) f (g) f^{-1}

20. (a) $\left\{ (s, r) \,\middle|\, r = \dfrac{16 - s}{3} \right\}$

(c) $\{(b, a) \,|\, a = \pm \sqrt{2 - b}\}$

(e) $\{(t, s) \,|\, s = t^2 - 2t + 1 \text{ and } t \geq 1\}$

(g) $\{(k, m) \,|\, m = (k - 12)/5\}$

(i) $\{(e, f) \,|\, f = e^2 + 2, e \geq 0\}$

(k) $\{(t, s) \,|\, s = \pm t, t \geq 0\}$

(m) $\{(r, y) \,|\, y = (-3 \pm \sqrt{r})/2\}$

(o) $\{(\beta, \alpha) \,|\, \alpha = \pm \sqrt{4 \pm \beta}, \beta \geq 0\}$

(q) $\{(b, a) \,|\, b = 0, a \in [0, \infty)\} \cup \{(b, a) \,|\, a = b/2, b < 0\}$

Problems

21. No, contains many pairs of ordered pairs with the same first element (e.g., Jay Smith, Tom Smith) and (Jay Smith, Sally Miller); domain $= \{x \,|\, x$ is a living person$\}$, range $= \{x \,|\, x$ is the name of a parent of someone now alive$\}$, $\{(x, y) \,|\, y$ is the name of a parent of x and x is a living human$\}$.

23. No, n is not a function, since each state has two senators (hence many cases of two ordered pairs with the same first element); m is a function since each senator comes from only one state (officially). $D_n = R_m = \{x \,|\, x$ is a state in the USA$\}$ $D_m = R_n = \{x \,|\, x$ is a U.S. senator$\}$.

25. A few samples would be:
    (a) Historians associate important events with key personalities (not a function) and particular wars with various causative factors (not a function).
    (b) Mechanics associate certain cars with their owners (may be a function) and particular models of car with various mechanical quirks (not a function either).
    (c) A lawyer tries to discover relations that relate clients with facts surrounding their legal problems.

27. (a) $C(w) = 28w$ (b) $C(w) = 36w$

29. $v(x) = \dfrac{5x}{29} (20 - 3x^2)$

31. $L(d) = 4\sqrt{100^2 + d^2}$

33. $A(b) = 35 + \dfrac{5}{2} b + \dfrac{60}{b}$

35. $V(x) = 4x^3 - 120x^2 + 875x$

37. (a) $d(c) = \sqrt{900 + 36c^2/25}$ (b) $L(c) = c/5$

39. $T(x) = \dfrac{\sqrt{x^2 + 1/4}}{30} + \dfrac{10 - x}{70}$

41. (a) Yes
    (b) Yes
    (c) $\{(x, z) \,|\, x$ is a student from the USA and z is x's home state$\}$.
    (d) $\{x \,|\, x$ is a student from the USA$\}$.
    (e) $\{s \,|\, s$ is a state in the USA$\}$ (assuming the universe of discourse is large enough).

•43. (b) The relation that associates with each living person the names of each living person in his immediate family (himself included) ; the relation associating with each real number the digits used in the decimal representation of the number.

•45.

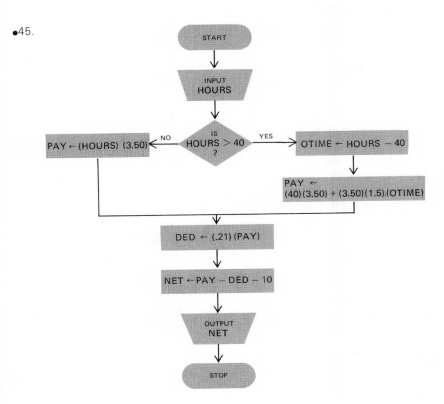

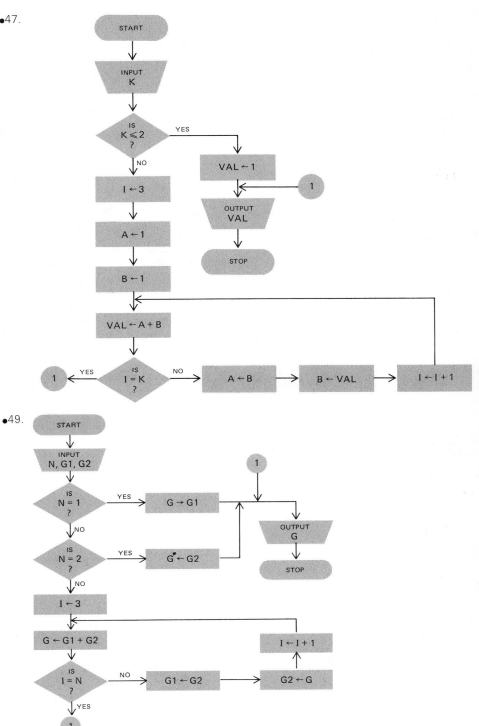

●51.

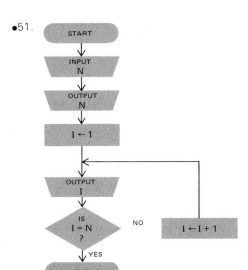

●53.

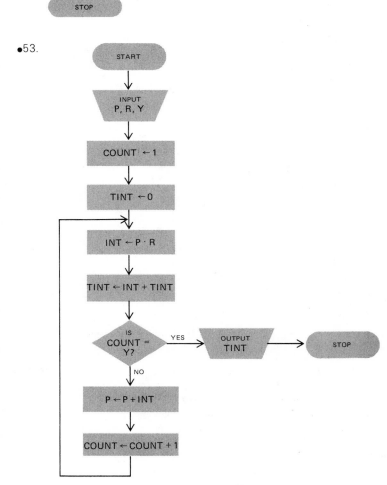

55. Quadrant I $= \{(x, y) \mid x > 0 \text{ and } y > 0\}$
Quadrant II $= \{(x, y) \mid x < 0 \text{ and } y > 0\}$
Quadrant III $= \{(x, y) \mid x < 0 \text{ and } y < 0\}$
Quadrant IV $= \{(x, y) \mid x > 0 \text{ and } y < 0\}$

•57. (a) (i)

(ii)

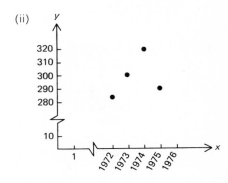

(b) (i)

(ii)

(c) (i)

(ii)

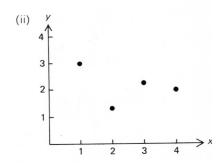

59. If $R_g \nsubseteq D_f$, then some $x \in D_g$ cannot be associated with a $z \in R_f$ because g will not accept $f(x)$ as an argument.

61. (a) The set of ordered pairs common to the two functions
(b) The set of simultaneous solutions of the two equations
(c) The set of points of intersection of the graphs of f and g

69. No

71. The constant function 1

75. (a) Function   (b) Function

(c) Not a function
(d) Not a function
(e) Function
(f) Function
(g) Function
(h) Not a function
(i) Function
(j) Not a function
(k) Not a function
(l) Not a function
(m) Not a function
(n) Not a function
(o) Not a function
(p) Function
(q) Not a function
(r) Not a function

77. No

81. None exists. Consider $f(x) = x$; $g(x) = x^3$.

83. $C = \{(a, s), (b, n), (c, v), (d, f), (e, r), (f, g), (g, h),$
        $(h, j), (i, o), (j, k), (k, l), (l, ;), (m, .), (n, m),$
        $(o, p), (p,]), (q, w), (r, t), (s, d), (t, y), (u, i),$
        $(v, b), (w, e), (x, c), (y, u), (z, x)\}$

   $C^{-1} = \{(b, v), (c, x), (d, s), (e, w), (f, d), (g, f), (h, g),$
       $(i, u), (j, h), (k, j), (l, k), (m, n), (o, i), (p, o),$
       $(r, e), (s, a), (t, r), (u, y), (v, c), (w, q), (x, z),$
       $(y, t), (., m), (], p), (:, 1), (n, b)\}$

87. (a) $R_f = [4, 8]$ (b) $[0, 2]$

89. (a)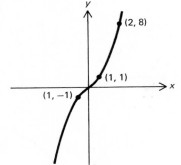

(b)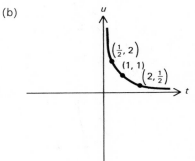

(c)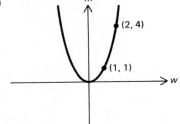

(d)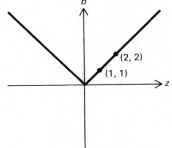

(e)

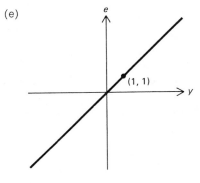

(f)

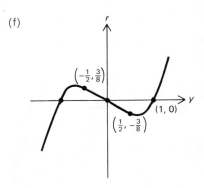

(g)

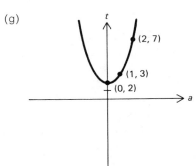

(h)

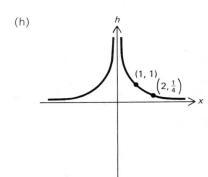

(i)

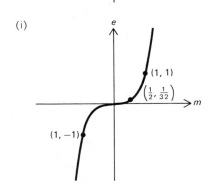

(j)

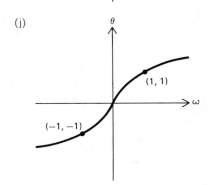

(k)

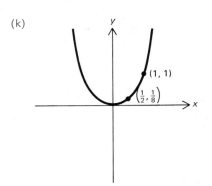

(l)

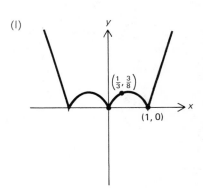

91. (a) Even   (b) Even   (c) Odd
    (d) Even   (e) Even   (f) Neither
    (g) Odd   (h) Odd   (i) Even
    (j) Neither   (k) Odd   (l) Odd
95. (a) No   (b) Yes   (c) No
    (d) Yes   (e) Yes   (f) Yes
    (g) No   (h) Yes   (i) No
    (j) Yes   (k) No

CHAPTER 2

Practice Exercises

1. (a)

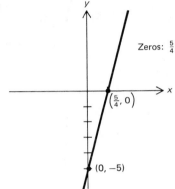

Zeros: $\frac{5}{4}$

$\left(\frac{5}{4}, 0\right)$

$(0, -5)$

(c)

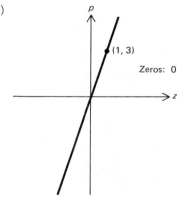

$(1, 3)$

Zeros: 0

(e)

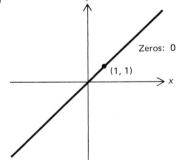

Zeros: 0

$(1, 1)$

(g)

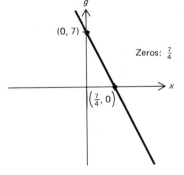

$(0, 7)$

Zeros: $\frac{7}{4}$

$\left(\frac{7}{4}, 0\right)$

(i)

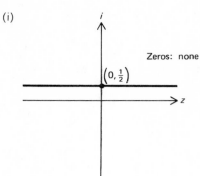

Zeros: none

(0, ½)

(k)

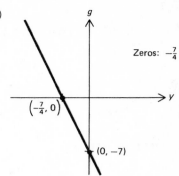

Zeros: $-\frac{7}{4}$

$\left(-\frac{7}{4}, 0\right)$

(0, −7)

(m)

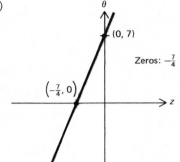

(0, 7)

Zeros: $-\frac{7}{4}$

$\left(-\frac{7}{4}, 0\right)$

(o)

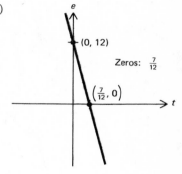

(0, 12)

Zeros: $\frac{7}{12}$

$\left(\frac{7}{12}, 0\right)$

(q)

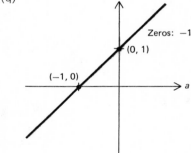

Zeros: −1

(0, 1)

(−1, 0)

2. (a) $\dfrac{1 + \sqrt{5}}{2}, \dfrac{1 - \sqrt{5}}{2}$

(c) No real zeros

(e) $\dfrac{-2 + \sqrt{10}}{3}, \dfrac{-2 - \sqrt{10}}{3}$

(g) No real zeros

(i) No real zeros

(k) $\dfrac{3 + \sqrt{13}}{2}, \dfrac{3 - \sqrt{13}}{2}$

(m) $\dfrac{-2 + \sqrt{19}}{3}$, $\dfrac{-2 - \sqrt{19}}{3}$

(o) $\dfrac{3 + \sqrt{29}}{2}$, $\dfrac{3 - \sqrt{29}}{2}$

(q) No real zeros

(s) $\dfrac{15 + \sqrt{229}}{2}$, $\dfrac{15 - \sqrt{229}}{2}$

(u) $\frac{2}{5}$, repeated

3. (a)

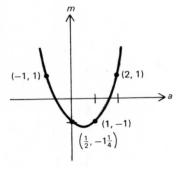

(c)

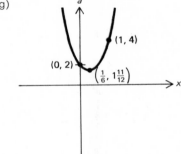

(e)

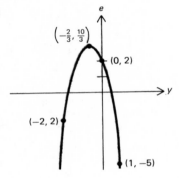

(g)

(i)

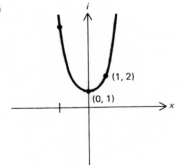

(k)

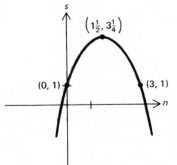

(m)

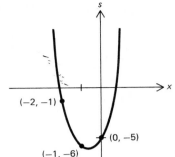

(−2, −1)

(0, −5)

(−1, −6)

(o)

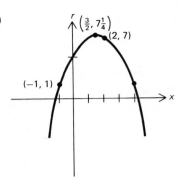

$\left(\frac{3}{2}, 7\frac{1}{4}\right)$

(2, 7)

(−1, 1)

(q)

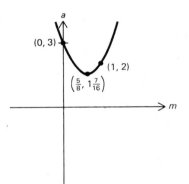

(0, 3)

(1, 2)

$\left(\frac{5}{8}, 1\frac{7}{16}\right)$

(s)

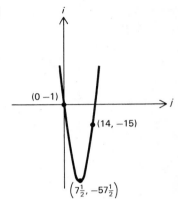

(0 −1)

(14, −15)

$\left(7\frac{1}{2}, -57\frac{1}{2}\right)$

(u)

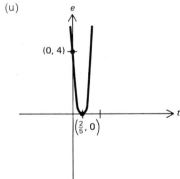

(0, 4)

$\left(\frac{2}{5}, 0\right)$

4. (a) Yes, 10   (c) No, square root
   (e) Yes, 2   (g) Yes, 1
   (i) No, absolute value   (k) Yes, 1
   (m) Yes, 5   (o) Yes, 6
   (q) Yes, 100

5. (a) 1, 3, −2   (c) +1, −1   (e) −4, −$\frac{1}{2}$, $\frac{2}{5}$
   (g) 1   (i) None   (k) $\frac{2}{3}$
   (m) 1, −1

6. (a) 0, 2, 3
   (e) 0, 6, −6
   (i) $-\sqrt{2}, \sqrt{2}$
   (m) $\sqrt{9/5}, -\sqrt{9/5}$

   (c) 1, 2, 3, 4
   (g) −1, +1, +2
   (k) $-\frac{4}{3}, +\sqrt{5/2}, -\sqrt{5/2}$

7. (a) $\left(x - \dfrac{5 + \sqrt{17}}{2}\right)\left(x - \dfrac{5 - \sqrt{17}}{2}\right)$

   (e) $(x + 2)^2(x - 3)(x + 1)$

   (i) $\left(x + \dfrac{3 - \sqrt{17}}{2}\right)\left(x + \dfrac{3 + \sqrt{17}}{2}\right)$

   (c) $(x + 5)(x - 3)(3x^2 - 3x + 1)$

   (g) $(x + 2)(x + 3)(x + 4)$

8. (a) $\sqrt{3}, -\sqrt{3}, 3$
   (e) 2, 2, 2, 2
   (i) −1 of multiplicity 3, $-\frac{1}{2}$ of multiplicity 1

   (c) 2, $\sqrt{5}, -\sqrt{5}$
   (g) 1, 1
   (k) $-\frac{1}{3}, \sqrt{3}, -\sqrt{3}$

9. (a) $3x^2 + 11x + 70$ with remainder 418
   (c) $x^4 - 2x^3 + 2x^2 - 4x + 9$ with remainder −19
   (e) $x^3 - x^2 + x - 1$ with remainder 2
   (g) $x^2 + 5x + 17$ with remainder 24
   (i) $x^2 - 3x + 9$ with remainder 0
   (k) $x^3 - 8x^2 - 23x + 210$ with remainder 0

10. (a) 229   (c) −8   (e) −333
    (g) 2079   (i) 0

11. (a) $x^2 - 2x + 2$ (c) $x^2 - x - 20$
    (e) $x^3 - 6x^2 + 12x - 8$

12. (a)

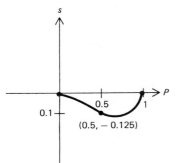

    (c)

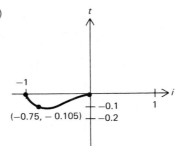

13. (a)   (c)   (e)

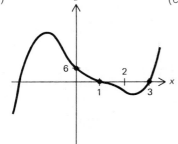
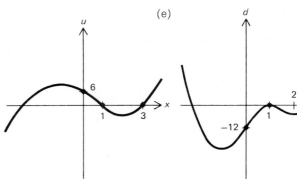

(g)

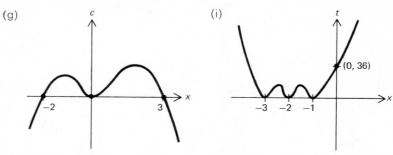

(i)

14. (a) The fifth midpoint is 1.40625. (c) The fifth midpoint is 2.0625.

Problems

17. (b) $-b/a$, c/a

●23.

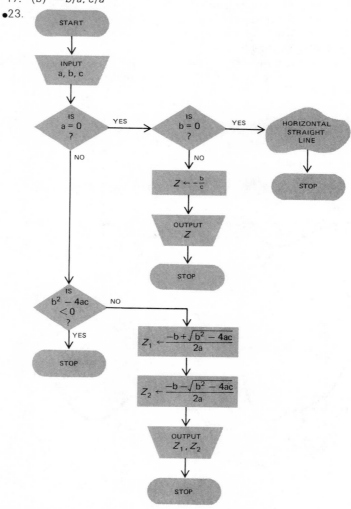

●33.

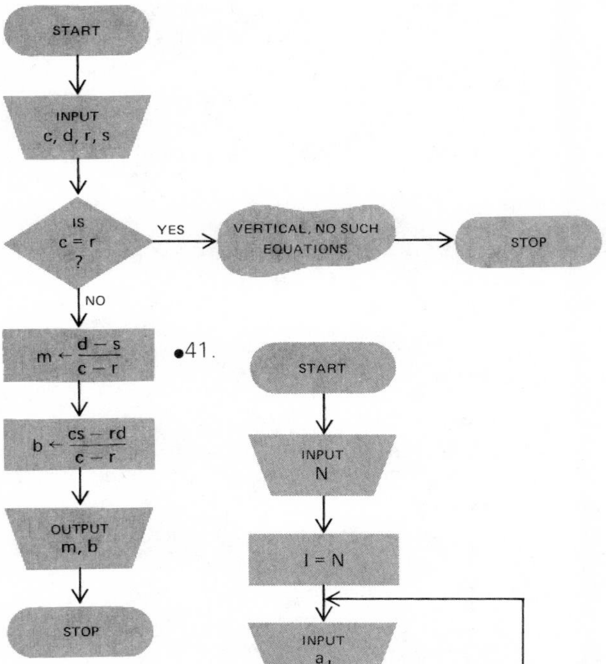

START

INPUT
c, d, r, s

IS
c = r
?

YES → VERTICAL, NO SUCH EQUATIONS → STOP

NO

$m \leftarrow \dfrac{d - s}{c - r}$

$b \leftarrow \dfrac{cs - rd}{c - r}$

OUTPUT
m, b

STOP

●41.

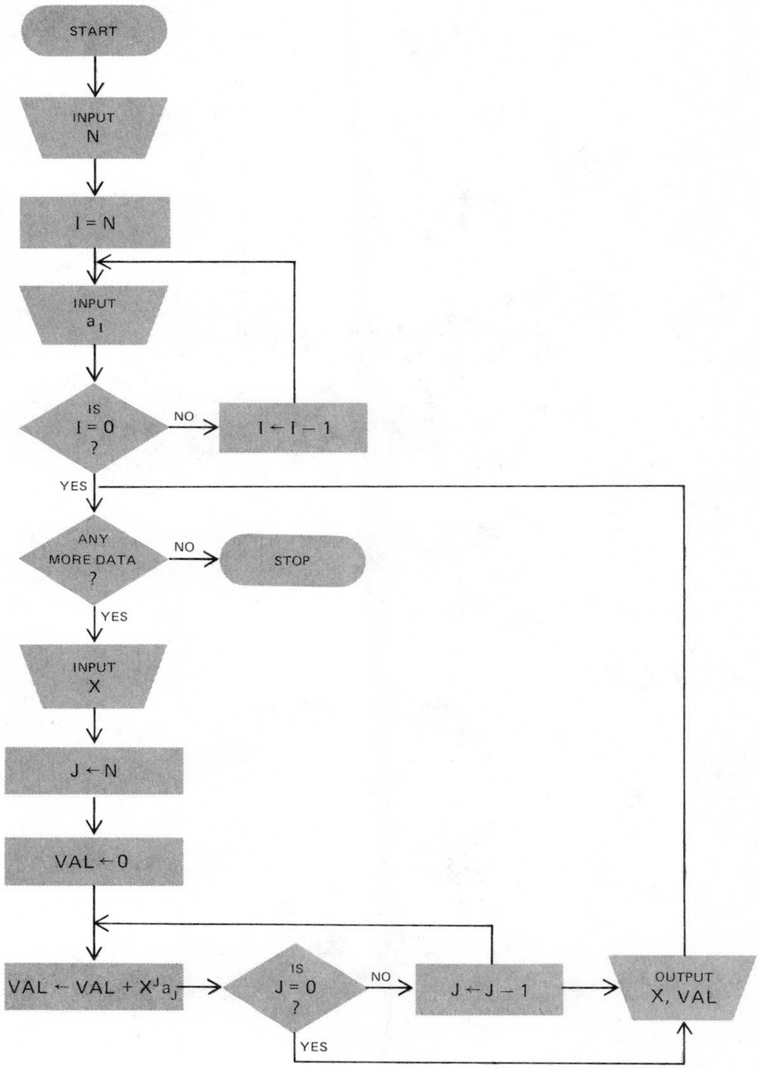

START

INPUT
N

I = N

INPUT
a_I

IS
I = 0
?

NO → $I \leftarrow I - 1$

YES

ANY
MORE DATA
?

NO → STOP

YES

INPUT
X

$J \leftarrow N$

$VAL \leftarrow 0$

$VAL \leftarrow VAL + X^J a_J$

IS
J = 0
?

NO → $J \leftarrow J - 1$ → OUTPUT X, VAL

YES

51. (a) 1, 4, -2, -6
    (b) -1, -5, 3
    (c) 3, -3
    (d) -4, $-\frac{1}{2}$, $\frac{1}{3}$
    (e) 6. $-\frac{1}{2}$, $\frac{1}{6}$
    (f) $-\frac{1}{2}$, -1 of multiplicity 3
    (g) -1, $\frac{1}{2}$, $\sqrt{5}$, $-\sqrt{5}$
    (h) $+2$ of multiplicity 5
    (i) 3, -2 each of multiplicity 2
    (j) $-\frac{1}{2}$, $\frac{1}{2}$, $\dfrac{-3 + \sqrt{21}}{2}$, $\dfrac{-3 - \sqrt{21}}{2}$
    (k) $+2$ of multiplicity 4
    (l) $\sqrt{3}$, $-\sqrt{3}$

•53.

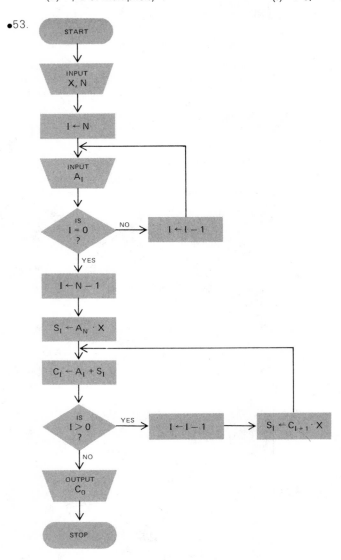

●55.

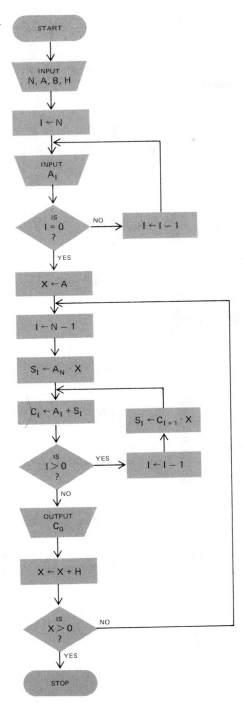

START

INPUT
N, A, B, H

$I \leftarrow N$

INPUT
A_I

IS
$I = 0$
?

NO → $I \leftarrow I - 1$

YES

$X \leftarrow A$

$I \leftarrow N - 1$

$S_I \leftarrow A_N \cdot X$

$C_I \leftarrow A_I + S_I$

$S_I \leftarrow C_{I+1} \cdot X$

IS
$I > 0$
?

YES → $I \leftarrow I - 1$

NO

OUTPUT
C_0

$X \leftarrow X + H$

IS
$X > 0$
?

NO

YES

STOP

•59. (a)

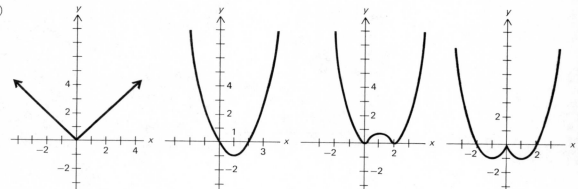

(b)

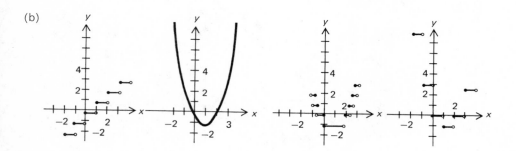

(c)

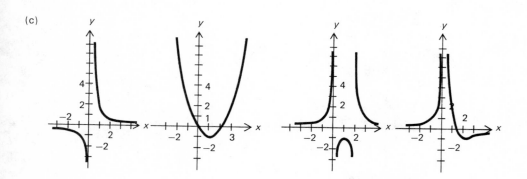

61. 50 persons.
63. (a) 196 ft   (b) $t = 1\frac{1}{2}$ sec   (c) $t = 5$ sec
65. (a) $9\frac{3}{4}$ weeks   (b) $4\frac{1}{2}$ weeks
67. Six units
69. Lengths of $\frac{1}{3}$ through $\frac{1}{2}$ mile and $\geq \frac{3}{4}$ mile
71. $-\frac{1}{4}$
73. $C_g(L) = 330 + 0.0158L$
    $C_e(L) = 290 + 2.1L$

CHAPTER 3

Practice Exercises

1.

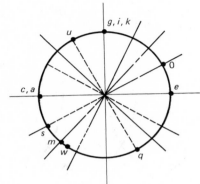

2. (a) 0   (c) 1   (e) 0   (g) −1
   (i) 0   (k) −1   (m) 0   (o) 1
   (q) −1

3.

	(a)	(c)
(i)	$0.8333\overline{3}$	0.84147
(ii)	$-0.8333\overline{3}$	−0.84147
(iii)	$0.9203\overline{3}$	0.92323
(iv)	$-0.9203\overline{3}$	−0.92323
(v)	0	0
(vi)	$-15.83\overline{3}$	−5.293
(vii)	$15.83\overline{3}$	5.923

4.

	(a)	(b)
(i)	0.5	0.54028
(ii)	0.5	0.54028
(iii)	−0.125	0.07012
(iv)	−0.125	0.07012
(v)	1	1
(vi)	−11.5	−7.1598
(vii)	−11.5	−7.1598

5. (a) 0.1494   (c) 0.8076   (e) 0.1395   (g) −0.7578
    (i) −0.2764   (k) 0.6594   (m) −0.9757   (o) 0.7033
    (q) 0.5480   (s) 0.6518   (u) −0.5480   (w) 0.4169
6. (a) $-\sqrt{3}/2$ (c) $\sqrt{2}/2$ (e) $\sqrt{3}/2$ (g) $-\frac{1}{2}$
    (i) $-\sqrt{2}/2$ (k) $-\frac{1}{2}$ (m) $\sqrt{2}/2$ (o) $-\sqrt{3}/2$
    (q) −1   (s) $\frac{1}{2}$ (u) $\frac{1}{2}$ (w) $\sqrt{2}/2$
    (y) $\frac{1}{2}$
7. (a) $\sqrt{3}$ (c) $-\sqrt{2}$ (e) $-\sqrt{3}$ (g) $-\sqrt{2}$
    (i) 1   (k) $-\sqrt{3}$ (m) −1   (o) $-\sqrt{3}/3$
    (q) 1   (s) 0   (u) $-\frac{1}{2}$ (w) $2\sqrt{3}/3$
8. (a) 0.8415   (c) −2.395   (e) −0.6552   (g) −0.1315
    (i) 3.602   (k) −0.4596   (m) −0.9959   (o) 1.101
    (q) 1.557
9. (a) $\sqrt{3}/2$ (c) −0.9505   (e) −6.657   (g) 0.8360
    (i) 7.096   (k) 0.9819   (m) −1.090   (o) −0.5062
    (q) Undefined   (s) 0.07074   (u) −0.6442

10. (a)

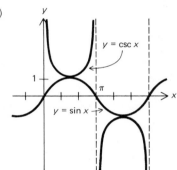
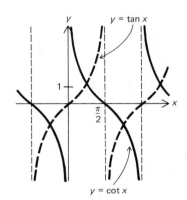

11. (a) sec 2t   (c) sin 2a   (e) 6   (g) 0
    (i) 2 tan² x   (k) csc² A   (m) 1   (o) sin x
    (q) tan x
13. (i) (a) $\frac{1}{2}(\sin 7x + \sin x)$ (c) $-\frac{1}{2}(\cos 16x - 1)$
        (e) $\frac{1}{2}(\sin 5x - \sin x)$ (g) $\frac{1}{2}(1 + \cos 4x)$
        (i) $\frac{1}{2}(\sin 6x)$

    (ii) (a) $2 \sin \dfrac{7x}{2} \cos \dfrac{x}{2}$ (c) 0

         (e) $2 \sin \dfrac{5x}{2} \sin \dfrac{x}{2}$ (g) $-2 \sin \dfrac{x}{2} \sin \dfrac{5x}{2}$

         (i) $2 \cos \dfrac{5x}{2} \sin \dfrac{11x}{2}$

14. (a) 0
    (c) $2\pi/3$
    (e) $\pi/4$
    (g) $\pi/6$

(i) $2\pi/3$

(k) $-\pi/2$

(m) $\{-\pi/6 + k2\pi \mid k \in \mathbb{Z}\} \cup \{-5\pi/6 + k2\pi \mid k \in \mathbb{Z}\}$

(o) $\{\pi/4 + k\pi \mid k \in \mathbb{Z}\}$

(q) $\{\pi + k2\pi \mid k \in \mathbb{Z}\}$

15. (a) 0.31

(c) 0.72

(e) 1.21

(g) $\{1.44 + k2\pi \mid k \in \mathbb{Z}\} \cup \{1.70 + k2\pi \mid k \in \mathbb{Z}\}$

(i) $\{-0.65 + k\pi \mid k \in \mathbb{Z}\}$

(k) $\{0.51 + k2\pi \mid k \in \mathbb{Z}\} \cup \{2.63 + k2\pi \mid k \in \mathbb{Z}\}$

(m) $\{0.9823 + k\pi \mid k \in \mathbb{Z}\}$

(o) $\{0.2806 + k2\pi \mid k \in \mathbb{Z}\} \cup \{2.8610 + k2\pi \mid k \in \mathbb{Z}\}$

(q) $\{-0.90 + k2\pi \mid k \in \mathbb{Z}\} \cup \{-2.24 + k2\pi \mid k \in \mathbb{Z}\}$

16. (a) $\pi/36$ rad.   (c) $\pi/8$ rad.   (e) $5\pi/18$ rad.   (g) $\pi/180$ rad.

(i) $\pi/5$ rad.

17. (a) $15°$ (c) $(4140/\pi)°$ (e) $\dfrac{180°}{7}$ (g) $\dfrac{5400°}{\pi}$

(i) $36°$

18. (a) 0.2588   (c) 0.9004   (e) 0.2250   (g) 1.001

(i) 4.705   (k) 0.8098   (m) 0.1932   (o) 2.162

(q) 13.81

19. (a) $17°$ (c) $55°$ (e) $21°$ (g) $9°30'$

(i) $38°39'$ (k) $78°34'$

20. (a) $a = 3.64$ (c) $\beta = 80°$ (e) $a = 59.6$ ft

$c = 10.6$ ft $a = 0.88$ in. $c = 77.8$ ft

$\beta = 70°$ $b = 4.9$ in. $\alpha = 50°$

(g) $c = 14.1$ ft   (i) $b = 13.7$ ft   (k) $a = 5.48$ in.

$\alpha = 45°$ $\alpha = 51°77'$ $\alpha = 60°31'$

$\beta = 45°$ $\beta = 38°31'$ $\beta = 29°29'$

(m) $\beta = 67°$ (o) $\beta = 29°$

$a = 8.9$ ft $a = 3.61$ ft

$b = 19.5$ ft $c = 4.13$ ft

21. (a) $b = 4.2$ (c) $b = 7.8$ (e) $A = 17°8'$

$A = 14°22'$ $B = 101°25'$ $B = 45°$

$C = 155°38'$ $C = 48°35'$ $C = 117°52'$

(g) $a = 9.8$ (i) $c = 29.4$

$c = 9.8$ $B = 17°27'$

$B = 100°$ $C = 132°33'$

23. (a) - (d)

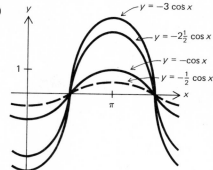

25. (a) - (d)

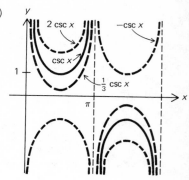

27. (a) - (d)

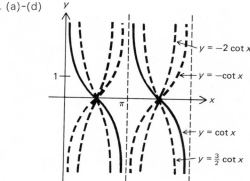

28. (a)

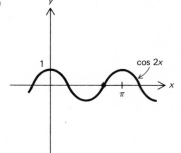

(c)

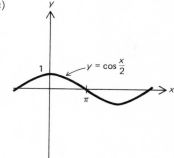

(e)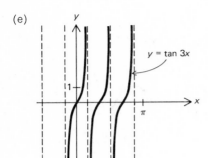

$y = \tan 3x$

(g)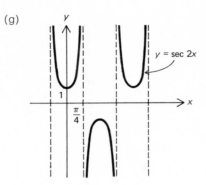

$y = \sec 2x$

(i)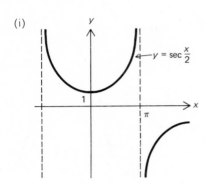

$y = \sec \dfrac{x}{2}$

(k)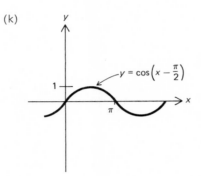

$y = \cos\left(x - \dfrac{\pi}{2}\right)$

(m)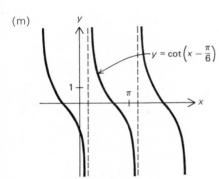

$y = \cot\left(x - \dfrac{\pi}{6}\right)$

(o)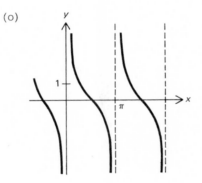

(q)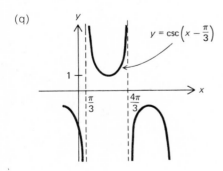

$y = \csc\left(x - \dfrac{\pi}{3}\right)$

29. (a)

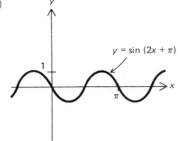

$y = \sin(2x + \pi)$

(c)

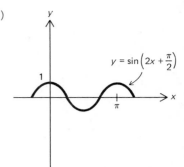

$y = \sin\left(2x + \dfrac{\pi}{2}\right)$

(e)

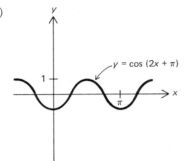

$y = \cos(2x + \pi)$

(g)

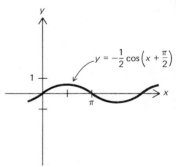

$y = 4\cos(x + \pi)$

(i)

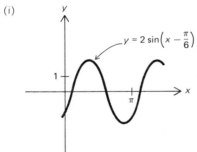

$y = 2\sin\left(x - \dfrac{\pi}{6}\right)$

(k)

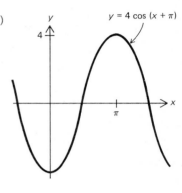

$y = -\dfrac{1}{2}\cos\left(x + \dfrac{\pi}{2}\right)$

(m)

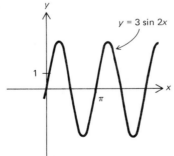

$y = 3\sin 2x$

(o)

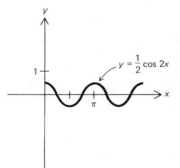

$y = \dfrac{1}{2}\cos 2x$

(q)

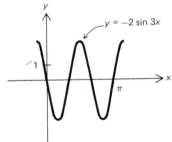

$y = -2 \sin 3x$

30. (a)

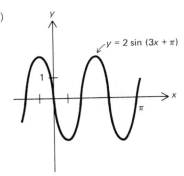

$y = 2 \sin (3x + \pi)$

(c)

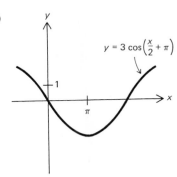

$y = 3 \cos \left(\frac{x}{2} + \pi\right)$

(e)

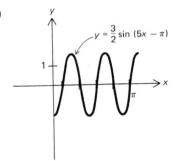

$y = \frac{3}{2} \sin (5x - \pi)$

(g)

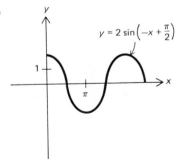

$y = 2 \sin \left(-x + \frac{\pi}{2}\right)$

(i)

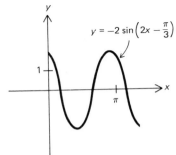

$y = -2 \sin \left(2x - \frac{\pi}{3}\right)$

(k)

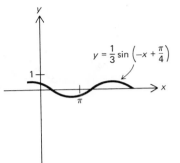

$y = \frac{1}{3} \sin \left(-x + \frac{\pi}{4}\right)$

(m)

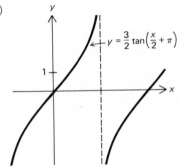

$y = \frac{3}{2} \tan\left(\frac{x}{2} + \pi\right)$

(o)

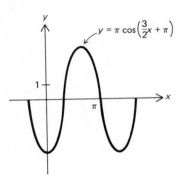

$y = \pi \cos\left(\frac{3}{2}x + \pi\right)$

(q)

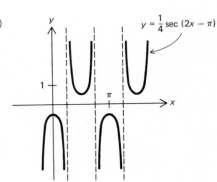

$y = \frac{1}{4} \sec(2x - \pi)$

Problems

31. (a) II   (b) III   (c) I   (d) IV   (e) I, III   (f) II, IV

35.

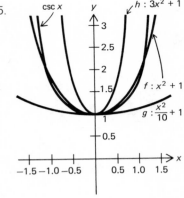

csc x $h : 3x^2 + 1$

$f : x^2 + 1$

$g : \frac{x^2}{10} + 1$

37. (a)

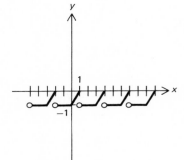

(b)

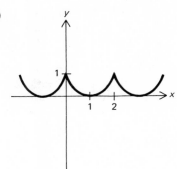

(c)

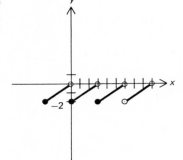

(d)

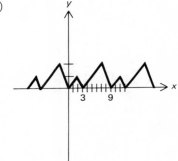

(e)

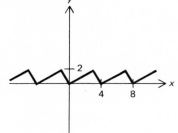

(f)

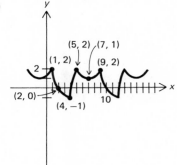

41. Sine, odd ; cosine, even ; tangent, odd ; cotangent, odd ; secant, even ; cosecant, odd.

43. (a) $\frac{3}{5}$ (b) $-\frac{3}{5}$ (c) $-\frac{12}{13}$ (d) $\frac{12}{13}$ (e) $\pm\frac{3}{4}$

45. (a) $\cdots[-4\pi, -2\pi], [0, 2\pi], [4\pi, 6\pi], [8\pi, 10\pi], \cdots$

     (b) $\cdots[-2\pi, 0], [2\pi, 4\pi], [6\pi, 8\pi], [10\pi, 12\pi], \cdots$

     (c) $\cdots[-\pi, \pi], [3\pi, 5\pi], [7\pi, 9\pi], [11\pi, 13\pi], \cdots$

     (d) $\cdots[-3\pi, -\pi], [\pi, 3\pi], [5\pi, 7\pi], [9\pi, 11\pi], \cdots$

51. (a) Need not be periodic   (b) Need not be periodic

     (c) Need not be periodic   (d) Need not be periodic

53. (a) $\dfrac{\sqrt{6}-\sqrt{2}}{4} \left(\text{or } \dfrac{\sqrt{2-\sqrt{3}}}{2} \right)$ (b) $\dfrac{\sqrt{6}+\sqrt{2}}{4} \left(\text{or } \dfrac{\sqrt{2+\sqrt{3}}}{2} \right)$

     (c) $2-\sqrt{3}$ (d) $\dfrac{\sqrt{2-\sqrt{2}}}{2}$

     (e) $\dfrac{\sqrt{2+\sqrt{2}}}{2}$ (f) $3-2\sqrt{2}$

     (g) $\dfrac{\sqrt{6}+\sqrt{2}}{4}$ (h) $\dfrac{\sqrt{2}-\sqrt{6}}{4}$

     (i) $-(2+\sqrt{3})$ (j) $(\sqrt{6}+\sqrt{2})$

     (k) $\sqrt{6}-\sqrt{2}$ (l) $2-\sqrt{3}$

59. 30.5 ft

61. 103.5 ft

63. (a) 44.6 ft   (b) 178.2π cu yd

65. 638.4 ft

67. 228.6 yd

69. 12.8 miles

71. 1.1 miles

73. 564 sq ft

75. 50 light years

77. (a) 255.5 m

     (b) Could be either 21°55′ or 29°55′

●81. (a) $2\sin 2x$ (b) $2\sec 2x$ (c) $\cot\left(x+\dfrac{\pi}{2}\right)$

     (d) $-\cot x$ (e) $-\cos 2x$ (f) $\dfrac{1}{2}\csc\left(\dfrac{3x}{2}\right)$

     (g) $\dfrac{1}{2}\sin\left(x+\dfrac{\pi}{6}\right)$ (h) $\tan\left(\dfrac{x}{2}\right)$

●85.

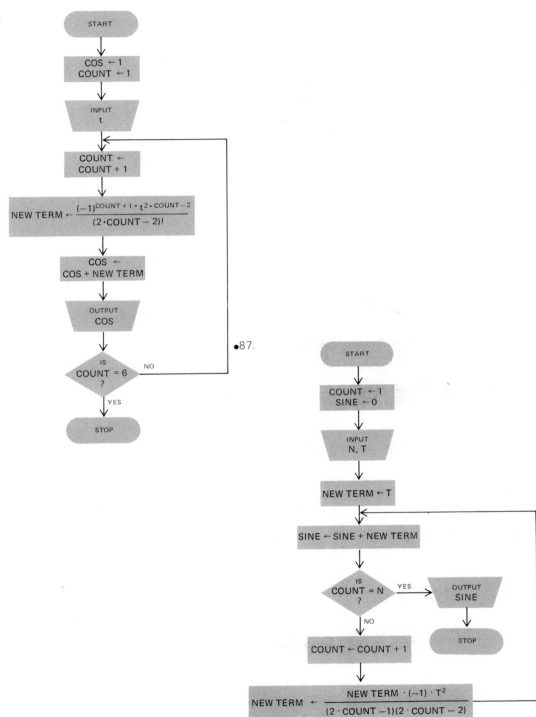

START

COS ← 1
COUNT ← 1

INPUT
t

COUNT ←
COUNT + 1

NEW TERM ← $\dfrac{(-1)^{COUNT + 1} \cdot t^{2 \cdot COUNT - 2}}{(2 \cdot COUNT - 2)!}$

COS ←
COS + NEW TERM

OUTPUT
COS

IS
COUNT = 6
?

NO

YES

STOP

●87.

START

COUNT ← 1
SINE ← 0

INPUT
N, T

NEW TERM ← T

SINE ← SINE + NEW TERM

IS
COUNT = N
?

YES

OUTPUT
SINE

STOP

NO

COUNT ← COUNT + 1

NEW TERM ← $\dfrac{\text{NEW TERM} \cdot (-1) \cdot T^2}{(2 \cdot COUNT - 1)(2 \cdot COUNT - 2)}$

93. (a) $\{\pi/6 + 2k\pi \,|\, k \in \mathbb{Z}\} \cup \{5\pi/6 + 2k\pi \,|\, k \in \mathbb{Z}\}$

(b) $\varnothing$

(c) $\{\pi/4 + 2k\pi \,|\, k \in \mathbb{Z}\} \cup \{-\pi/4 + 2k\pi \,|\, k \in \mathbb{Z}\}$

(d) $\{-\pi/6 + 2k\pi \,|\, k \in \mathbb{Z}\} \cup \{-5\pi/6 + 2k\pi \,|\, k \in \mathbb{Z}\}$

(e) $\{-2 + k\pi \,|\, k \in \mathbb{Z}\}$

(f) $\{-2 + \pi/6 + 2k\pi \,|\, k \in \mathbb{Z}\} \cup \{-2 + 5\pi/6 + 2k\pi \,|\, k \in \mathbb{Z}\}$

(g) $\{(2k+1)\pi/8) \,|\, k \in \mathbb{Z}\}$

(h) $\{1.107 + \pi/2 + k\pi \,|\, k \in \mathbb{Z}\}$ (approximately)

(i) $\{1.107 + k\pi \,|\, k \in \mathbb{Z}\}$ (approximately)

(j) $\{2\pi/3 + k\pi \,|\, k \in \mathbb{Z}\}$

(k) $\{\pi/8 + k\pi \,|\, k \in \mathbb{Z}\} \cup \{-\pi/8 + k\pi \,|\, k \in \mathbb{Z}\}$

(l) $\{3\pi/2 + 2k\pi \,|\, k \in \mathbb{Z}\}$

95. (a) $2 \sin^2 \dfrac{3x}{2}$ (b) $\frac{1}{2} \sin 8x$

(c) $\cos B$ (d) $\dfrac{\sqrt{2}}{8}$

(e) $\dfrac{1}{128} (3 - 2 \cos 2\theta + \cos 4\theta)$ (f) $8 \sin^2 \dfrac{5x}{2}$

(g) $2 \sin^2 \theta$ (h) $2(\cos^2 x - \sin^2 x)^2$

(i) 1 (j) $\dfrac{1 - \cos 8x}{8}$

(k) $\frac{1}{2}$ (l) 3

(m) -1 (n) $\cos 2t$

(o) $8 \sin^2 \theta \cos^2 \theta$ (p) $\frac{1}{8}(3 - 4 \cos 2w + \cos 4w)$

(q) $-\tan \dfrac{t}{2}$

CHAPTER 4

Practice Exercises

1. (a)

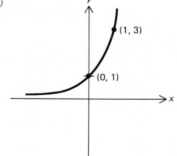

(c)

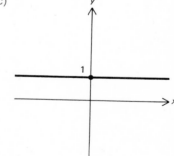

(e)

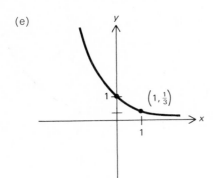

2. (a)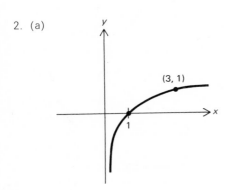

(c)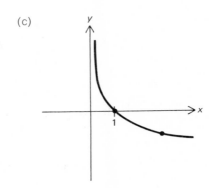

3. 14,400,000
5. $4 \log_{1/2}(1/10)$ months
7. $59 \log_e 2$ years
9. $4 \log_e 9$ hr
11. $10^{10} \log_e 2$ years
12. (a) 0.0828   (d) 4.5911   (e) 0
    (g) 7900   (i) 934.4   (k) 4.3373
    (m) $5.5224 - 10$ (o) 0.577

13 and 14. Answers as listed are for Exercise 13. Those for Exercise 14 would be the same numbers but with fewer significant digits.

    (a) 307,800   (c) 10,650
    (e) 53,600   (g) 0.0277
    (i) 10.53   (k) 707.5
    (m) 1.468   (o) 1477
    (q) 0.02938   (s) 7,037,000
    (u) 1,716,000,000   (w) 0.02491
    (y) $1.241 \cdot 10^{12}$

Problems

•21. The expansion could consist of adding $\boxed{\text{VAL} \leftarrow \text{A}^x}$ immediately after the most recent term X of the sequence in Problem 20 has been determined, and outputting VAL at the same time X is outputted.

23. (a)

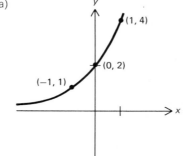

(b)

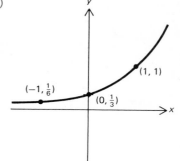

(c)

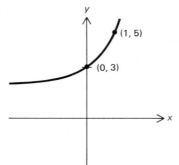

(d)

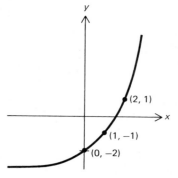

(e)

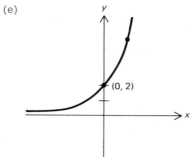

(f)

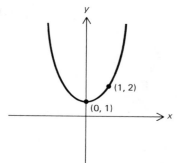

(g)

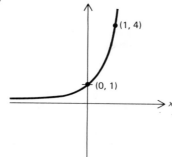

●31.

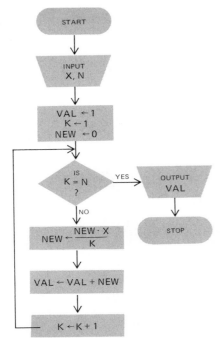

●33. The modification could consist of replacing an instruction such as the following and adding a loop to generate the values of **x** desired in the table.

37. The characteristic of $\log_a x$ is $\lfloor \log_a x \rfloor$. The mantissa of $\log_a x$ is $\log_a x - \lfloor \log_a x \rfloor$.

CHAPTER 5

Practice Exercises

1. (a) $1, \frac{1}{4}, \frac{1}{9}, \frac{1}{10}, \frac{1}{25}, \frac{1}{36}$ (c) $\frac{1}{2}, \frac{2}{3}, \frac{3}{4}, \frac{4}{5}, \frac{5}{6}, \frac{6}{7}$
   (e) $\frac{1}{2}, \frac{4}{3}, \frac{9}{4}, \frac{16}{5}, \frac{25}{6}, \frac{36}{7}$ (g) $\frac{1}{2}, 1, \frac{3}{2}, 2, \frac{5}{2}, 3$

●2. (a) $\langle 7 + 3n \rangle$ (c) $\langle (n+1)/4 \rangle$ (e) $\langle 2 + n \rangle$
   (g) $\langle 26 - 7n \rangle$ (i) $\langle 2 - n/2 \rangle$ (k) $\langle 113 - 6n \rangle$

●3. (a) $\langle \frac{1}{2} \rangle$ (c) $\langle -6(-\frac{1}{2})^{n-1} \rangle$

   (e) $\langle \frac{9}{4}(\frac{4}{3})^{n-1} \rangle$ (g) $\left\langle \dfrac{3}{32}(2)^{n-1} \right\rangle$

   (i) $\left\langle -\dfrac{1}{686}(-7)^{n-1} \right\rangle$ (k) $\langle -\frac{5}{2}(-\frac{1}{2})^{n-1} \rangle$

4. (a) 27   (c) 62   (e) $\frac{665}{8}$
   (g) 12,727,500 (i) 1,111,110   (k) $\frac{7381}{2520}$
   (m) 21   (o) $\frac{19}{4}$ (q) 13
   (s) $\frac{1}{2}$ (t) 484

Problems

•7. (a) 1, 1, 2, 3, 5, 8, 13, 21, 34, 55

•(b)

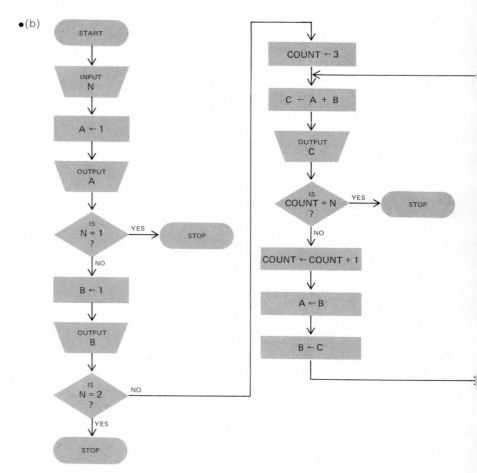

9. •(b)

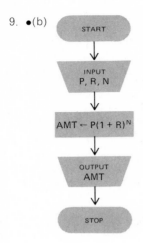

11. No
•13. $\langle 0 \rangle$
15. (a) $f_n(x) = x^n$
•17. $\langle h_0(\frac{2}{3})^n \rangle$

•19.

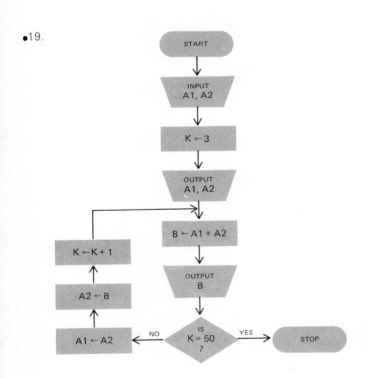

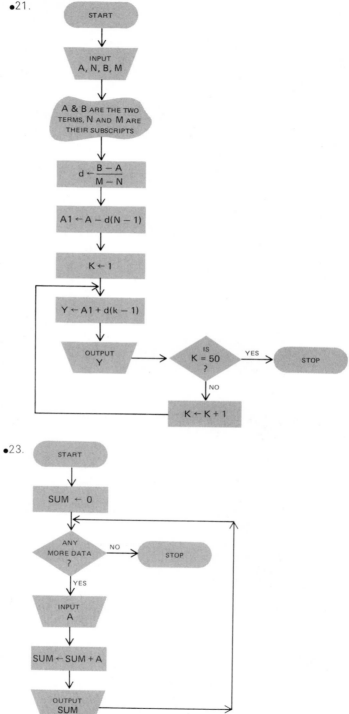

•21.

START

INPUT
A, N, B, M

A & B ARE THE TWO
TERMS, N AND M ARE
THEIR SUBSCRIPTS

$d \leftarrow \dfrac{B - A}{M - N}$

$A1 \leftarrow A - d(N - 1)$

$K \leftarrow 1$

$Y \leftarrow A1 + d(k - 1)$

OUTPUT
Y

IS
K = 50
?

YES

STOP

NO

$K \leftarrow K + 1$

•23.

START

$SUM \leftarrow 0$

ANY
MORE DATA
?

NO

STOP

YES

INPUT
A

$SUM \leftarrow SUM + A$

OUTPUT
SUM

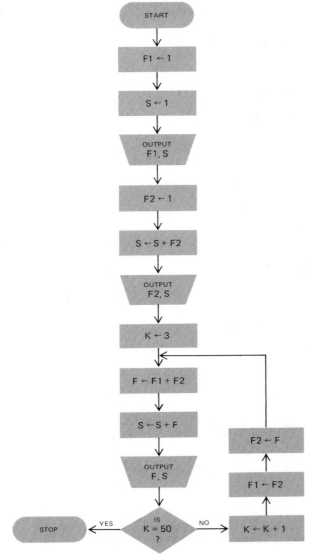

●25.

START

F1 ← 1

S ← 1

OUTPUT
F1, S

F2 ← 1

S ← S + F2

OUTPUT
F2, S

K ← 3

F ← F1 + F2

S ← S + F

OUTPUT
F, S

F2 ← F

F1 ← F2

IS
K = 50
?

STOP   YES

NO   K ← K + 1

CHAPTER 6

Practice Exercises

1. (a) $2\sqrt{3}$ (c) $\sqrt{10}$ (e) $5\sqrt{3}$
    (g) $2\sqrt{37}$ (i) $\sqrt{29}$ (k) 2
    (m) $\frac{1}{2}\sqrt{205}$ (o) 7   (q) $2b$

(s) $\sqrt{a^2 + b^2}$ (u) $(a - b)\sqrt{2}$

2. (a) $(4, 4)$ (c) $\left(-\frac{7}{2}, \frac{7}{2}\right)$ (e) $\left(\frac{3}{2}, 1\right)$

    (g) $(0, 1)$ (i) $\left(-\frac{1}{2}, -8\right)$ (k) $(0, -7)$

    (m) $\left(0, -\frac{3}{2}\right)$ (o) $\left(0, \frac{7}{2}\right)$ (q) $(a, 0)$

    (s) $(3a/2, 3b/2)$ (u) $\left(\dfrac{a + b}{2}, \dfrac{a + b}{2}\right)$

3. (a) 2 (c) -3 (e) $-\frac{8}{3}$

    (g) 6 (i) $\frac{2}{5}$ (k) 0

    (m) $\frac{14}{3}$ (o) Undefined   (q) Undefined

    (s) b/a (u) -1

4. (a) $63°26'$ (c) $108°27'$ (e) $110°34'$

    (g) $80°32'$ (i) $21°48'$ (k) $0°$

    (m) $77°54'$ (o) $90°$ (q) $90°$

    (s) $\text{Tan}^{-1}(b/a)$ (u) $135°$

5. (a) $y = x - 2$ (c) $y = -3x - 7$

    (e) $3y = -8x - 1$ (g) $y = 6x + 1$

    (i) $5y = 2x - 39$ (k) $y = -7$

    (m) $6y = 28x - 7$ (o) $x = 0$

    (q) $x = a$ (s) $ay = bx$

    (u) $y = -x + b + u$

6. (a) $2y = x$ (c) $y = -x - 1$

    (e) $y = 4$ (g) $4y + 12x = -3$

    (i) $x = 0$ (k) $y = 10x - 3$

7. (a)

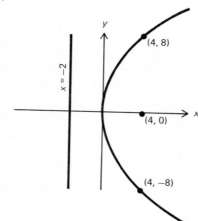

(c)

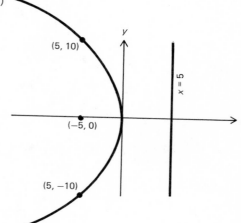

(e)

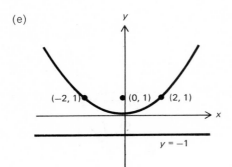

(g)

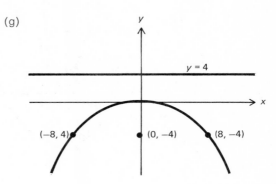

8. (a)

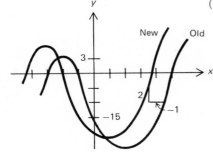

(c)

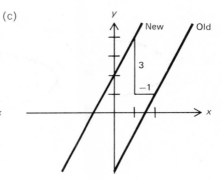

(e)

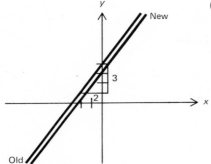

(g)

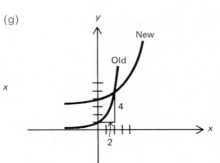

(i)   (k)

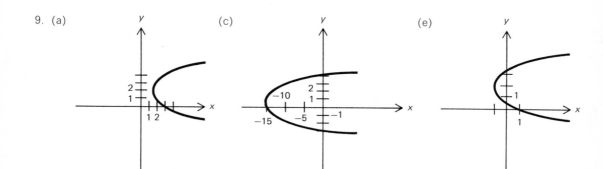

9. (a)   (c)   (e)

10. (a) $y + 3 = (x - 2)^2$ (c) $(y - \frac{5}{4}) = (-1)(x - \frac{1}{2})^2$
    (e) $y + 3 = (x + 2)^2$

Problems

13. (a) Neither   (b) Perpendicular   (c) Perpendicular
    (d) Neither   (e) Neither   (f) Perpendicular
    (g) Perpendicular   (h) Parallel   (i) Parallel
    (j) Neither   (k) Parallel   (l) Parallel
    (m) Neither   (n) Perpendicular   (o) Neither

15.   (i)   (ii)
    (a) -6 $\frac{3}{2}$
    (b) 6 $-\frac{2}{3}$
    (c) None   None
    (d) None   None
    (e) $\pm\sqrt{21}$ None
    (f) None   Any real number

17. $(y - k)^2 = \pm 4a(x - h)$

23. B/A

•25.

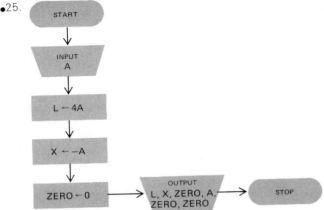

31. $2(r-a)x + 2(s-b)y = r^2 + s^2 - a^2 - b^2$
33. (a) $x^2 + y^2 = r^2$ (b) $(x-h)^2 + (y-k)^2 = r^2$
35. $3x^2 - 8ax + 3y^2 + a^2 = 0$
•37. (a) If a parabola is determined by the point F and the line L, F is called the focus of the parabola.
   (b) If a parabola is determined by the point F and the line L, L is called the directrix of the parabola.
   (c) The line perpendicular to the directrix of a parabola and containing the focus of the parabola is called the axis of the parabola.
   (d) The point of intersection of a parabola and the line segment from the focus perpendicular to the directrix is called the vertex of the parabola.
   (e) The latus rectum of a parabola is the line segment parallel to the directrix, through the focus, with endpoints on the parabola.

CHAPTER 7

Practice Exercises

1.

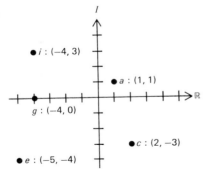

2. (a) $10 - i$ (c) $-4 + 9i$ (e) $-32 + 14i$
   (g) $\frac{11}{4} + 5i/3$ (i) $-5 + 10i$ · (k) $\frac{1}{2} + i/2$
   (m) $-i$ (o) i (q) $-i$
   (s) $-i$ (u) i (w) 5
   (y) $2\sqrt{221}$ (aa) $5 - 6i$ (cc) $-34 - 31i$
   (ee) 5

3. (a) $-\frac{24}{25} + 7i/25$ (c) $-\frac{3}{4} + i$
   (e) -2 (g) $-\frac{2}{3} - i$
   (i) -1

4. (a) $\frac{3}{2} \pm (\sqrt{3}/2)i$ (c) $\frac{1}{2} \pm (\sqrt{3}/2)i$ (e) $-1 \pm 2\sqrt{3}$
   (g) $-\frac{1}{4} \pm (\sqrt{3}/4)i$ (i) $-\frac{1}{2} \pm (\sqrt{3}/2)i$

5. (a) $-1, \frac{3}{2} + (\sqrt{3}/2)i, \frac{3}{2} - (\sqrt{3}/2)i$
   (c) $3, -1, (\sqrt{5}/3)i, -(\sqrt{5}/3)i$
   (e) $\pm i$ each of multiplicity 2

6. (a) $5(\cos 53°8' + i \sin 53°8')$
   (c) $10(\cos 216°52' + i \sin 216°52')$
   (e) $\sqrt{13}(\cos(-56°18') + i \sin(-56°18'))$
   (g) $6(\cos 0° + i \sin 0°)$
   (i) $\sqrt{13}(\cos 56°18' + i \sin 56°18')$
   (k) $\sqrt{13}(\cos 236°19' + i \sin 236°19')$
   (m) $\sqrt{97}(\cos(-23°57') + i \sin(-23°57'))$
   (o) $\sqrt{85} \cdot (\cos(-77°28') + i \sin(-77°28'))$

7. (a) $5\sqrt{3}/2 + 5i/2$ (c) $-1 + i$
   (e) $-6i$ (g) $-0.08685 + 0.4924i$
   (i) $-3.7558 - 1.2680i$ (k) $6\sqrt{2} + 6\sqrt{2}i$
   (m) $\frac{3}{8} - (3\sqrt{3}/8)i$ (o) $-3.420 - 9.397i$

8. Your checking should provide confirmation of your work.

9. (a) $\sqrt{2}, \sqrt{2}(\cos 90° + i \sin 90°), \sqrt{2}(\cos 180° + i \sin 180°), \sqrt{2}(\cos 270° + i \sin 270°)$

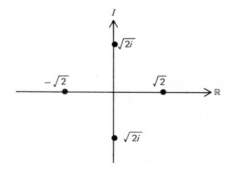

(c) 2, 2(cos 72° + i sin 72°), 2(cos 144° + i sin 144°),
   2(cos(−72°) + i sin(−72°)), 2(cos(−144°) + i sin(−144°))

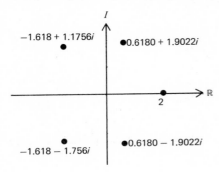

(e) −2, 2(cos 60° + i sin 60°), 2(cos 300° + i sin 300°)

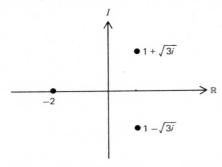

(g) cos 11¼° + i sin 11¼°, cos 101¼° + i sin 101¼°,
   cos 191¼° + i sin 191¼°, cos 281¼° + i sin 281¼°

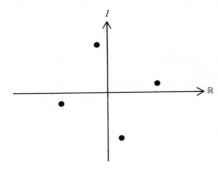

•11.

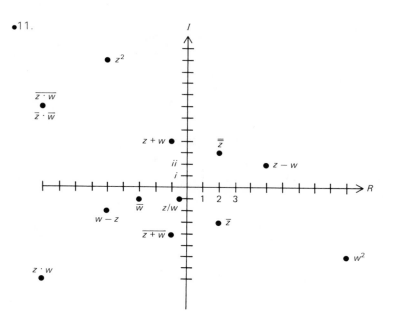

21. $w = \frac{19}{15} - \frac{13}{15}i$, $z = -\frac{3}{10} - \frac{9}{10}i$

•23.

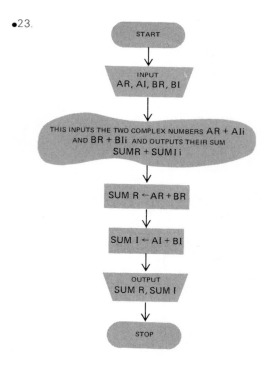

•33.

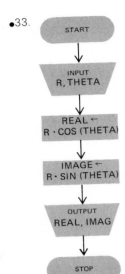

INDEX